T0202801

Lecture Notes in Computer Science

Lecture Notes in Artificial Intelligence   **14197**

Founding Editor

Jörg Siekmann

Series Editors

Randy Goebel, *University of Alberta, Edmonton, Canada*
Wolfgang Wahlster, *DFKI, Berlin, Germany*
Zhi-Hua Zhou, *Nanjing University, Nanjing, China*

The series Lecture Notes in Artificial Intelligence (LNAI) was established in 1988 as a topical subseries of LNCS devoted to artificial intelligence.

The series publishes state-of-the-art research results at a high level. As with the LNCS mother series, the mission of the series is to serve the international R & D community by providing an invaluable service, mainly focused on the publication of conference and workshop proceedings and postproceedings.

Murilo C. Naldi · Reinaldo A. C. Bianchi
Editors

Intelligent Systems

12th Brazilian Conference, BRACIS 2023
Belo Horizonte, Brazil, September 25–29, 2023
Proceedings, Part III

 Springer

Editors
Murilo C. Naldi (ID)
Federal University of São Carlos
São Carlos, Brazil

Reinaldo A. C. Bianchi (ID)
Centro Universitario da FEI
São Bernardo do Campo, Brazil

ISSN 0302-9743   ISSN 1611-3349 (electronic)
Lecture Notes in Artificial Intelligence
ISBN 978-3-031-45391-5   ISBN 978-3-031-45392-2 (eBook)
https://doi.org/10.1007/978-3-031-45392-2

LNCS Sublibrary: SL7 – Artificial Intelligence

This Springer imprint is published by the registered company Springer Nature Switzerland AG
The registered company address is: Gewerbestrasse 11, 6330 Cham, Switzerland

Paper in this product is recyclable.

Preface

The 12th Brazilian Conference on Intelligent Systems (BRACIS 2023) was one of the most important events held in Brazil in 2023 for researchers interested in publishing significant and novel results related to Artificial and Computational Intelligence. The Brazilian Conference on Intelligent Systems (BRACIS) originated from the combination of the two most important scientific events in Brazil in Artificial Intelligence (AI) and Computational Intelligence (CI): the Brazilian Symposium on Artificial Intelligence (SBIA, 21 editions) and the Brazilian Symposium on Neural Networks (SBRN, 12 editions). The Brazilian Computer Society (SBC) supports the event, the Special Committee of Artificial Intelligence (CEIA), and the Special Committee of Computational Intelligence (CEIC). The conference aims to promote theoretical aspects and applications of Artificial and Computational Intelligence and exchange scientific ideas among researchers, practitioners, scientists, engineers, and industry.

In 2023, BRACIS took place in Belo Horizonte, Brazil, from September 25th to 29th, 2023, in the Campus of the Universidade Federal de Minas Gerais. The event was held in conjunction with two other events: the National Meeting on Artificial and Computational Intelligence (ENIAC) and the Symposium on Information and Human Language Technology (STIL).

BRACIS 2023 received 242 submissions. All papers were rigorously reviewed by an international Program Committee (with a minimum of three double-blind peer reviews per submission), followed by a discussion phase for conflicting reports. After the review process, 89 papers were selected for publication in three volumes of the Lecture Notes in Artificial Intelligence series (an acceptance rate of 37%).

The topics of interest included, but were not limited to, the following:

- Agent-based and Multi-Agent Systems
- Bioinformatics and Biomedical Engineering
- Cognitive Modeling and Human Interaction
- Combinatorial and Numerical Optimization
- Computer Vision
- Constraints and Search
- Deep Learning
- Distributed AI
- Education
- Ethics
- Evolutionary Computation and Metaheuristics
- Forecasting
- Foundation Models
- Foundations of AI
- Fuzzy Systems
- Game Playing and Intelligent Interactive Entertainment
- Human-centric AI

- Hybrid Systems
- Information Retrieval, Integration, and Extraction
- Intelligent Robotics
- Knowledge Representation and Reasoning
- Knowledge Representation and Reasoning in Ontologies and the Semantic Web
- Logic-based Knowledge Representation and Reasoning
- Machine Learning and Data Mining
- Meta-learning
- Molecular and Quantum Computing
- Multidisciplinary AI and CI
- Natural Language Processing
- Neural Networks
- Pattern Recognition and Cluster Analysis
- Planning and Scheduling
- Reinforcement Learning

We want to thank everyone involved in BRACIS 2023 for helping to make it a success: we are very grateful to the Program Committee members and reviewers for their volunteered contribution to the reviewing process; we would also like to thank all the authors who submitted their papers and laboriously worked to have the best final version possible; the General Chairs and the Local Organization Committee for supporting the conference; the Brazilian Computing Society (SBC); and all the conference's sponsors and supporters. We are confident that these proceedings reflect the excellent work in the artificial and computation intelligence communities.

September 2023   Murilo C. Naldi
                                                             Reinaldo A. C. Bianchi

Organization

General Chairs

Gisele Lobo Pappa   Universidade Federal de Minas Gerais, Brazil
Wagner Meira Jr.   Universidade Federal de Minas Gerais, Brazil

Program Committee Chairs

Murilo Coelho Naldi   Universidade Federal de São Carlos, Brazil
Reinaldo A. C. Bianchi   Centro Universitário FEI, Brazil

Steering Committee

Aline Paes   Universidade Federal Fluminense, Brazil
André Britto   Universidade Federal do Sergipe, Brazil
Anna H. R. Costa   Universidade de São Paulo, Brazil
Anne Canuto   Universidade Federal do Rio Grande do Norte, Brazil
Arnaldo Cândido Jr.   Universidade Estadual Paulista, Brazil
Felipe Meneguzzi   University of Aberdeen, UK
Filipe Saraiva   Universidade Federal do Pará, Brazil
Gina M. B. Oliveira   Universidade Federal Uberlandia, Brazil
Helida Santos   Universidade Federal do Rio Grande, Brazil
Leliane N. de Barros   Universidade de São Paulo, Brazil
Livy Real   B2W Digital, Brazil
Maria V. de Menezes   Universidade Federal do Ceará, Brazil
Marlo Souza   Universidade Federal da Bahia, Brazil
Renato Tinos   Universidade de São Paulo, Brazil
Ricardo Marcacini   Universidade de São Paulo, Brazil
Tatiane Nogueira   Universidade Federal da Bahia, Brazil
Thiago Pardo   Universidade de São Paulo - São Carlos, Brazil

Program Committee

Adenilton da Silva	Universidade Federal de Pernambuco, Brazil
Adriane Serapião	Universidade Estadual Paulista, Brazil
Adrião Duarte D. Neto	Universidade Federal do Rio Grande do Norte, Brazil
Alexandre Salle	Universidade Federal do Rio Grande do Sul, Brazil
Aline Neves	Universidade Federal do ABC, Brazil
Aline Paes	Universidade Federal Fluminense, Brazil
Alneu Lopes	Universidade de São Paulo - São Carlos, Brazil
Aluizio Araújo	Universidade Federal de Pernambuco, Brazil
Alvaro Moreira	Universidade Federal do Rio Grande do Sul, Brazil
Amedeo Napoli	LORIA, France
Ana Bazzan	Universidade Federal do Rio Grande do Sul, Brazil
Ana Carolina Lorena	Instituto Tecnológico de Aeronáutica, Brazil
Ana C. B. K. Vendramin	Universidade Tecnológica Federal do Paraná, Brazil
Anderson Soares	Universidade Federal de Goiás, Brazil
André Britto	Universidade Federal do Sergipe, Brazil
André P. L. F. Carvalho	Universidade de São Paulo - São Carlos, Brazil
André Rossi	Universidade Estadual Paulista, Brazil
André Ruela	Marinha do Brasil, Brazil
André Takahata	Universidade Federal do ABC, Brazil
Andrés E. C. Salazar	Universidade Tecnológica Federal do Paraná, Brazil
Anna H. R. Costa	Universidade de São Paulo, Brazil
Anne Canuto	Universidade Federal do Rio Grande do Norte, Brazil
Araken Santos	Universidade Federal Rural do Semi-árido, Brazil
Artur Jordão	Universidade de São Paulo, Brazil
Aurora Pozo	Universidade Federal do Paraná, Brazil
Bernardo Gonçalves	Universidade de São Paulo, Brazil
Bruno Masiero	Universidade Estadual de Campinas, Brazil
Bruno Nogueira	Universidade Federal de Mato Grosso do Sul, Brazil
Bruno Souza	Universidade Federal do Maranhão, Brazil
Bruno Veloso	Universidade Portucalense, Portugal
Carlos Ribeiro	Instituto Tecnológico de Aeronáutica, Brazil
Carlos Silla	Pontifícia Universidade Católica do Paraná, Brazil

Carlos Thomaz	Centro Universitário FEI, Brazil
Carlos A. E. Montesco	Universidade Federal de Sergipe, Brazil
Carlos E. Pantoja	Centro Federal de Educação Tecnológica - RJ, Brazil
Carolina P. de Almeida	Universidade E. do Centro-Oeste do Paraná, Brazil
Celia Ralha	Universidade de Brasilia, Brazil
Claudio Bordin Jr.	Universidade Federal do ABC, Brazil
Claudio Toledo	Universidade de São Paulo, Brazil
Cleber Zanchettin	Universidade Federal de Pernambuco, Brazil
Cristiano Torezzan	Universidade Estadual de Campinas, Brazil
Daniel Araújo	Universidade Federal do Rio Grande do Norte, Brazil
Daniel Dantas	Universidade Federal de Sergipe, Brazil
Danilo Perico	Centro Universitário FEI, Brazil
Danilo Sanches	Universidade Tecnológica Federal do Paraná, Brazil
Debora Medeiros	Universidade Federal do ABC, Brazil
Denis Mauá	Universidade de São Paulo, Brazil
Dennis B. Aranibar	Universidad Católica San Pablo, Peru
Diana Adamatti	Universidade Federal do Rio Grande, Brazil
Diego Furtado Silva	Universidade de São Paulo, Brazil
Donghong Ji	Wuhan University, China
Eder M. Gonçalves	Universidade Federal do Rio Grande, Brazil
Edson Gomi	Universidade de São Paulo, Brazil
Edson Matsubara	Universidade Federal de Mato Grosso do Sul, Brazil
Eduardo Costa	Corteva Agriscience, Brazil
Eduardo Goncalves	Escola Nacional de Ciências Estatísticas, Brazil
Eduardo Palmeira	Universidade Estadual de Santa Cruz, Brazil
Eduardo Spinosa	Universidade Federal do Paraná, Brazil
Edward H. Haeusler	Pontifícia Universidade Católica do R. de J., Brazil
Elaine Faria	Universidade Federal Uberlandia, Brazil
Elizabeth Goldbarg	Universidade Federal do Rio Grande do Norte, Brazil
Emerson Paraiso	Pontificia Universidade Catolica do Paraná, Brazil
Eric Araújo	Universidade Federal de Lavras, Brazil
Evandro Costa	Universidade Federal de Alagoas, Brazil
Fabiano Silva	Universidade Federal do Paraná, Brazil
Fábio Cozman	Universidade de São Paulo, Brazil
Felipe Leno da Silva	Lawrence Livermore National Lab., USA
Felipe Meneguzzi	University of Aberdeen, UK

Felix Antreich	Instituto Tecnológico de Aeronáutica, Brazil
Fernando Osório	Universidade de São Paulo, São Carlos, Brazil
Flavia Bernardini	Universidade Federal Fluminense, Brazil
Flavio Tonidandel	Centro Universitário FEI, Brazil
Flávio S. C. da Silva	Universidade de São Paulo, Brazil
Francisco Chicano	University of Málaga, Spain
Francisco De Carvalho	Universidade Federal de Pernambuco, Brazil
Gabriel Ramos	Universidade do Vale do Rio dos Sinos, Brazil
George Cavalcanti	Universidade Federal de Pernambuco, Brazil
Gerson Zaverucha	Universidade Federal do Rio de Janeiro, Brazil
Giancarlo Lucca	Universidade Federal do Rio Grande, Brazil
Gisele Pappa	Universidade Federal de Minas Gerais, Brazil
Gracaliz Dimuro	Universidade Federal do Rio Grande, Brazil
Guilherme Barreto	Universidade Federal do Ceará, Brazil
Guilherme Coelho	Universidade Estadual de Campinas, Brazil
Guilherme Derenievicz	Universidade Federal do Paraná, Brazil
Guilherme D. Pelegrina	Universidade Estadual de Campinas, Brazil
Guillermo Simari	Universidad Nacional del Sur in B. B., Argentina
Gustavo Giménez-Lugo	Universidade Tecnológica Federal do Paraná, Brazil
Heitor Gomes	Victoria University of Wellington, New Zealand
Helena Caseli	Universidade Federal de São Carlos, Brazil
Helida Santos	Universidade Federal do Rio Grande, Brazil
Heloisa Camargo	Universidade Federal de São Carlos, Brazil
Huei Lee	Universidade Estadual do Oeste do Paraná, Brazil
Isaac da Silva	Centro Universitário FEI, Brazil
Ivandré Paraboni	Universidade de São Paulo, Brazil
Ivette Luna	Universidade Estadual de Campinas, Brazil
Jaime S. Sichman	Universidade de São Paulo, Brazil
Jean Paul Barddal	Pontifícia Universidade Católica do Paraná, Brazil
João Papa	Universidade Estadual Paulista, Brazil
João C. Xavier-Júnior	Universidade Federal do RN (UFRN), Brazil
João Paulo Canário	Stone Co., Brazil
Jomi Hübner	Universidade Federal de Santa Catarina, Brazil
Jonathan Andrade Silva	Universidade Federal de Mato Grosso do Sul, Brazil
José Antonio Sanz	Universidad Publica de Navarra, Spain
José A. Baranauskas	Universidade de São Paulo, Brazil
Jose E. O. Luna	Universidad Católica San Pablo, Peru
Julio Nievola	Pontifícia Universidade Católica do Paraná, Brazil
Karla Roberta Lima	Universidade de São Paulo, Brazil

Karliane Vale	Universidade Federal do Rio Grande do Norte, Brazil
Kate Revoredo	Humboldt-Universität zu Berlin, Germany
Krysia Broda	Imperial College, UK
Laura De Miguel	Universidad Pública de Navarra, Spain
Leila Bergamasco	Centro Universitário FEI, Brazil
Leliane Nunes de Barros	Universidade de São Paulo, Brazil
Leonardo Emmendorfer	Universidade Federal do Rio Grande, Brazil
Leonardo Matos	Universidade Federal de Sergipe, Brazil
Leonardo T. Duarte	Universidade Estadual de Campinas, Brazil
Leonardo F. R. Ribeiro	Amazon, USA
Levy Boccato	Universidade Estadual de Campinas, Brazil
Li Weigang	Universidade de Brasilia, Brazil
Livy Real	B2W Digital, Brazil
Lucelene Lopes	Universidade de São Paulo - São Carlos, Brazil
Luciano Digiampietri	Universidade de São Paulo, Brazil
Luis Garcia	Universidade de Brasília, Brazil
Luiz H. Merschmann	Universidade Federal de Lavras, Brazil
Marcela Ribeiro	Universidade Federal de São Carlos, Brazil
Marcelo Finger	Universidade de São Paulo, Brazil
Marcilio de Souto	Université d'Orléans, France
Marcos Domingues	Universidade Estadual de Maringá, Brazil
Marcos Quiles	Universidade Federal de São Paulo, Brazil
Maria Claudia Castro	Centro Universitario FEI, Brazil
Maria do C. Nicoletti	Universidade Federal de São Carlos, Brazil
Marilton Aguiar	Universidade Federal de Pelotas, Brazil
Marley M. B. R. Vellasco	Pontifícia Universidade Católica do R. de J., Brazil
Marlo Souza	Universidade Federal da Bahia, Brazil
Marlon Mathias	Universidade de São Paulo, Brazil
Mauri Ferrandin	Universidade Federal de Santa Catarina, Brazil
Márcio Basgalupp	Universidade Federal de São Paulo, Brazil
Mário Benevides	Universidade Federal Fluminense, Brazil
Moacir Ponti	Universidade de São Paulo, Brazil
Murillo Carneiro	Universidade Federal de Uberlândia, Brazil
Murilo Loiola	Universidade Federal do ABC, Brazil
Murilo Naldi	Universidade Federal de São Carlos, Brazil
Myriam Delgado	Universidade Tecnológica Federal do Paraná, Brazil
Nuno David	Instituto Universitário de Lisboa, Portugal
Patrícia Tedesco	Universidade Federal de Pernambuco, Brazil
Paula Paro Costa	Universidade Estadual de Campinas, Brazil

Paulo Cavalin	IBM Research, Brazil
Paulo Pirozelli	Universidade de São Paulo, Brazil
Paulo Quaresma	Universidade de Évora, Portugal
Paulo Santos	Flinders University, Australia
Paulo Henrique Pisani	Universidade Federal do ABC, Brazil
Paulo T. Guerra	Universidade Federal do Ceará, Brazil
Petrucio Viana	Universidade Federal Fluminense, Brazil
Priscila Lima	Universidade Federal do Rio de Janeiro, Brazil
Priscila B. Rampazzo	Universidade Estadual de Campinas, Brazil
Rafael Giusti	Universidade Federal do Amazonas, Brazil
Rafael G. Mantovani	Universidade Tecnológica Federal do Paraná, Brazil
Rafael Parpinelli	Universidade do Estado de Santa Catarina, Brazil
Reinaldo A. C. Bianchi	Centro Universitario FEI, Brazil
Renato Krohling	UFES - Universidade Federal do Espírito Santo, Brazil
Renato Tinos	Universidade de São Paulo, Brazil
Ricardo Cerri	Universidade Federal de São Carlos, Brazil
Ricardo Marcacini	Universidade de São Paulo - São Carlos, Brazil
Ricardo Prudêncio	Universidade Federal de Pernambuco, Brazil
Ricardo Rios	Universidade Federal da Bahia, Brazil
Ricardo Suyama	Universidade Federal do ABC, Brazil
Ricardo A. S. Fernandes	Universidade Federal de São Carlos, Brazil
Roberta Sinoara	Instituto Federal de C., E. e T. de São Paulo, Brazil
Roberto Santana	University of the Basque Country, Spain
Rodrigo Wilkens	Université Catholique de Louvain, Belgium
Romis Attux	Universidade Estadual de Campinas, Brazil
Ronaldo Prati	Universidade Federal do ABC, Brazil
Rosangela Ballini	Universidade Estadual de Campinas, Brazil
Roseli A. F. Romero	Universidade de São Paulo - São Carlos, Brazil
Rosiane de Freitas R.	Universidade Federal do Amazonas, Brazil
Sandra Sandri	Instituto Nacional de Pesquisas Espaciais, Brazil
Sandro Rigo	Universidade do Vale do Rio dos Sinos, Brazil
Sílvia Maia	Universidade Federal do Rio Grande do Norte, Brazil
Sílvio Cazella	Universidade Federal de Ciências da S. de P. A., Brazil
Silvia Botelho	Universidade Federal do Rio Grande, Brazil
Solange Rezende	Universidade de São Paulo - São Carlos, Brazil
Tatiane Nogueira	Universidade Federal da Bahia, Brazil
Thiago Covoes	Wildlife Studios, Brazil
Thiago Homem	Instituto Federal de E., C. e T. de São Paulo, Brazil

Thiago Pardo	Universidade de São Paulo - São Carlos, Brazil
Tiago Almeida	Universidade Federal de São Carlos, Brazil
Tiago Tavares	Insper, Brazil
Valdinei Freire	Universidade de São Paulo, Brazil
Valerie Camps	Paul Sabatier University, France
Valmir Macario	Universidade Federal Rural de Pernambuco, Brazil
Vasco Furtado	Universidade de Fortaleza, Brazil
Vinicius Souza	Pontifícia Universidade Católica do Paraná, Brazil
Viviane Torres da Silva	IBM Research, Brazil
Vladimir Rocha	Universidade Federal do ABC, Brazil
Washington Oliveira	Universidade Estadual de Campinas, Brazil
Yván Túpac	Universidad Católica San Pablo, Peru
Zhao Liang	Universidade de São Paulo, Brazil

Additional Reviewers

Alexandre Alcoforado
Alexandre Lucena
Aline Del Valle
Aline Ioste
Allan Santos
Ana Ligia Scott
Anderson Moraes
Antonio Dourado
Antonio Leme
Arthur dos Santos
Brenno Alencar
Bruna Zamith Santos
Bruno Labres
Carlos Caetano
Carlos Forster
Carlos José Andrioli
Caroline Pires Alavez Moraes
Cedric Marco-Detchart
Cinara Ghedini
Daiane Cardoso
Daniel da Silva Junior
Daniel Guerreiro e Silva
Daniela Vianna
Diego Cavalca
Douglas Meneghetti
Edilene Campos

Edson Borin
Eduardo Costa Lopes
Eduardo Max
Elias Silva
Eliton Perin
Emely Silva
Estela Ribeiro
Eulanda Santos
Fabian Cardoso
Fabio Lima
Fagner Cunha
Felipe Serras
Felipe Zeiser
Fernando Pujaico Rivera
Guilherme Mello
Guilherme Santos
Israel Fama
Javier Fumanal
Jefferson Oliva
João Fabro
João Lucas Luz Lima Sarcinelli
Joelson Sartori
Jorge Luís Amaral
José Angelo Gurzoni Jr.
José Gilberto Medeiros Junior
Juan Colonna

Leandro Lima
Leandro Miranda
Leandro Stival
Leonardo da Silva Costa
Lucas Alegre
Lucas Buzuti
Lucas Carlini
Lucas da Silva
Lucas Pavelski
Lucas Queiroz
Lucas Rodrigues
Lucas Francisco Pellicer
Luciano Cabral
Luiz Celso Gomes Jr.
Maëlic Neau
Maiko Lie
Marcelino Abel
Marcella Martins
Marcelo Polido
Marcos José
Marcos Vinícius dos Santos Ferreira
Marisol Gomez
Matheus Rocha
Mária Minárová
Miguel de Mello Carpi
Mikel Sesma
Murillo Bouzon
Murilo Falleiros Lemos Schmitt

Newton Spolaôr
Odelmo Nascimento
Pablo Silva
Paulo Rodrigues
Pedro Da Cruz
Rafael Berri
Rafael Gomes Mantovani
Rafael Krummenauer
Rafael Orsi
Ramon Abílio
Richard Gonçalves
Rubens Chaves
Sarah Negreiros de Carvalho Leite
Silvio Romero de Araújo Júnior
Tatiany Heiderich
Thiago Bulhões da Silva Costa
Thiago Carvalho
Thiago Dal Pont
Thiago Miranda
Thomas Palmeira Ferraz
Tiago Asmus
Tito Spadini
Victor Varela
Vitor Machado
Weber Takaki
Weverson Pereira
Xabier Gonzalez

Contents – Part III

xvi   Contents – Part III

Language and Models

Graph Neural Networks

Pattern Recognition

AI Applications

Evolutionary Algorithms

Multiobjective Evolutionary Algorithms Applied to the Optimization of Expanded Genetic Codes

Maísa de Carvalho Silva, Paulo Guilherme Pinheiro Pereira,
Lariza Laura de Oliveira⬤, and Renato Tinós$^{(\boxtimes)}$⬤

Department of Computing and Mathematics, FFCLRP,
University of São Paulo, Ribeirão Preto, Brazil
rtinos@ffclrp.usp.br

Abstract. There is great interest in the creation of genetically modified organisms that use amino acids different from the naturally encoded amino acids. Unnatural amino acids have been incorporated into genetically modified organisms to develop new drugs, fuels and chemicals. When incorporating new amino acids, it is necessary to change the standard genetic code. Expanded genetic codes have been created without considering the robustness of the code. In this work, multi-objective genetic algorithms are proposed for the optimization of expanded genetic codes. Two different approaches are compared: weighted and Pareto. The expanded codes are optimized in relation to the frequency of replaced codons and two measures based on robustness (for polar requirement and molecular volume). The experiments indicate that multi-objective approaches allow to obtain a list of expanded genetic codes optimized according to combinations of the three objectives. Thus, specialists can choose an optimized solution according to their needs.

Keywords: Genetic Algorithms · Multi-objective Optimization · Expanded Genetic Codes

1 Introduction

Proteins are vital macromolecules in living organisms [3]. They are composed of amino acids joined by covalent bonds forming series with different sizes and constitutions. Changes in the amino acid sequence can cause modifications in the three-dimensional structure of the protein and consequently in its function. Each amino acid is encoded in the DNA (*Deoxyribonucleic Acid*) by a sequence

This work was partially supported by São Paulo Research Foundation - FAPESP (under grants #2021/09720-2 and #2013/07375-0), National Council for Scientific and Technological Development - CNPq (under grant #306689/2021-9), and Center for Artificial Intelligence - C4AI (supported by FAPESP, under grant #2019/07665-4, and IBM Corporation).

M. C. Naldi and R. A. C. Bianchi (Eds.): BRACIS 2023, LNAI 14197, pp. 3–16, 2023.
https://doi.org/10.1007/978-3-031-45392-2_1

of three nucleotides, called codon. Sixty-one codons specify amino acids and three codons indicate the end of protein sequencing (stop codons), during its synthesis. There are 20 types of amino acids that are generally used in proteins (natural amino acids). Since there are $4^3 = 64$ possible combinations of the 4 nucleotides (A, C, G, T) in a codon, some amino acids are encoded by more than one codon.

Living beings share the same *standard genetic code*, with rare exceptions. There are approximately 1.4×10^{70} hypothetical genetic codes, i.e., ways to associate codons to the natural amino acids. When compared to all codes, the standard genetic code is very robust [5], which is explained by two main factors. First, when the organization of the standard genetic code is examined, one can see that many amino acids are encoded by similar codons (see Fig. 1). That is, small changes in the nucleotide sequence sometimes generate no change in the sequence of amino acids of a protein. Second, most of the modifications result in changing the amino acid to one with similar physical-chemical properties [14].

Recently, there has been great interest in creating genetically modified organisms that use unnatural amino acids, i.e., amino acids other than the 20 amino acids naturally encoded in the standard genetic code. These amino acids can be interesting for many reasons. For example, they may contain heavy atoms that facilitate some crystallographic studies involving X-rays. New amino acids have been incorporated into genetically modified organisms to produce drugs, fuels and chemicals of great economic interest [11].

When adding new amino acids to genetically modified organisms, it is necessary to modify the standard genetic code. Expanded genetic codes can be created by using codons with four nucleotides instead of three [1]. Another possibility is to add synthetic nucleotides to create new codons [15]. However, the most attractive way to modify the standard genetic code is by replacing the association of some codons with their respective amino acids. In the new genetic code, these codons are now associated to the new amino acids. In general, the codons chosen to have the association replaced are those that are least frequently used in the organism.

The robustness is not taken in account when creating the expanded genetic codes. According to the authors' knowledge, optimization methods have not been used for creating expanded genetic codes, with exception for a previous paper of our group [13]. In [13], we proposed a single-objective *Genetic Algorithm* (GA) for optimizing expanded genetic codes. GAs have been previously employed for creating hypothetical genetic codes with the aim of investigating the optimality of the standard genetic code. In [12], single-objective GAs were used to find robust genetic codes considering only one robustness measure each time. In [9, 10], more than one robustness measure are simultaneously optimized in a multi-objective evolutionary approach. That is, instead of comparing the codes using a single measure of robustness based on a given physical-chemical property, the codes are compared using two or more measures concurrently. Using more than one objective results in hypothetical codes more similar to the standard genetic code.

In this paper, we propose the use of multi-objective GAs for optimizing expanded genetic codes. The expanded codes are optimized in relation to the frequency of replaced codons and two measures based on robustness (for polar requirement and molecular volume). There are two main approaches for multi-objective GAs [6]: i) the weighted approach, where the multi-objective problem is transformed into a single-objective problem; ii) the Pareto approach, where a set of non-dominated solutions are considered. Here, the Pareto multi-objective approach is proposed for optimizing expanded genetic codes. The weighted approach was proposed in our previous work [13], but with only two objectives (robustness for polar requirement and frequency of replaced codons); here we consider the weighted approach with the same three objectives employed in the Pareto approach. An advantage of the Pareto approach over the weighted approach for this problem is that it is not necessary to set weights for each objective. In addition, the multi-objective Pareto approach can generate a list of solutions with different properties, rather than just one solution.

The rest of this paper is organized as follows. In Sect. 2, the proposed GAs for optimizing expanded genetic codes are presented. We present in Sect. 3 experiments where the GAs are compared in a problem where the genetic code should incorporate the codification for a hypothetical amino acid. Finally, the conclusions and discussion about future work are presented in Sect. 4.

	U			C			A			G			
U	UUU	Phe	1.9	UCU	Ser	1.1	UAU	Tyr	1.6	UGU	Cys	0.4	U
	UUC	Phe	1.8	UCC	Ser	1.0	UAC	Tyr	1.4	UGC	Cys	0.6	C
	UUA	Leu	1.0	UCA	Ser	0.7	UAA	Stop	0.2	UGA	Stop	0.1	A
	UUG	Leu	1.1	UCG	Ser	0.8	UAG	Stop	0.03	UGG	Trp	1.4	G
C	CUU	Leu	1.0	CCU	Pro	0.7	CAU	His	1.2	CGU	Arg	2.4	U
	CUC	Leu	0.9	CCC	Pro	0.4	CAC	His	1.1	CGC	Arg	2.2	C
	CUA	Leu	0.3	CCA	Pro	0.8	CAA	Gln	1.3	CGA	Arg	0.3	A
	CUG	Leu	5.2	CCG	Pro	2.4	CAG	Gln	2.9	CGG	Arg	0.5	G
A	AUU	Ile	2.7	ACU	Thr	1.2	AAU	Asn	1.6	AGU	Ser	0.7	U
	AUC	Ile	2.7	ACC	Thr	2.4	AAC	Asn	2.6	AGC	Ser	1.5	C
	AUA	Ile	0.4	ACA	Thr	0.1	AAA	Lys	3.8	AGA	Arg	0.2	A
	AUG	Met	2.6	ACG	Thr	1.3	AAG	Lys	1.2	AGG	Arg	0.2	G
G	GUU	Val	2.0	GCU	Ala	1.8	GAU	Asp	3.3	GGU	Gly	2.8	U
	GUC	Val	1.4	GCC	Ala	2.3	GAC	Asp	2.3	GGC	Gly	3.0	C
	GUA	Val	1.2	GCA	Ala	0.1	GAA	Glu	4.4	GGA	Gly	0.7	A
	GUG	Val	2.4	GCG	Ala	3.2	GAG	Glu	1.9	GGG	Gly	0.9	G

Fig. 1. Standard genetic code. The codon is a three-letter sequence from an alphabet with 4 nucleotides (A, C, G, T). The names of the amino acids are abbreviated in the figure (examples: *Phe* is Phenylalanine and *Leu* is Leucine). The numbers represent the frequency (F) of codons in *E. coli*. Source: adapted from [8].

2   Proposed Genetic Algorithms

In the proposed GAs, individuals represent hypothetical expanded genetic codes that incorporate the codification of a new unnatural amino acid. Here, we consider that only one new amino acid is incorporated. However, the algorithms can be easily adapted for problems where more than one amino acid are incorporated[1]. The two approaches based on GAs for the optimization of expanded genetic codes are presented in Sects. 2.3 and 2.4. Before, the common elements to all the proposed GAs are introduced in Sects. 2.1 and 2.2. Here, the optimization problem is considered a minimization problem.

2.1   Codification and Operators

The binary codification is used for representing an expanded genetic code in the chromosome of an individual of the GA (Fig. 2). The chromosome has 61 elements, each one representing a specific codon of the genetic code. The three stop codons (UAA, UAG and UGA) are not represented in the chromosome. An element equal to 1 in the i-th position of the chromosome means that the respective codon in the standard genetic code will now be related to the new amino acid. An element equal to 0 means that the amino acid in the respective codon is not changed.

All amino acids (natural and incorporated) must be present in the expanded genetic codes. In this way, the initial population of the GA is created ensuring that all amino acids are presented in the solutions (expanded genetic codes) associated to the individuals. In addition, when reproduction operators result in the removal of one of the amino acids from the expanded genetic code, the individual is penalized by adding a value of 10,000 to its fitness, making the selection of that individual very rare. The value of 10,000 was determined according to the theoretical maximum and minimal that can be obtained for each objective.

Here, the two GAs use the same reproduction and selection operators. Tournament selection and elitism are used to select individuals. In tournament selection, the best of s_t randomly chosen individuals is selected. In elitism, the best individual of the population is copied to the new population. Two-point crossover and flip mutation are used to transform individuals. Crossover is applied with rate p_c, while each element of the chromosome is transformed with rate p_m.

2.2   Objectives

Three objectives are considered. Two objectives are based on robustness and one objective is the frequency of replaced codons. Here, robustness of a genetic code C is defined as the inverse of $M_{st}(C)$, that is the mean square change for the values of a given property regarding mistranslation and base position errors [7,9]. The equation for $M_{st}(C)$ is:

$$M_{st}(C) = \frac{\sum_{ij} a(i,j)(X(i,C) - X(j,C))^2}{\sum_{ij} N(i,j)} \tag{1}$$

---

[1] The binary encoding of the chromosome is used here because we consider only one unnatural amino acid. If more than one new amino acid is considered, the integer encoding must be used, where integer $i > 0$ represents the i-th new amino acid. The only modification needed in this case is in the way the new chromosomes are generated and mutated.

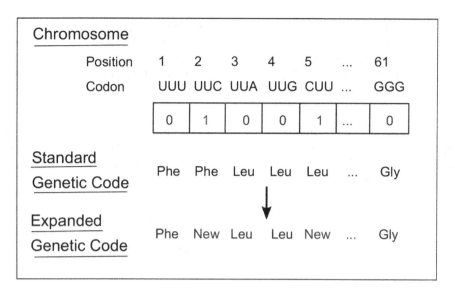

Fig. 2. Example of representation of the expanded genetic code in the chromosome of an individual of the GA. In the expanded genetic code, the amino acid *New* replaces the naturally encoded amino acids in the positions with elements equal to 1 in the chromosome of the GA's individual.

where $X(i, C)$ is the amino acid property value for the amino acid codified by the i-th codon of the genetic code C, $N(i, j)$ is the number of possible replacements between codons i and j, and $a(i, j)$ is a weight for the change between amino acids codified by the i-th and j-th codons. By minimizing $M_{st}(C)$ for a given amino acid property, we are selecting a more robust genetic code regarding mistranslation and base position errors.

The three objectives are:

$f_F(\mathbf{x})$: given by the sum of frequency (codon usage) of the codons that encode the new amino acid in the genetic code C codified by chromosome $\mathbf{x}$, i.e.,:

$$f_F(\mathbf{x}) = \sum_{i=1}^{61} x(i)\phi(i) \tag{2}$$

where $\phi(i)$ is the codon usage (frequency) of the i-th codon, considering organism *E. coli*. The codon usage of *E. coli* is given in Fig. 1. *E. coli* is a prokaryotic model organism very important in applications of biotechnology. This objective is considered in order to avoid codes with many replacements, specially in codons that are often used. Additional codon replacements incur in higher economic cost because new biological molecules must be designed and utilized. Besides, codon replacements can lead to unwanted biological effects.

$f_{PR}(\mathbf{x})$: given by $M_{st}(C)$ (Eq. 1) considering polar requirement of the amino acids (Table 1). When a new amino acid is incorporated, the polar requirement of this amino acid is also used in Eq. 2. Polar requirement is a very important property of amino acids regarding structure and function of proteins.

$f_{MV}(\mathbf{x})$: given by $M_{st}(C)$ (Eq. 1) considering molecular volume of the amino acids (Table 1). Molecular volume is also an important property of amino acids regarding structure and function of proteins.

In the experiments, values of f_F, f_{PR}, and f_{MV} are presented. In the weighted approach, the evaluation of each objective is given by the normalization by maximum values. For simplicity, we use the same symbols here for the original and normalized values.

Table 1. Polar requirement and molecular volume for the natural amino acids [7].

Amino Acid	Polar Requirement (PR)	Molecular Volume (MV)
Ala	7	31
Arg	9.1	124
Asp	13	54
Asn	10	56
Cys	4.8	55
Glu	12.5	83
Gln	8.6	85
Gly	7.9	3
His	8.4	96
Ile	4.9	111
Leu	4.9	111
Lys	10.1	119
Met	5.3	105
Phe	5	132
Pro	6.6	32.5
Ser	7.5	32
Thr	6.6	61
Trp	5.2	170
Tyr	5.4	136
Val	5.6	84

2.3   Weighted Approach

In this approach, each evaluation of objective is calculated separately. Then, weights are assigned according to the importance of each objective. Thus, the multi-objective problem is transformed into a single-objective problem. Here, the fitness of a solution (genetic code) codified by chromosome $\mathbf{x}$ is given by:

$$f(\mathbf{x}) = w_F f_F(\mathbf{x}) + w_{PR} f_{PR}(\mathbf{x}) + w_{MV} f_{MV}(\mathbf{x}) \tag{3}$$

where w_F, w_{PR}, and w_{MV} are the weights respectively associated to $f_F(\mathbf{x})$, $f_{PR}(\mathbf{x})$, and $f_{MV}(\mathbf{x})$.

The GA used in the weighted approach is here called *WGA*. Three versions of WGA are tested, each one with different values for the weights (Table 2). In WGA1, all weights are equal, i.e., no objective is prioritized. In WGA2, w_{PR} is higher, i.e., $f_{PR}(\mathbf{x})$ is prioritized. In WGA3, w_F is higher, i.e., $f_F(\mathbf{x})$ is prioritized.

Table 2. Weights for the different versions of WGA.

Weight	Version		
	WGA1	WGA2	WGA3
w_F	1/3	1/3	1/2
w_{PR}	1/3	1/2	1/3
w_{MV}	1/3	1/6	1/6

2.4 Pareto Approach

This approach uses the concept of Pareto dominance in order to obtain a subset of non-dominated solutions to a multi-objective problem. According to the Pareto dominance concept, a solution $\mathbf{x_A}$ dominates a solution $\mathbf{x_B}$ if it is better in at least one of the objectives, and it is not worse in any of the objectives. There are many multi-objective GAs that uses the Pareto approach [2]. Here, the *Non-dominated Sorting Genetic Algorithm II* (NSGA-II) [4] is used because of two main reasons. First, NSGA-II has good performance when the number of objectives is not high. NSGA-II presents a worst-case time complexity of $O(mn^2)$ per generation for problems with n objectives and when the population size is equal to m. In addition, it has a mechanism for maintenance of population diversity. Second, it was used in [9] for investigating the genetic code adaptability, a problem similar to the expanded genetic code optimization problem. Despite being similar, the chromosome codification used in [9] is different from the codification used here. The reproduction operators, objectives, and other characteristics are different too.

The NSGA-II algorithm used here can be summarized as follows [9]:

i. A population $P(0)$ is randomly generated and sorted in layers (fronts) according to the Pareto dominance. Thus, the first layer is formed by solutions which are not dominated by other solutions, i.e., the best Pareto optimal solution set found so far.

ii. At iteration t, the population $P(t)$ is transformed into population $Q(t)$ by using selection and reproduction operators. The next step is to sort the union population, $P(t) + Q(t)$, according to the Pareto dominance.

iii. A new population $P(t+1)$ is created by merging the layers of $P(t)$ and $Q(t)$. When the number of individuals in the last layer exceeds the population size, the crowding distance is used to select the most diverse individuals.

Here, the objectives of the NSGA-II are the same used in the weighted approach: $f_F(\mathbf{x})$, $f_{PR}(\mathbf{x})$, and $f_{MV}(\mathbf{x})$.

3   Experiments

3.1   Experimental Design

Experiments were performed considering the insertion of one hypothetical amino acid, named here *new*. The values of polar requirement and molecular volume for the new amino acid were obtained by averaging all respective values for the natural amino acids (see Table 1). Experiments, not shown here, with other hypothetical amino acids were also done.

Most parameters of the GAs are equal to those used in [9]; preliminary tests were carried out in order to adjust the other parameters. The GAs (in the two approaches) have population size equal to $m = 100$ individuals. The number of runs is 10, each one with a different random seed. The number of generations is 120.

For all GAs, the same parameters for tournament selection and reproduction are used: $s_t = 3$, $p_c = 0.6$, and $p_m = 0.01$. The results of the best solutions (expanded genetic codes) obtained in the runs of the different versions of WGA are presented. For NSGA-II, the results of the Pareto front obtained by applying the dominance criterion in the union of first front solutions for each run is presented.

3.2   Experimental Results

The results for the evaluation of the best solutions obtained by the 3 versions of WGA (Sect. 2.3) are presented in Table 3. When the different versions of WGA are compared, the best solutions for f_F are those obtained by WGA3, as expected. The best solutions for f_{PR} and f_{MV} were obtained by WGA1, where the weights are equal. It is interesting that the solution with best f_{PR} is obtained by WGA1 and not by WGA2, that has a higher weight w_{PR}. However, the difference in f_{PR} for the best solutions obtained by WGA1 and WGA2 is small.

The number of replacements, i.e., codons where the association to the amino acid changed, are high for the genetic codes obtained by WGA. Figure 3 shows the best solutions obtained by WGA1, WGA2, and WGA3. When the number of replacements is high, the robustness is also high. This occurs because most of

Table 3. Evaluation of the best solutions obtained by the weighted approach. The best results for each algorithm are in bold.

Property	WGA1	WGA2	WGA3
F	35.60	33.10	**18.40**
PR	**1.13**	1.16	1.45
MV	**617.71**	654.02	842.09
Replacements	34	32	21

Table 4. Evaluation of the best solutions obtained by NSGA-II regarding each objective. The mean for all solutions of the Pareto front are also presented.

Property	Mean	Best F	Best PR	Best MV
F	16.32	**0.20**	48.10	48.10
PR	1.74	2.58	**1.08**	**1.08**
MV	996.95	1667.63	**562.54**	**562.54**
Replacements	17.52	1	39	39

the changes in the codons will result in the codification of the same amino acid, i.e., the *new* amino acid.

Unlike the weighted approach, NSGA-II allows to obtain a list of expanded genetic codes. The evaluation of the subset of solutions obtained by applying the dominance criterion to the union of the first front obtained in different runs of NSGA-II is shown in Fig. 4. The best solutions obtained by the weighted approaches are also presented in the figure. One can observe that the best solutions obtained by the weighted approaches are in the Pareto front obtained by NSGA-II. Table 4 shows the evaluation of the best solutions for each objective obtained by NSGA-II. It is interesting to observe that, for the values of the properties of the *new* hypothetical amino acid, the best solutions for f_{PR} and f_{MV} are the same. In additional experiments, not shown here, with other values for the properties of the hypothetical amino acid, this does not necessarily happen.

One can observe that, while the genetic code with best f_F obtained by NSGA-II replaces only one codon, 21 replacements are generated by WGA3. Codes that results in many replacements, specially in codons that are frequently used, incur in higher economic cost and can lead to unwanted biological effects. It is important to observe that less replacements could be obtained by setting w_f to values much higher than the values of the other two weights in WGA. However, this would result in manually testing many different settings for the weights. The Pareto approach is interesting because it is not necessary to define weights or priority for each objective. Besides, it allows to obtain a list of genetic codes, that could be offered to the specialist for a particular selection, given a real-world application.

Codon	Amino Acid	Codon	Amino Acid	Codon	Amino Acid	Codon	Amino Acid
UUU	Phe	UCU	New	UAU	Tyr	UGU	New
UUC	Phe	UCC	New	UAC	New	UGC	Cys
UUA	New	UCA	Ser	UAA	Stop	UGA	Stop
UUG	New	UCG	New	UAG	Stop	UGG	Trp
CUU	New	CCU	New	CAU	His	CGU	New
CUC	New	CCC	New	CAC	New	CGC	New
CUA	New	CCA	New	CAA	New	CGA	Arg
CUG	Leu	CCG	Pro	CAG	Gln	CGG	New
AUU	Ile	ACU	Thr	AAU	New	AGU	New
AUC	Ile	ACC	Thr	AAC	Asn	AGC	New
AUA	New	ACA	New	AAA	Lys	AGA	New
AUG	Met	ACG	Thr	AAG	New	AGG	New
GUU	Val	GCU	Ala	GAU	Asp	GGU	New
GUC	New	GCC	Ala	GAC	New	GGC	Gly
GUA	New	GCA	New	GAA	Glu	GGA	New
GUG	Val	GCG	Ala	GAG	New	GGG	New

Codon	Amino Acid	Codon	Amino Acid	Codon	Amino Acid	Codon	Amino Acid
UUU	Phe	UCU	Ser	UAU	Tyr	UGU	New
UUC	Phe	UCC	New	UAC	New	UGC	Cys
UUA	Leu	UCA	New	UAA	Stop	UGA	Stop
UUG	Leu	UCG	New	UAG	Stop	UGG	Trp
CUU	New	CCU	New	CAU	His	CGU	New
CUC	New	CCC	New	CAC	New	CGC	New
CUA	New	CCA	New	CAA	New	CGA	Arg
CUG	Leu	CCG	Pro	CAG	Gln	CGG	New
AUU	Ile	ACU	Thr	AAU	New	AGU	New
AUC	Ile	ACC	Thr	AAC	Asn	AGC	New
AUA	New	ACA	New	AAA	Lys	AGA	New
AUG	Met	ACG	Thr	AAG	New	AGG	New
GUU	Val	GCU	Ala	GAU	Asp	GGU	New
GUC	New	GCC	Ala	GAC	New	GGC	Gly
GUA	New	GCA	New	GAA	Glu	GGA	New
GUG	Val	GCG	Ala	GAG	New	GGG	New

Codon	Amino Acid	Codon	Amino Acid	Codon	Amino Acid	Codon	Amino Acid
UUU	Phe	UCU	Ser	UAU	Tyr	UGU	New
UUC	Phe	UCC	Ser	UAC	New	UGC	Cys
UUA	Leu	UCA	Ser	UAA	Stop	UGA	Stop
UUG	Leu	UCG	Ser	UAG	Stop	UGG	Trp
CUU	Leu	CCU	Pro	CAU	His	CGU	Arg
CUC	Leu	CCC	New	CAC	New	CGC	New
CUA	New	CCA	Pro	CAA	Gln	CGA	New
CUG	Leu	CCG	Pro	CAG	Gln	CGG	New
AUU	Ile	ACU	Thr	AAU	New	AGU	New
AUC	Ile	ACC	Thr	AAC	Asn	AGC	New
AUA	New	ACA	New	AAA	Lys	AGA	New
AUG	Met	ACG	Thr	AAG	New	AGG	New
GUU	Val	GCU	Ala	GAU	Asp	GGU	Gly
GUC	Val	GCC	Ala	GAC	New	GGC	Gly
GUA	Val	GCA	New	GAA	Glu	GGA	New
GUG	Val	GCG	Ala	GAG	New	GGG	New

Fig. 3. Expanded genetic codes for the best solutions respectively obtained by WGA1 (top), WGA2 (middle), and WGA3 (bottom).

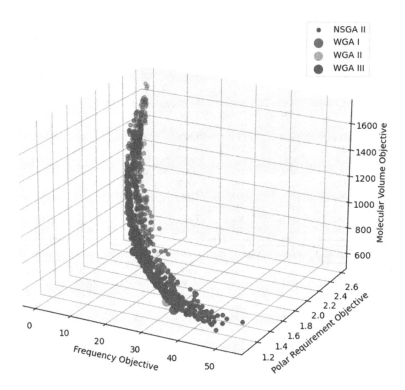

Fig. 4. Pareto front for NSGA-II and best solutions for WGA1, WGA2, and WGA3.

This advantage of NSGA-II is illustrated in Fig. 5, that shows the expanded genetic codes for three solutions of the Pareto front: solution with best f_F, solution with best f_{PR} for 2 replacements, and solution with best f_{PR} for 3 replacements. The two last solutions are those of the Pareto front with best robustness for polar requirement among those with two and three replacements. In this way, the specialist can define that she/he wants a list of solutions with best robustness for a given amino acid property and with a given number of replacements.

Codon	Amino Acid	Codon	Amino Acid	Codon	Amino Acid	Codon	Amino Acid
UUU	Phe	UCU	Ser	UAU	Tyr	UGU	Cys
UUC	Phe	UCC	Ser	UAC	Tyr	UGC	Cys
UUA	Leu	UCA	Ser	UAA	Stop	UGA	Stop
UUG	Leu	UCG	Ser	UAG	Stop	UGG	Trp
CUU	Leu	CCU	Pro	CAU	His	CGU	Arg
CUC	Leu	CCC	Pro	CAC	His	CGC	Arg
CUA	Leu	CCA	Pro	CAA	Gln	CGA	Arg
CUG	Leu	CCG	Pro	CAG	Gln	CGG	Arg
AUU	Ile	ACU	Thr	AAU	Asn	AGU	Ser
AUC	Ile	ACC	Thr	AAC	Asn	AGC	Ser
AUA	Ile	ACA	Thr	AAA	Lys	AGA	Arg
AUG	Met	ACG	Thr	AAG	Lys	AGG	New
GUU	Val	GCU	Ala	GAU	Asp	GGU	Gly
GUC	Val	GCC	Ala	GAC	Asp	GGC	Gly
GUA	Val	GCA	Ala	GAA	Glu	GGA	Gly
GUG	Val	GCG	Ala	GAG	Glu	GGG	Gly

Codon	Amino Acid	Codon	Amino Acid	Codon	Amino Acid	Codon	Amino Acid
UUU	Phe	UCU	Ser	UAU	Tyr	UGU	New
UUC	Phe	UCC	Ser	UAC	Tyr	UGC	Cys
UUA	Leu	UCA	Ser	UAA	Stop	UGA	Stop
UUG	Leu	UCG	Ser	UAG	Stop	UGG	Trp
CUU	Leu	CCU	Pro	CAU	His	CGU	Arg
CUC	Leu	CCC	Pro	CAC	His	CGC	Arg
CUA	Leu	CCA	Pro	CAA	Gln	CGA	Arg
CUG	Leu	CCG	Pro	CAG	Gln	CGG	Arg
AUU	Ile	ACU	Thr	AAU	Asn	AGU	Ser
AUC	Ile	ACC	Thr	AAC	Asn	AGC	Ser
AUA	Ile	ACA	Thr	AAA	Lys	AGA	Arg
AUG	Met	ACG	Thr	AAG	Lys	AGG	New
GUU	Val	GCU	Ala	GAU	Asp	GGU	Gly
GUC	Val	GCC	Ala	GAC	Asp	GGC	Gly
GUA	Val	GCA	Ala	GAA	Glu	GGA	Gly
GUG	Val	GCG	Ala	GAG	Glu	GGG	Gly

Codon	Amino Acid	Codon	Amino Acid	Codon	Amino Acid	Codon	Amino Acid
UUU	Phe	UCU	Ser	UAU	Tyr	UGU	New
UUC	Phe	UCC	Ser	UAC	Tyr	UGC	Cys
UUA	Leu	UCA	Ser	UAA	Stop	UGA	Stop
UUG	Leu	UCG	Ser	UAG	Stop	UGG	Trp
CUU	Leu	CCU	Pro	CAU	His	CGU	Arg
CUC	Leu	CCC	Pro	CAC	His	CGC	Arg
CUA	Leu	CCA	Pro	CAA	Gln	CGA	Arg
CUG	Leu	CCG	Pro	CAG	Gln	CGG	New
AUU	Ile	ACU	Thr	AAU	Asn	AGU	Ser
AUC	Ile	ACC	Thr	AAC	Asn	AGC	Ser
AUA	Ile	ACA	Thr	AAA	Lys	AGA	Arg
AUG	Met	ACG	Thr	AAG	Lys	AGG	New
GUU	Val	GCU	Ala	GAU	Asp	GGU	Gly
GUC	Val	GCC	Ala	GAC	Asp	GGC	Gly
GUA	Val	GCA	Ala	GAA	Glu	GGA	Gly
GUG	Val	GCG	Ala	GAG	Glu	GGG	Gly

Fig. 5. Three expanded genetic codes obtained by NSGA-II: solution with best f_F (top), solution with best f_{PR} for 2 replacements (middle), and solution with best f_{PR} for 3 replacements (bottom).

4   Conclusions

We propose the use of multi-objective GAs for the optimization of expanded genetic codes. Three objectives are considered: robustness regarding polar requirement, robustness regarding molecular volume, and frequency of use of replaced codons. Two approaches were investigated: weighted (WGA) and Pareto (NSGA-II).

Experiments with a hypothetical amino acid indicated that WGA found codes with many replacements of codons. NSGA-II allowed to obtain codes with only one replacement, while the best codes for WGA resulted respectively in 21 replacements. Replacing many codons is not interesting in many aspects. Both approached obtained robust codes. Another advantage of the Pareto approach is that a list of genetic codes is offered to the specialist, that can select a genetic code according to the characteristics of an application.

It is important to highlight that this is a theoretical work, without taking into account restrictions that may occur from technological and biological points of view. In practice, more information about the biological application is necessary to choose an expanded genetic code. However, the work shows a computational approach for optimizing expanded genetic codes that can be useful, when used in conjunction to other strategies, for helping specialists.

A possible future work is to investigate the introduction of new amino acids through the creation of synthetic nucleotides [15]. In this case, the standard genetic code is not modified; it is only expanded to accommodate the new codons related to the new synthetic nucleotides. For example, assuming that a synthetic nucleotide Y is created, we would have the possibility of associating the new codons that have Y, i.e., $AAY, ACY, \dots, AYA, \dots, YGG$, to the new amino acids. Usually, not all new codons are associated with amino acids. In this case, optimization via GAs shows a promising approach.

Another possible future work, from a technological point of view, is to use other algorithms in the calculation of the Pareto Set, such as SPEA-II (Strength Pareto Evolutionary Algorithm 2). Other optimization techniques could also be considered if the number of replacements is constrained. For example, if the maximum number of replacements is small, exhaustive search can be used to find the best codes. Finally, the investigation of new objectives that may be interesting from an technological, experimental, and/or biological point of view could also be investigated.

References

1. Anderson, J.C., et al.: An expanded genetic code with a functional quadruplet codon. Proc. Natl. Acad. Sci. **101**(20), 7566–7571 (2004)
2. Coello, C.A.C., Lamont, G.B.: Applications of Multi-objective Evolutionary Algorithms, vol. 1. World Scientific, London (2004)
3. Cox, M.M., Nelson, D.L.: Lehninger Principles of Biochemistry, vol. 5. WH Freeman, New York (2008)

4. Deb, K., Pratap, A., Agarwal, S., Meyarivan, T.: A fast and elitist multiobjective genetic algorithm: NSGA-II. IEEE Trans. Evol. Comput. **6**(2), 182–197 (2002)
5. Freeland, S.J., Hurst, L.D.: The genetic code is one in a million. J. Mol. Evol. **47**(3), 238–248 (1998)
6. Freitas, A.A.: A critical review of multi-objective optimization in data mining: a position paper. ACM SIGKDD Explor. Newsl. **6**(2), 77–86 (2004)
7. Haig, D., Hurst, L.D.: A quantitative measure of error minimization in the genetic code. J. Mol. Evol. **33**(5), 412–417 (1991). https://doi.org/10.1007/BF02103132
8. Maloy, S.R., Stewart, V.J., Taylor, R.K., Miller, S.I.: Genetic analysis of pathogenic bacteria. Trends Microbiol. **4**(12), 504 (1996)
9. Oliveira, L.L., Freitas, A.A., Tinós, R.: Multi-objective genetic algorithms in the study of the genetic code's adaptability. Inf. Sci. **425**, 48–61 (2018)
10. Oliveira, L.L., Oliveira, P.S.L., Tinós, R.: A multiobjective approach to the genetic code adaptability problem. BMC Bioinform. **16**(1), 1–20 (2015)
11. Rovner, A.J., et al.: Recoded organisms engineered to depend on synthetic amino acids. Nature **518**(7537), 89–93 (2015)
12. Santos, J., Monteagudo, Á.: Simulated evolution applied to study the genetic code optimality using a model of codon reassignments. BMC Bioinform. **12**(1), 1–8 (2011)
13. Silva, M.C., Oliveira, L.L., Tinós, R.: Optimization of expanded genetic codes via genetic algorithms. In: Anais do XV Encontro Nacional de Inteligência Artificial e Computacional, pp. 473–484 (2018)
14. Yockey, H.P.: Information Theory, Evolution, and the Origin of Life. Cambridge University Press, Cambridge (2005)
15. Zhang, Y., et al.: A semi-synthetic organism that stores and retrieves increased genetic information. Nature **551**(7682), 644–647 (2017)

Genetic Algorithms with Optimality Cuts to the Max-Cut Problem

Pablo Luiz Braga Soares[1]([✉]) and Carlos Victor Dantas Araújo[2]

[1] NEMO Research Laboratory, Federal University of Ceará - Campus Russas,
Russas, CE 62900-420, Brazil
`pablo.soares@ufc.br`
[2] Institute of Computing, University of Campinas, Campinas, SP 13083-970, Brazil

Abstract. The MAX-CUT Problem involves dividing a set of n vertices in a weighted graph $G = (V, E)$ into two subsets $(S, \bar{S})$ in such a way that the sum of the weights between the subsets is maximized. This research introduces two heuristic methods that combine Genetic Algorithm, Tabu Search, and a set of optimality cuts, which are also proven in this work. To the best of our knowledge, we are the first to utilize these inequalities in conjunction with the genetic algorithm methodology to solve the MAX-CUT problem. Computational experiments using a benchmark set of 54 instances, ranging from 800 to 3000 vertices, demonstrate that the incorporation of optimality cuts is a crucial factor for our methodologies to compete effectively with six state-of-the-art approaches for the MAX-CUT problem and our genetic algorithm that incorporated optimality cuts in the population generation was able to improve the state-of-the-art value for the G51 instance and find the same solutions as the literature in 31 other instances.

Keywords: Max-Cut Problem · Optimality Cuts · Genetic Algorithm

1 Introduction

The MAX-CUT problem is a combinatorial optimization problem that, due to its vast domain of applications such as social networks, where the MAX-CUT value is generally a measure of network robustness and structural balance originating from social behaviors [1,5,11,18], statistical physics, image segmentation, design of Very Large Scale Integrated (VLSI) Circuits and communications network design [4], has attracted the scientific interest of several researchers in the areas of graph theory, mathematical optimization and discrete mathematics [6]. This high applicability motivates the study and development of algorithms that obtain good results in a viable computational time.

The MAX-CUT problem can be described as follows: given an undirected graph $G = (V, E)$ where V is a set with n nodes, E is a set with m edges

---

This work is partially supported by Universal CNPq [422912/2021-2].

M. C. Naldi and R. A. C. Bianchi (Eds.): BRACIS 2023, LNAI 14197, pp. 17–32, 2023.
https://doi.org/10.1007/978-3-031-45392-2_2

and a cost $c_{ij} = c_{ji} \in \mathbb{R}$ for each edge $(i, j) \in E$. Any partition of V, represented by $(S, \bar{S})$, defines a cut of G. The cut value is the sum of all c_{ij}, such that $i \in S$ and $j \in \bar{S}$. It is important to note that S or $\bar{S}$ can be empty. The MAX-CUT problem consists in finding a cut of G with the largest value. Figure 1 a) show a weighted undirected graph G with $V = \{1, 2, 3, 4, 5\}$ and $E = \{(1, 2), (1, 4), (2, 3), (2, 4), (2, 5), (3, 4), (3, 5), (4, 5)\}$. Figure 1 b) presents a possible partition $(S, \bar{S})$, where $S = \{1, 2, 3\}$, $\bar{S} = \{4, 5\}$ and its associated cut value $c_{14} + c_{24} + c_{25} + c_{34} + c_{35} = 10 + 10 + 5 + 10 + 5 = 40$.

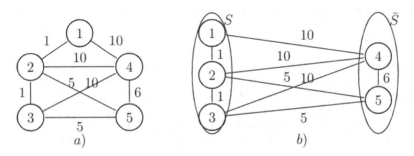

Fig. 1. a) Example of a weighted undirected graph G with 5 vertices and 8 edges. b) Example of a $(S, \bar{S})$ partition of V.

Classified as an NP-hard problem [14], it is considered a computational challenge to solve it optimally, even for moderately sized instances [19]. Therefore, for these instances, it is necessary to use approaches that provide results close to the optimal solution, such heuristics, meta-heuristics and approximation algorithms [10].

For large instances, heuristic and metaheuristic methods are commonly used to find "goodenough" sub-optimal solutions. In particular, for the very popular max-cut problem, many heuristic algorithms have been proposed, including simulated annealing [23], tabu search [17], tabu search with local search procedure [2], scatter search with Greedy Randomized Adaptive Search Procedure - GRASP [21], memetic algorithm [27] and multiple search operator heuristic [20]. These six procedures were tested using the benchmark of instances, called *G set* and the results obtained (only cut value), as far as is known, represents the state-of-the-art to MAX-CUT.

In this context, a genetic algorithm is one of the population-based meta-heuristic [22] that has been used in several studies, such as test data generation and selection in a regression testing environment [24], for the job shop scheduling problem [26], for designing the weights to combine Recurrent Neural Network (RNN) and long short-term memory (LSTM) [7] and for the MAX-CUT problem [10, 15, 16].

Considering the literature, it is notorious an extensive use of heuristics and meta-heuristics to obtain solutions for the MAX-CUT, a fact that motivates the creation of hybrid and non-exact guided methodologies to improve the results of the state-of-the-art. Heuristic approaches using optimality cuts proved to be an interesting possibility to reduce the search space and realize less random modifications to the solutions. Generally, valid inequalities, optimality cuts and optimality conditions are used in exact approaches [8,9,25]. On the other hand, it is difficult to find and prove these cuts. Also, considering inequalities that already exist, it is difficult to efficiently apply them to optimization problems.

In this paper, we developed two variations of genetic algorithms that use optimally cuts in their composition. The first one uses the optimally cuts in the generation of the initial population and also as a modification (a kind of mutation) procedure. The second uses the optimally cuts in population generation and also uses them as criterion for choosing candidates to compose a tabu list. The main difference between the two versions is the effort spent on generating new individuals. The first variant puts more effort into generating a good population, using the mutation procedure and local search, while the second version puts more effort into improving the value of individuals through the tabu search.

The main contributions of this paper are thus (1) the use of optimality cuts obtained through the study of the objective function of a quadratic programming model for the maximum cut problem; (2) development of genetic algorithms that incorporate optimality cuts to guide the search for better solutions; and (3) the genetic algorithm that incorporated optimality cuts in the population generation was able to improve the state-of-the-art value for the G51 instance and find the same solutions as the literature in 31 other instances.

The remaining sections of this paper are organized as follows. The optimality cuts used are presented and proved in Sect. 2. In Sect. 3, we present how we used the optimally cuts in the composition of the two variants of the genetic algorithm. Section 4 contains the computational results. Finally, Sect. 5 concludes the paper, with our final remarks.

2   Optimality Cuts for Max-Cut

Let $(S, \bar{S})$ be a partition represented by the binary vector $x \in \{0,1\}^n$ with $x_i = 0$ if $i \in S$ and $x_i = 1$ otherwise, $\forall i = \{1, \ldots, n\}$ and let $\bar{x}_j = 1 - x_j$, $\forall j = \{1, \ldots, n\}$, then the max-cut problem can be formulated as an unrestricted problem, with a quadratic objective function and its variables assuming binary values [3]. The constant $c_{ij} = c_{ji}, \forall i = \{1, \ldots, n\}$, $j = \{1, \ldots, n\}$ and $i \neq j$, represent the edge weight; in cases where the edge does not exist its value is zero. Let (1) be a mathematical formulation as follows:

$$\left\{ \max \sum_{i=1}^{n-1} \sum_{j=i+1}^{n} c_{ij}(x_i \bar{x}_j + \bar{x}_i x_j) : x \in \{0,1\}^n, \quad \bar{x} = 1 - x \right\}. \qquad (1)$$

Let $Q(x) = \sum_{i=1}^{n-1} \sum_{j=i+1}^{n} c_{ij}(x_i \bar{x}_j + \bar{x}_i x_j)$, notice that

$$Q(x) = \sum_{i=1}^{n-1} \sum_{j=i+1}^{n} c_{ij}(x_i\bar{x}_j + \bar{x}_i x_j) = \sum_{i=1}^{n} \sum_{\substack{j=1 \\ j\neq i}}^{n} c_{ij}x_i\bar{x}_j, \text{(with } c_{ij} = c_{ji}) \quad (2)$$

Let $x \in \{0,1\}^n$, $k \in \{1,2,\ldots,n\}$ and given $x^{\bar{k}} \in \{0,1\}^n$ be a vector such that: $x_i^{\bar{k}} = x_i$, if $i \neq k$ and $x_i^{\bar{k}} = 1 - x_i$ if $i = k$. We are interested in studying the variation $\Delta_k(x)$ in the value of the objective function when we modify only the component k of a feasible vector x.

Lemma 1. *Let* $\Delta_k(x) = Q(x^{\bar{k}}) - Q(x) = \sum_{j\neq k} c_{kj}(1 - 2x_k)(1 - 2x_j)$, *for all*

$x \in \{0,1\}^n$ *and for all* $k \in \{1,2,\ldots,n\}$.

Proof. Using the equation (2) we have $\Delta_k(x)$

$$= \sum_{j\neq k} c_{kj}\left[x_k^{\bar{k}}(1 - x_j^{\bar{k}}) - x_k(1 - x_j)\right] + \sum_{i\neq k} c_{ik}\left[x_i^{\bar{k}}(1 - x_k^{\bar{k}}) - x_i(1 - x_k)\right]$$

$$= \sum_{j\neq k} c_{kj}\left[(1 - x_k)(1 - x_j) - x_k(1 - x_j)\right] + \sum_{i\neq k} c_{ik}\left[x_i x_k - x_i(1 - x_k)\right]$$

$$= \sum_{j\neq k} c_{kj}(1 - 2x_k)(1 - x_j) - \sum_{i\neq k} c_{ki}x_i(1 - 2x_k) = \sum_{j\neq k} c_{kj}(1 - 2x_k)(1 - 2x_j)$$

∎

Let $C_k = \sum_{\substack{j=1 \\ j\neq k}}^{n} c_{kj}$ and $C_k(x) = \sum_{\substack{j=1 \\ j\neq k}}^{n} c_{kj}x_j$, Corollary 1 present properties of an optimal solution that are based on Lemma 1.

Corollary 1. *Let* x^* *be an optimal solution for formulation* (1) *and let* $k \in \{1,2,\ldots,n\}$. *Then, the following holds:*

1. *If* $x_k^* = 1$, *then* $C_k(x^*) \leq C_k/2$ *and If* $C_k(x^*) < C_k/2$, *then* $x_k^* = 1$.
2. *If* $x_k^* = 0$, *then* $C_k(x^*) \geq C_k/2$ *and If* $C_k(x^*) > C_k/2$ *then* $x_k^* = 0$.

Proof. Let $\bar{x} = (x^*)^{\bar{k}}$. We have that $Q(\bar{x}) \leq Q(x^*)$ by optimality of x^*. First, consider $x_k^* = 1$. By Lemma (1) we have

$$0 \geq Q(\bar{x}) - Q(x^*) = \sum_{j\neq k} -c_{kj}(1 - 2x_j^*) = -\sum_{j\neq k} c_{kj} + 2\sum_{j\neq k} c_{kj}x_j^* = -C_k + 2C_k(x^*).$$

Showing the first implication. To confirm the second claim of item 1 in the Corollary 1, consider $x_k^* = 0$. Again, by Lemma 1 we have

$$Q(\bar{x}) - Q(x^*) = \sum_{j\neq k} c_{kj}(1 - 2x_j) = C_k - 2C_k(x^*) \leq 0.$$

∎

The proof of item 2 in the Corollary 1 will be omitted as it utilizes the same arguments as item 1.

3 Developed Algorithms

In this section, we describe the strategy for encoding the optimality cuts and how they will be used in conjunction with the genetic algorithms, as well as the main ideas and pseudocodes of the algorithms. To exemplify the use of the optimality cuts, consider the graph G of Fig. 1. Each cut in G is a binary vector where each position represents a vertex and the value at the position represents the set to which the vertex belongs: if $v \in S, x_v = 0$, otherwise $v \in \bar{S}$ and $x_v = 1$. Let $x^* = [0, 0, 0, 1, 1]$ be the binary vector that represents the cut $(S, \bar{S})$ on G in Fig. 1. It is worth noting that the vector x^* represents an optimal solution for G. We have $C_4 = c_{41} + c_{42} + c_{43} + c_{45} = 36$, $C_4(x^*) = c_{41}x_1^* + c_{42}x_2^* + c_{43}x_3^* + c_{45}x_5^* = 6$, $x_4^* = 1$ and $C_4(x^*) \leq C_4/2$.

For this example, it is easy to see that x^* is satisfied by the optimality cut. We know that these inequalities hold to any optimal solution, and if we modify, for example, the vertex 5 in set $\bar{S}$ to the set S, the inequalities of vertices $\{1, 2, 3, 4\}$ still satisfy the optimality cuts, however the new inequality that represents vertex 5 it is infeasible, ensuring that the new cut is not optimal.

The Algorithm 1 presents the procedure that uses optimality cuts to modify a vector x. The computation of C_k values is performed only once and their values do not change during the execution of the algorithm; this allows the calculation to be done at the instance loading step, prior to Algorithm 1. However, $C_k(x)$ values are calculated dynamically and it must be computed according to the values of the current x vector.

The computation of $C_k(x)$ for each vertex k in lines (2)–(6) has complexity $O(n)$ and can be improved to $\Theta(1)$ after the first iteration by saving the values in a vector and updating after each change in the cut. The lines (7) and (10) run in constant time, and the procedure to change the vertex between sets can be made in $O(n)$-time. Assuming α as a constant indicating the number of iterations in loop (1)–(11), the time complexity of the Algorithm 1 can be limited to $O(\alpha n^2)$.

Algorithm 1: Modification with Optimality Cuts

Input: Individual (x), Vector (C)
Output: x modified by optimality cuts
1 **repeat**
2   **for** $k = 1, 2, \ldots, |\ V\ |$ **do**
3         $C_k(x) \leftarrow 0$
4   **for** $j = 1, 2, \ldots, n$ **do**
5   **if** $k \neq j$ **then**
6                 $C_k(x) \leftarrow C_k(x) + (c_{kj} * x_j)$
7   **if** $C_k(x) < C_k/2$ *and* $x_k = 0$ **then**
8            $x_k \leftarrow 1$
9   **else if** $C_k(x) > C_k/2$ *and* $x_k = 1$ **then**
10            $x_k \leftarrow 0$
11 **until** *each vertex respect the optimality cut*
12 **return** x

3.1   Genetic Algorithm with Optimality Cuts

The motivation to use Genetic Algorithms (GA) is the fact there are many different solutions to which the cuts can be applied, along with the fact that genetic operators make it possible to maintain diversity in solutions, especially after applying the inequalities which often result in similar solutions. This section describes the first developed genetic algorithm that uses the optimality cuts, referenced in the rest of this document as GA-OC. The complete version of GA-OC is presented in the Algorithm 2.

Algorithm 2: Genetic Algorithm with Optimality Cuts

Input: Population Size (ps), Number of Individuals with Optimality Cuts ($nioc$), Number of New Individuals per Generation ($nnig$), Mutation Rate (mr), Number of Individuals to Modification (nim).

Output: $x^{fittest}$

1 $P \leftarrow \emptyset$; // Population
2 **for** $j = 1, 2, \ldots, ps$ **do**
3      $\hat{x} \leftarrow$ NewIndividual( ) // randomly generated
4   **if** $j \leq nioc$ **then**
5          $\hat{x} \leftarrow$ ModificationWithOptimalityCuts($\hat{x}$) // Algorithm 1
6      $P \leftarrow P \cup \{\hat{x}\}$
7 $x^{fittest} \leftarrow$ best individual $\in P$
8 **while** *unsatisfied stop criterion* **do**
9      $NI \leftarrow \emptyset$; // Set that represents the new individuals
10   **for** $1, 2, \ldots, nnig$ **do**
11         $x^{new} \leftarrow$ Crossover(Tournament(P), Tournament(P)) // Algorithm 4
12         $x^{new} \leftarrow$ LocalOptimization(x^{new}) // Algorithm 3
13         $NI \leftarrow NI \cup \{x^{new}\}$
14   **for** $j = 1, 2, \ldots, nim$ **do**
15         $x^{any} \leftarrow$ randomly and not visited individual $\in P$
16         $P \leftarrow P \backslash \{x^{any}\}$
17         $x^{any} \leftarrow$ ModificationWithOptimalityCuts(x^{any})// Algorithm 1
18         $P \leftarrow P \cup \{x^{any}\}$
19   **for** $j = 1, 2, \ldots, nim$ **do**
20         $x^{any} \leftarrow$ randomly and not visited individual $\in P$
21         $P \leftarrow P \backslash \{x^{any}\}$
22         $x^{any} \leftarrow$ DefaultMutation(x^{any}, mr) // Algorithm 5
23         $P \leftarrow P \cup \{x^{any}\}$
24     $P \leftarrow$ Selection(P, NI) // Algorithm 6
25     $x^{fittest} \leftarrow$ best individual of P
26 **return** $x^{fittest}$

Each individual in the population is a binary vector that represents a cut in the graph and the fitness function is the cut value. The first step of GA-OC lines (1)–(6) randomly generates the initial population, where (ps) and ($nioc$) are parameters referring to the size of the population and the amount of randomly generated individuals that are modified by the optimality cuts in the initial population. It is important to mention that we chose a vertex that did not have its value modified when using the Algorithm 1. New individuals are generated through crossover (Algorithm 4) and local optimization (Algorithm 3) in lines (9)–(13). The selection of parents is performed through tournaments, which selects the fittest individual from a random subset of the population. The

---

Algorithm 3: Local Optimization

Input: Individual (x) // if $v \in S, x_v = 0$,   if $v \in \bar{S}, x_v = 1$
Output: Individual x

1 **for** $i \in S$ **do**

2 | $\Delta_k \leftarrow \sum_{j \in \bar{S}} c_{kj} - \sum_{j \in S} c_{kj}$ // Randomly choose an unvisited vertex $k \in S$

3 |

4 | **if** $\Delta_k > 0$ **then**

5 | | $x_k \leftarrow 1 - x_k$, mark k as visited

6 **for** $i \in \bar{S}$ **do**

7 | $\Delta_k \leftarrow \sum_{j \in S} c_{kj} - \sum_{j \in \bar{S}} c_{kj}$ // Randomly choose an unvisited vertex $k \in \bar{S}$

8 |

9 | **if** $\Delta_k > 0$ **then**

10 | | $x_k \leftarrow 1 - x_k$, mark k as visited

11 **return** x

Algorithm 4: Crossover

Input: Individual ($\hat{x}$), Individual ($\bar{x}$)
Output: x^{new}

1 **for** $i = 1, 2, \ldots, |V|$ **do**

2 | **if** $\hat{x}_i = \bar{x}_i$ **then**

3 | | $x_i^{new} \leftarrow \hat{x}_i$

4 | **else**

5 | | Let $j \in \mathbb{R}$ such that $0 \leq j \leq 1$

6 | | **if** $j < \frac{fitness(\bar{x})}{(fitness(\bar{x}) + fitness(\hat{x}))}$ **then**

7 | | | $x_i^{new} \leftarrow \bar{x}_i$

8 | | **else**

9 | | | $x_i^{new} \leftarrow \hat{x}_i$

10 **return** x^{new}

---

crossover function creates a new individual from a pair returned from tournaments. The next step in lines (14)–(18), is to select a subset of individuals from the population P to apply the modification procedure using the optimality cuts, which consists of fixing a random vertex and applying the Algorithm 1. The same occurs in the lines (19)–(23), where a different subset from the population P is selected to apply the default mutation, as shown in Algorithm 5. In the end, new individuals are inserted into the population after applying a procedure of selection, presented on Algorithm 6.

Algorithm 5: Default Mutation

Input: Individual (x), Mutation Rate (r)
Output: Individual x

1 **for** $k = 1, 2, \ldots, |V|$ **do**

2 | Choose a real number j such that $0 \leq j < 1$

3 | **if** $j < r$ **then**

4 | | $x_k \leftarrow 1 - x_k$

5 **return** x

Algorithm 6: Selection

Input: Population (P), Set of New Individuals (NI)
Output: New Population P
1 **for** *each individual* $x \in NI$ **do**
2   **if** $x \notin P$ **then**
3        $w \leftarrow$ individual with worst fitness $\in P$
4   **if** *fitness*$(x) \geq$ *fitness*(w) **then**
5           $P \leftarrow P \cup \{x\}$, $P \leftarrow P \backslash \{w\}$
6 **return** P

3.2 Genetic Algorithm with Perturbation-Based on Tabu Search

The Genetic Algorithm with Perturbation-Based on Tabu Search (GA-TS) pseudocode is presented on (Algorithm 7). First, the initial population is created in the same way as in GA-OC lines (1)–(6), then the new individuals are computed using crossover (Algorithm 4) and perturbation-based on tabu search (Algorithm 8). We discard the default mutation procedure and in the end, new individuals are inserted into the population after applying a procedure of selection, presented on Algorithm 6.

It is important to emphasize that in Algorithm 8, the Tabu List (TL) size is dynamically and that size changes following the rule presented by Galinier [12]. The condition to shuffle the solution was based on Wu [27], according to their results and some preliminary tests that we made. Also, values above 500 iterations do not show significant difference. The Shuffle procedure in the line (8) of the Algorithm 8 selects 150 vertices from the current solution and changes the set where they are in.

Algorithm 9 is based on the principle of best improvement, in which the movement candidate list is generated based on two properties: the cut improvement of moving the vertex k to the other set of the partition, and if this vertex, according to the optimality cuts, must be in the opposite set. Thus, the movement that results in the greatest cut gain for the solution is made. This procedure is called on Algorithm 8 until the maximum number of iterations is reached, always updating the *Incumbent*, i.e., the best solution found until the moment.

The choice of a Perturbation-Based on Tabu Search (TS) in conjunction with the GA is due to the decision to use more time optimizing the new individuals generated, allowing for changes only when they improve the cut value. The main difference between GA-TS and the GA-OC is the fact that GA-TS makes a stronger search to improve the new individuals.

Algorithm 7: GA with Pertubation-Based on Tabu Search

Input: Population Size (ps), Number of Individuals with Optimality Cuts ($nioc$), Number of New Individuals per Generation ($nnig$), Number of Iterations of Perturbation-Based Tabu Search (ni)

Output: $x^{fittest}$

1   $P \leftarrow \emptyset$ // Population
2   **for** $j = 1, 2, \ldots, ps$ **do**
3      $\hat{x} \leftarrow$ NewIndividual( ) // randomly generated
4   **if** $j \leq nioc$ **then**
5        $\hat{x} \leftarrow$ ModificationWithOptimalityCuts($\hat{x}$) // Algorithm 1
6      $P \leftarrow P \cup \{\hat{x}\}$
7   $x^{fittest} \leftarrow$ best individual $\in P$
8   **while** *unsatisfied stop criterion* **do**
9      $NI \leftarrow \emptyset$; // Set that represents the new individuals
10   **for** $1, \ldots, nnig$ **do**
11        $x^{new} \leftarrow$ Crossover(Tournament(P), Tournament(P)) // Algorithm 4
12        $x^{new} \leftarrow$ PerturbationBasedTabuSearch(x^{new},ni) // Algorithm 8
13        $NI \leftarrow NI \cup \{x^{new}\}$
14   P $\leftarrow$ Selection(P, NI) // Algorithm 6
15      $x^{fittest} \leftarrow$ best individual $\in P$
16   **return** $x^{fittest}$

---

4 Computational Results

In this section, we describe the computational experiments that were performed to test the efficiency of the two heuristics developed in this work. The algorithms were implemented using the C++11 programming language and are available at https://github.com/cvaraujo/max-cut-hybrid-ga-ts. The experiments were carried out using a machine with Intel(R) Xeon(R) Silver 4114 (2.20 GHz) × 10 with 32 GB RAM and Linux Ubuntu 16.04 64 bits operating system using a sample of 54 instances. This set of benchmark instances, called *G set*, is available at http://www.grafo.etsii.urjc.es/optsicom/maxcut and it was generated by Helmberg et al. [13].

Algorithm 8: Perturbation-Based on Tabu Search

Input: Individual (x), Number of Iterations (ni)

Output: $x^{Incumbent}$

1   $x^{Incumbent} \leftarrow x$
2   TL $\leftarrow \emptyset$; // The Tabu List of movements
3   **for** $1, \ldots, ni$ **do**
4      $x \leftarrow$ NeighborhoodMove(x, TL) // Algorithm 9
5   **if** *fitness(x) > fitness($x^{Incumbent}$)* **then**
6        $x^{Incumbent} \leftarrow x$
7   **if** *500 iterations in a row without changing $x^{Incumbent}$* **then**
8   Shuffle(x)
9   **return** $x^{Incumbent}$

To set these values and parameters we consider an empirical analysis. We tested the algorithms with different values for some parameters and the results did not present significant differences. We observed that it is preferable that

Algorithm 9: Neighborhood Move

Input: Individual (x), Tabu List (TL), Vector (C)
Output: Individual x

1 $CL \leftarrow \emptyset$; // The list of vertex candidate movements
2 $\Delta_{best} \leftarrow -\infty$, $bv \leftarrow 0$; // bv best vertex
3 **for** $k = 1, 2, \ldots, |V|$ **do**
4 $\quad x^{new} \leftarrow x$, $x_k^{new} \leftarrow 1 - x_k^{new}$, $\Delta_k \leftarrow fitness(x^{new}) - fitness(x)$
5 $\quad$ **if** $(C_k(x^{new}) < C_k/2$ **and** $x_k^{new} = 0)$ **or** $(C_k(x^{new}) > C_k/2$ **and** $x_k^{new} = 1)$ **or** $\Delta_k > 0$
$\quad\quad$ **then**
6 $\quad\quad\quad CL \leftarrow CL \cup \{k\}$
7 Shuffle(CL)
8 **for** *each vertex* $k \in CL$ **do**
9 $\quad x^{new} \leftarrow x$, $x_k^{new} \leftarrow 1 - x_k^{new}$, $\Delta_k \leftarrow fitness(x^{new}) - fitness(x)$
10 $\quad$ **if** $k \notin TL$ **and** $\Delta_k > \Delta_{best}$ **then**
11 $\quad\quad\quad bv \leftarrow k$, $\Delta_{best} \leftarrow \Delta_k$
12 **if** $bv \neq 0$ **then**
13 $\quad x_{bv} \leftarrow 1 - x_{bv}$, $CL \leftarrow CL \setminus \{bv\}$, $TL \leftarrow TL \cup \{bv\}$
14 **return** x

GA-TS manages fewer individuals per iteration since it takes more time optimizing each one. For GA-OC, creating more individuals and renewing part of the population facilitates the variety and the quality of solutions that are optimized, either by the inequalities or by the local optimization procedure. The operator and parameter values to algorithms GA-OC and GA-TS are: **Representation:** GA-OC and GA-TS use binary vector representation; **Population:** The initial population size for GA-OC is 300, with 10% NIOCs and 50 new individuals are created in each generation; In GA-TS, the initial population size is 50, with 10% NIOCs and a one individual in each iteration; **Tournament:** for both algorithms the tournaments use 4 randomly selected individuals and the best of them is returned; **Crossover:** for both algorithms, the crossover used is uniform with two individuals. If the vertex value of the parents is different, the child has a chance of inheriting from the fittest parent. The crossover procedure also generates only one child; **Mutation:** the mutation on GA-OC is applied to 20% of the population, excluding the fittest. Each gene of the selected individuals has a probability of 10% to change. On GA-TS, the mutation function is not used; **TS Iterations:** the number of iterations is 10^6; **Stopping Criterion:** time limit of 1800 s for both algorithms.

To show the difference in the quality of individuals created with the use of optimality cuts, we select a sample of 12 instances, where 3 are from each size of vertices, and created a population of 50 individuals through Algorithm 1, referred here as NIOC, another population of 50 Randomly Generated Individuals, referring here to RCI. In the population of RCIs, the local optimization procedure (Algorithm 3) was also used on all individuals. The results are in Table 1, the first and second columns (Graph, n) are the instance name and the number of vertices. For each algorithm (NIOC, RCI and the RCI with local optimization RCI-LO) there are two columns (avg, dev) that inform the average and standard deviation of the cut values, respectively. The highest averages are in bold.

In all 12 sample instances, the NIOC population consistently exhibits higher mean values compared to RCI and RCI-LO. This superiority is also observed in the average standard deviation, except for instance G46 where NIOC did not achieve the best result. The results of RCI-LO demonstrate the effectiveness of the local optimization procedure in improving the quality of randomly generated individuals across all tested instances. These findings highlight the positive impact of incorporating optimality cuts on the overall output quality of the algorithm, enabling the generation of solutions with higher cut values within a shorter timeframe. However, it is important to acknowledge that relying solely on optimality cuts for optimization may lead to individuals getting trapped in local maxima more quickly than other methods like RCIs. To address this limitation, we have implemented additional measures such as an alternative local optimization procedure, a high mutation rate, and a preference for maximum diversity during the selection process. These measures aim to mitigate the risk of premature convergence and enhance the algorithm's exploration capabilities, enabling it to search for more optimal solutions across a wider solution space.

The second experiment compares two versions of our algorithm, GA-OC and GA-TS, with a standard genetic algorithm implementation. We conducted 30 runs of GA-OC, GA-TS, and the standard GA algorithm for each of the 54 instances. Table 3 presents the average, standard deviation, and minimum values obtained by each of the algorithms. Considering the results, the hybrid approaches outperform the default version of the GA for all instances. Considering only the two versions of GA using the optimality cuts, the algorithm GA-OC outperforms GA-TS with respect to the averages and lowest values. For all instances, GA-OC has 28 highest values of column *min*, and in 6 cases the value is the same, while for the 20 remaining the value of GA-OC is less than GA-TS. Considering columns *avg*, to 29 instances GA-OC is the highest, 4 were draws, and GA-TS has higher values in others 21. For column *dev*, GA-OC has a smaller variation in solutions, such that for 30 instances it has the standard deviation value less than GA-TS, to 4 instances both the algorithms have value 0 of standard deviation and for the remaining 20 instances GA-TS has smaller values. Based on these results, the probability associated with the t-test was calculated using the mean values (columns 4 and 7) obtained by the GA-OC and GA-TS algorithms across the 54 instances, using a two-tailed distribution and a 95% confidence interval as parameters. The p-value of 0.009 indicates that the difference between the means of the algorithms is statistically significant.

We compared the results of our algorithms with the most effective heuristics currently in the literature. It is important to emphasize that the conditions and configuration parameters of our algorithms were set under different circumstances than those reported in the literature for the compared heuristics, such as the programming languages used, termination criteria, and hardware configuration. Therefore, no implementation of the algorithms from the compared papers was conducted. Only the results presented by the authors were used for the purpose of comparison with our algorithms.

Table 4 compares our approaches with 6 state-of-the-art algorithms. The first two columns contain the instance identifier (Graph) and the number of vertices (n). Columns (GA-OC) and (GA-TS) exhibit the best values obtained by our algorithms and from the fifth to the tenth column are the best results found by

Table 1. The difference between NIOC, RCI and RCI-LO

Graph	n	NIOC avg	dev	RCI avg	dev	RCI-LO avg	dev
G1	800	**11310.40**	26.75	9598.40	61.46	10977.40	46.73
G2	800	**11363.80**	28.10	9582.80	63.68	10975.30	47.60
G3	800	**11357.95**	39.12	9575.55	70.05	10976.70	37.66
G22	2000	**12822.80**	26.60	10004.55	72.61	12258.70	43.62
G23	2000	**12730.75**	18.97	9999.60	68.60	12261.20	44.92
G24	2000	**12788.00**	18.98	9969.65	57.33	12267.95	39.94
G45	1000	**6368.75**	20.35	5009.95	50.42	6122.90	26.71
G46	1000	**6404.65**	22.04	4994.65	54.23	6121.65	19.12
G47	1000	**6382.60**	16.81	4989.40	39.18	6120.95	30.43
G48	3000	**5974.40**	45.39	2997.30	38.40	4648.20	46.07
G49	3000	**5976.70**	36.61	2989.00	41.99	4651.10	43.87
G50	3000	**5860.80**	26.40	2999.60	39.58	4632.90	31.73

Table 2. Comparison of GA-OC and GA-TS with the six algorithms from the literature

GA-OC	TS [17]	SS [21]	SA [23]	MA [27]	MSOP [20]	SILS [2]	GA-TS	TS [17]	SS [21]	SA [23]	MA [27]	MSOP [20]	SILS [2]
Better	22	44	38	31	20	2	Better	10	36	34	31	8	8
Equal	23	9	11	15	26	32	Equal	25	11	10	13	28	28
Worse	9	1	5	8	8	20	Worse	19	7	10	10	18	18

these reference algorithms. The "-" symbol represents that no value was made available for that instance and values in bold font are the actual best-known results. Table 2 show the summary of the comparison with the best result of our algorithms. The rows 2, 3, 4 for GA-OC in Table 2 respectively denote: the number of instances in which our algorithms obtain better, equal, and worse cut values when compared to the corresponding reference algorithm.

From Table 2, we observe that the GA-OC algorithm outperforms 5 out of 6 reference algorithms in terms of the number of wins, i.e., the number of instances in which each algorithm achieved the best cut value. Compared to the SILS method, GA-OC won in instances G37 and G51, with G51 being the instance where it was able to improve the state-of-the-art and achieved the same results in 32 other instances. To the 20 remaining instances, the $gap = \frac{STA-OA}{STA} \times 100$, between the best of our algorithms (OA) and the state-of-the-art (STA) is less than 1%, more specifically the gap of one instance is 1% while for the other 19 this value is less than 0.3%. On the other hand, the GA-TS algorithm outperforms 3 out of 6 reference algorithms. These results confirm the effectiveness of our genetic algorithms that use our set of optimality cuts to deliver high-quality solutions for the 54 instances belonging to G set.

Table 3. Comparison of algorithms GA-OC, GA-TS and GA

(Graph, n)	GA-OC			GA-TS			GA		
	min	avg	dev	min	avg	dev	min	avg	dev
(G1, 800)	**11624**	**11624.0**	0.0	11607	11622.3	5.1	11565	11560.4	10.8
(G2, 800)	**11620**	**11620.0**	0.0	11607	11616.3	4.7	11555	11554.8	13.0
(G3, 800)	**11622**	**11622.0**	0.0	11622	11622.0	0.0	11577	11571.1	23.1
(G4, 800)	**11646**	**11646.0**	0.0	11641	11644.5	2.3	11583	11580.4	16.4
(G5, 800)	**11631**	**11631.0**	0.0	11627	11630.5	1.2	11575	11568.4	13.6
(G6, 800)	**2178**	**2178.0**	0.0	2166	2176.5	3.6	2105	2094.6	18.0
(G7, 800)	**2006**	**2006.0**	0.0	1999	2002.7	3.4	1983	1965.5	32.4
(G8, 800)	**2005**	**2005.0**	0.0	1998	2001.7	3.3	1917	1912.4	8.2
(G9, 800)	**2054**	**2054.0**	0.0	2038	2049.2	5.8	1952	1938.5	6.2
(G10, 800)	1999	1999.8	0.4	**2000**	**2000.0**	0.0	1904	1901.0	6.9
(G11, 800)	**560**	562.2	1.9	560	**562.6**	1.3	560	549.8	25.6
(G12, 800)	552	**555.2**	1.3	**554**	555.0	1.0	550	540.0	25.7
(G13, 800)	574	**579.2**	2.4	**580**	580.4	0.8	572	562.8	26.0
(G14, 800)	3052	3057.9	3.0	**3058**	**3060.6**	1.5	3009	3005.9	12.0
(G15, 800)	3033	3041.7	5.0	**3046**	**3048.1**	1.1	3015	3006.8	19.7
(G16, 800)	3038	3045.8	4.6	**3047**	**3050.0**	1.7	2992	2989.9	15.2
(G17, 800)	3035	**3040.3**	3.2	**3040**	3044.1	2.3	2986	2983.6	10.1
(G18, 800)	987	988.5	1.7	**990**	**991.1**	0.7	959	955.5	20.4
(G19, 800)	**905**	**905.9**	0.3	904	905.4	0.9	886	876.8	21.6
(G20, 800)	**941**	**941.0**	0.0	**941**	**941.0**	0.0	932	921.5	23.9
(G21, 800)	**930**	930.4	0.5	**930**	**930.5**	0V5	899	892.8	17.7
(G22, 2000)	**13353**	**13355.9**	1.7	13257	13291.3	15.9	13316	13268.1	110.6
(G23, 2000)	**13324**	**13332.0**	4.3	13297	13312.6	9.9	13281	13229.3	87.3
(G24, 2000)	**13323**	**13329.9**	3.2	13257	13287.5	13.1	13254	13218.9	92.2
(G25, 2000)	**13327**	**13333.0**	3.9	13260	13281.3	15.7	13270	13226.1	95.2
(G26, 2000)	**13307**	**13317.7**	5.2	13266	13285.1	11.4	13234	13202.5	88.8
(G27, 2000)	**3324**	**3330.9**	4.1	3250	3288.8	19.4	3146	3123.1	50.2
(G28, 2000)	**3284**	**3289.1**	4.0	3241	3257.7	17.0	3158	3122.0	76.3
(G29, 2000)	**3374**	**3396.2**	8.7	3335	3358.5	16.9	3238	3207.4	66.9
(G30, 2000)	**3397**	**3407.1**	5.0	3335	3370.7	20.5	3165	3149.4	39.5
(G31, 2000)	**3298**	**3301.4**	2.9	3243	3255.3	10.8	3142	3115.4	63.2
(G32, 2000)	1384	1394.8	7.8	**1400**	**1401.8**	1.9	1368	1340.3	78.8
(G33, 2000)	1354	1367.0	5.6	**1368**	**1371.6**	2.7	1350	1320.8	76.6
(G34, 2000)	1362	1371.4	5.2	**1372**	**1375.2**	2.0	1356	1325.3	77.7
(G35, 2000)	7635	7646.1	7.6	**7657**	**7667.2**	6.5	7536	7521.4	49.3
(G36, 2000)	7615	7638.0	9.9	**7636**	**7655.0**	10.7	7542	7524.8	52.8
(G37, 2000)	7634	7648.3	8.5	**7657**	**7667.2**	5.4	7568	7547.1	56.8
(G38, 2000)	7636	7644.4	9.3	**7652**	**7661.3**	7.9	7553	7533.0	52.9
(G39, 2000)	**2385**	**2396.2**	6.4	2372	2385.7	8.3	2302	2275.5	65.3
(G40, 2000)	**2383**	**2390.9**	4.2	2358	2371.1	8.9	2239	2227.5	51.1
(G41, 2000)	**2384**	**2395.4**	5.4	2356	2368.6	9.1	2278	2256.1	59.0
(G42, 2000)	**2462**	**2468.4**	5.5	2421	2438.4	11.3	2379	2350.4	68.5
(G43, 1000)	**6660**	**6660.0**	0.0	6656	6659.4	1.2	6589	6581.6	18.8
(G44, 1000)	**6649**	**6649.9**	0.3	6639	6647.2	4.4	6571	6566.6	22.6
(G45, 1000)	**6649**	**6652.9**	1.6	6641	6647.1	4.8	6571	6565.3	20.1
(G46, 1000)	**6645**	**6647.9**	1.4	6637	6645.1	3.8	6575	6573.3	21.2
(G47, 1000)	**6655**	**6656.5**	0.7	6648	6652.2	3.5	6569	6567.6	20.4
(G48, 3000)	**6000**	**6000.0**	0.0	**6000**	**6000.0**	0.0	5880	5807.5	240.4
(G49, 3000)	**6000**	**6000.0**	0.0	**6000**	**6000.0**	0.0	5940	5854.0	251.0
(G50, 3000)	5878	5879.8	0.6	**5880**	**5880.0**	0.0	5872	5785.3	232.6
(G51, 1000)	3827	3838.3	5.5	**3843**	**3845.4**	1.2	3789	3780.0	20.6
(G52, 1000)	3830	3835.9	3.9	**3846**	**3849.0**	1.6	3780	3773.9	16.6
(G53, 1000)	3828	3837.9	4.8	**3844**	**3846.5**	1.7	3783	3777.0	18.7
(G54, 1000)	3829	3837.7	4.4	**3843**	**3848.0**	2.7	3785	3778.5	21.2

Table 4. Best results obtained by GA-OC, GA-TS and the current state-of-the-art

Graph	n	GA-OC	GA-TS	TS [17]	SS [21]	SA [23]	MA [27]	MSOP [20]	SILS [2]
G1	800	**11624**	**11624**	**11624**	**11624**	11621	**11624**	**11624**	**11624**
G2	800	**11620**	**11620**	**11620**	**11620**	11612	**11620**	**11620**	**11620**
G3	800	**11622**	**11622**	**11622**	**11622**	11618	**11622**	**11622**	**11622**
G4	800	**11646**	**11646**	**11646**	**11646**	11644	-	**11646**	**11646**
G5	800	**11631**	**11631**	**11631**	**11631**	11628	-	**11631**	**11631**
G6	800	**2178**	**2178**	**2178**	2165	**2178**	-	**2178**	**2178**
G7	800	**2006**	**2006**	**2006**	1982	**2006**	-	**2006**	**2006**
G8	800	**2005**	**2005**	**2005**	1986	**2005**	-	**2005**	**2005**
G9	800	**2054**	**2054**	**2054**	2040	**2054**	-	**2054**	**2054**
G10	800	**2000**	**2000**	**2000**	1993	1999	-	**2000**	**2000**
G11	800	**564**	**564**	**564**	562	**564**	**564**	**564**	**564**
G12	800	**556**	**556**	**556**	552	554	**556**	**556**	**556**
G13	800	**582**	**582**	580	578	580	**582**	580	**582**
G14	800	**3064**	3063	3061	3060	3063	**3064**	3061	**3064**
G15	800	**3050**	**3050**	**3050**	3049	3049	**3050**	**3050**	**3050**
G16	800	**3052**	**3052**	3052	3045	3050	**3052**	**3052**	**3052**
G17	800	**3047**	**3047**	3046	3043	3045	-	3046	**3047**
G18	800	**992**	**992**	991	988	990	-	**992**	**992**
G19	800	**906**	**906**	904	903	904	-	904	**906**
G20	800	**941**	**941**	**941**	**941**	**941**	-	**941**	**941**
G21	800	**931**	**931**	**931**	930	927	-	**931**	**931**
G22	2000	13358	13311	**13359**	13346	13158	13358	**13359**	**13359**
G23	2000	13340	13325	13342	13317	13116	**13344**	13342	**13344**
G24	2000	13334	13307	**13337**	13303	13125	**13337**	**13337**	**13337**
G25	2000	**13340**	13311	13332	13320	13119	-	13332	**13340**
G26	2000	13326	13304	**13328**	13294	13098	-	**13328**	**13328**
G27	2000	3337	3319	3336	3318	**3341**	-	3336	**3341**
G28	2000	3296	3285	3295	3285	**3298**	-	3295	**3298**
G29	2000	**3405**	3385	3391	3389	3394	-	3391	**3405**
G30	2000	3412	3396	3403	3403	3412	-	3403	**3413**
G31	2000	3307	3281	3288	3288	**3309**	-	3288	**3310**
G32	2000	1408	1406	1406	1398	**1410**	**1410**	1406	**1410**
G33	2000	1378	1376	1378	1362	1376	**1392**	1378	1382
G34	2000	1382	1378	1378	1364	1382	**1384**	1378	**1384**
G35	2000	7676	7675	7678	7668	7485	**7686**	7678	7682
G36	2000	7667	7675	7670	7660	7473	**7679**	7670	7672
G37	2000	7672	7675	7682	7664	7484	**7690**	7482	7484
G38	2000	7674	7674	**7683**	7681	7479	-	**7683**	**7683**
G39	2000	2404	2393	2397	2393	2405	-	2397	**2407**
G40	2000	2396	2388	2390	2374	2378	-	2390	**2400**
G41	2000	**2405**	2380	2400	2386	**2405**	-	2400	**2405**
G42	2000	2480	2458	2469	2457	2465	-	2469	**2481**
G43	1000	**6660**	**6660**	**6660**	6656	6658	**6660**	**6660**	**6660**
G44	1000	**6650**	**6650**	6639	6648	6646	**6650**	**6650**	**6650**
G45	1000	**6654**	**6654**	6652	6642	6652	**6654**	**6654**	**6654**
G46	1000	**6649**	**6649**	**6649**	6634	6647	-	**6649**	**6649**
G47	1000	6657	6657	**6665**	6649	6652	-	**6665**	6657
G48	3000	**6000**	**6000**	**6000**	**6000**	**6000**	**6000**	**6000**	**6000**
G49	3000	**6000**	**6000**	**6000**	**6000**	**6000**	**6000**	**6000**	**6000**
G50	3000	**5880**	**5880**	**5880**	**5880**	5858	5800	**5880**	**5880**
G51	1000	**3848**	3847	3847	3846	3841	-	3847	3844
G52	1000	3850	**3851**	3849	3849	3845	-	3849	**3851**
G53	1000	3849	**3850**	3848	3846	3845	-	3848	**3850**
G54	1000	3851	3851	3851	3846	3845	-	3851	**3852**

5 Conclusion

This work presents a new set of optimality cuts and two heuristics based on genetic algorithms that use these inequalities in their composition. Besides the proposal to present an approach to improve the efficiency of these heuristics, as far as is known nothing of the kind was reported to MAX-CUT problem. Analysis using the benchmark set G considered the best cut value obtained by our heuristics in each instance and we were able to improve the best-known value for the instance G51 and be strongly competitive with current state of the art algorithms, presenting results with a maximum of 1% gap of the best-known values for all instances. Although the experiments consider a limited execution time, increasing it would not have a significant impact on the results provided by our algorithms. Thus, it is possible to conclude that the use of the proposed optimality cuts in GA and TS results in a good improvement to obtain solutions to the MAX-CUT Problem. For future work, we hope to find new ways to explore the optimality cuts presented and apply them to different heuristics and meta-heuristics, such as Simulated Annealing (SA) and Scatter Search (SS). It is possible to use these optimality cuts in Math-heuristics, that are approaches which use mathematical models and heuristics. Also, improvements can be sought in the cuts, searching for more efficient and quick ways to apply them.

References

1. Agrawal, R., Rajagopalan, S., Srikant, R., Xu, Y.: Mining newsgroups using networks arising from social behavior. In: Proceedings of the 12th international conference on World Wide Web, pp. 529–535 (2003)
2. Alidaee, B., Sloan, H., Wang, H.: Simple and fast novel diversification approach for the UBQP based on sequential improvement local search. Comput. Ind. Eng. **111**, 164–175 (2017)
3. Barahona, F.: The max-cut problem on graphs not contractible to K5. Oper. Res. Lett. **2**(3), 107–111 (1983)
4. Barahona, F., Grötschel, M., Jünger, M., Reinelt, G.: An application of combinatorial optimization to statistical physics and circuit layout design. Oper. Res. **36**(3), 493–513 (1988)
5. Bramoullé, Y.: Anti-coordination and social interactions. Games Econom. Behav. **58**(1), 30–49 (2007)
6. Burer, S., Monteiro, R., Zhang, Y.: Rank-two relaxation heuristic for max-cut and other binary quadratic problems. SIAM J. Optim. **12**(2), 503–521 (2001/2002)
7. Chui, K.T., Gupta, B.B., Vasant, P.: A genetic algorithm optimized RNN-LSTM model for remaining useful life prediction of turbofan engine. Electronics **10**(3), 285 (2021)
8. De Simone, C., Diehl, M., Jünger, M., Mutzel, P., Reinelt, G., Rinaldi, G.: Exact ground states of Ising spin glasses: new experimental results with a branch-and-cut algorithm. J. Stat. Phys. **80**(12), 487–496 (1995)
9. De Simone, C., Rinaldi, G.: A cutting plane algorithm for the max-cut problem. Optim. Methods Softw. **3**(13), 195–214 (1994)

10. Dunning, I., Gupta, S., Silberholz, J.: What works best when? A systematic evaluation of heuristics for Max-Cut and QUBO. INFORMS J. Comput. **30**(3), 608–624 (2018)
11. Facchetti, G., Iacono, G., Altafini, C.: Computing global structural balance in large-scale signed social networks. Proc. Natl. Acad. Sci. **108**(52), 20953–20958 (2011)
12. Galinier, P., Boujbel, Z., Fernandes, M.: An efficient memetic algorithm for the graph partitioning problem. Annals OR **191**, 1–22 (2011)
13. Helmberg, C., Rendl, F.: A spectral bundle method for semidefinite programming. SIAM J. Optim. **10**(3), 673–696 (2000)
14. Karp, R.M.: Reducibility among combinatorial problems. In: Miller, R.E., Thatcher, J.W., Bohlinger, J.D. (eds.) Complexity of Computer Computations, pp. 85–103. Springer, Cham (1972). https://doi.org/10.1007/978-1-4684-2001-2_9
15. Kim, S.H., Kim, Y.H., Moon, B.R.: A hybrid genetic algorithm for the MAX CUT problem. In: Proceedings of the 3rd Annual Conference on Genetic and Evolutionary Computation, pp. 416–423. Morgan Kaufmann Publishers Inc. (2001)
16. Kim, Y.H., Yoon, Y., Geem, Z.W.: A comparison study of harmony search and genetic algorithm for the MAX-CUT problem. Swarm Evol. Comput. **44**, 130–135 (2019)
17. Kochenberger, G.A., Hao, J.K., Lü, Z., Wang, H., Glover, F.: Solving large scale Max Cut problems via tabu search. J. Heuristics **19**(4), 565–571 (2013)
18. Kolli, N., Narayanaswamy, B.: Influence maximization from cascade information traces in complex networks in the absence of network structure. IEEE Trans. Comput. Soc. Syst. **6**(6), 1147–1155 (2019)
19. Krislock, N., Malick, J., Rouoin, F.: Improved semidefinite bounding procedure for solving Max-Cut problems to optimality. Math. Program. **143**(1–2), 61–86 (2014)
20. Ma, F., Hao, J.K.: A multiple search operator heuristic for the max-k-cut problem. Ann. Oper. Res. **248**(1–2), 365–403 (2017)
21. Martí, R., Duarte, A., Laguna, M.: Advanced scatter search for the Max-Cut problem. INFORMS J. Comput. **21**(1), 26–38 (2009)
22. Mirjalili, S.: Genetic algorithm. In: Evolutionary Algorithms and Neural Networks. SCI, vol. 780, pp. 43–55. Springer, Cham (2019). https://doi.org/10.1007/978-3-319-93025-1_4
23. Myklebust, T.G.: Solving maximum cut problems by simulated annealing. arXiv preprint (2015)
24. Pandey, A., Banerjee, S.: Test suite optimization using firefly and genetic algorithm. Int. J. Softw. Sci. Comput. Intell. (IJSSCI) **11**(1), 31–46 (2019)
25. Rendl, F., Rinaldi, G., Wiegele, A.: A branch and bound algorithm for Max-Cut based on combining semidefinite and polyhedral relaxations. In: Fischetti, M., Williamson, D.P. (eds.) IPCO 2007. LNCS, vol. 4513, pp. 295–309. Springer, Heidelberg (2007). https://doi.org/10.1007/978-3-540-72792-7_23
26. Sun, L., Cheng, X., Liang, Y.: Solving job shop scheduling problem using genetic algorithm with penalty function. Int. J. Intell. Inf. Process. **1**(2), 65–77 (2010)
27. Wu, Q., Hao, J.-K.: A memetic approach for the Max-Cut problem. In: Coello, C.A.C., Cutello, V., Deb, K., Forrest, S., Nicosia, G., Pavone, M. (eds.) PPSN 2012. LNCS, vol. 7492, pp. 297–306. Springer, Heidelberg (2012). https://doi.org/10.1007/978-3-642-32964-7_30

Assessment of Robust Multi-objective Evolutionary Algorithms on Robust and Noisy Environments

Mateus Clemente de Sousa[1,4(✉)], Ivan Reinaldo Meneghini[2,4],
and Frederico Gadelha Guimarães[3,4]

[1] Instituto Federal de Minas Gerais, Bambuí, Minas Gerais, Brazil
mateus.clemente@ifmg.edu.br
[2] Instituto Federal de Minas Gerais, Ibirité, Minas Gerais, Brazil
ivan.reinaldo@ifmg.edu.br
[3] Universidade Federal de Minas Gerais, Belo Horizonte, Minas Gerais, Brazil
fredericoguimaraes@ufmg.br
[4] Machine Intelligence and Data Science – MINDS Lab, Universidade Federal de
Minas Gerais, Belo Horizonte, Brazil

Abstract. Robust optimization considers uncertainty in the decision
variables while noisy optimization concerns with uncertainty in the evaluation of objective and constraint functions. Although many evolutionary
algorithms have been proposed to deal with robust or noisy optimization problems, the research question approached here is whether these
methods can deal with both types of uncertainties at the same time.
In order to answer this question, we extend a test function generator
available in the literature for multi-objective optimization to incorporate
uncertainties in the decision variables and in the objective functions. It
allows the creation of scalable and customizable problems for any number of objectives. Three evolutionary algorithms specifically designed for
robust or noisy optimization were selected: RNSGA-II and RMOEA/D,
which utilize Monte Carlo sampling, and the C-RMOEA/D, which is a
coevolutionary MOEA/D that uses a deterministic robustness measure.
We did experiments with these algorithms on multi-objective problems
with (i) uncertainty in the decision variables, (ii) noise in the output, and
(iii) with both robust and noisy problems. The results show that these
algorithms are not able to deal with simultaneous uncertainties (noise
and perturbation). Therefore, there is a need for designing algorithms to
deal with simultaneously robust and noisy environments.

Keywords: Robust Optimization · Noisy Optimization · Evolutionary
Algorithm · Test functions

1 Introduction

Uncertainties exist in the real world, such as in the measurement system or final
control element. The search for optimal solutions in the presence of uncertainties

M. C. Naldi and R. A. C. Bianchi (Eds.): BRACIS 2023, LNAI 14197, pp. 33–48, 2023.
https://doi.org/10.1007/978-3-031-45392-2_3

is often referred to as robust optimization in the literature [3,4,9,11]. However, the term robust optimization can be used to encompass uncertainties in both the decision variables and the parameters of the problem, as discussed by Ben-Tal, El Ghaoui, and Nemirovski in their book on Robust Optimization [2]. This includes considering uncertainty in the input of the mathematical model of the optimization problem. In this article, our focus is specifically on uncertainties in the decision variables, referred to as perturbations within the context of robust optimization. On the other hand, uncertainties in the objective functions, which pertain to the output of the process, are referred to as noise within the context of noisy optimization. Figure 1 illustrates these definitions and the distinctions between noisy and robust optimization.

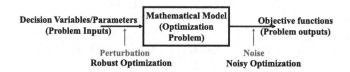

Fig. 1. Robust Optimization (perturbation in the decision variables/parameters) versus Noisy Optimization (noise in the function output).

Real optimization problems can have different types of uncertainties. These uncertainties can prevent the implementation of practical solutions if not taken into account. The designer (optimizer) therefore faces the challenge of finding solutions that are less sensitive to uncertainties. Optimization with uncertainty is a relatively new and rapidly growing research field that has gained significant attention in the past decade [10]. Some works with applications using optimization under uncertainties are [1,7,12,17,20,25]. Examples of uncertainties include noise, model inaccuracies, time variations, measurement inaccuracies, disturbances, and other uncontrolled factors that can degrade the performance of the designed solutions [19].

Recently, Shaaban Sahmoud and Haluk Rahm [22] worked on noisy and dynamic optimization. We aim to fill a gap in the literature by conducting a study that tests evolutionary algorithms in the presence of both types of uncertainties; in decision variables and objective functions. In order to answer this research question, we extend a test function generator available in the literature for multi-objective optimization. The goal is to provide an examination of the behavior of these algorithms under such conditions.

The generator of benchmark problems from [18] uses a bottom-up approach and generates problems with various features, the objective space, and the decision space separately. It allows the creation of scalable and customizable problems for any number of objectives and Pareto fronts with different shapes and topologies, and can incorporate different features such as dissimilarity, robustness, and modality. The objective space can always be represented as a vector $(\mathbf{x}_p, \mathbf{x}_d)$, where $\mathbf{x}_p$ responsible for the spatial location of the points in the objective space and $\mathbf{x}_d$ governs the convergence.

The extension introduces different types of noise (Gaussian, Uniform, and Cauchy) for the objective functions. These noises have different properties, such as being additive or multiplicative, and can have different intensities. The noise intensities were taken from the work of [8]. Proposed intensities were implemented, along with a third intensity we specifically proposed for the Gaussian and Uniform noises.

Then, three evolutionary algorithms designed for robust or noisy optimization are selected: RNSGA-II, RMOEA/D, and C-RMOEA/D. The results of the tests show that the algorithms were not able to handle simultaneous uncertainties in decision variables and objective functions, leading to a degradation in the quality of the solutions compared to only one type of uncertainty (either perturbation or noise). Thus, we conclude that there is a need for designing algorithms specifically for handling simultaneous uncertainties.

In this work, the preliminary definitions and concepts such as uncertainties and robustness measures are presented in Sect. 2. In Sect. 3 presents the extension of the function generator proposed in this work. The results of the computational experiment, including a brief description of the three algorithms used and the experimental setup, are presented in Sect. 4. Finally, the conclusion of the work is presented in Sect. 5.

2   Preliminary Concepts

Some of the main concepts covered in this section include types of uncertainties and robustness measures. These concepts provide the foundation for the later discussions in the paper about robust and noisy multi-objective optimization.

2.1   Types of Uncertainties

There are several classifications of uncertainties in the literature. Three prevalent classifications of uncertainties are presented in Table 1. The parameter δ represents uncertainties in the decision variables (perturbation), $f(\mathbf{x} + \delta)$, or in the objective functions (noise), $f(\mathbf{x}) + \delta$.

According to [4], uncertainties are categorized based on their sources, with some sources of different types of uncertainties being identified in the design and optimization process. Ong et al. [21] consider which elements of the model are affected, such as the objective function, variables, or environment. Finally, Jin and Branke [13] classify uncertainties in optimization problems according to how they impact the optimization process.

We observed that there is an equivalence between the classifications of uncertainties mentioned in the literature. This equivalence is depicted in Table 2. To illustrate this equivalence, consider Type B, which classifies the source of uncertainty as production tolerance. For example, a final control element has uncertainty (a hypothetical example would be a drill that aims to drill 5 m, but may have a $\pm 10\%$ error). This error in the final control element generates a perturbation in the decision variables, thus making it both Type II and Category II.

Table 1. Classifications of types of uncertainties, according to [4,13,21].

Authors	Classification
Jin and Branke (2005) [13]	Type I: Uncertainty in the objective function
	Type II: Uncertainty in decision variables
	Type III: Objective function approximation
	Type IV: Time-varying objective function
Ong et al. (2006) [21]	Category I: $f(\mathbf{x}) + \delta$
	Category II: $f(\mathbf{x} + \delta)$
	Category III: variations operating conditions
Beyer and Sendhoff (2007) [4]	Type A: Environment and operating conditions
	Type B: Tolerances in production
	Type C: Output system uncertainty
	Type D: Feasibility Uncertainty

Table 2. Equivalence between classifications of uncertainties, according to [4,13,21].

Classification	Equivalence
Type I: Uncertainty in the objective function	Type C and Cat. I
Type II: Uncertainty in decision variables	Type B and Cat. II
Type III: Objective function approximation	Type C
Type IV: Time-varying objective function	–
Category I: $f(\mathbf{x}) + \delta$	Type C and Type I
Category II: $f(\mathbf{x} + \delta)$	Type B and Type II
Category III: variations operating conditions	Type A
Type A: Environment and Operating Conditions	Cat. III
Type B: Production Tolerances	Type II and Cat. II
Type C: Output System Uncertainty	Type I, IV and Cat. I
Type D: Feasibility Uncertainty	–

2.2 Robustness Measurements

Mathematically, there are different ways to quantify the uncertainties classified above. One paper that defined robustness measures was by [4]. According to this work, uncertainties can basically be modeled deterministically, probabilistically, or possibilistically:

1. The deterministic type defines domains of uncertainty parameters. An example is the worst-case measure:

$$R(\mathbf{x}, \delta) = \max_{\delta \in \Delta} f(\mathbf{x}, \delta) \tag{1}$$

2. The probabilistic type defines probability measures that describe the probability by which a given event can occur. An example is the Monte Carlo integration followed by the mean:

$$\hat{R}(\mathbf{x}, \delta) = \frac{1}{k} \sum_{i=1}^{k} [f(\mathbf{x} + \delta^i)] \qquad (2)$$

3. The possibilistic type defines fuzzy rules that describe the degree of adherence by which a given event may be acceptable. An example is the treatment of uncertainty sets using fuzzy logic and the theory of evidence [14].

3   Function Generator Extension - Robust Optimization And/or Noisy Optimization

A contribution of this work is the proposal of an extension to the function generator described in [18]. The main aspect of this extension is the integration of concepts from robust and noisy optimization. The Gaussian, Uniform, and Cauchy noise models present in the BBOB functions [8] are incorporated into the function generator, resulting in an optimization problem formulation with uncertainty in both the decision variables and the objective functions. Mathematically:

$$f(\mathbf{x}, \delta_x, \delta_f) = f(\mathbf{x} + \delta_x) + \delta_f \qquad (3)$$

where δ_x indicates the uncertainties in the decision variables and δ_f indicates the uncertainties in the objective functions. In Eq. 3, the noise is considered additive, but it can also be multiplicative:

$$f(\mathbf{x}, \delta_x, \delta_f) = f(\mathbf{x} + \delta_x) \times \delta_f \qquad (4)$$

The main objective of the proposed extension is to evaluate the behavior of algorithms in problems with uncertainty in both the decision variables and the objective functions. To date, the algorithms have only been tested in either robust optimization or noisy optimization. They were designed specifically for one type of uncertainty, so the question is whether they will perform well in the presence of two types of uncertainty simultaneously.

The test function used in this study is called **GPD** and is taken from [18]. This problem allows for changing the format of the Pareto front through a parameter p. The parameter p defines the norm ($p \geq 1$) or quasi-norm ($0 < p < 1$) used in the function $h(\mathbf{x}) = ||T(\mathbf{x})||_p$. If $0 < p < 1$, the Pareto front is convex. If $p = 1$, it is a hyperplane, and if $p > 1$, it is concave. In this article, it was defined as follows: $p = 0.5$ or $p = 1$ or $p = 2$.

The focus of this work is to assess the robustness and noise tolerance of the algorithms. To achieve this, unrestricted and bi-objective problems were chosen. The problem **GPD** is evaluated according to the methodology described in [18].

As previously stated, this work utilizes three models of noise: Gaussian, Uniform, and Cauchy (δ_f of Eqs. 3 and 4). These noises will be incorporated into the function generator. These noise models were adopted from the work of [8] (BBOB Functions). Finck et al. [8] defined two levels of intensity for the noise: moderate and severe. The mathematical description of the three noise models is presented below.

– **Gaussian Noise**

$$f_{GN}(f,\beta) = f \times \exp(\beta\mathcal{N}(0,1)) \tag{5}$$

where β controls the intensity of the noise and $\mathcal{N}(0,1)$ is a random variable with a normal distribution with a mean of 0 and a variance of 1.

– **Uniform noise**

$$f_{UN}(f,\alpha,\beta) = f \times \mathcal{U}(0,1)^{\beta} \max\left(1,\left(\frac{10^9}{f+\epsilon}\right)^{\alpha\mathcal{U}(0,1)}\right) \tag{6}$$

where $\mathcal{U}(0,1)$ represents a random, uniformly distributed number in the interval $(0,1)$ and α is a parameter that controls the intensity of the noise in conjunction with the parameter β, resulting in two random factors. The first factor is uniformly distributed in the interval [0,1] for $\beta = 1$. The second factor (max) is greater than or equal to 1. Finally, the factor ϵ is introduced to avoid division by zero, with a value of 10^{-99} being adopted. The Uniform noise is considered to be more severe than Gaussian noise.

– **Cauchy Noise**

$$f_{CN}(f,\alpha,\mathrm{p}) = f + \alpha\left(1000 + \mathbb{I}_{\{\mathcal{U}(0,1)<\mathrm{p}\}}\frac{\mathcal{N}(0,1)}{|\mathcal{N}(0,1)|+\epsilon}\right) \tag{7}$$

where α defines the intensity of the noise and p determines the frequency of the noise disturbance. The value of ϵ is 10^{-199}. This noise model has two important characteristics. First, only a small percentage of the function values are affected by noise. Second, the noise distribution is considered to be unusual, with outliers occurring from time to time.

The Gaussian and Uniform noises are multiplicative, so the intensity remains constant regardless of the range of values in the objective function. For example, if the maximum noise value is 5, the effect on the objective function will be to increase its value by 5 times, whether the value ranges from [0,1] or [0,100]. Cauchy's noise differs from the previous two in that it is additive. In this case, the scale of the objective function must be considered. For example, if the range of values for one problem is [0,1] and for another problem is [0,100], and the noise peak has an intensity of 1, the intensity for the first problem will be 100%

since the noise is additive, whereas, for the second problem, it will only be 1%. Therefore, in this article, Eq. 7 will not include the 1000 summation term. In this article, the intensity values from [8] are kept.[1]

4   Computational Experiment

In this section, the computational experiments will be presented. A brief description of the three selected algorithms and the testing setup will be provided. Subsequently, the solutions that the algorithms found for problems with perturbation and/or noise will be displayed. Finally, a discussion of the results will be made.

4.1   Algorithms

Typically, robustness measures (discussed in Sect. 2.2) are employed to account for uncertainties, resulting in a new objective function. As a result, existing algorithms in the literature are transformed into robust versions. For instance, the RMOEA/D employs MOEA/D [27] and the RNSGA-II employs NSGA-II [5], with the addition of Monte Carlo sampling in the uncertainty space. The average of this sampling is then calculated [6]. The RMOEA/D selects a set of vectors with the best weighting.

C-RMOEA/D [19] is a coevolutionary algorithm for robust multi-objective optimization that employs the worst-case minimization methodology and the decomposition/aggregation strategy. The decomposition approach facilitates the implementation of a coevolutionary approach [19]. The algorithm uses a competitive strategy between two populations of candidate solutions, with X being a set of vectors from the decision space and Δ being a set of perturbation vectors.

The RNSGA-II and RMOEA/D employ a probabilistic robustness measure, whereas C-RMOEA/D is a coevolutionary algorithm that uses a deterministic robustness measure. Other recent strategies for robust optimization or noisy optimization can be found at [15, 16, 26].

4.2   Experimental Setup

The Function Generator Extension (Sect. 3) will be utilized in the experiments. The Pareto front will be tested with convex, linear, and concave formats with simultaneous uncertainties. Problems with only perturbations or noise will be

---

[1] For Gaussian noise: moderate intensity considers $\beta = 0.01$, which corresponds to a variation of up to 3% (either multiplying the function by 1.03 or multiplying the function by 0.97); severe intensity considers $\beta = 1$, resulting in a variation of up to 20 times (either multiplying the function by 20 or dividing the function by 20). For uniform noise: moderate intensity considers $\beta = 0.01$ and $\alpha = 0.01(0.49 + 1/D)$, where D represents the number of decision variables (always considered as 24 in this work), resulting in a variation of up to 12%; severe intensity considers $\beta = 1$ and $\alpha = 0.49 + 1/D$, resulting in a variation of up to tens of thousands. For Cauchy noise: moderate intensity considers $\alpha = 0.01$ and $p = 0.05$; severe intensity considers $\alpha = 1$ and $p = 0.2$.

demonstrated using the concave Pareto front. The primary aim of this study is to evaluate algorithms in the presence of simultaneous uncertainties. To achieve this, the tests will be conducted with bi-objective and unconstrained scenarios. As mentioned earlier, the function generator is adaptable to these parameters. Tests with perturbations only (Robust Optimization) are conducted as shown in [18].

The following parameters were used for all tests: a population size of 210; a total of 24 decision variables; the stopping criterion is the maximum number of evaluations of the objective function, equal to 150,000. The RNSGA-II and RMOEA/D algorithms used 100 samples for averaging. The C-RMOEA/D used the PBI[2] (Penalty Boundary Intersection) aggregation function and a subpopulation size of 15. The RMOEA/D also utilized a subpopulation size of 15.

4.3   Results

The results will be presented in three parts: (i) robust optimization (with only perturbation); (ii) noisy optimization (with only noise); (iii) robust and noisy optimization (with both perturbation and noise - simultaneous uncertainties).

Robust Optimization. Figure 2 shows the results for $\delta_x = 0.1$, which is considered a high perturbation as it represents an uncertainty of $\pm 10\%$. The RNSGA-II and RMOEA/D (using a mean-based robustness measure) mapped the Robust Front. The algorithm based on a worst-case measure, C-RMOEA/D, also mapped the Robust Front. The results with convex and linear fronts are not displayed due to page limitations. The results are similar to those obtained with a concave front.

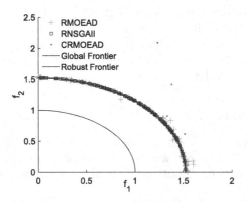

Fig. 2. Result with $\delta_x = 0.1$.

Noisy Optimization. The C-RMOEA/D does not incorporate coevolution with noise only. Hence, it becomes the MOEA/D algorithm, without a second evolutionary cycle. The results were compared between MOEA/D and

---

[2] Further details on the decomposition algorithm and methods can be found in [23].

RMOEA/D and between NSGA-II and RNSGA-II. This comparison was made between the original algorithm without any uncertainty handling, and the algorithm that uses sampling followed by the mean metric.

Figure 3 presents the solutions obtained from RNSGA-II and NSGA-II. The results were similar for Gaussian and Uniform noise. The algorithms showed good convergence and dispersion from the Global Front with moderate intensity. However, they did not produce any mapping for severe intensity. Finally, both RNSGA-II and NSGA-II mapped the Global Front well for Cauchy noise (single additive noise) at both intensities. Due to page limitations, the results of MOEA/D and RMOEA/D will not be shown. These results were similar to those of NSGA-II and RNSGA-II from the Fig. 3.

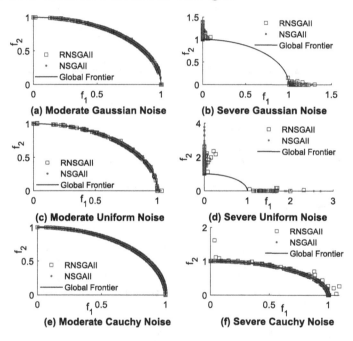

Fig. 3. Result for function generator extension with noise for NSGAII and RNSGAII.

Robust and Noisy Optimization. Figure 4 displays the outcome of extending the function generator with Gaussian noise and $\delta_x = 0.1$. The three algorithms failed to converge for high-intensity noise. This outcome was predictable because the severe noise greatly impacted the original function. However, the C-RMOEA/D solutions exhibited a tendency towards the Robust Front. For moderate-intensity noise, the algorithms achieved convergence and good dispersion. Nonetheless, the RMOEA/D had some poor solutions, but, generally, the solutions converged. The same can be said for the C-RMOEA/D algorithm with concave PF.

Figure 5 displays the results of extending the function generator with Uniform noise and $\delta_x = 0.1$. The analysis of results for Uniform noise is similar to

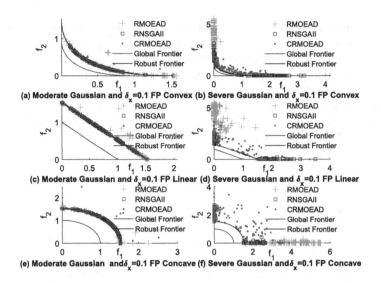

Fig. 4. Result for function generator extension - Gaussian noise and $\delta_x = 0.1$.

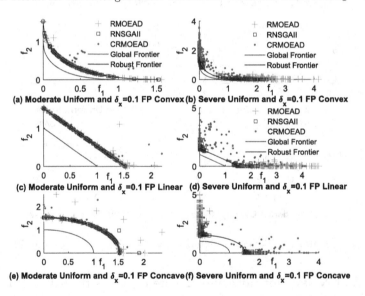

Fig. 5. Result for function generator extension - Uniform noise and $\delta_x = 0.1$.

Gaussian noise. The three algorithms did not converge for high-intensity noise. The algorithms converged and had good dispersion for moderate-intensity noise. However, the RMOEA/D and C-RMOEA/D had some poor solutions. RNSGA-II had some poor solutions for concave PF. As a result, it becomes evident that these algorithms with simultaneous uncertainties begin to deteriorate (even with small noise).

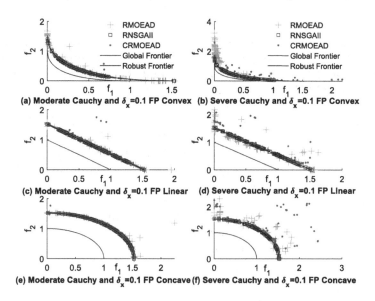

Fig. 6. Result for function generator extension - Cauchy noise and $\delta_x = 0.1$.

Figure 6 displays the result of extending the function generator with Cauchy noise and $\delta_x = 0.1$. This noise differs from the previous ones in that it is additive and not multiplicative. The C-RMOEA/D and RNSGA-II showed good convergence and dispersion for all PF formats and intensities. For high-intensity noise, the C-RMOEA/D had some solutions without convergence, and for the convex PF it failed to map the robust PF at the extremes. The RMOEA/D algorithm showed good convergence and dispersion for moderate-intensity noise. Only some final solutions were poor. Finally, RMOEA/D produced some solutions that mapped small parts of the PF to high-intensity noise. Therefore, the solutions did not have good dispersion and most of them did not converge. The worst convergence result was for convex PF.

As stated in the previous results, the original function is severely contaminated by high Gaussian and Uniform noise. Therefore, we created an intermediate-intensity noise for further testing. The value was $\beta = 0.5$ for Gaussian noise. For Uniform noise, the parameters were $\beta = 0.5$ and $\alpha = 0.2 * (0.49 + 1/D)$. The effect of intermediate-intensity noise showed that the trend of the original function remained. Consequently, EAs must be able to eliminate the effect of noise at this intensity.

For this test set (intermediate-intensity), 30 simulations were performed for each front and algorithm. The IGD (Inverted Generational Distance) evaluation metric [24] was calculated for each simulation, using the Robust Front as the reference set. The solution set that performed best according to the IGD metric (closest to zero) was chosen, and Fig. 7 shows these results for tests of the function generator extension with intermediate noise intensity and $\delta_x = 0.1$.

The algorithms did not achieve good convergence, regardless of the geometry of the PF (convex, linear, or concave) or the type of noise (Gaussian or uniform). This indicates that, when the noise has a greater effect (intermediate intensity), the tested algorithms were unable to eliminate its effect with perturbation.

Specifically, RNSGA-II started to converge at the extremes and in the middle of the PF for Gaussian noise and convex PF. The same happened for linear PF, but with a smaller number of solutions. Some RMOEA/D solutions also converged in the central region of the robust PF (linear PF and Gaussian noise). These algorithms did not show convergence when the PF was concave. The solutions of the C-RMOEA/D algorithm showed a trend of robust PF shapes for Gaussian noise, mainly the convex PF.

Uniform noise had a more significant detrimental effect (compared to Gaussian) on the original function. The results (Fig. 7) were consistent for all PF formats and Uniform noise. The algorithms did not converge. The C-RMOEA/D showed a tendency towards solutions on the border.

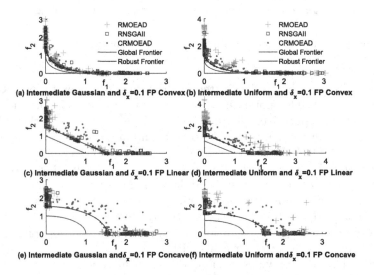

Fig. 7. Result for the extension of the function generator - intermediate noise and $\delta_x = 0.1$ (perturbation in the objective functions and decision variables).

Table 3 contains the average of the IGD metric for the 30 simulations, considering the Robust Front as a reference. The previous results qualitatively demonstrated that the algorithms were not able to deal with simultaneous uncertainties. The quantitative results (Table 3) reinforce this conclusion. The average IGD was high for all algorithms (keep in mind that a lower IGD value indicates better performance), except for the CRMOEA/D algorithm with convex shape. The CRMOEA/D can be considered the superior algorithm regardless of the format of the Pareto Front. Only for the Concave Front and Uniform noise, the RNSGA-II algorithm showed slightly better performance.

Table 3. Results of the IGD metric for intermediate intensity noise and $\delta_x = 0.1$.

Noise	Shape	RMOEA/D	RNSGA-II	CRMOEA/D
Gaussian	Convex	0.9128 ± 0.3981	0.5089 ± 0.0499	0.1707 ± 0.0566
	Linear	1.0129 ± 0.4174	0.7047 ± 0.0670	0.4531 ± 0.1089
	Concave	0.9997 ± 0.4843	0.7339 ± 0.0663	0.5839 ± 0.1448
Uniform	Convex	2.0650 ± 0.2778	0.6528 ± 0.0672	0.2813 ± 0.0763
	Linear	2.2647 ± 0.3573	0.7195 ± 0.0672	0.6580 ± 0.1490
	Concave	1.2511 ± 0.2363	0.7383 ± 0.0416	0.7246 ± 0.3342

Table 3 displays the variability of the IGD metric across 30 simulations, using the Robust Front as the reference, as indicated by the standard deviation. The RNSGA-II algorithm demonstrated superiority in this test across different noise and Border Shape settings, with consistently low standard deviation. The standard deviation test indicates that the RNSGA-II algorithm exhibited minimal variations in IGD across the 30 simulations. The CRMOEA/D demonstrated superior performance compared to the average IGD. However, it is worth noting that the standard deviation was high, suggesting the presence of simulations with exceptionally poor results. Conversely, the RMOEA/D exhibited unsatisfactory outcomes in terms of both the mean and standard deviation.

4.4 Discussion

Algorithms with a robustness measure had no difficulties in finding the Robust Front when the problem has only perturbation.

The evolutionary algorithms were able to eliminate the effect of noise for moderate intensity (noisy optimization), including algorithms without any robustness measure (NSGA-II and MOEA/D). However, the algorithms did not achieve convergence when the effect of Gaussian and Uniform noise was high (severe intensity). One possible explanation is that the intensity of these noises severely deteriorates the original function, causing the function's trend to be lost.

Cauchy noise is different from the previous ones. This noise has peaks of high intensity. As a result, the trend of the original function remains even with severe Cauchy noise. The RMOEA/D algorithm had difficulties in mapping the Global Front to severe Cauchy noise. The reason is that noise has contaminated the average value (noise spikes have increased the average). The original MOEA/D algorithm managed to map the Global Front for this noise. One explanation is that the evolutionary process eliminated the effect of noise on solutions. The NSGA-II and RNSGA-II algorithms achieved excellent mapping of the Global Front for Cauchy noise. This shows that the characteristics of the NSGA-II (archive, elitism, dominance) eliminated the effect of this noise, even when the average was used (RNSGA-II).

Finally, for robust noisy problems, the results demonstrate that the tested algorithms had difficulties in this new class of problems. Some solutions started to deteriorate with moderate noise and perturbation. We performed some tests by

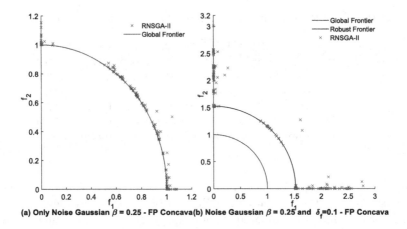

Fig. 8. Result for the extension of the function generator - $\beta = 0.25$ Gaussian Noise).

increasing the noise intensity and observed that the algorithms presented greater difficulty for problems with simultaneous uncertainties. Thus, we created an intermediate noise. The algorithms did not solve the problems with perturbation and intermediate noise.

To confirm the conclusion presented in this discussion, results are provided for Gaussian noise with a value of $\beta = 0.25$, as depicted in Fig. 8. The RNSGA-II algorithm successfully mapped the Global Frontier when the problem contained only noise. However, when both perturbation and noise were present, the majority of algorithm solutions did not converge.

5   Conclusion

In the real world, simultaneous uncertainties are common. A practical example is a drilling process, where the sensor measuring the depth is subject to uncertainty (resulting in noise in the objective function), and the drill (actuator) is also subject to uncertainty (resulting in perturbations in the decision variables). The authors identified a deficiency in the field of optimization with uncertainty, as no previous research had dealt with simultaneous uncertainties.

The results indicated that the algorithms tested (RNSGA-II, RMOEA/D, and CRMOEA/D) were unable to effectively handle simultaneous uncertainties. Consequently, there is a need to formulate algorithms that are capable of handling both robust and noisy environments concurrently.

Acknowledgment. This work has been supported by the Brazilian agencies (i) National Council for Scientific and Technological Development (CNPq), Grant no. 312991/2020-7; (ii) Coordination for the Improvement of Higher Education Personnel (CAPES) through the Academic Excellence Program (PROEX) and (iii) Foundation for Research of the State of Minas Gerais (FAPEMIG, in Portuguese), Grant no. APQ-01779-21. MINDS Laboratory https://minds.eng.ufmg.br/

References

1. Balouka, N., Cohen, I.: A robust optimization approach for the multi-mode resource-constrained project scheduling problem. Eur. J. Oper. Res. **291**(2), 457–470 (2021)
2. Ben-Tal, A., El Ghaoui, L., Nemirovski, A.: Robust Optimization, vol. 28. Princeton University Press, Princeton (2009)
3. Ben-Tal, A., El Ghaoui, L., Nemirovski, A.: Robust Optimization in Applied Mathematics. Princeton Series, Princeton (2009)
4. Beyer, H.G., Sendhoff, B.: Robust optimization – a comprehensive survey. Comput. Methods Appl. Mech. Eng. **196**(33), 3190–3218 (2007). https://doi.org/10.1016/j.cma.2007.03.003
5. Deb, K., Pratap, A., Agarwal, S., Meyarivan, T.: A fast and elitist multiobjective genetic algorithm: NSGA-II. IEEE Trans. Evol. Comput. **6**(2), 182–197 (2002)
6. Deb, K., Sindhya, K., Hakanen, J.: Introducing robustness in multi-objective optimization. Evol. Comput. **14**(4), 463–494 (2006)
7. Duan, J., He, Z., Yen, G.G.: Robust multiobjective optimization for vehicle routing problem with time windows. IEEE Trans. Cybern. **52**(8), 8300–8314 (2021)
8. Finck, S., Hansen, N., Ros, R., Auger, A.: Real-parameter black-box optimization benchmarking 2010: presentation of the noisy functions. Technical report. Citeseer (2010)
9. Gaspar-Cunha, A., Covas, J.A.: Robustness in multi-objective optimization using evolutionary algorithms. Comput. Optim. Appl. **39**(1), 75–96 (2007). https://doi.org/10.1007/s10589-007-9053-9
10. Goerigk, M., Schöbel, A.: Algorithm Engineering in Robust Optimization. Springer, Cham (2016). https://doi.org/10.1007/978-3-319-49487-6_8
11. Gorissen, B.L., Yanıkoğlu, İ, den Hertog, D.: A practical guide to robust optimization. Omega **53**, 124–137 (2015). https://doi.org/10.1016/j.omega.2014.12.006
12. Häse, F., et al.: Olympus: a benchmarking framework for noisy optimization and experiment planning. Mach. Learn. Sci. Technol. **2**(3), 035021 (2021)
13. Jin, Y., Branke, J.: Evolutionary optimization in uncertain environments – a survey. Trans. Evol. Comput. **9**(3), 303–317 (2005)
14. Klir, G.J., Folger, T.A.: Fuzzy Sets, Uncertainty, and Information. Prentice-Hall, Englewood Cliffs (1998)
15. Liu, J., Liu, Y., Jin, Y., Li, F.: A decision variable assortment-based evolutionary algorithm for dominance robust multiobjective optimization. IEEE Trans. Syst. Man Cybern. Syst. **52**(5), 3360–3375 (2021)
16. Liu, R., Li, Y., Wang, H., Liu, J.: A noisy multi-objective optimization algorithm based on mean and Wiener filters. Knowl.-Based Syst. **228**, 107215 (2021)
17. Lu, Y., Xu, Y., Herrera-Viedma, E., Han, Y.: Consensus of large-scale group decision making in social network: the minimum cost model based on robust optimization. Inf. Sci. **547**, 910–930 (2021)
18. Meneghini, I.R., Alves, M.A., Gaspar-Cunha, A., Guimaraes, F.G.: Scalable and customizable benchmark problems for many-objective optimization. Appl. Soft Comput. **90**, 106139 (2020)
19. Meneghini, I.R., Guimaraes, F.G., Gaspar-Cunha, A.: Competitive coevolutionary algorithm for robust multi-objective optimization: the worst case minimization. In: IEEE Congress on Evolutionary Computation (CEC), pp. 586–593 (2016). https://doi.org/10.1109/CEC.2016.7743846

20. Mou, W., Wang, Q., Peng, J.: Accelerating gradient-based optimization via importance sampling. J. Mach. Learn. Res. **22**(22), 1–29 (2021)
21. Ong, Y.S., Nair, P.B., Lum, K.Y.: Max-min surrogate-assisted evolutionary algorithm for robust design. IEEE Trans. Evol. Comput. **10**(4), 392–404 (2006). https://doi.org/10.1109/TEVC.2005.859464
22. Sahmoud, S., Topcuoglu, H.R.: Dynamic multi-objective evolutionary algorithms in noisy environments. Inf. Sci. **634**, 650–664 (2023)
23. Trivedi, A., Srinivasan, D., Sanyal, K., Ghosh, A.: A survey of multiobjective evolutionary algorithms based on decomposition. IEEE Trans. Evol. Comput. **21**(3), 440–462 (2016)
24. Van Veldhuizen, D.A., Lamont, G.B.: Multiobjective evolutionary algorithm research: a history and analysis. Technical report. Citeseer (1998)
25. Yang, J., Su, C.: Robust optimization of microgrid based on renewable distributed power generation and load demand uncertainty. Energy **223**, 120043 (2021)
26. Yang, Y.: Robust multi-objective optimization based on the idea of multi-tasking and knowledge transfer. In: Proceedings of the 14th International Conference on Computer Modeling and Simulation, pp. 257–265 (2022)
27. Zhang, Q., Li, H.: MOEA/D: a multiobjective evolutionary algorithm based on decomposition. IEEE Trans. Evol. Comput. **11**(6), 712–731 (2007)

Optimization Strategies

Binary Flying Squirrel Optimizer for Feature Selection

Luiz Fernando Merli de Oliveira Sementille, Douglas Rodrigues[(✉)],
André Nunes de Souuza, and João Paulo Papa

São Paulo State University, Bauru, SP, Brazil
{lf.sementille,d.rodrigues,andre.souza,joao.papa}@unesp.br

Abstract. Bio-inspired optimization algorithms aim to address the most diverse problems without the need for derivatives, and they are independent of the shape of the search space. The Flying Squirrel Optimizer belongs to the family of bio-inspired algorithms and simulates the movement of flying squirrels from tree to tree in search of food. This paper proposes a binary version of the flying squirrel optimizer for feature selection problems. To elucidate the performance of the proposed algorithm, we employed six other well-known bio-inspired algorithms for comparison purposes in sixteen benchmark datasets widely known in the literature. Furthermore, we employ the binary flying squirrel optimizer in selecting gas concentrations to identify faults in power transformers. The results expressed that Binary Flying Squirrell Optimizer can either find compact feature sets or improve classification effectiveness, corroborating its robustness.

Keywords: Flying Squirrel Optimizer · Metaheuristic · Feature Selection

1 Introduction

Finding the most cost-efficient route for transporting goods, the ideal amount of raw material to manufacture a product, or even the best way to drill oil wells are just a few examples of decision-making many companies face. In these cases and several other complex problems, mathematical optimization can be used to find feasible and optimal solutions.

Through mathematical models, optimization aims for the optimal solution among the feasible candidates in a search space guided by the objective function and the problem's constraints. Given the nature of the objective function, classical mathematical optimization algorithms can be inefficient due to the function's dependence on being continuous and differentiable in the search space. Furthermore, classical algorithms may lose performance in cases where the function is multimodal.

Bio- or nature-inspired algorithms emerge as an elegant alternative for solving complex optimization problems, mitigating the disadvantages of classical

M. C. Naldi and R. A. C. Bianchi (Eds.): BRACIS 2023, LNAI 14197, pp. 51–64, 2023.
https://doi.org/10.1007/978-3-031-45392-2_4

algorithms. Through the idea that natural processes are essentially optimal, many algorithms were created following metaphors of nature, such as Particle Swarm Optimization [10], Bat Algorithm [19], Flower Pollination Algorithm [18], Whale Optimization Algorithm [13], Butterfly Optimization Algorithm [2], and Jellyfish Search [4], among others. This group of bio-inspired algorithms offers near-optimal solutions in a reasonable time, an alternative to solving NP-hard combinatorial problems.

Artificial intelligence has benefited from using bio-inspired algorithms to optimize neural networks, support vector machines, or optimum-path forest models, among many others. In addition to hyperparameter adjustments to minimize misclassification, the dimensionality of datasets can negatively affect the classifier's performance during training, increasing the computational burden and even decreasing accuracy.

One way to solve this problem is through dimensionality reduction techniques, i.e., selecting the most relevant features for the classification task. In other words, we want to remove irrelevant and redundant features to reduce the computational cost required during training and maximize the classifier's hit rate. Over the last few years, bio-inspired algorithms have successfully addressed feature selection problems, for they can obtain good solutions in a reasonable time, even if the problem is complex.

The literature is vast and with promising results. Nakamura et al. [14] proposed the Binary Bat Algorithm (BBA) for feature selection. Although BA is effective for such a purpose, it lacks efficiency compared to other metaheuristics. Rodrigues et al. [16] introduced the Binary Cuckoo Search (BCS) to the same context, and Rodrigues et al. [16] designed a binary-valued Flower Pollination Algorithm (BFPA) to select the most critical sensors to person identification using electroencephalogram signals.

This paper proposes a binary version of the Flying Squirrel Optimizer (FSO) for feature selection, called BFSO. The FSO was proposed by Azizyan et al. [3] inspired by flying squirrels moving from one tree to another in search of food. This work's main contributions lie in using a wrapper-based approach employing BFSO and the Naive Bayes classifier for the feature selection task. However, any other supervised classifier can be used. We considered 16 benchmark datasets to evaluate the proposed approach's performance. The performance of the BFSO was also validated to select relevant gas concentrations to identify faults in power transformers.

In short, this paper figures the following contributions:

- To propose the Binary Flying Squirrel Optimizer;
- To evaluate the proposed approach for feature selection; and
- To employ BFSO for fault diagnosis in power transformers using gas concentration.

The remainder of this paper is organized as follows. Section 2 presents the theoretical background concerning the flying squirrel optimizer, while Sects. 3 and 4 discuss the methodology and experimental results, respectively. Section 5 states conclusions and future works.

2   Flying Squirrel Optimizer

Flying Squirrel Optimizer [3] is a flying squirrel swarm-based algorithm whose population is given by $\mathcal{X} = \{\mathbf{x}_1, \mathbf{x}_2 \ldots, \mathbf{x}_m\}$, where m is the number of squirrels and $\mathbf{x}_i \in \Re^n$ denotes a single possible solution. The proposed model follows two movements squirrels perform in their pursuit of food. The first simulates the awkward gait of squirrels when they are on the ground. This motion is modeled as random walks following a normal distribution $\mathbf{r} \sim \mathcal{N}(\mu, \sigma^2)$ in which μ is the mean position of all flying squirrels and σ is given as follows:

$$\sigma = -\ln\left(1 - \frac{1}{\sqrt{t+2}}\right)^2, \tag{1}$$

where t corresponds to the current iteration.

In the second movement, the squirrels fly from one tree to another through a Lévy flight distribution, as follows:

$$L(\lambda, s, \alpha) = \frac{\lambda \cdot \Gamma(\lambda) \cdot \sin(\lambda)}{\pi} \cdot \frac{\alpha}{s^{1+\lambda}}, \quad |s| \to \infty, \tag{2}$$

where $\Gamma(\lambda)$ stands for the gamma function with index λ, α is a control parameter for the tail distribution ($\alpha = 1$), and s is the step size. According to Mantegna [12], for large steps $s \gg s_0 > 0$, s can be computed through a linear transformation as follows:

$$s = \frac{\psi}{\eta^{\lambda-1}}, \tag{3}$$

where ψ and η are drawn from a Gaussian distribution with zero mean and standard deviations σ_ψ and σ_η computed as follows

$$\sigma_\psi = \left[\frac{\Gamma(1+\lambda)}{\lambda\Gamma((1+\lambda)/2)} \frac{\sin(\frac{\pi\lambda}{2})}{2^{(\lambda-1)/2}}\right], \sigma_\eta = 1. \tag{4}$$

In this algorithm, λ grows linearly at each iteration, increasing the length of steps of the Lévy flight. Moreover, λ is computed as follows:

$$\lambda = \beta + (2 - \beta) * ((t + 1)/T), \tag{5}$$

in which β is a user-configured parameter, and T corresponds to the maximum number of iterations. The position of each squirrel is updated as follows:

$$\mathbf{x}_i(t + 1) = \mathbf{x}_i(t) + L(\lambda, s, \alpha) \cdot \mathbf{r} \otimes (\mathbf{x}_i(t) - \mathbf{x}_{best}), \tag{6}$$

where $\mathbf{x}_{best}$ is the best solution found so far, and $\otimes$ denotes the pointwise multiplication.

2.1   Algorithmic Analysis

Algorithm 1 describes how FSO works in more detail. Lines 1–4 initialize each possible solution (squirrel) with random values uniformly distributed within $[0, 1]$. The fitness value of each solution is set to a great value in Line 4. This loop takes $\theta(mn)$ operations. The main loop in Lines 6–24 controls the FSO mechanism, taking $\theta(T)$ steps. As mentioned earlier, squirrels follow two main movements. The parameter σ of the first (Eq. 1) is computed in Line 7, and variable λ, which concerns the second movement, is obtained in Line 8.

The inner loop in Lines 10–19 computes a new position for each squirrel (Lines 13–14), evaluates its fitness value (Line 15), and updates its position if it has a better solution (Lines 16–19). These steps take $\theta(mn)$ calculations. Last but not least, Lines 20–24 update the best solution whithin the swarm. This loop takes $O(mn)$ operations. The overall complexity is given by $\theta(Tmno)$ operations in which $\theta(o)$ represents the complexity to compute the fitness function.

3   Methodology

Table 1 presents the datasets used to evaluate the robustness of BFSO in the context of feature selection. Note that for this work, we used publicly available datasets in the UCI repository[1] and six datasets, i.e., the last six items presented in Table 1, which will be used to validate the BFSO in the context of identifying faults in power transformers. The data used to form these six data sets were obtained from IEC TC10 [1] and scientific papers [5]. One can observe the different scenarios, i.e., datasets with varying size, from different domains, and with a wide range concerning the number of features. Each dataset was split into training, validation, and testing sets using a $2 : 1 : 1$ ratio. The pipeline of the proposed approach is depicted in Fig. 1, in which Fig. 1a ilustrates the optimization process to find the best subswt of features and Fig. 1b ilustrates the final evaluation step, where the selected features are employed to classify the test subset.

At each iteration, the agent's position in the search space is updated, and the fitness value is computed by evaluating the objective function. Given that agents move through the search space with real values, we propose to use a transfer function to convert real to binary values before evaluating the objective function. The transfer function is given as follows:

$$\mathbf{x}_i = \begin{cases} 1 \text{ if } T(\mathbf{x}_i) > \phi, \\ 0 \text{ otherwise} \end{cases} \tag{7}$$

in which $\mathbf{x}_i$ represents the i-th agent, $\phi \sim \mathcal{U}(0, 1)$, and $T(\cdot)$ stands for a transfer function described as follows:

$$T(\mathbf{z}) = \frac{1}{1 + e^{-\mathbf{z}}}. \tag{8}$$

---

[1] https://archive.ics.uci.edu/.

Algorithm 1: Flying Squirrel Optimizer

---

input : Variable β, $\alpha = 1$, number of flying squirrels m, dimensions n, and iterations T.
output : Global best position $\mathbf{x}_{best}$.
auxiliaries: Fitness value $\mathbf{f} \in \Re^m$ for each individual, auxiliary array $\mathbf{p} \in \Re^n$, and variables acc, and $globalfit$.

1 **for** *each flying squirrel i ($\forall i = 1, \ldots, m$)* **do**
2   **for** *each dimension j ($\forall j = 1, \ldots, n$)* **do**
3       $x_i^j \leftarrow U\{0, 1\}$;
4    $f_i \leftarrow \infty$;
5 $globalfit \leftarrow \infty$;
6 **for** *each iteration t ($t = 1, \ldots, T$)* **do**
7   Update σ according to Eq. 1;
8   Update s according to Eq. 3;
9   Update λ according to Eq. 5;
10   **for** *each flying squirrel i ($\forall i = 1, \ldots, m$)* **do**
11   Calculate its mean position $\mu \in \Re^n$;
12       $\mathbf{r} \leftarrow N\{\mu, \sigma\}$;
13   **for** *each dimension j ($\forall j = 1, \ldots, n$)* **do**
14          $p^j \leftarrow x_i^j + (L(\lambda, s, \alpha) \cdot r^j \cdot (x_i^j - x_{best}^j))$;
15       $acc \leftarrow f(\mathbf{x}_i)$;
16   **if** *($acc < f_i$)* **then**
17          $f_i \leftarrow acc$;
18   **for** *each dimension j ($\forall j = 1, \ldots, n$)* **do**
19             $x_i^j \leftarrow p^j$;

20   **for** *each flying squirrel i ($\forall i = 1, \ldots, m$)* **do**
21   **if** *($f_i < globalfit$)* **then**
22          $globalfit \leftarrow f_i$;
23   **for** *each dimension j ($\forall j = 1, \ldots, n$)* **do**
24             $x_{best}^j \leftarrow x_i^j$;

After applying the transfer function, a binary vector will originate a new subset, selecting only the representative characteristics of the original set. This process is accomplished by multiplying the binary vector that represents the agent's position by the feature vector of the data set, i.e., the position of the vector that has the value 1 indicates that the characteristic corresponding to that position in the feature vector will be selected to compose the new subset. On the other hand, a 0 value indicates the feature will not be selected.

Table 1. Datasets used in the comparative study.

Dataset	Task	Samples	Features
Arcene	Mass Spectrometry	200	10,000
BASEHOCK	Text	1,993	4,862
Caltech101	Image Silhouettes	8,671	784
COIL20	Face Image	1,540	1,024
Isolet	Spoken Letter Recognition	1,560	617
Lung	Biological	203	3,312
Madelon	Artificial	2,600	500
Mushrooms	Biological	8,124	22
ORL	Face Image	400	1,024
PCMAC	Text	1,943	3,289
Phishing	Network Security	11,055	68
Segment	Image Segmentation	2,310	19
Semeion	Handwritten Digits	1,593	256
Sonar	Signal	208	60
Spambase	Network Security	4,601	48
Vehicle	Image Silhouettes	846	18
Wine	Chemical	178	13
1069_5gt	Engeneering	1,069	5
1069_7gt	Engeneering	1,069	7
1086_5ge	Engeneering	1,086	5
1086_7ge	Engeneering	1,086	7
1143_5gte	Engeneering	1,143	5
1143_7gte	Engeneering	1,143	7

Therefore, with a new training and validation subsets in hand, the classifier is trained on the new training set and classifies the samples of the new validation set. The classifier's hit rate is a fitness value to guide the optimization process. The feature subset that achieves the highest accuracy in this process will be stored and used later to train the classifier once again and classify the test subset to determine the real accuracy of the model. We used Naïve Bayes classifier in this paper, for it is parameterless and fast for training[2].

---

[2] It is worthy to say that any other supervised classifier can be used. We recommend models that figure a reasonably efficient training step, for the fitness function might be evaluated several times during the optimization process.

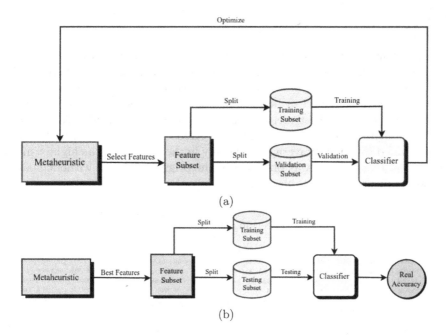

Fig. 1. Proposed approach pipeline: (a) feature optimization and (b) final evaluation.

Regarding FSO running parameters, we used the same recommended by Azizyan et al. [3], i.e., $\beta = 0.5$, and the number of agents and iterations were set to 30 and 60, respectively. Once metaheuristics are stochastic processes, we performed 20 independent runs to allow the calculation of the mean and standard deviation. Besides, different seeds were generated for each run so that distinct samples constituted the division of training, validation, and test subsets.

To provide robust analysis, we chose five widely known bio-inspired algorithms[3] for comparison purposes with the BFSO:

- Binary Aquila Optimizer (BAO) [11];
- Binary Bat Algorithm (BBA) [14];
- Binary Firefly Algorithm (BFA) [6];
- Binary Flower Pollination Algorithm (BFPA) [16]; and
- Binary Particle Swarm Optimization (BPSO) [9].

It is worth noting that the values of the hyperparameters of each algorithm are the same proposed by their respective authors.

---

[3] The algorithms used for comparison purposes and FSO are part of Opytimizer library, which contains several implementations of metaheuristics in Python. The Opytimizer library is available in: https://github.com/gugarosa/opytimizer.

4   Experimental Results

Table 2 presents the F1-score obtained using the methodology described in Sect. 3 with regard to the final evaluation, as illustrated in Fig. 1b. The values are the mean and standard deviation for each of the seventeen benchmark datasets. Note that the values highlighted in bold represent the best results for each dataset. BFSO achieved the highest mean values in twelve datasets, followed by BFA, which obtained the highest mean values in Arcene and BASEHOCK datasets. BAO obtained the highest average value only in the Wine set. The highest mean value in the PCMAC set was obtained in the original set, i.e., without feature selection, indicated in the table using the Naïve Bayes (NB) classifier.

Table 2. Results concerning the test set over all datasets.

Datasets	BAO	BBA	BFA	BFPA	BFSO	BPSO	NB
Arcene	59.3088 ± 8.3389	57.5877 ± 10.3192	**59.8683 ± 9.2318**	57.7523 ± 8.5524	59.1910 ± 10.5680	59.2751 ± 11.3801	53.6850 ± 9.5646
Basehock	94.0338 ± 0.9792	94.0227 ± 0.8793	**95.7893 ± 0.9784**	94.0722 ± 1.1839	94.3733 ± 1.0879	93.9473 ± 0.9834	**95.7893 ± 0.9784**
Caltech101	34.1234 ± 1.5458	34.0947 ± 1.3933	34.7767 ± 1.6077	34.8583 ± 1.6466	**35.1594 ± 1.9759**	34.3723 ± 1.6767	34.8920 ± 1.5331
Coil20	84.3091 ± 4.9616	83.9465 ± 4.9509	82.8525 ± 4.9616	84.9327 ± 5.0247	**86.9788 ± 4.6495**	84.7001 ± 4.8842	79.5231 ± 4.6923
Isolet	80.7023 ± 2.0034	78.8507 ± 3.0662	77.6084 ± 3.0457	81.0067 ± 2.0086	**83.5065 ± 2.6442**	80.0914 ± 3.0755	75.4493 ± 2.8987
Lung	74.0387 ± 7.7944	71.8917 ± 6.7042	71.8445 ± 6.6693	73.5910 ± 8.7942	**74.9191 ± 6.9269**	72.2351 ± 7.0388	67.2273 ± 7.1381
Madelon	58.6742 ± 1.8451	58.4740 ± 1.7233	58.1306 ± 1.7026	58.7479 ± 1.7006	**59.7056 ± 1.6511**	58.3765 ± 1.7613	58.6320 ± 1.4970
Mushrooms	99.9108 ± 0.0790	99.6492 ± 0.2913	99.3877 ± 0.3679	99.9354 ± 0.0532	**99.9446 ± 0.0642**	99.7969 ± 0.2523	98.4284 ± 1.0558
Orl	39.0554 ± 4.7198	38.7282 ± 5.1377	38.6586 ± 5.3897	39.3422 ± 5.6624	**40.7033 ± 4.6595**	38.8620 ± 4.9223	35.4178 ± 4.7992
Pcmac	83.5293 ± 2.4724	82.4870 ± 1.8595	85.0110 ± 2.1555	82.8450 ± 2.1896	83.7857 ± 2.0898	82.4825 ± 2.1978	**85.2949 ± 2.0053**
Phishing	91.7404 ± 0.5551	91.3935 ± 0.5903	91.1606 ± 0.6236	91.8376 ± 0.5977	**92.0855 ± 0.5028**	91.6013 ± 0.6132	69.1982 ± 1.0777
Segment	88.6052 ± 1.9314	85.7717 ± 4.1481	84.6891 ± 3.9508	89.5156 ± 1.4329	**89.6811 ± 1.7589**	86.9927 ± 3.9002	77.2281 ± 2.7683
Semeion	78.9357 ± 2.2782	78.3433 ± 2.9065	76.2091 ± 2.3609	78.1723 ± 2.8818	**80.2570 ± 3.0613**	78.3726 ± 2.9118	75.4282 ± 3.1367
Sonar	70.1847 ± 7.0922	70.0928 ± 6.9859	69.6273 ± 5.7411	71.3052 ± 6.4080	**72.5106 ± 5.6098**	69.5813 ± 6.7136	69.5458 ± 4.2817
Spambase	89.2693 ± 0.9433	88.1176 ± 1.6737	87.7436 ± 1.6078	89.5733 ± 1.2327	**90.1784 ± 0.8849**	88.8250 ± 1.9227	78.9545 ± 3.7119
Vehicle	53.4430 ± 5.8986	51.3056 ± 6.8372	50.3072 ± 6.2957	**53.8514 ± 5.2129**	53.6630 ± 3.8678	51.4325 ± 7.1698	43.5802 ± 5.7693
Wine	**94.8283 ± 4.3541**	94.0176 ± 5.1421	93.7128 ± 4.9796	94.5585 ± 4.5397	93.5664 ± 4.5140	93.9975 ± 3.9848	97.0611 ± 3.1347

Figure 2 illustrates the average number of features selected for the Arcene, BASEHOCK, Coil20, ORL, Segment, and Spambase datasets. BFSO selected the lowest number of features in the Spambase set compared to the other bio-inspired algorithms and was the second-best technique in the ORL and Coil20 datasets, second only to BAO. Furthermore, it is worth the caveat of the size of the standard deviation of the BAO in these datasets. In the Arcene and BASEHOCK datasets, BFSO was not among the best feature selection algorithms.

Moreover, we performed the Friedman test [7,8] with a significance of 0.05 (5%) to assess whether the average F1-score obtained using each bio-inspired algorithm is similar, i.e., null hypothesis H_0 for this test. Then we employed the Nemenyi post-hoc test [15]. Figure 3c illustrates the test performed for the Coil20 dataset, where it can be seen that BFSO performed similarly to BFPA. In the Segment set, illustrated in Fig. 3e, we notice the similarity in the performance of BFSO with BFPA and BAO. Finally, in the Spambase set, illustrated in Fig. 3f, note the similarity between BFSO, BFPA, BAO, and BPSO.

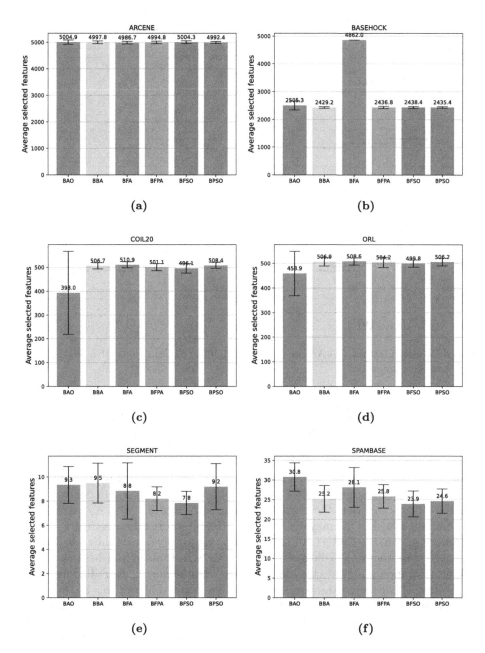

Fig. 2. Average selected features considering each metaheuristic algorithm on: (a) Arcene (b) BASEHOCK (c) Coil20 (d) ORL (e) Segment datasets. (f) Spambase datasets.

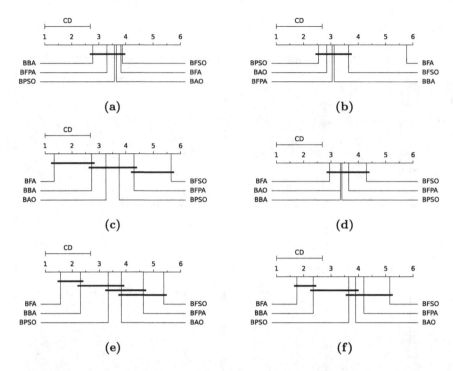

Fig. 3. Nemenyi test on: **(a)** Arcene **(b)** BASEHOCK **(c)** Coil20 **(d)** ORL **(e)** Segment **(f)** Spambase datasets.

Additionally, the Wilcoxon Signed-Rank test [17] with a significance of 0.05 (5%) was performed to validate the data obtained more robustly. Table 3 presents the test result considering F1-score metric obtained during the 20 independent runs. The symbol = indicates that the result obtained after feature selection using the bio-inspired algorithms performed statistically similar to that obtained using the original dataset, i.e., we accept the null hypothesis H_0. In contrast, the symbol $\neq$ indicates that the results were statistically different, i.e., we rejected the null hypothesis. One may observe that BFSO performed statistically similar to the NB only in the Caltech101 set. Based on the BFSO results shown in Table 2 together with the Wilcoxon test obtained, BFSO obtained the best result by significantly increasing the classifier's hit rate when removing degrading features.

Following, we tested the performance of bio-inspired algorithms in Dissolved Gas Analysis (DGA) datasets for the task of gas concentration selection for fault identification in power transformers. Table 4 presents the mean values and the standard deviation for each of the six datasets. Also, note that the top performers are highlighted in bold. NB obtained the highest average values in the datasets 1069_5gt, 1069_7gt, 1143_5gte, and 1143_7gte. BBA and BFA obtained the highest mean values in the 1086_5ge and 1086_7ge datasets, respectively.

Table 3. Results concerning the Wilcoxon Signed-Rank test set over all datasets.

Datasets	BAO	BBA	BFA	BFPA	BFSO	BPSO
Arcene	$\neq$	$\neq$	$\neq$	$\neq$	$\neq$	$\neq$
BASEHOCK	$\neq$	$\neq$	$=$	$\neq$	$\neq$	$\neq$
Caltech101	$\neq$	$\neq$	$=$	$=$	$=$	$\neq$
Coil20	$\neq$	$\neq$	$\neq$	$\neq$	$\neq$	$\neq$
Isolet	$\neq$	$\neq$	$\neq$	$\neq$	$\neq$	$\neq$
Lung	$\neq$	$\neq$	$\neq$	$\neq$	$\neq$	$\neq$
Madelon	$=$	$=$	$=$	$=$	$\neq$	$=$
Mushrooms	$\neq$	$\neq$	$\neq$	$\neq$	$\neq$	$\neq$
ORL	$\neq$	$\neq$	$\neq$	$\neq$	$\neq$	$\neq$
PCMAC	$\neq$	$\neq$	$=$	$\neq$	$\neq$	$\neq$
Phishing	$\neq$	$\neq$	$\neq$	$\neq$	$\neq$	$\neq$
Segment	$\neq$	$\neq$	$\neq$	$\neq$	$\neq$	$\neq$
Semeion	$\neq$	$\neq$	$=$	$\neq$	$\neq$	$\neq$
Sonar	$=$	$=$	$=$	$=$	$\neq$	$=$
Spambase	$\neq$	$\neq$	$\neq$	$\neq$	$\neq$	$\neq$
Vehicle	$\neq$	$\neq$	$\neq$	$\neq$	$\neq$	$\neq$
Wine	$\neq$	$\neq$	$\neq$	$\neq$	$\neq$	$\neq$

Figure 4 illustrates the mean of the selected features for the datasets 1069_5gt, 1069_7gt, 1086_5ge, 1086_7g, and 1143_5gte, and 1143_7gte. It is important to note that the bio-inspired algorithms selected, on average, half of the features of the original set, i.e., even if there is a little loss in the hit rate, the computational cost for training can compensate in these cases, as well as the cost for feature extraction.

Table 5 present the Wilcoxon Signed-Rank test on DGA datasets. All bio-inspired algorithms performed similarly to NB statistically except in the 1143_5gte set and the BPSO in the 1143_7gte set. The difference in the hit rate obtained by the NB to the bio-inspired algorithms was insignificant, which further emphasizes the advantage of training in reduced datasets.

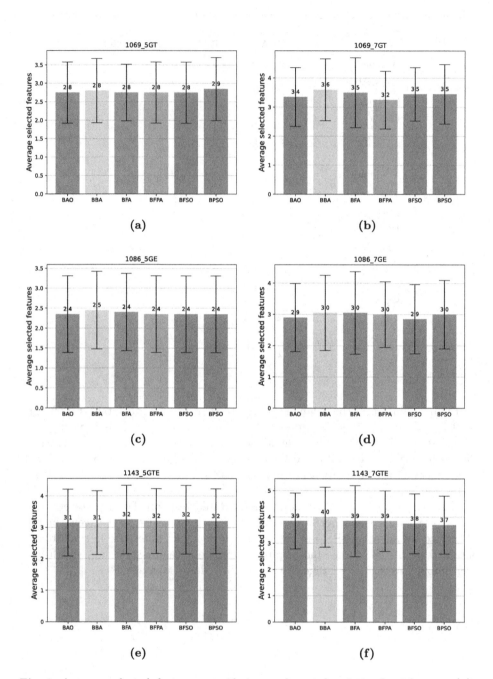

Fig. 4. Average selected features considering each metaheuristic algorithm on: (a) 1069_5gt (b) 1069_7gt (c) 1086_5ge (d) 1086_7ge (e) 1143_5gte (f) 1143_7gte datasets.

Table 4. Average F1-score values considering the comparison among the bio-inspired algorithms on DGA datasets.

Datasets	BAO	BBA	BFA	BFPA	BFSO	BPSO	NB
1069_5gt	97.3243 ± 1.1641	97.2737 ± 1.2112	97.2157 ± 1.1842	97.3243 ± 1.1641	97.3243 ± 1.1641	97.3068 ± 1.1621	**97.3384 ± 1.0960**
1069_7gt	97.2012 ± 1.1321	97.2287 ± 1.1183	97.1757 ± 1.0382	97.2217 ± 1.1281	97.1813 ± 1.1174	97.2011 ± 1.1369	**97.2909 ± 1.0006**
1086_5ge	98.3119 ± 0.4669	**98.4436 ± 0.5092**	98.4188 ± 0.4892	98.3119 ± 0.4669	98.3119 ± 0.4669	98.3119 ± 0.4669	98.1526 ± 0.7254
1086_7ge	98.2952 ± 0.5423	98.3414 ± 0.4875	**98.3807 ± 0.4773**	98.2952 ± 0.5423	98.2898 ± 0.5531	98.2952 ± 0.5423	98.1104 ± 0.8147
1143_5gte	97.1861 ± 0.9976	97.2358 ± 0.8860	97.2347 ± 0.9116	97.2587 ± 0.9162	97.2587 ± 0.9162	97.2587 ± 0.9162	**97.5426 ± 0.8362**
1143_7gte	97.0704 ± 0.8791	97.1016 ± 0.7974	96.9667 ± 0.9500	97.1130 ± 0.9799	97.0212 ± 0.9171	96.9823 ± 0.8714	**97.3678 ± 0.7821**

Table 5. Wilcoxon Signed-Rank test for the bio-inspired algorithms in comparison to Naïve Bayes classifier on DGA datasets.

Datasets	BAO	BBA	BFA	BFPA	BFSO	BPSO
1069_5gt	=	=	=	=	=	=
1069_7gt	=	=	=	=	=	=
1086_5ge	=	=	=	=	=	=
1086_7ge	=	=	=	=	=	=
1143_5gte	≠	≠	≠	≠	≠	≠
1143_7gte	=	=	=	=	=	≠

5 Conclusions

In this work, we propose the binary version of the Flying Squirrel optimizer. Considering the feature selection task, we validated its robustness and performance on sixteen benchmark datasets. Next, we employ the BFSO to select gaseous concentrations for fault identification in power transformers. For BFSO performance comparison purposes, we used the bioinspired algorithms BAO, BBA, BCS, BFA, BFPA, and BPSO.

The results showed that the BFSO can greatly reduce the set of features in all used datasets. Furthermore, in some cases, it achieved better predictive performance than the other bioinspired algorithms used for comparison.

Thus, the performance of the BFSO demonstrated in the feature selection task makes it a viable tool for its performance and low complexity. For future work, one idea would be to use chaotic maps for initializing the flying squirrels and employ opposition-based learning further to improve the exploration and exploitation of the algorithm.

References

1. IEC 60599:2022 Mineral oil-filled electrical equipment in service - Guidance on the interpretation of dissolved and free gases analysis. IEC, Geneva, Switzerland, 4 edn. (2022)
2. Arora, S., Singh, S.: Butterfly optimization algorithm: a novel approach for global optimization. Soft. Comput. **23**(3), 715–734 (2019). https://doi.org/10.1007/s00500-018-3102-4

3. Azizyan, G., Miarnaeimi, F., Rashki, M., Shabakhty, N.: Flying squirrel optimizer (FSO): A novel SI-based optimization algorithm for engineering problems. Iranian J. Optimiz. **11**(2), 177–205 (2019)
4. Chou, J.S., Truong, D.N.: A novel metaheuristic optimizer inspired by behavior of jellyfish in ocean. Appl. Math. Comput. **389**, 125535 (2021)
5. Equbal, M.D., Khan, S.A., Islam, T.: Transformer incipient fault diagnosis on the basis of energy-weighted dga using an artificial neural network. Turk. J. Electr. Eng. Comput. Sci. **26**(1), 77–88 (2018)
6. Falcón, R., Almeida, M., Nayak, A.: Fault identification with binary adaptive fireflies in parallel and distributed systems. In: Proceedings of the IEEE Congress on Evolutionary Computation, pp. 1359–1366. IEEE (2011)
7. Friedman, M.: The use of ranks to avoid the assumption of normality implicit in the analysis of variance. J. Am. Stat. Assoc. **32**(200), 675–701 (1937)
8. Friedman, M.: A comparison of alternative tests of significance for the problem of m rankings. Ann. Math. Stat. **11**(1), 86–92 (1940)
9. Kennedy, J., Eberhart, R.C.: A discrete binary version of the particle swarm algorithm. In: IEEE International Conference on Systems, Man, and Cybernetics, vol. 5, pp. 4104–4108 (1997)
10. Kennedy, J., Eberhart, R.: Particle swarm optimization. In: Proceedings of ICNN 1995-International Conference on Neural Networks, vol. 4, pp. 1942–1948. IEEE (1995)
11. Li, L., Pan, J.S., Zhuang, Z., Chu, S.C.: A novel feature selection algorithm based on aquila optimizer for covid-19 classification. In: Shi, Z., Zucker, J.D., An, B. (eds.) Intelligent Information Processing XI, pp. 30–41. Springer International Publishing, Cham (2022). https://doi.org/10.1007/978-3-031-03948-5_3
12. Mantegna, R.N.: Fast, accurate algorithm for numerical simulation of lévy stable stochastic processes. Phys. Rev. E **49**, 4677–4683 (1994)
13. Mirjalili, S., Lewis, A.: The whale optimization algorithm. Adv. Eng. Softw. **95**, 51–67 (2016). https://doi.org/10.1016/j.advengsoft.2016.01.008
14. Nakamura, R.Y.M., Pereira, L.A.M., Costa, K.A., Rodrigues, D., Papa, J.P., Yang, X.S.: BBA: a binary bat algorithm for feature selection. In: 2012 25th SIBGRAPI Conference on Graphics, Patterns and Images, pp. 291–297 (2012)
15. Nemenyi, P.: Distribution-free Multiple Comparisons. Princeton University (1963)
16. Rodrigues, D., et al.: BCS: a binary cuckoo search algorithm for feature selection. In: IEEE International Symposium on Circuits and Systems, pp. 465–468 (2013)
17. Wilcoxon, F.: Individual comparisons by ranking methods. Biometrics Bull. **1**(6), 80–83 (1945)
18. Yang, X.S.: Flower pollination algorithm for global optimization. In: International conference on Unconventional Computing and Natural Computation, pp. 240–249. Springer (2012). https://doi.org/10.1007/978-3-031-03948-5_3
19. Yang, X.S., Gandomi, A.H.: Bat algorithm: a novel approach for global engineering optimization. Eng. Comput. (2012)

Fitness Landscape Analysis of TPOT Using Local Optima Network

Matheus Cândido Teixeira[1]([envelope])[iD] and Gisele Lobo Pappa[2][iD]

[1] Instituto Federal de Mato Grosso (IFMT), Cuiaba, Brazil
matheus.candido@ifmt.edu.com
[2] Universidade de Minas Gerais (UFMG), Belo Horizonte, Brazil
glpappa@dcc.ufmg.br

Abstract. AutoML addresses the challenge of automatically configuring machine learning pipelines for specific data analysis tasks. These pipelines encompass techniques for preprocessing and classifying data. Numerous approaches exist for discovering the optimal pipeline configuration, with most focusing on optimization methods such as Bayesian optimization and evolutionary algorithms. Nevertheless, limited knowledge exists regarding the structure of the search space that these methods operate within. What is certain is that these spaces incorporate categorical, continuous, and conditional hyperparameters, and effectively handling them is not straightforward. To shed light on this matter, the present study conducts an examination of AutoML search spaces generated by the Tree-based Pipeline Optimization Tool (TPOT) algorithm utilizing local optimal networks (LON). The goal is to gain deeper insights into the overall characteristics of the search space, enhancing comprehension of the search strategy employed and the algorithm's limitations. This investigation aids in understanding the search strategy and constraints of the algorithm, ultimately contributing to the advancement of novel optimization algorithms or the refinement of existing ones within the scientific literature. The findings have implications for enhancing optimization algorithms by illuminating how the search space is explored and the consequent impact on the discovered solutions.

Keywords: TPOT · Local Optima Network · Fitness Landscape

1 Introduction

In the last decade, Automated Machine Learning (AutoML) has experienced significant advancements in the development of algorithms and strategies for generating tailored machine learning pipelines for specific problems [8]. A machine learning pipeline consists of a sequence or parallel execution of operations, including data preprocessing, classification, and post-processing. While classification is commonly used, other machine learning tasks like regression or clustering can also be accommodated. These methods aim to improve accuracy while considering computational constraints, as evaluating these pipelines requires substantial

M. C. Naldi and R. A. C. Bianchi (Eds.): BRACIS 2023, LNAI 14197, pp. 65–79, 2023.
https://doi.org/10.1007/978-3-031-45392-2_5

computational resources. Major cloud platforms, such as Google, Microsoft, and Amazon, have incorporated AutoML methods into their offerings [4].

The generation of these pipelines primarily relies on Bayesian optimization, evolutionary computation, or hybrid techniques [8]. In recent years, the focus has shifted towards optimizing the hyperparameters of artificial neural networks, a process known as Neural Architecture Search (NAS) [3]. However, little is known about the structure of the search space in both AutoML and NAS domains, which could provide insights into more effective and efficient exploration methods.

Given the success of evolutionary computation in exploring these pipeline spaces, which started with basic methods like TPOT [12] and Recipe [16] and progressed to more advanced approaches like CMA-ES [7], it is natural to investigate the fitness landscape that these methods traverse. The fitness landscape of a problem is defined by three components: the set of valid solutions ($\mathcal{X}$), the neighborhood of a solution ($\mathcal{N}$), and a fitness function ($f : \mathcal{X} \rightarrow \mathbb{R}$) that evaluates the solutions in $\mathcal{X}$. By assessing solutions using f and considering the notion of neighborhood, the fitness landscape can be characterized.

Previous research has explored the fitness landscape of AutoML problems using Local Optimal Networks (LON) [10], which offers a global perspective compared to traditional metrics such as FDC. LONs provide a comprehensive view of the fitness landscape, while FDC focuses on local features of the search space. However, previous work employed an artificial search space generated using brute force, with a fixed number of neighbors for each solution, which is unrealistic since many algorithms employ efficient strategies to explore the search space. To address this limitation, we construct the search space using TPOT, an AutoML algorithm.

Essentially, our grammar and search space represent 69,960 valid machine learning pipelines that comprise a classification algorithm and potentially a preprocessing algorithm[1]. Although our grammar is smaller compared to the default TPOT configuration space, which can represent an enormous number of different ML pipelines, this simplification allows us to evaluate every pipeline represented by our grammar and determine the global optimum in the search space. The results reveal that in many cases, TPOT fails to optimize the problem, frequently evaluating suboptimal solutions instead of the global optimum in numerous datasets, indicating TPOT's struggle to optimize in certain search spaces and its tendency to get trapped in local optima.

This work contributes to the field by constructing a new search space for AutoML problems using a real optimization algorithm (TPOT), proposing a novel approach for assigning weights to edges in constructing the Local Optima Network (LON), and conducting an analysis of the constructed LONs.

The remainder of this paper is structured as follows: Sect. 2 reviews related work, Sect. 3 presents our methodology, including details about the grammar, TPOT configuration, search space construction, and LON generation. Section 4

---

[1] In this work, the terms machine learning pipeline, solution, and configuration are used interchangeably, referring to one or more preprocessing/classification algorithms and their parameters.

describes our experimental setup, including the datasets used, statistical information, and hardware specifications. Section 5 presents the results and ensuing discussion. Finally, Sect. 6 concludes the paper.

2   Related Work

Few works have looked so far at the fitness landscape of AutoML problems. Garciarena et al. [6] were the first to perform the analysis of AutoML landscapes considering a subspace of TPOT. Their objective was to identify the local characteristics of the space close to optimal solutions by using metrics such as slope, roughness and neutrality. Their results suggest that many regions of high fitness exist in the space, but these are prone to overfitting. In this same direction, Pimenta et al. [13] looked at fitness landscape metrics to better understand the search space of a huge space of machine learning pipelines. They looked at fitness distance correlation (FDC) and metrics of neutrality, and concluded FDC was a poor metric for performing the analyses.

Turning from AutoML complete pipelines to analyses of loss function in neural architecture search (NAS), Rodrigues et al. [15] characterized fitness landscapes of meta-heuristics for neuroevolution of Convolutional Neural Networks using autocorrelation and the entropic measure of ruggedness. Nunes et al. [9] also analyzed the fitness landscape of NAS in the context of graph neural network architectures. They used FDC together with the dispersion metric (which measures the dispersion between the funnels in the landscape), and have also looked at the neutrality of the space.

In the realm of AutoML problems, there has been ongoing research on the fitness landscape. Pushak et al. [14] conducted an assessment of the landscape of algorithm configuration, specifically focusing on the modality and convexity of parameter responses. Their approach involved defining parameter response slices, where a specific parameter, denoted as p, was varied within a defined window centered around an optimum solution identified by SMAC. By keeping all other parameters constant and measuring the algorithm's performance as a function of p, they evaluated various algorithms for typical optimization problems. The findings revealed that many of the parameter slices exhibited unimodal and convex characteristics, both within instance sets and on individual instances.

One of the few works analyzing fitness landscapes based on LONs and related to our research is presented in [17]. The authors adapted LONs to analyze the overall structure of parameter configuration spaces. They examined the metrics derived from LONs and fitness distance correlation (FDC), observing significant discrepancies when tuning the same algorithm for different problem instances. Notably, in complex scenarios, a substantial number of sub-optimal funnels were identified, while simpler problems displayed a single global funnel. Similarly, Cleghorn et al. [2] investigated parameter spaces for Particle Swarm Intelligence (PSO) with a similar objective. Their analysis unveiled that the macro-level view of PSO's parameter landscapes appeared relatively simple, yet the micro-level view revealed a much higher level of complexity. This complexity posed challenges for parameter tuning, exceeding the initial assumptions.

3 Methodology

The problem of AutoML can be formally defined as a generalization of the Combined Algorithm Selection and Hyperparameter optimization (CASH) problem [5]. In its original definition, given a set $\mathcal{A} = \{A^{(1)}, A^{(2)}, \ldots, A^{(k)}\}$ of learning algorithms, where each algorithm $A^{(j)}$ has a hyperparameter space $\Lambda^{(j)} = \{\lambda^{(1)}, \ldots, \lambda^{(S)}\}$, defined from the full set of algorithm's hyper-parameters Ω, the CASH problem is defined as in Eq. 1[2].

$$A^*_{\lambda^*} = \underset{A^{(j)} \in \mathcal{A}, \lambda \in \Lambda^{(i)}}{\operatorname{argmax}} \frac{1}{k} \sum_{i=1}^{k} \mathcal{F}\left(A_\lambda^{(j)}, \mathcal{D}_{train}^{(i)}, \mathcal{D}_{valid}^{(i)}\right) \tag{1}$$

where $\mathcal{F}(A_\lambda^{(j)}, \mathcal{D}_{train}^{(i)}, \mathcal{D}_{valid}^{(i)})$ is the gain achieved when a learning algorithm A, with hyperparameters Λ, is trained and validated on disjoint training and validation sets $\mathcal{D}_{train}^{(i)}$ and $\mathcal{D}_{valid}^{(i)}$, respectively, on each partition $1 \leq i \leq k$ of a k-fold cross-validation procedure.

A generalization can be made if we replace $\mathcal{A}$ by a set of pipelines $\mathcal{P} = \{P^{(1)}, \ldots, P^{(V)}\}$, which includes a subset of algorithms from $\mathcal{A}$ and their respective set of hyperparameters $\Gamma^{(i)} = \{\Lambda^{(1)}, \ldots, \Lambda^{(S)}\}$, represented by the full set Ψ, as defined in Eq. 2

$$\mathbf{P}^*_{\mathbf{\Gamma}^*} = \underset{\mathcal{P}^{(i)} \subseteq \mathbf{P}, \mathbf{\Gamma}^{(i)} \subseteq \Psi}{\operatorname{argmax}} \frac{1}{K} \cdot \sum_{j=1}^{K} \mathcal{F}(\mathbf{P}^{(i)}_{\mathbf{\Gamma}^{(i)}}, D_{train}^{(j)}, D_{valid}^{(j)}) \tag{2}$$

Figure 1 illustrates a fitness landscape, where the horizontal axis represents the configuration space and the vertical axis represents the fitness of a given configuration. As the landscape fitness concept is multidisciplinary (it can be applied to several problems), it is necessary to define what the configuration space represents and how the fitness is calculated. Therefore, in this article, the configurations space is formed by ML pipelines and the fitness was defined as the F1-score of the pipeline.

Fig. 1. Illustration of a simple fitness landscape

Fig. 2. Examples of pipelines

ML pipelines can be represented through a tree as illustrated in Fig. 2. Each configuration (represented by the horizontal axis) in Fig. 1 corresponds

---

[2] The original definition casts the problem as a minimization one. Here we replace the loss function by a gain function.

to a pipeline according to the examples illustrated in Fig. 2. Pipelines are built according to a context-free grammar that defines that every grammatically correct pipeline has 0 or 1 preprocessing algorithms and a classification algorithm. In the figure, the leaf nodes correspond to the algorithms and their respective hyperparameters. Although some algorithms have continuous hyperparameters, a set of discrete values have been selected and can be consulted through grammar[3].

3.1   TPOT: Tree-Based Pipeline Optimization Tool

TPOT [12] is an automatic optimization tool for ML pipelines (ie, AutoML) that uses Genetic Programming (GP) as optimization heuristics. GP, as well as several algorithms based on the theory of evolution, use crossover, mutation, recombination and selection operators to evolve individuals. In the case of TPOT, individuals are ML pipelines, whose algorithms and hyperparameters are defined through a configuration file.

3.2   Construction of the Fitness Landscape

To analyze how TPOT explores the space, we first select a relatively small configuration space composed of 4 options of preprocessing algorithms (one option is not to use a preprocessing algorithm) and 5 classification algorithms. Each algorithm has a set of hyperparameters and, considering algorithms and hyperparameters, it is possible to form about 70,000 different pipelines. Each pipeline is made up of a classification algorithm and (optionally) a preprocessing algorithm. We evaluate the fitness (f1-score) of each of these solutions.

Then, we run the TPOT in that same space to be able to analyze how the algorithm explores the space, that is, if the TPOT, i.e. the GP, is an algorithm capable of effectively optimizing the AutoML problem. We run TPOT 30 times on each dataset using a different seed for statistical purposes. To build the fitness landscape, it is necessary to define the concept of neighborhood. We define the neighborhood through 3 different types of operators: mutation, recombination and reproduction. When a solution u is mutated and generates another solution v, we say that v is a neighbor of u, that is, $v \in \mathcal{N}(u)$. The same process occurs in the case of recombination. As for the crossover, the process is different. Two solutions u_1 and u_2 are selected for reproduction and a solution v is generated as a product of this operation. Figure 3 illustrates the process.

3.3   Compressing Neutral Nodes

The LON – shown below – is affected by the number of local optima in space. By definition, a locally optimal solution is a solution whose fitness is greater than that of its neighbors, therefore, a local optimum can be verified by comparing its fitness with that of its neighbors or through a local search algorithm like Hill

---

[3] Link to grammar.

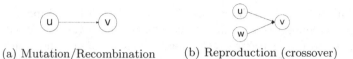

(a) Mutation/Recombination   (b) Reproduction (crossover)

Fig. 3. Edges

Climbing, where the local optimum is the solution where the algorithm does not find any other optimization points. Regardless of the method used, both methods should achieve the same result, however, after some experiments, a divergence was observed in the nodes identified as local optima by both methods. After analyses, it was identified that the order in which the local search algorithm visits neutral regions of the fitness landscape can generate a greater amount of optima because the first solution visited in the neutral region is reported as an LO – there is no solution with fitness greater than current. Although dealing with the Hill Climbing acceptance criteria seems to be a solution to the problem – seeking to advance in a solution with fitness greater than or equal to the current one – the problem still persists if there is no solution with fitness greater than the one already found in the region neutral.

Therefore, the solution adopted in this work was to compress neutral solutions into a single node and insert an attribute that counts the number of merged nodes for later analysis purposes. The strategy used to find neutral solutions was to use Breadth-first search (BFS) to enter neutral regions through the neighborhood of a solution. After this adjustment, the graph with clustered neutral regions is used to construct the LON.

3.4   Local Optima Networks (LON)

A LON is represented by a directed graph $LON = (V,E)$, where V are the vertices that represent the local optima of the fitness landscape and E are the edges that connect the vertices, which can represent basin-transition, escape or permutation edges, as detailed below. By extracting a set of features from this network, we obtain a view of the global structure of the fitness landscape of a problem. Here, we use the original LON model [11] to perform our analysis, along with other models more adapted to neutral spaces, such as Compressed LON (CLON), Monotonic LON (MLON), and Compressed Monotonic LON (CMLON) [1,10].

3.5   Nodes

A local optimum is defined as a solution x such that $\forall x' \in \mathcal{N}(x)$, $f(x) \geq f(x')$, where $\mathcal{N}(\cdot)$ is the neighborhood operator and $f(\cdot)$ is the fitness function. The local optima are defined by a local search algorithm. In this work, the local search looks for solutions with fitness greater than or equal to the current fitness.

Note that the f-measure varies from 0 to 1, and some machine learning pipelines may differ only on the third or fourth decimal point. These small differences make little to no effect on the results, and hence we could say solutions with these differences in fitness are neutral. Because of that, a pipeline p_i is defined as better than a pipeline p_j if the fitness value of p_i is greater than the fitness value of $p_j + \delta$, where δ is a specified tolerance value given by the standard deviation of the mean fitness value of 30 independent random samples.

The literature defines a variety of local search algorithms to find the local optima of a LON, including Iterated Local Search (ILS) and Tabu Search. We have used a classical hill-climbing, as the search space is combinatorial and enumerable.

3.6   Edges

Once the LON nodes are determined, we define whether there is an edge e_{ij} between two local optima LO_i and LO_j by defining a edge weight w_{ij}. An edge exists when $w_{ij} > 0$. The literature defines at least three ways of assigning weights to LON edges [10,11,18]: basin-transition edges, escape edges, and perturbation edges. Each of these methods brings new information about the relationship between local optima, but basin-transition is more useful for combinatorial problems because of its probabilistic nature. A problem with this method is that it takes a long time to evaluate the weight of every edge (in the order of $O(n^3)$), so we propose a new methodology: common ancestor, as described below, and use the basin-transition as a baseline.

Basin-Transition (BT): As the local search defines a mapping from the search space to a set of locally optimal solutions, it also generates basins of attraction. The basin of attraction b_i of a local optimum LO_i is composed by all solutions s in the search space that satisfy $LS(s) = LO_i$, that is, $b_i = \{v \in V(G) \,|\, LS(v) = LO_i\}$. Therefore, in the basin transition method [11], the weight of the edge that connects two local optima LO_i and LO_j is given by:

$$w(e_{ij}) = \frac{1}{|b_i|} \sum_{s \in b_i} \sum_{s' \in b_j} p(s \to s') \tag{3}$$

where $p(s \to s')$ is the probability of a mutation in s generates s'.

Common Ancestor (CA): This method assigns weights to edges proportionally to the number of common ancestors. A node v is an ancestor of u if there exists a path that connects v to u, i.e., $u \rightsquigarrow v$.

3.7   Network Statistics

Number of Nodes: Indicates the total number of nodes (or vertices).

Number of Edges: Indicates the total number of edges present in the graph.

Number of Self-loops: Indicates the number of simple loops in the graph, where an edge's destination is equal to its origin (e.g., (u, u)).

Number of Isolated nodes: Indicates the number of nodes without any incoming or outgoing edges.

Degree Centrality: Measures the importance of a node in the graph based on the number of edges incident to that node. It quantifies the node's connectivity within the network.

In-Degree: The number of edges that terminate at a given vertex. It indicates the importance of a solution in the search space and identifies frequently visited solutions during exploration.

Out-Degree: The number of edges that originate from a given vertex. It indicates the importance of a solution in the search space and identifies solutions that are challenging for the search algorithm to reach.

Density: Measures the proportion of edges present in the graph relative to the total number of possible edges.

Mean Clustering: The average clustering coefficient of each node in the graph, which captures the degree of interconnectedness among a node's neighbors.

Total Weight: The sum of the weights of all edges present in the graph.

Increasing Weights: The sum of the weights of edges that connect nodes where the fitness of the destination node is *greater* than the fitness of the origin node.

Decreasing Weights: The sum of the weights of edges that connect nodes where the fitness of the destination node is *smaller* than the fitness of the origin node.

Neutral Weights: The sum of the weights of edges that connect nodes where the fitness of the destination node is *equal* to the fitness of the origin node.

4   Experimental Setup

4.1   Characterization of the Fitness Landscapes

The fitness landscape of a problem depends directly on the data being analyzed. In this work the pipelines were evaluated in 20 datasets selected from UCI Machine Learning Repository[4] and from Kaggle[5]. The selection criteria were: (i) popularity, (ii) numerical or categorical features and (iii) the task intended was classification.

Considering the search space defined in Sect. 3, we generate all the solutions and evaluated them for each of the 20 datasets, generating 20 different fitness landscapes.

Table 1 presents some features of the datasets used to generate the fitness landscape. The "Code" column indicates the code used to reference each dataset, the "Instances" column indicates the number of instance, the "Features" column indicates the number of features, the "Classes" column indicates the number of

---

[4] https://archive.ics.uci.edu/ml/index.php.
[5] https://www.kaggle.com/datasets.

classes present in the target feature. Following, the "#Optimum" column indicates the number of solutions that achieve the value of optimal fitness. The "Var.", "Mean", "Max." and "Min." columns indicate the variance, mean, highest and lowest fitness value of the evaluated pipelines.

Table 1. Summary of the fitness value of the pipelines evaluated in each dataset

Code	Dataset	Size	Features	Classes	# Optimum	Var	Mean	Max	Min
DS01	abalone	4,177	8	28	1	0	0.22	0.28	0.05
DS02	bank	11,162	16	2	8	0.01	0.72	0.84	0.48
DS03	car-evaluation	1,728	6	4	8	0.01	0.69	0.94	0.37
DS04	diabetes	768	8	2	8	0	0.69	0.79	0.57
DS05	dry-bean	13,611	16	7	32	0.02	0.76	0.93	0.12
DS06	fire	17,442	6	2	8	0	0.89	0.95	0.59
DS07	fruit	898	34	7	1	0.02	0.65	0.92	0.12
DS08	heart	303	13	2	96	0.01	0.64	0.82	0.48
DS09	ml-prove	6,118	51	6	21	0	0.38	0.45	0.16
DS10	mushrooms	8,124	22	7	10	0.01	0.55	0.67	0.31
DS11	nursery	12,960	8	5	1	0.03	0.69	0.99	0.3
DS12	pistachio-28	2,148	28	2	6	0.01	0.79	0.93	0.42
DS13	pumpkin	2,500	12	2	3	0.02	0.74	0.88	0.43
DS14	raisin	900	7	2	1	0	0.81	0.88	0.31
DS15	statlog-segment	2,310	19	7	24	0.03	0.81	0.97	0.27
DS16	texture	5,500	40	11	6	0.06	0.83	1	0.16
DS17	vehicle	846	18	4	7	0.01	0.59	0.79	0.23
DS18	water-potability	3,276	9	2	1	0	0.58	0.66	0.39
DS19	wilt	4,839	5	2	2	0	0.93	0.99	0.41
DS20	wine-quality-red	1,599	11	6	16	0	0.53	0.64	0.26

Further, Fig. 4 shows the box-plots of the fitness distribution of the pipelines generated for each dataset. Note that, for some datasets, the fitness of the solutions is predominantly high or low, while for others they are better distributed. Observe that this distribution does not affect the FLA, but gives an insight on the difficulty of the problem.

Table 1 shows more detailed statistics of the fitness of the pipelines evaluated for each dataset, summarized in the box-plots of Fig. 4. The variance is relatively low since the fitness (F1-score) can vary from 0.00 to 1.00.

Figure 5 shows the total time required for evaluating the entire configuration space. The total time is equal to the sum of the time needed to train and evaluate each solution individually. Some factors that justify the variation in total time are the number of samples and the size of each dataset. For example, DS09 is the largest dataset (51 features). The experiments were run on an Intel(R) Xeon(R) CPU E5-2620 v2 @ 2.10 GHz and approximately 65 GB of RAM.

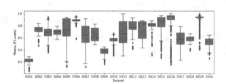

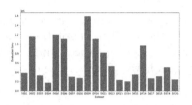

Fig. 4. Box-plot of the fitness of the pipelines in different datasets.

Fig. 5. Evaluation time for the complete search space defined by the grammar.

The configuration space generated by TPOT is composed of pipelines that have a feature selection algorithm, a preprocessing algorithm and a classification algorithm. There are approximately 3.94×10^{10}, 2.66×10^{10} and 1.50×10^{38} combinations of selection, preprocessing and classification algorithms, respectively, considering the different combinations of hyperparameters.

The TPOT configurations are: population_size is 10, generations is 30, mutation rate is 0.9, crossover rate is 0.1 and the scoring function is f1-weighted. The experiments were run on 20 datasets for selected classification problems from the Kaggle platform.

5 Results

Our experiments aim to verify how the TPOT explores the configuration space. The first experiments analyze the performance of TPOT until finding the global optimum (GO) and the number of times the algorithm was able to find the GO. This helps to verify the difficulty faced by the algorithm in the search spaces of the experiment. As the entire space was evaluated by brute force and the space was conditioned to only a few solutions (about 70 thousand), it is possible to detect whether the algorithm found the GO.

The results of the fitness landscape of TPOT show that some solutions are evaluated frequently, that is, even when changing the seed of the algorithm, a subset of solutions (in some cases only one) are evaluated in most executions (more than 70% of the times). The analysis of these solutions indicates that in all cases, they are not the global optimum and that the average error (average of the difference between their fitness and the global optimum) is 0.06 with a standard deviation of 0.07.

Also, TPOT is not able to optimize the datasets in several cases (even in this small search space). For example, it is possible to verify through the column "Hits (%)" of Table 2 that TPOT does not find the global optimum in all executions in most datasets. For example, the global optimum was only found in 16,67% of the runs in the raisin dataset, while in the bank and nursery dataset it was found in all TPOT runs.

However, regardless of the number of hits, in all cases, the average error (between the highest fitness found and the global optimum in space) occurs only

Table 2. Summarization of 30 runs of TPOT using the same

Code	Error	Fitness	Global	Hits (%)
DS01	0	0.284	0.284	100.000
DS02	0	0.838	0.838	100.000
DS03	0	0.938	0.938	73.333
DS04	0.001	0.789	0.790	66.667
DS05	0	0.931	0.931	23.333
DS06	0	0.954	0.954	66.667
DS07	0.001	0.914	0.916	86.667
DS08	0.002	0.819	0.822	50.000
DS09	0.002	0.446	0.448	30.000
DS10	0.001	0.667	0.668	86.667
DS11	0	0.994	0.994	100.000
DS12	0.001	0.929	0.930	73.333
DS13	0.001	0.882	0.883	73.333
DS14	0.004	0.877	0.881	13.333
DS15	0	0.968	0.968	93.333
DS16	0	0.998	0.998	90.000
DS17	0	0.788	0.788	100.000
DS18	0.001	0.664	0.664	93.333
DS19	0.001	0.988	0.989	86.667
DS20	0.001	0.639	0.640	66.667

Table 3. Performance do TPOT: regarding the number of generations until finding the GO

Code	Gen	# Found	# Global
DS01	4.6	1	1
DS02	10.5	7.4	8
DS03	9.64	5.7	8
DS04	10.35	4.87	8
DS05	4.57	6.43	32
DS06	14.35	3.83	8
DS07	5.85	0.87	1
DS08	9.73	18.87	96
DS09	11.78	4.73	21
DS10	11.92	8.53	10
DS11	5	1	1
DS12	6.36	3.9	6
DS13	7.68	1.43	3
DS14	4	0.13	1
DS15	10.46	14.53	24
DS16	4.48	5.33	6
DS17	2.57	5.47	7
DS18	9.36	0.93	1
DS19	4.15	1.5	2
DS20	5.7	9.17	16

from the third decimal place. This result indicates that the TPOT was close to the global optimum in all runs, since a difference of 0.004 in the f1-score between different ML pipelines is not a significant gain or loss in most applications.

Table 3 presents the generation in which the TPOT finds the global optimum (GO). The column "Gen." indicates which generation the first GO was found. The other two columns indicate the amount that was found and the amount that the dataset has. Finding the GO in a large generation indicates that the algorithm found some difficulty in exploring the space and this may be an indication of the difficulty of the problem. Another aspect is the amount of GOs found (especially in multimodal problems) in the space, since the exploration of the search space is an important feature because of the diversity of solutions, which is an important feature for AutoML, as it allows the choice between different algorithms. For example, in the DS08 dataset, the number of GOs is 96, while the average of optimals found by TPOT is only 18.87, that is, a large part of the space diversity remains unexplored.

Figure 6 presents the centrality, density and clustering statistics of the graphs generated by the TPOT exploration. Reported results are the average of 30 runs of the experiment. It is possible to observe that, when we create the LON using the CA method, the results are closer to the original graph, while the BT method presents more discrepant results.

In the three metrics shown in the figure, the BT method has values greater than those calculated in the other graphs. This allows us to conclude that the BT results in a LON with greater centrality and denser than the others.

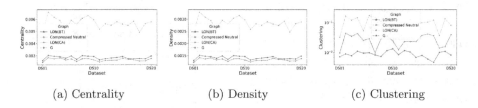

(a) Centrality   (b) Density   (c) Clustering

Fig. 6. Performance of TPOT in different runs on the same dataset

Figure 7 shows several runs of the algorithm on the same dataset, but with different seeds. It is possible to verify that in the first case, the algorithm was not able to find the GO and the exploration was concentrated in only one region — represented by the densest region in the Figure. In the second case, the algorithm found 75% of the global optima in space and the graph has many explored regions and the concentrations are scattered around several solutions. In the third case, the algorithm was able to find all GOs and the Figure indicates a concentration of the exploration around some solutions (visually less than in the second case). The explanation for this is that some hyperparameters do not affect the fitness of the solutions (in certain spaces), therefore, all solutions that differ only in the values of such hyperparameters, obtain the same fitness. Therefore, some solutions are close to each other due to this phenomenon.

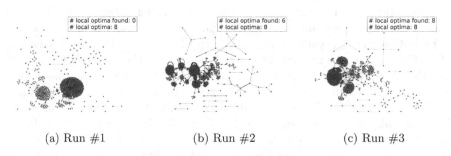

(a) Run #1   (b) Run #2   (c) Run #3

Fig. 7. Performance of TPOT in different runs on the same dataset

Figure 8 presents the results of various graph statistics computed on the LON CA and LON BT graphs. For comparison purposes, the value of the statistics

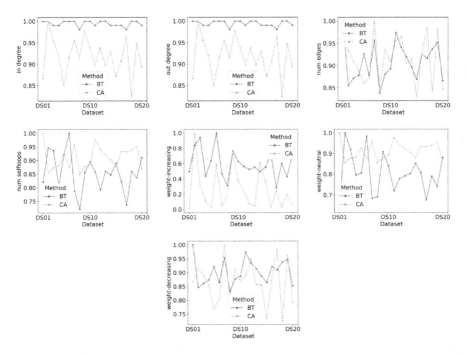

Fig. 8. Several metrics calculated using the LON constructed using the BT and CA method. The results serve as a basis for comparison between methods.

were normalized by the maximum value obtained in each method, that is, we divided the value of each statistic by the maximum value found in datasets. It is possible to observe that for both the in-degree and the out-degree, the statistic evaluated in LON BT is smoother (and higher) than the one evaluated in LON CA. In the other statistics, both vary significantly as the datasets vary. Therefore, we initially analyzed the correlation between the statistics of the two methods.

Considering the relationship between the LON BT and CA, it is possible to verify that the correction between the average degree of entry and exit is 0.051, that is, there is no correlation between the two types. Likewise, the correlation between the weight of the edges that form loops in the LON is -0.335, that is, the correlation is equally weak. The correlation between the number of edges and the weight of the graph is 0.5653 and 0.5414, respectively. Although in these last two cases the correlation is greater, it is still not possible to consider that there is a strong relationship between them. Thus, it is possible to observe that the structure of the LON are relatively different from each other.

In order to compare the difference between both methods, the following statistical test was performed: The correlation between the metrics of the two LONs were correlated with the features of the TPOT exploration graph (From Table 2: Hits (%), Error, Fitness, Global/ From Table 3: Generation, # Found, # Global) and the t-test[6] was used to compare whether there is a difference between the two populations. The resulting p-value was 0.57, which allows accepting the null hypothesis that both means are equal. In this case, we consider that the equality of the mean is a measure of the similarity between the two populations.

6 Conclusion

The results show that TPOT suffers somewhat from local optima, as indicated by the fact that suboptimal solutions remain for many generations. However, the solutions have a fitness relatively close to the global optimum (difference less than 10×10^{-4}).

Another observation is the importance of repeating the experiments, as the algorithm is affected by the position where it was initialized. TPOT can quickly converge to suboptimal solutions and fail to explore space. Through several iterations of the algorithm, it is possible to verify and 'force' the space exploration through different initialization points.

The results also show that LON CA has structural differences in relation to LON BT. The correlations between several metrics between them are low, which suggests that the two methods create "different" LONs. However, both methods produce a graph that has a similar correlation with the metrics of the original graph, with the advantage that the CA method is more efficient. It can be computed through the adjacency matrix of the graphs (cost $O(N^2)$), while the BT method needs to calculate the combinations between pairs of adjacent nodes (cost $O(N^3)$).

Acknowledgements. We would like to thank IFMT for the financial support for transportation and accommodation during the participation in the Conference.

References

1. Adair, J., Ochoa, G., Malan, K.M.: Local optima networks for continuous fitness landscapes. In: Proceedings of the Genetic and Evolutionary Computation Conference Companion, pp. 1407–1414. ACM, New York (Jul 2019). https://doi.org/10.1145/3319619.3326852, https://dl.acm.org/doi/10.1145/3319619.3326852
2. Cleghorn, C.W., Ochoa, G.: Understanding parameter spaces using local optima networks. In: Proceedings of the Genetic and Evolutionary Computation Conference Companion, pp. 1657–1664. ACM, New York (Jul 2021). https://doi.org/10.1145/3449726.3463145, https://dl.acm.org/doi/10.1145/3449726.3463145
3. Elsken, T., Metzen, J.H., Hutter, F.: Neural architecture search: a survey. J. Mach. Learn. Res. **20**(1), 1997–2017 (2019)
4. Erickson, N., et al.: Autogluon-tabular: robust and accurate automl for structured data. arXiv preprint arXiv:2003.06505 (2020)
5. Feurer, M., Klein, A., Eggensperger, K., Springenberg, J., Blum, M., Hutter, F.: Efficient and robust automated machine learning. In: Advances in Neural Information Processing Systems, pp. 2962–2970 (2015)

---

[6] The same variance was not assumed for the two populations.

6. Garciarena, U., Santana, R., Mendiburu, A.: Analysis of the complexity of the automatic pipeline generation problem. In: 2018 IEEE Congress on Evolutionary Computation (CEC). pp. 1–8. IEEE (jul 2018). https://doi.org/10.1109/CEC.2018.8477662, https://ieeexplore.ieee.org/document/8477662/

7. G. Shala, Biedenkapp, A., N.Awad, Adriaensen, S., M.Lindauer, Hutter, F.: Learning step-size adaptation in cma-es. In: Proceedings of the Sixteenth International Conference on Parallel Problem Solving from Nature (PPSN 2020) (Sep 2020)

8. Hutter, F., Kotthoff, L., Vanschoren, J.: Automated machine learning: methods, systems, challenges. Springer Nature (2019). https://doi.org/10.1007/978-3-030-05318-5

9. Nunes, M., Fraga, P.M., Pappa, G.L.: Fitness landscape analysis of graph neural network architecture search spaces. In: Proceedings of the Genetic and Evolutionary Computation Conference, pp. 876–884. ACM, New York (jun 2021). https://doi.org/10.1145/3449639.3459318, https://dl.acm.org/doi/10.1145/3449639.3459318

10. Ochoa, G., Chicano, F.: Local optima network analysis for MAX-SAT. In: Proceedings of the Genetic and Evolutionary Computation Conference Companion, pp. 1430–1437. ACM, New York (jul 2019). https://doi.org/10.1145/3319619.3326855, https://dl.acm.org/doi/10.1145/3319619.3326855

11. Ochoa, G., Verel, S., Daolio, F., Tomassini, M.: Local Optima Networks: A New Model of Combinatorial Fitness Landscapes, pp. 233–262 (2014). https://doi.org/10.1007/978-3-642-41888-4_9, http://link.springer.com/10.1007/978-3-642-41888-4_9

12. Olson, R.S., Moore, J.H.: Tpot: a tree-based pipeline optimization tool for automating machine learning. In: Workshop on Automatic Machine Learning, pp. 66–74 (2016)

13. Pimenta, C.G., de Sá, A.G.C., Ochoa, G., Pappa, G.L.: Fitness Landscape Analysis of Automated Machine Learning Search Spaces, pp. 114–130 (2020). https://doi.org/10.1007/978-3-030-43680-3_8, http://link.springer.com/10.1007/978-3-030-43680-3_8

14. Pushak, Y., Hoos, H.: Algorithm configuration landscapes: In: Auger, A., Fonseca, C.M., Lourenço, N., Machado, P., Paquete, L., Whitley, D. (eds.) PPSN 2018. LNCS, vol. 11102, pp. 271–283. Springer, Cham (2018). https://doi.org/10.1007/978-3-319-99259-4_22

15. Rodrigues, N.M., Silva, S., Vanneschi, L.: A Study of Fitness Landscapes for Neuroevolution (jan 2020), http://arxiv.org/abs/2001.11272

16. de Sá, A.G.C., Pinto, W.J.G.S., Oliveira, L.O.V.B., Pappa, G.L.: RECIPE: a grammar-based framework for automatically evolving classification pipelines. In: McDermott, J., Castelli, M., Sekanina, L., Haasdijk, E., García-Sánchez, P. (eds.) EuroGP 2017. LNCS, vol. 10196, pp. 246–261. Springer, Cham (2017). https://doi.org/10.1007/978-3-319-55696-3_16

17. Treimun-Costa, G., Montero, E., Ochoa, G., Rojas-Morales, N.: Modelling parameter configuration spaces with local optima networks. In: Proceedings of the 2020 Genetic and Evolutionary Computation Conference, pp. 751–759 (2020)

18. Yafrani, M.E., Martins, M.S.R., Krari, M.E., Wagner, M., Delgado, M.R.B.S., Ahiod, B., Lüders, R.: A fitness landscape analysis of the travelling thief problem. In: Proceedings of the Genetic and Evolutionary Computation Conference, pp. 277–284. ACM, New York (jul 2018). https://doi.org/10.1145/3205455.3205537, https://dl.acm.org/doi/10.1145/3205455.3205537

Optimization Strategies for BERT-Based Named Entity Recognition

Monique Monteiro[(✉)] and Cleber Zanchettin

Informatics Center - Federal University of Pernambuco (UFPE), Recife, Brazil
{mlbm,cz}@cin.ufpe.br

Abstract. Transfer learning through language modeling achieved state-of-the-art results for several natural language processing tasks such as named entity recognition, question answering, and sentiment analysis. However, despite these advancements, some tasks still need more specific solutions. This paper explores different approaches to enhance the performance of Named Entity Recognition (NER) in transformer-based models that have been pre-trained for language modeling. We investigate model soups and domain adaptation methods for Portuguese language entity recognition, providing valuable insights into the effectiveness of these methods in NER performance and contributing to the development of more accurate models. We also evaluate NER performance in few/zero-shot learning settings with a causal language model. In particular, we evaluate diverse BERT-based models trained on different datasets considering general and specific domains. Our results show significant improvements when considering model soup techniques and in-domain pretraining compared to within-task pretraining.

Keywords: NER · BERT · transfer learning · model soups

1 Introduction

Named entity recognition (NER) is an important task for meaning extraction of textual content. In recent years, advances in transfer learning with deep neural networks based on Transformers [15] enabled substantial performance improvements for NLP models, especially in low-resource languages. In this context, transfer learning enables training by leveraging previous knowledge of a particular language's grammatical structure - that knowledge is embedded in models previously trained on unlabeled data through self-supervision. Language models are the most classic examples among these pre-trained models.

In particular, for the Portuguese language, the most commonly used model is BERTimbau [12], a neural network that replicates with minor changes the architecture and training procedures for BERT [4], by using brWaC [16] as training corpus. BERTimbau was validated in specific tasks such as named entity recognition, textual similarity, and recognizing textual entailment, surpassing previously published results. Currently, a classical solution for constructing entity recognizers, classifiers, and other discriminative models for low-resource languages

© The Author(s), under exclusive license to Springer Nature Switzerland AG 2023
M. C. Naldi and R. A. C. Bianchi (Eds.): BRACIS 2023, LNAI 14197, pp. 80–94, 2023.
https://doi.org/10.1007/978-3-031-45392-2_6

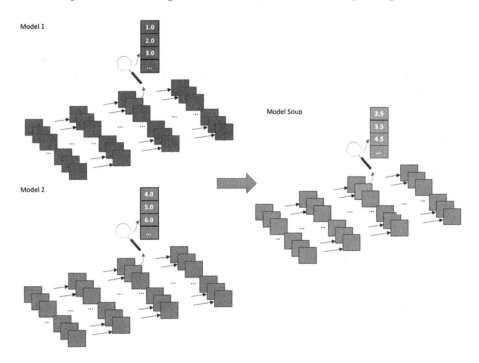

Fig. 1. Model soup - model creation by averaging weights of multiple models trained with different hyperparameter sets. Here, two pre-trained neural networks with identical architecture but different parameters are used to create a third model ("Model Soup"), whose weights are calculated by simply averaging the respective weights of Model 1 and Model 2, without any training.

requires fine-tuning a pre-trained language model - PLM (e.g., BERTimbau for Portuguese) - in the target task by adding linear layers and adjusting the network weights by retraining with a new optimization objective. Such an approach has achieved satisfactory results for most situations at a low training cost [7,8,11].

However, there are still not yet explored techniques, at least for entity recognition tasks in the Portuguese language: domain adaptation, one-shot/few-shot learning, and recent techniques such as *model soups* [17].

This paper applies model soups and domain adaptation techniques to named entity recognition (NER) task for Brazilian Portuguese language. We assess the pertinence of such techniques for this specific language and evaluate multiple ways of applying them, finding, for example, that certain configurations of uniform model soups worked best. We present a study reusing existing models for a relatively low-resource language, and we show that techniques proposed for other problems can effectively transfer to NER.

So, from the viewpoint of practical applications in natural language understanding, our main contributions can be summarized as 1) experiments on a medium-to-low resource language (Brazilian Portuguese, as most gold standard datasets such as HAREM [10] were conceived for the Portuguese variant); 2) further investigation on model soups on NLP, specifically for NER; 3) further investigation on domain adaptation focused on entity recognition, while previous research was more commonly conducted on document classification; and 4) evaluation of NER performance in few/zero-shot learning setups with a causal large language model (LLM).

This paper is organized in the following way: 2 introduces the techniques to be analyzed and other related works, while 3 and 4 present the detailed experiments and results, respectively. Source code is available at https://github.com/monilouise/opt-bert-ner.

2   Related Work

Wortsman et al. [17] propose a technique that consists of generating a model by averaging the weights of two or more trained models, in opposition to the traditional approach, which is based on 1) training multiple models with several hyperparameters and 2) choosing the model with the best performance on a validation set. Also, the idea is to combine multiple models without the additional inference and memory costs related to traditional ensemble learning. The authors refer to the technique as model soups, illustrated in Fig. 1. It supports three variations: 1) construction of a model by averaging all models (*uniform soup*); 2) *greedy soup*, in which models are added sequentially to the soup as they improve the model accuracy in the validation dataset; and 3) *learned soup*, in which the interpolation weights for each model are optimized through gradient descent. According to the authors, the greedy strategy showed the best results. At the same time, learned soups require loading in memory all the models simultaneously, generating more extensive networks and leading to little gain in accuracy. Their experiments were conducted on image and text classification. In this work, we evaluate the application of this technique to the NER task.

Sun, Qiu, Xu and Huang [13] conducted experiments on fine-tuning BERT pre-trained models on document classification. In particular, they analyzed domain adaptation - a training approach with training data from the target task or the same domain. Here, we apply domain adaptation to the NER task.

Houlsby et al. [6] propose transfer learning with adapter modules. According to the authors, adapter modules add only a few trainable parameters per task. New tasks can be added without revisiting previous ones, keeping the parameters of the original network fixed, with a high degree of parameter sharing. They transferred BERT Transformer model to 26 text classification tasks (including the GLUE benchmark), attaining near state-of-the-art performance. The following tasks were evaluated: document classification on several datasets, linguistic acceptability, sentiment analysis, paraphrase detection, semantic textual similarity, textual entailment, and question answering. Only English datasets were used, and the paper does not report performance on the named entity recognition task. Compared to our work, they focus on parameter efficiency and flexibility for different tasks, while we focus on entity recognition performance in a low-resource language. Even so, we intend to investigate Adapters architecture performance for NER tasks in future research.

Regarding Portuguese language, Rodrigues et al. [9] introduced Albertina, a Transformer-based foundation model both for European (Portugal) and American (Brazil) Portuguese variants. Compared to BERTimbau, it uses 128-token sequence truncation instead of 512. The authors report results on tasks such as text entailment, semantic textual similarity and disambiguation tasks, but to our knowledge, they did not evaluate the model on entity recognition.

Finally, BERT was one of the first transformer-based models; up to our knowledge, the current main Portuguese PLMs were based on it. It takes some concerns about its potential compared to more recent PLMs. However, experiments conducted by Tänzer et al. [14] show that BERT can reach near-optimal performance even when many of the training set labels have been corrupted.

3   Methodology and Experiments

3.1   Model Soups Experiments

Creating a model by averaging other models' weights was validated on image and document classification tasks [17].

Formally, let $\theta = FineTune(\theta_0, h)$ the set of parameters obtained by fine-tuning the pre-trained initialization θ_0 with the hyperparameters configuration h. The technique uses a mean value for θ_i, i.e., $\theta_S = \frac{1}{|S|} \sum_{i \in S} \theta_i$, where $S \subset \{1, ..., k\}$ and k is the number of hyperparameter configurations (or models) to be used.

Initially, we reproduced the idea using a uniform average. The greedy approach was not analyzed due to our low number of available candidate models. So, we created a model whose weights are the average of the weights of different models:

(A) the entity recognition model developed to validate BERTimbau [12], trained on an adapted version from the First HAREM[1] [10];
(B) model analogous to (A), but trained with data augmentation[2] and;
(C) model adapted to the same corpus for the target task (First HAREM), as described in Sect. 3.2.

We denote this model as the average model or model soup.

For the combinations cited above, we evaluated the BERT base and large variations, which differ in the number of parameters. Also, we evaluated settings with and without an additional layer for conditional random fields (CRF), which we used to reduce the probability of generating invalid label sequences.

There is a difference between our experiments and the original proposal by Wortsman et al. (2022): the authors assume that the models were independently

---

[1] The adapted version refers to a setting called "selective" by the authors, in which only 5 classes are used (PERSON, ORGANIZATION, LOCAL, VALUE and DATE).
[2] Here, we used label-wise token replacement (LwTR) [3].

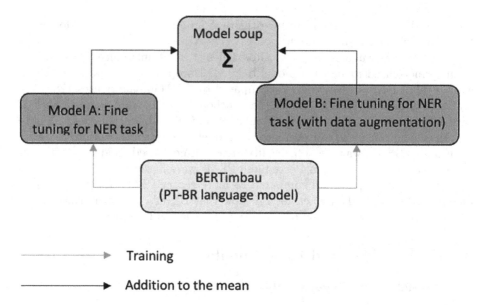

Fig. 2. Model soup - the original idea (adapted to the setting of two models trained on a NER task from a common initialization - in this case, the language model).

optimized from the same initialization, which would lead them to the same region or valley on the error surface. Such strategy is represented in Fig. 2. But here, according to Fig. 3, only models A and B start from the same initialization (BERTimbau language model). In contrast, model C was finetuned on the First HAREM textual content.

The experiments' results are shown in Sect. 4.1.

3.2   Domain Adaptation Experiments

Sun et al. [13] conducted experiments on different fine-tuning options, including multi-task learning. However, they concluded that the benefits from multi-task learning tend to be inferior to those obtained from additional pretraining. So, we did not experiment with multi-task learning.

To conduct the experiments on domain adaptation, we use as start points:

(A) a NER model for (long) documents related to public accounts auditing[3][4];
(B) the NER model trained for BERTimbau evaluation [12];
(C) an entity recognition model [11] trained on a public news dataset [7].

In (A), during domain adaptation, the original language model - BERTimbau - received additional training on a dataset different from the one used to train the

---

[3] We used the hyperparameters for learning rate and batch size suggested by Silva et al. [11].

[4] This dataset contains non-public data and cannot be made publicly available.

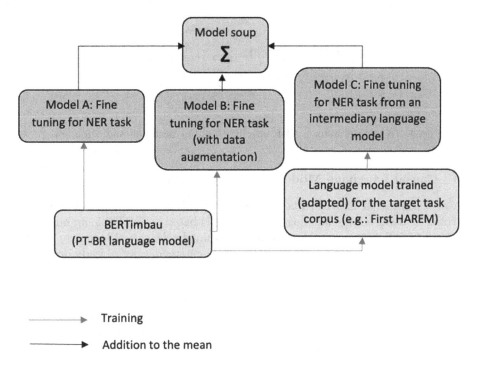

Fig. 3. Model soup - alternative version using two models trained from a common parameter set (BERTimbau language model) and a third model trained on an intermediary language model.

target NER model. However, this dataset came from the same domain and origin (documents related to public accounts auditing), leading to an intermediary and domain-optimized language model. As the training task was Masked Language Model (MLM), such dataset does not contain any label and can be considered a kind of "superset" for the entity recognition dataset: for example, the dataset for training the domain-adapted language model has 52,912 documents, against 431 documents for the labeled dataset used during entity recognition model training. The complete flow is described in Fig. 4.

For (B) and (C), during the construction of the intermediary language models, the respective datasets were used: First HAREM (the same train split used by Souza et al. [12]) and the media news dataset (the same train split as the original dataset [7]). The labels were discarded for both datasets in the phase of MLM training.

The three resulting language models were used as base models for retraining the three cited entity recognizers to measure domain adaptation impact. Learning rate and batch size hyperparameters for all intermediary language models training were the same as reported by Souza et al. [12].

Section 4.2 shows the experiments' results.

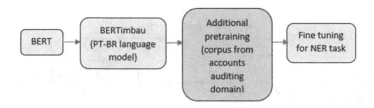

Fig. 4. Training flow for a NER specific to public accounts audit domain.

3.3   Causal Language Modeling - Few/zero-Shot Learning

We also conducted experiments on entity recognition with few and zero-shot learning by using a large language model (LLM) based on GPT-3.5, an improved version of GPT-3 [1], a language model pre-trained with causal language modeling objective. We used the same Brazilian public news dataset already described [7] and compared the following settings:

(A) a NER model [8] based on finetuning BERTimbau to the news dataset
(B) GPT 3.5 few-shot learning with instruction prompt and examples
(C) GPT 3.5 few-shot learning with no instruction prompt (only examples)
(D) GPT 3.5 zero-shot learning

The results are shown in Sect. 4.3.

4   Results and Discussions

4.1   Model Soup Results Analysis

Table 1 summarizes the results from the main experiments on the model soup technique. The bolded rows refer to the baseline (original) models, with mean and standard deviation from 5 training executions. Rows prefixed as "M1:", "M2:", ..., and "M12:" are the respective "best" models for each variant among five training runs. These models are the model soups components. Finally, rows related to the model soups - labeled with + (plus) signs - do not involve training (only two or three models averaging), so they were not randomized (this is why standard deviation values are not shown in these rows, except for the setup with additional fine-tuning, as explained below). We report precision, recall, and F1 metrics for all the experiments.

In the first experiment, we evaluate the direct application of the model soup technique. We used a uniform average from a set composed of the respective best models (i.e., best training) among the variations described in 3.1.

Later, we evaluated the second setup, in which the model soup received additional fine-tuning for the NER task. As shown in Table 1, for the smaller variant in model size (base), the first model, without additional fine-tuning, shows better results for precision and F1 metrics.

Table 1. Experiments with model soups for NER BERTImbau ("BERT$_{\text{BASE}}$" or "BERT$_{\text{LARGE}}$") trained on First HAREM

Model	Precision (%)	Recall (%)	F1 (%)	↑
Original BERT$_{\text{BASE}}$	82.1 ($\sigma = \pm1.3$)	82.4 ($\sigma = \pm0.6$)	82.3 ($\sigma = \pm0.9$)	–
M1: BERT$_{\text{BASE}}$: original (best F1)	83.9	83.1	83.5	–
M2: BERT$_{\text{BASE}}$: (HAREM)	82.9	83.0	82.9	–
M3: BERT$_{\text{BASE}}$ (data augm.)	82.2	82.4	82.3	–
M1 + M3	**85.5**	82.9	**84.2**	+1.9
M1 + M2 + M3	**86.1**	82.0	84.0	+1.7
Original BERT$_{\text{LARGE}}$	81.8 ($\sigma = \pm1.2$)	82.4 ($\sigma = \pm0.5$)	82.1 ($\sigma = \pm0.5$)	–
M4: BERT$_{\text{LARGE}}$: original (best F1)	82.8	82.3	82.6	–
M5: BERT$_{\text{LARGE}}$: (HAREM)	82.7	83.3	83.0	–
M6: BERT$_{\text{LARGE}}$ with data augm	84.2	82.8	83.5	–
M4 + M6	**86.0**	82.6	**84.3**	+2.2
M4 + M5 + M6	**87.1**	81.1	**84.0**	+1.9
Original BERT$_{\text{BASE}}$ + CRF	84.1 ($\sigma = \pm1.2$)	82.0 ($\sigma = \pm0.6$)	83.0 ($\sigma = \pm0.8$)	–
M7: BERT$_{\text{BASE}}$ + CRF: original	85.0	82.3	83.6	–
M8: BERT$_{\text{BASE}}$ + CRF (HAREM)	82.7	83.3	83.0	–
M9: BERT$_{\text{BASE}}$ + CRF (data augm.)	86.2	81.8	83.9	–
M7 + M9	**87.5**	81.5	84.4	+1.4
M7 + M8 + M9	**87.8**	80.4	83.9	+0.9
Original BERT$_{\text{LARGE}}$ + CRF	83.9 ($\sigma = \pm1.2$)	82.0 ($\sigma = \pm0.8$)	83.0 ($\sigma = \pm0.6$)	–
M10: BERT$_{\text{LARGE}}$ + CRF: original	84.7	82.8	83.7	–
M11: BERT$_{\text{LARGE}}$ + CRF (HAREM)	84.8	83.0	83.9	–
M12: BERT$_{\text{LARGE}}$ + CRF (data augm.)	86.1	82.0	84.0	–
M10 + M12	**87.2**	81.1	84.1	+1.1
M10 + M11 + M12	86.3	79.3	82.7	−0.3

Experiments with model soups: the strategy shown in Fig. 2 (i.e., two models initialized from the same weights and independently optimized) leads to better results than the alternative strategy shown in Fig. 3. The base variant shows better results for precision and F1 metrics without additional fine-tuning, making the extra training step unnecessary. Values shown in boldface are above one standard deviation from the mean value for each metric.

As already described in 3.1, for each model size (base/large), we added the following components to each combination: (A) original BERTimbau NER; (B) BERTimbau NER retrained with data augmentation; and (C) BERTimbau NER retrained after domain adaptation to First HAREM (original BERTimbau language model fine-tuned to First HAREM text set), as described in 3.2.

Table 2. Domain adaptation for documents related to public accounts audit.

Model	Precision (%)	Recall (%)	**F1 (%)**	↑
BERT$_{BASE}$: original	82.2 ($\sigma = \pm0.4$)	89.1 ($\sigma = \pm0.3$)	86.0 ($\sigma = \pm0.1$)	-
BERT$_{BASE}$: domain adapted	82.4 ($\sigma = \pm0.3$)	89.4 ($\sigma = \pm0.8$)	**86.5** ($\sigma = \pm0.4$)	**+0.5**

Domain adaptation for documents related to public accounts audit: here, we used in-domain adaptation and achieved the best result among the experiments, in contrast to within-task adaptation. For Masked F1, the improvement was exactly on the standard deviation frontier. Precision, recall, and Masked F1 values are the mean values from 5 runs with random initialization.

Later, the third variant (C) was removed from each combination, leading to the original schema shown in Fig. 2. Finally, we observed that the combinations based only on (A) (original NER) and (B) (data augmented NER) led to better values for precision and F1 metrics, confirming Wortsman's (2022) original hypothesis of using independently optimized models from the same initialization.

When we compare our methodology with the one used by the authors [17], they use accuracy in most image and text classification. For example, the authors do not refer explicitly to recall, which shows worse results in our experiments.

So, further investigation is needed about the reason recall is worsening by the use of the model soup technique, which at the moment makes us believe that such a method could be more suitable to situations in which precision is more important than getting a high number of entities or false positive cases.

On the other hand, given known limitations in using precision-recall-F1 for entity recognition, better and more interpretable metrics for this task are a current research topic [5].

Finally, according to Wortsman et al. [17], preliminary experiments show that improvement in the textual corpus, although present, is less profound than in image classification. The authors stress the need for additional investigation into this subject. They used ImageNet and several variations, including illustrations and other representations beyond real photos (e.g., drawings, arts, cartoons, tattoos, origami, sculptures, etc.). But for textual classification, they used general domain datasets for paraphrase detection, coherence vs. contradiction, linguistic acceptability, and sentiment analysis. Preliminary and qualitative analysis of the different data for images vs. texts shows more variability and larger data size for the first case (e.g., ImageNet contains millions of images), which could have led to a more significant impact on image classification.

4.2 Domain Adaptation Results Analysis

In this subsection, we report results achieved by NER models when trained over intermediary language models, as described in 3.2.

The results of the experiment with the NER model trained on documents related to public accounts auditing are shown in Table 2.

Table 3. Domain adaptation for media news.

Model	Precision (%)	Recall (%)	F1 (%)	↑
BERT$_{BASE}$: original	85.1 ($\sigma = \pm0.7$)	87.6 ($\sigma = \pm1.2$)	86.9 ($\sigma = \pm0.7$)	–
BERT$_{BASE}$: domain adapted	85.3 ($\sigma = \pm0.7$)	87.7 ($\sigma = \pm1.0$)	87.2 ($\sigma = \pm0.6$)	+0.3

Domain adaptation for media news: within-task domain adaptation led to little improvement for Masked F1 (below one standard deviation).

Table 4. Domain adaptation for First HAREM.

Model	Precision (%)	Recall (%)	F1 (%)	↑
BERT$_{BASE}$: original	82.1 ($\sigma = \pm1.3$)	82.4 ($\sigma = \pm0.6$)	82.3 ($\sigma = \pm0.9$)	–
BERT$_{BASE}$ + CRF: original	84.1 ($\sigma = \pm1.2$)	82.0 ($\sigma = \pm0.6$)	83.0 ($\sigma = \pm0.8$)	–
BERT$_{LARGE}$: original	81.8 ($\sigma = \pm1.2$)	82.4 ($\sigma = \pm0.5$)	82.1 ($\sigma = \pm0.5$)	–
BERT$_{LARGE}$ + CRF: original	83.9 ($\sigma = \pm1.2$)	82.0 ($\sigma = \pm0.8$)	83.0 ($\sigma = \pm0.6$)	–
BERT$_{BASE}$: domain adapted	81.9 ($\sigma = \pm1.0$)	82.2 ($\sigma = \pm0.8$)	82.0 ($\sigma = \pm0.9$)	–0.3
BERT$_{BASE}$ + CRF: domain adapted	85.1 ($\sigma = \pm0.7$)	81.6 ($\sigma = \pm2.0$)	83.3 ($\sigma = \pm1.1$)	+0.3
BERT$_{LARGE}$: domain adapted	81.0 ($\sigma = \pm2.3$)	82.8 ($\sigma = \pm0.9$)	81.9 ($\sigma = \pm1.5$)	- -0.2
BERT$_{LARGE}$ + CRF: domain adapted	84.9 ($\sigma = \pm0.8$)	81.7 ($\sigma = \pm0.8$)	83.3 ($\sigma = \pm0.4$)	+0.3

Domain adaptation for First HAREM: within-task domain adaptation led to little improvement for Masked F1 (below one standard deviation) in some setups and to worse results in others.

As a comparison metric, Masked F1 [11] was used. This metric is F1 calculated over post-processed output, correcting invalid transitions according to the IOB2 schema instead of using the raw output directly. Based on in-domain adaptation, this setup led to the most pronounced improvements.

In the experiment conducted on the NER model for media news [7,8,11], the results are shown in Table 3.

The experiment was conducted only with variants based on a pre-trained Brazilian Portuguese language model (BERTimbau) because multi-language models gave an inferior performance, according to Silva et al. [11] and Chen et al. [2]. Masked F1 was also used as the main comparison metric. Table 4 shows the results achieved with the NER model used in BERTimbau evaluation [12].

As can be noted in Tables 3 and 4, experiments conducted with the media news NER and BERTimbau NER did not reveal significant differences after domain adaptation. Such results confirm observations by Sun et al. [13]: the domain adaptation made on media news and First HAREM is "within-task" (the texts used are the same as the training texts for the target task). In general,

Table 5. Domain adaptation for First HAREM - qualitative analysis.

Target Classification	BERT$_{BASE}$+ CRF	Dom. Adapt. BERT$_{BASE}$ + CRF
"... que costumavam comer na noite de Consoada TIME" (...who used to eat at **Consoada** night)	"... que costumavam comer na noite de Consoada LOC"	"... que costumavam comer na noite de Consoada TIME"
"Gostei muito da feira do Soajo LOC" (I enjoyed **Soajo** Fair a lot)	"Gostei muito da feira do Soajo PERSON"	"Gostei muito da feira do Soajo LOC"
"E qual a lembrança mais antiga da cidade? São João do Souto LOC" (And what's the oldest city's memory? **São João do Souto**)	"E qual a lembrança mais antiga da cidade? São João do Souto"	"E qual a lembrança mais antiga da cidade? São João do Souto LOC"
"... comecei a fazer trabalhos na Abade da Loureira ORG" (... I started to work for **Abade da Loureira**)	"... comecei a fazer trabalhos na Abade da Loureira LOC"	"... comecei a fazer trabalhos na Abade da Loureira ORG"
"Tirei o curso de formação no Centro de Formação de Informática do Minho ORG" (I took the graduation course at **Minho Informatics Center**)	"Tirei o curso de formação no Centro de Formação de Informática do Minho LOC"	"Tirei o curso de formação no Centro de Formação de Informática do Minho ORG"

The second column shows outputs generated by the baseline NER, while the third column shown outputs from the NER trained on the intermediary language model (domain-adapted NER). While the first one misclassifies Portuguese expressions, the second one labels them correctly. All the examples belong to First HAREM test dataset.

"in-domain" pretraining - using texts from the same domain which are different from the texts in the training dataset for the target task - gives superior results. We suspect that within-task pretraining could lead to overfitting because it uses the same texts from the target task dataset.

However, after error analysis, we realized that some European expressions, organizations, and local names could be correctly classified only after BERTimbau Brazilian language model domain adaptation on (European) HAREM. At the same time, they were misclassified when NER was trained directly from the raw BERTimbau language model. The results are shown in Table 5, where we have the following examples:

- 1st row: In Portugal, "Consoada" refers to Christmas Night, which should be classified as a temporal (TIME) mention.
- 2nd row: "Soajo" is the name of a Portuguese village.
- 3rd row: "Abade da Loureira" refers to both an organization (ORG) and a street (LOC). But given the specific context in the sentence, it should be classified as an organization (ORG).

Table 6. Sample outputs by a generic domain LM vs. specific domain LM.

LM trained on First Harem	LM trained on Accounts Auditing
QUESTION 1: ESTE É UM GRANDE (THIS IS A BIG) <?>	
Este é um grande desafio	Este é um grande administrador
(This is a big challenge)	(This is a great administrator)
Este é um grande passo	Este é um grande empresário
(This is a big step)	(This is a great enterpreneur)
Este é um grande tesouro	Este é um grande produtor
(This is a big treasure)	(This is a great producer)
QUESTION 2: ESTA É UMA GRANDE (THIS IS A BIG) <?>	
Esta é uma grande vitória	Esta é uma grande empresa
(This is a big victory)	(This is a big company)
Esta é uma grande oportunidade	Esta é uma grande economia
(This is a great opportunity)	(This is a huge economy)
Esta é uma grande conquista	Esta é uma grande concorrência
(This is a great achievement)	(This is a big bid)

Table 7. Few/zero-shot learning

Model	Precision (%)	Recall (%)	F1 (%)
BERT$_{BASE}$	86	91	88
GPT 3.5 few-shot w/ instruction prompt	77.45	60.68	68.04
GPT 3.5 few-shot w/o instruction prompt	70.96	57.42	63.47
GPT 3.5 zero-shot	–	48.10	–

- 4th row: "Centro de Formação de Informática do Minho" is a local educational institution.
- 5th row: "São João do Souto" refers to an extinct Portuguese parish.

It makes sense because First HAREM is a Portuguese corpus, different from the Brazilian corpus used to train BERTimbau. These results show semantic gains from domain adaptation, although quantitative performance differences are not statistically significant.

Further, the results shown here were obtained in the NER task context. For document classification, within-task pretraining has been commonly used.

Finally, we showed that domain adaptation in experiment (A) - by training an intermediary language model with a larger, same-domain dataset - led to a higher impact on F1 metrics when compared to the experiments with within-task domain adaptation ((B) and (C)). Furthermore, qualitative analysis for the intermediary language model shows example outputs for predicting masked term tasks, i.e., the public accounts auditing language model can generate texts related to themes such as contracts and biddings, as seen in Table 6.

4.3   Causal Language Modeling - Few/zero-Shot Learning

This subsection summarizes experiments conducted on GPT 3.5, a large language model pre-trained with a causal language modeling objective. The results are shown in Table 7.

In the few-shot setting, we gave the model some examples. These examples could be accompanied by an instruction prompt telling the model the kinds of entities expected. We also tested sending only the examples without any further instruction. In the zero-shot setting, we only asked the question and let the model be free to return the information in any format, giving no examples. So it returned the results in a conversational unstructured format, making it difficult to measure the output precision and F1. Therefore, we only show recall.

Despite the recent impressive reasoning skills shown by GPT 3.5 and "Chat-GPT" models, it is interesting to note its quantitative performance still lags behind traditionally fine-tuned models. We hypothesize that BERT bidirectional masked language model objective may be more suitable to discriminative tasks, while causal language models - concerned only with predicting the future words - may lose information from past tokens. On the other hand, we believe that more research is necessary to investigate prompt engineering practices more suitable to NER and similar tasks. Finally, we realized impressive qualitative results on GPT 3.5 zero-shot learning: although its recall is considerably worse than the BERT-based baseline, it returns the entities and the relationships between them.

5   Conclusions

Among the analyzed techniques, model soups achieved the best results for the NER task. In the experiments conducted on domain adaptation, the best results were achieved with in-domain adaptation. We did not observe significant improvements in within-task domain adaptation. However, we realized the model could learn domain-specific terms with the First HAREM corpus. The lack of both quantity and diversity of "golden" labeled datasets for Portuguese when compared to English or Chinese is a substantial limitation to research on several tasks or multitask learning, as done, for example, by Houlsby at al. [6]. We believe advances in few-shot learning and prompt engineering could solve this limitation through synthetic data generation. So, besides investigating fine-tuning with Adapters, we expect to conduct further experiments on few-shot learning. Finally, we intend to investigate the fine-tuning of the recent Albertina LLM to the NER task and check if its larger capacity can compensate for its smaller context window (128 tokens) compared to BERTimbau.

References

1. Brown, T., et al.: Language models are few-shot learners. In: Larochelle, H., Ranzato, M., Hadsell, R., Balcan, M., Lin, H. (eds.) Advances in Neural Information Processing Systems, vol. 33, pp. 1877–1901. Curran Associates, Inc. (2020), https://proceedings.neurips.cc/paper_files/paper/2020/file/1457c0d6bfcb4967418bfb8ac142f64a-Paper.pdf

2. Chen, Y., Mikkelsen, J., Binder, A., Alt, C., Hennig, L.: A comparative study of pre-trained encoders for low-resource named entity recognition. In: Gella, S., et al (eds.) Proceedings of the 7th Workshop on Representation Learning for NLP, RepL4NLP@ACL 2022, Dublin, Ireland, 26 May 2022, pp. 46–59. Association for Computational Linguistics (2022). https://doi.org/10.18653/v1/2022.repl4nlp-1.6
3. Dai, X., Adel, H.: An analysis of simple data augmentation for named entity recognition. In: Scott, D., Bel, N., Zong, C. (eds.) Proceedings of the 28th International Conference on Computational Linguistics, COLING 2020, Barcelona, Spain (Online), 8-13 December 2020, pp. 3861–3867. International Committee on Computational Linguistics (2020). https://doi.org/10.18653/v1/2020.coling-main.343
4. Devlin, J., Chang, M., Lee, K., Toutanova, K.: BERT: pre-training of deep bidirectional transformers for language understanding. CoRR arXiv: 1810.04805
5. Fu, J., Liu, P., Neubig, G.: Interpretable multi-dataset evaluation for named entity recognition. In: Webber, B., Cohn, T., He, Y., Liu, Y. (eds.) Proceedings of the 2020 Conference on Empirical Methods in Natural Language Processing, EMNLP 2020, Online, 16-20 November 2020, pp. 6058–6069. Association for Computational Linguistics (2020). https://doi.org/10.18653/v1/2020.emnlp-main.489
6. Houlsby, N., et al.: Parameter-efficient transfer learning for NLP. In: Chaudhuri, K., Salakhutdinov, R. (eds.) Proceedings of the 36th International Conference on Machine Learning, ICML 2019, 9-15 June 2019, Long Beach, California, USA. Proceedings of Machine Learning Research, vol. 97, pp. 2790–2799. PMLR (2019), http://proceedings.mlr.press/v97/houlsby19a.html
7. Monteiro, M.: Extrator de entidades mencionadas em notícias da mídia. https://github.com/SecexSaudeTCU/noticias_ner (2021), (Accessed 21 May 2022)
8. Monteiro, M.: Riskdata brazilian portuguese ner. https://huggingface.co/monilouise/ner_news_portuguese (2021), (Accessed 21 May 2022)
9. Rodrigues, J., et al.: Advancing neural encoding of portuguese with transformer albertina PT-. CoRR https://doi.org/10.48550/arXiv.2305.06721, https://doi.org/10.48550/arXiv.2305.06721 (2023)
10. Santos, D., Seco, N., Cardoso, N., Vilela, R.: HAREM: an advanced NER evaluation contest for portuguese. In: Calzolari, N., et al. (eds.) Proceedings of the Fifth International Conference on Language Resources and Evaluation, LREC 2006, Genoa, Italy, 22-28 May 2006, pp. 1986–1991. European Language Resources Association (ELRA) (2006), http://www.lrec-conf.org/proceedings/lrec2006/summaries/59.html
11. Silva, E.H.M.D., Laterza, J., Silva, M.P.P.D., Ladeira, M.: A proposal to identify stakeholders from news for the institutional relationship management activities of an institution based on named entity recognition using BERT. In: Wani, M.A., Sethi, I.K., Shi, W., Qu, G., Raicu, D.S., Jin, R. (eds.) 20th IEEE International Conference on Machine Learning and Applications, ICMLA 2021, Pasadena, CA, USA, 13–16 December 2021, pp. 1569–1575. IEEE (2021). https://doi.org/10.1109/ICMLA52953.2021.00251
12. Souza, F., Nogueira, R., Lotufo, R.: BERTimbau: pretrained BERT models for Brazilian Portuguese. In: Cerri, R., Prati, R.C. (eds.) BRACIS 2020. LNCS (LNAI), vol. 12319, pp. 403–417. Springer, Cham (2020). https://doi.org/10.1007/978-3-030-61377-8_28
13. Sun, C., Qiu, X., Xu, Y., Huang, X.: How to fine-tune BERT for text classification? In: Sun, M., Huang, X., Ji, H., Liu, Z., Liu, Y. (eds.) CCL 2019. LNCS (LNAI), vol. 11856, pp. 194–206. Springer, Cham (2019). https://doi.org/10.1007/978-3-030-32381-3_16

14. Tänzer, M., Ruder, S., Rei, M.: Memorisation versus generalisation in pre-trained language models. In: Proceedings of the 60th Annual Meeting of the Association for Computational Linguistics (Volume 1: Long Papers), pp. 7564–7578. Association for Computational Linguistics, Dublin, Ireland (May 2022). https://doi.org/10.18653/v1/2022.acl-long.521

15. Vaswani, A., Shazeer, N., Parmar, N., Uszkoreit, J., Jones, L., Gomez, A.N., Kaiser, L.u., Polosukhin, I.: Attention is all you need. In: Guyon, I., et al. (eds.) Advances in Neural Information Processing Systems, vol. 30. Curran Associates, Inc. (2017). https://proceedings.neurips.cc/paper_files/paper/2017/file/3f5ee243547dee91fbd053c1c4a845aa-Paper.pdf

16. Wagner Filho, J.A., Wilkens, R., Idiart, M., Villavicencio, A.: The brWaC corpus: a new open resource for Brazilian Portuguese. In: Proceedings of the Eleventh International Conference on Language Resources and Evaluation (LREC 2018). European Language Resources Association (ELRA), Miyazaki, Japan (May 2018). https://aclanthology.org/L18-1686

17. Wortsman, M., et al.: Model soups: averaging weights of multiple fine-tuned models improves accuracy without increasing inference time. CoRR abs/2203.05482 (2022)

FlexCon-CE: A Semi-supervised Method with an Ensemble-Based Adaptive Confidence

Arthur Medeiros[1]([✉])[iD], Arthur C. Gorgônio[2][iD],
Karliane Medeiros Ovidio Vale[1][iD], Flavius L. Gorgônio[1][iD],
and Anne Magály de Paula Canuto[1][iD]

[1] Department of Computing and Technology (DCT), Federal University of Rio
Grande do Norte (UFRN), Caicó, RN, Brazil
`arthur.santos.017@ufrn.edu.br`,
{karliane.vale,flavius.gorgonio,anne.canuto}@ufrn.br
[2] Department of Informatics and Applied Mathematics (DIMAP), Federal University
of Rio Grande do Norte (UFRN), Natal, RN, Brazil
`arthur.gorgonio.099@ufrn.edu.br`

Abstract. Semi-supervised learning is characterized by a low number of
labeled instances and a high number of unlabeled instances. FlexCon-C
(Flexible Confidence Classifier) is a well-known semi-supervised method
that uses the self-training learning algorithm as basis to generate prediction models. The main difference between self-training and FlexCon-C is
that the former uses a fixed threshold to select the unlabeled instances,
while the latter has a dynamically adjusted confidence. FlexCon-C
applies a confidence adjustment equation based on the classifier performance. In this sense, the classifier performance is used to select and to
label unlabeled instances. In Machine Learning, it is well-known that the
classifier performance can be further improved through the use of classifier ensembles. Therefore, this study proposes the use classifier ensembles in the FlexCon-C confidence adjustment equation, aiming to provide
a more efficient measure to select and to label unlabeled instances. In
order to assess the viability of the proposed method (FlexCon-CE), an
empirical analysis will be conducted, using 20 datasets, three different
classification algorithms and five different configurations of initially unlabeled data. The results indicate that the proposed method outperformed
the traditional method, therewith proving itself promising for the task
of automatic data selection and labeling in the semi-supervised context.

Keywords: Classifier ensemble · Self-training algorithm · FlexCon-C

1 Introduction

Known as a field of study that has gained significant importance in recent years,
machine learning (ML) has emerged as a way of deducing a hypothesis based

M. C. Naldi and R. A. C. Bianchi (Eds.): BRACIS 2023, LNAI 14197, pp. 95–109, 2023.
https://doi.org/10.1007/978-3-031-45392-2_7

on past or lived experience [1]. In this context, machines are programmed to learn from previously stored data derived from previous experiences in order to generate knowledge. In ML, the learning algorithms can be broadly divided into supervised, semi-supervised and unsupervised algorithms. In this paper, semi-supervised learning (SSL) algorithms are assessed and these algorithms use labeled instances to build their initial hypothesis and combine the information obtained by these examples to label the unlabeled instances. In other words, it is possible to use partially supervised information to guide the learning process and increase the evidence of the target labels [2].

Nevertheless, we have observed that the automatic assignment of labels in an SSL algorithm is a difficult task, mainly related to the correct selection of the unlabeled instances to be labeled. An inefficient selection of the unlabeled instances and the consequent labeling of these instances can degrade enormously the performance of the SSL algorithms. A solution for this problem is proposed in [3] using a semi-supervised learning model called Flexible Confidence Classifier (FlexCon-C), which automatically adjusts the confidence parameter to select and label the unlabeled instance. This confidence is based solely on the accuracy of a classifier built in the previous iterations.

In the ML context, it is well-known that the accuracy of a single classifier can not be efficient for complex tasks. For this reason, combining different classifiers to improve the classifier performance has been emerged as an efficient alternative, as observed in several studies in the field [4–9]. In these studies, it has been noticed that combining information from various sources has resulted in systems with more reliability and performance than individual classifiers. Systems combining classifiers are known as classifier ensembles and they are composed of a set of base classifiers organized in a parallel way. These classifiers receive the input data and they produce several outputs that are used, in turn, as inputs by a combination method, which combines these outputs and generates the final ensemble output [7].

In this paper, we seek to improve the FlexCon-C [3] performance by introducing the classifier ensembles strategy to select and to label unlabeled instances and to include them in the labeled data set. Therefore, this paper proposes the Flexcon-CE (Flexible COnfidence with Classifier Ensembles) model, which uses an ensemble-based confidence adaptive confidence equation. In fact, two different versions of the Flexcon-CE method are proposed in this paper. The first one, called FlexCon-CE(Hom), uses a homogeneous ensemble (base classifiers of the same type) while the second one, named FlexCon-CE(Het), uses an heterogeneous ensemble. This study assesses the efficiency of the proposed methods when building representative classifier ensembles by analyzing the classification effectiveness of these methods in 20 well-known datasets, with three different classification algorithms and five different configurations of initially unlabeled data.

This article is organized as follows: Sect. 2 presents the theoretical aspects of semi-supervised learning algorithms as well as the FlexCon-C method and the classifier ensembles. Section 3 highlights some related work to this research while

Sect. 4 presents and details the SSL method proposed in this article. Section 5 describes some methodological aspects of the empirical analysis while Sect. 6 presents the obtained results. Finally, Sect. 7 highlights some final considerations and future possibilities of researches.

2 Theoretical Aspects

2.1 Semi-supervised Learning

According to the tasks to which they are applied, the machine learning algorithms can be broadly divided into three types: 1) supervised learning - predictive tasks; 2) unsupervised learning - descriptive tasks; 3) semi-supervised learning - combining predictive and descriptive tasks. In other words, supervised learning requires the training dataset to have input and output attributes (labels or classes) in order to find a model, based on the data used for training, that can be used to predict a label or class of a given instance [1]. In unsupervised learning, dataset attributes are not required to have an output or label to shape the instances' behavior pattern. This is possible due to the existence of tasks that look for behavior patterns of the instances based on the input attributes. Unsupervised or descriptive tasks are more related to exploring the dataset, with the aim of obtaining cluster patterns or association rules present in the data [2].

The existence of datasets that have a part of the instances with labels and another part without assigned labels became a motivation for the creation of a new classification model with the objective of improving the performance of learning tasks benefiting from the strategy of using labeled and unlabeled data, known as semi-supervised learning [10]. Following this concept, different algorithms were developed that propose the classification of unlabeled instances based on the ones with a label, including *self-training*, which will be used in this work.

The *self-training* method is characterized by using its own prediction results, obtained by a single classifier, and applying a new training to the model until all instances that did not have labels are properly labeled [11]. The Algorithm 1 depicts the flow of instructions that are executed in the *self-training* process. First, the database is divided into two sets: D_l - labeled instances and D_u - unlabeled instances (line 1). Then, a classifier f is generated through D_l (line 3). After assigning the new labels to the unlabeled data, these instances are inserted into the subset S which will be removed from D_u and added into D_l (lines 4–6). Then the process will restart until there is no more unlabeled data. Note that, based on the results generated by the classifier, the model is retrained in order to learn according to its assertiveness rate.

In [12], the authors addressed an exploratory analysis of the *self-training* algorithm, regarding its adaptation by inserting a new parameter, called confidence, which is used to select and to label unlabeled instances. The main aim is to select instances that have a prediction higher than a confidence threshold. However, the confidence threshold is initially defined and this value is used throughout the whole labeling process. In [3], the authors proposed the FlexCon-C method,

---

Algorithm 1: Self-Training

---

variables: labeled instances D_l, unlabeled instances D_u, classifier f, k value

1 initially we have $D_l = \{(X_i, Y_i) \mid i = 1 \ldots L\}$ e $D_u = \{(X_j) \mid j = L + 1 \ldots L + U\}$;

2 **while** $D_u \neq \emptyset$ **do**

3   Train classifier f based on D_l using supervised learning;

4   Apply f on the instances in D_u;

5   Remove a subset $S = \{s1, s2, \ldots, sn\}$ from D_u, which contains the first k instances with a higher confidence of their predictions;

6   Add subset $\{(x, f(x)) \mid x \in S\}$ to subset D_l;

7 **end**

Result: labeled data

---

which is an extension of the work proposed in [12], which applies an adaptive (or flexible) confidence threshold, which is calculated in every iteration of the semi-supervised process. The FlexCon-C algorithm is presented by algorithm 2.

Algorithm 2: Self-Training with Confidence Adjustment (FlexCon-C)

---

variables: labeled data D_l, unlabeled data D_u

1 initially we have $D_l = (x_i, y_i \mid i = 1 \ldots l$ e $D_u = x_j \mid l + 1 \ldots l + u$;

2 **while** $D_u \neq \emptyset$ **do**

3   Train classifier f based on D_l using supervised learning;

4   Apply f on the instances in D_u;

5   Calculate the new threshold value;

6   Select and label each instance in subset S;

7   Remove a subset $S = \{s_1, s_2, \ldots, s_n\}$ from D_u, the instances whose prediction is higher than or equal to the confidence threshold;

8   Add subset $\{(x, f(x)) \mid x \in S\}$ to set D_l;

9 **end**

Result: labeled data

In Algorithm 2, the lines marked in blue highlight the main differences between this algorithm and the original *self-training* algorithm. Initially, D_l represents the initially labeled set, while D_u stands for the set of unlabeled instances. At each iteration, a classifier f is trained using the set of labeled instances D_l as training set and classifies the D_u instances (lines 3 to 4). Then, a new confidence threshold value is calculated (line 5). Then, select all unlabeled instances whose prediction is higher than or equal to the confidence threshold. These instances are then labeled (line 6). These instances are removed from the D_u set and included to the D_l set. This process is then repeated until the D_u set is empty.

In [3], three SSL algorithms are proposed (FlexCon-G, FlexCon, and FlexCon-C). All three algorithms use adaptive confidence threshold and they are all based on the *self- training* method. In the cited article, the FlexCon-C algorithm adjusts the confidence parameter at each iteration based solely on the precision of a classifier built the previous iteration. In this method, the adjustment will be done by increasing or decreasing the confidence value based on a change rate cr. Equation 1 presents the confidence update equation for FlexCon-C.

$$conf(t_{i+1}) = \begin{cases} conf(t_i) - cr, \text{ if } acc \geq mp + e \\ conf(t_i), \text{ if } mp - e < acc < mp + e \\ conf(t_i) + cr, \text{ if } acc \leq mp - e \end{cases} \qquad (1)$$

where $conf(t_{i+1})$ is the confidence value of the current iteration, mp is the minimum acceptable precision; cr is the change rate it is a user-given hyperparameter, acc is the accuracy of the classifier. Finally, e is an acceptable precision variation that is allowed in order to define an stabilization in precision.

In this equation, if the accuracy is higher than a minimum acceptable precision (considering the acceptable variation e), mp, the confidence threshold is decreased by a change rate cr. If the accuracy is lower than a minimum acceptable precision, the confidence threshold is increased by a change rate cr. Otherwise, the confidence threshold does not change.

This method was further divide into two sub-methods with respect to the label definition of the unlabeled instances, FlexCon-C1 and FlexCon-C2. FlexCon-C1 uses classifier ensembles to define the label of each pattern, while FlexCon-C2 uses the label predicted by the classifier generated in the first iteration, which is stored by the algorithm. In FlexCon-C1, the classifier ensemble is composed of classifiers built in all iteration performed until a current iteration. In addition, these individual classifiers are combined by sum and majority voting, leading to two versions of this method, FlexCon-C1(v) and FlexCon-C1(s).

2.2   Classifier Ensembles

As previously mentioned, the increased complexity and wide applicability of classification systems has led to exploring many approaches and methodologies. Nevertheless, there is a perception that no classifier is considered completely satisfactory for a particular task; therefore, the idea of combining different methods to improve performance has emerged as a very promising possibility [13]. This combination is called classifier ensembles, also known as multi-classifier systems.

In classification tasks, an ensemble includes several sub-models called base classifiers, which are usually obtained by training a basic learning algorithm (decision tree, neural network, k nearest neighbors, among others). The ensembles can be built based on the same learning algorithm, producing homogeneous ensembles, or using different algorithms and generating heterogeneous ensembles [14].

The proposition of classifier ensembles is to create and combine several inductive models for the same domain, obtaining better prediction quality [9]. After generating the base classifier set, the next step is the choice for the methods to combine their outputs. There is a vast number of methods for combining classifiers in the literature. In this work, we will be used voting. The choice for this method was made due to the use of information from all classifiers. The voting method, often used to combine classifiers, performs the combination by voting on the results of each classifier when a new instance is presented.

3   Related Work

Over the years, several works have been developed using a confidence parameter in semi-supervised ML algorithms [3,12,15–18]. The goal is to investigate the behavior of these algorithms, including this parameter as a threshold for selecting new instances for the labeled dataset. On the other hand, researchers have explored the use of classifier ensembles to enhance the performance of various machine learning algorithms [4–6,8]. The present work aims to merge these two approaches, both using the confidence parameter as well as classifier ensembles, as a way to improve the performance of the semi-supervised learning algorithm FlexCon-C.

In [12,15] new versions of the self-training algorithm were proposed that included a fixed confidence parameter with the objective of minimizing inconsistencies in the classification choices during the labeling process. In [3,16–19] algorithms for the automatic adjustment of the confidence threshold were proposed as an extension of those researches. Therefore, the focus of research development was on building solutions that could adjust the confidence parameter during the data labeling process.

In [8] a study was conducted on the existing risks in railway safety inspections. These inspections are reported using text, which generates a large amount of textual data. As motivation for the authors, a predictive model was proposed that used classifier ensembles to predict risk situations that could be addressed proactively. The chosen classification model was the decision tree, which was applied to an ensemble of Bagging classifiers that combined the results of the classifiers by voting. Similarly, [5] used different models of classifiers in an ensemble to predict a credit score for customers of financial organizations and banks.

In [4] a new ensemble learning algorithm named E_RHFS was developed and applied as a tool to predict software failure. The authors used the voting combination method on the base classifiers to obtain the final ensemble classifier for software defect prediction. In [6] the proposal focused on two primary factors in creating a classifier ensemble, namely: accuracy and diversity. The paper aimed to propose a new diversity measure called Misclassification Diversity (MD) and an Incremental Layered Classifier Selection approach to build the ensemble.

4   The Proposed Method: FlexCon-CE

As already mentioned, the purpose of this paper is to propose a new FlexCon-C variation, called FlexCon-CE (Flexible Confidence with Classifier Ensemble). This variation uses an ensemble-based confidence threshold. Therefore, in order to develop the proposed method, it was necessary to modify the original FlexCon-C Algorithm 2 with lines marked in blue highlight. Algorithm 3 presents the FlexCon-CE step-by-step operation.

Algorithm 3: Flexible Confidence with Ensemble (FlexCon-CE)

---
variables: labeled instances D_l, unlabeled instances D_u, list of classifiers F_n

1   $D_l = (x_i, y_i \mid i = 1 \ldots l$ e $D_u = x_j \mid l+1 \ldots l+u$;

2   create an empty ensemble e;

3   **while** $D_u \neq \emptyset$ **do**

4   **for** $f \in F_n$ **do**

5   train classifier f based on D_l using supervised learning;

6   add f to e;

7   **end**

8   Apply e on the instances in D_u;

9   Calculate new confidence threshold;

10   Select and label each instance in subset S;

11   Remove a subset $S = \{s_1, s_2, \ldots, s_n\}$ from D_u, the instances whose values in $e(x)$ are greater or equal to the confidence threshold;

12   Add subset $\{(x, e(x)) \mid x \in S\}$ to set D_l;

13 **end**

   Result: labeled data

Initially, in Algorithm 3, the database is divided into two sets: a set of labeled instances D_l and a set of unlabeled instances D_u. Additionally, a pool of classifiers F_n is available. An ensemble is defined as empty (line 1). Then, for each classifier f in F_n, they are trained using D_l as training set. After that, the trained classifier f is added to the ensemble e (lines 2 to 7). Finally, the ensemble e will be applied to D_u in order to calculate the effectiveness generated by e (line 8). Then, the new confidence threshold is calculated (line 9), the unlabeled instances are selected, labeled, removed from the unlabeled set D_u and included in the labeled set D_l (lines 9 to 12). It is important to emphatize that the threshold is used to select unlabeled instances that will be labeled by the voting process of the ensemble.

It is important to emphasize that the calculation of the new confidence threshold is made using Eq. 1. The main difference is that the mp value is related to the ensemble accuracy (FlexCon-CE) instead of the classifier accuracy (FlexCon-C).

As previously described, two FlexCon-CE versions are proposed. In the first one, the classifier ensemble is constructed heterogeneously, using more than one type of classification method. In the second version, the classifier ensemble is built homogeneously, when using only one type of method. In this work, both heterogeneous and homogeneous ensemble approaches will be assessed.

5   Experimental Methodology

In order to assess the feasibility of the proposed method, an empirical analysis will be carried out. The next subsections will describe the main aspects of the experimental methodology.

5.1   Datasets

In this empirical analysis, the FlexCon-CE method is applied to 20 different databases. These datasets are available in platforms and repositories that maintain various experimental data, such as GitHub [20], UCI Machine Learning [21] and Kaggle Datasets [22]. Table 1 describes the characteristics of all datasets, regarding the number of attributes, instances, number of classes, data types (Categorical - C, Real - R and Integer - I), and data distribution (Balanced - B and Unbalanced - U).

Table 1. Datasets characteristics

Dataset	Instances	Atributes	Classes	Type(s)	Balanced
Btsc[1]	748	5	2	R	U
Car Evaluation	1728	6	4	C, I	U
Cnae-9	1080	857	9	I	B
Hill-valley	606	101	2	I	B
Image segmentation	2310	19	7	R	B
Kr-vs-kp[2]	3196	36	2	C	B
Mammographic Mass	961	6	2	I	B
Multiple Features	2000	64	10	I, R	B
Mushroom	8124	22	2	C	B
Ozone Level Detection	2536	73	2	R	U
Pima	768	9	2	I, R	U
Planning Relax	182	13	2	R	U
Seeds	210	7	3	R	B
Semeion	1593	256	10	I	B
Solar Flare	1484	8	10	R	U
SPECTF Heart	267	44	2	I	U
Tic Tac Toe Endgame	958	9	2	C	U
Vehicle	946	18	4	I	B
Waveform	5000	40	3	R	B
Wilt	4839	6	2	R	U

[1] Blood Transfusion Service Center.
[2] King-Rook vs King-Pawn.

5.2   Methods and Materials

All method and learning procedures of this paper are implemented using the scikit-learn package available in the Python language.

In order to create the algorithm proposed in this article, we decided to use the scikit-learn Python library [23], since it offers several learning methods for classification, regression, and clustering domains. It also integrates with other widely used Python libraries for machine learning experiments, such as Matplotlib [24], Numpy [25], Pandas [26], and others. In order to facilitate the comparative analysis of the results of the FlexCon-CE method with FlexCon-C, three classifiers are used, which are: Naive Bayes (NB), Decision Tree (Tree) – using a CART-based algorithm – and K-Nearest Neighbors (k-NN) – with k values from 4 to 8. Additionally, the percentage values of the initially labeled instances, n, are 5%, 10%, 15%, 20% and 25%, in the same way as it was performed by [16,19,27,28].

Each trained classifier was included to the classifier ensemble heterogeneously and homogeneously, for FlexCon-CE (Het) and FlexCon-CE (Hom), respectively. For the combination of all n classifiers, the voting criterion was applied to define the overall ensemble output. The voting method consists of a non-linear combination of the classifiers' outputs, and its process consists of determining the winning class, from a data input pattern, by the total number of votes counted for each classifier [4,8,29]. Therefore, the confidence of the classifier ensemble is defined by the number of classifiers that selected the winning class divided by the number of classifiers of the used ensemble.

The main aim of this empirical analysis is to evaluate the performance of the FlexCon-CE algorithm and to compare them to the FlexCon-C (previously explained: FlexCon-C1(v), FlexCon-C1(s) and FlexCon-C2) results.

In order to assess the obtained results, the Friedman statistical test was achieved. According to [30], this test compares related sample data, which means that the same object can be analyzed more than once. As it is non-parametric test, numerical values are not used directly, but rather their positions. After the ranking are performed for the groups separately, the hypothesis of similarity of the sum of the position of each group is verified. In this way, it is possible to compare the performance of different data classification algorithms in the scenarios created for each different dataset.

6   Experimental Results

This section presents the analysis of the obtained results. As previously mentioned, the analysis of the results will be conducted in a comparative way, evaluating the performance of both FlexCon-CE versions and three FlexCon-C versions, FlexCon-C1(v), FlexCon-C1(s), and FlexCon-C2 [3]. The next sections will present the analysis of the experimental results. The first one presents an analysis of the accuracy results presented by all analyzed methods.while the next section conducts a statistical analysis of the obtained results.

Table 2. Results of average accurary

Classifier	Method	5%	10%	15%	20%	25%
Naive Bayes	FlexCon-CE(Hom)	68.43 ± 8.33	**71.00 ± 7.82**	**71.30 ± 7.00**	**71.77 ± 7.25**	**71.62 ± 7.28**
	FlexCon-C1(v)	68.3 ± 18.26	68.57 ± 18.46	68.41 ± 18.13	68.57 ± 18.26	69.0 ± 18.09
	FlexCon-C1(s)	**69.18 ± 17.98**	68.36 ± 17.6	68.13 ± 18.35	68.72 ± 17.98	68.97 ± 18.95
	FlexCon-C2	68.29 ± 18.38	68.08 ± 18.47	69.16 ± 17.74	68.7 ± 18.38	68.38 ± 19.25
Tree	FlexCon-CE(Hom)	**74.93 ± 6.17**	**78.41 ± 5.86**	**79.75 ± 5.7**	**80.88 ± 4.82**	**81.84 ± 5.33**
	FlexCon-C1(v)	69.74 ± 19.35	75.2 ± 14.75	76.97 ± 14.12	77.92 ± 13.8	79.21 ± 13.06
	FlexCon-C1(s)	69.78 ± 19.0	74.36 ± 15.6	76.18 ± 14.88	78.27 ± 13.67	76.41 ± 17.4
	FlexCon-C2	67.21 ± 16.05	72.6 ± 12.31	73.73 ± 12.22	74.58 ± 12.4	75.91 ± 11.62
k-NN	FlexCon-CE(Hom)	70.70 ± 6.34	75.28 ± 6.21	76.97 ± 5.13	78.27 ± 4.91	79.03 ± 4.63
	FlexCon-C1(v)	**74.19 ± 13.34**	**76.98 ± 12.68**	**78.14 ± 12.95**	**79.48 ± 12.37**	79.65 ± 12.21
	FlexCon-C1(s)	74.06 ± 13.6	76.94 ± 12.36	77.93 ± 12.83	79.21 ± 12.17	**79.84 ± 12.35**
	FlexCon-C2	73.49 ± 14.1	76.91 ± 12.84	77.9 ± 13.0	78.86 ± 12.65	79.78 ± 12.61

Table 3. Results of best average accuracy

Classifiers	Method	5%	10%	15%	20%	25%
–	FlexCon-CE(Het)	**75.21 ± 6.21**	77.68 ± 5.4	79.16 ± 5.03	79.95 ± 5.58	80.41 ± 5.32
Naive Bayes	FlexCon-CE(Hom)	68.43 ± 8.33	71.00 ± 7.82	71.30 ± 7.00	71.77 ± 7.25	71.62 ± 7.28
Tree	FlexCon-CE(Hom)	74.93 ± 6.17	**78.41 ± 5.86**	**79.75 ± 5.7**	**80.88 ± 4.82**	**81.84 ± 5.33**
k-NN	FlexCon-C1(v)	74.19 ± 13.34	76.98 ± 12.68	78.14 ± 12.95	79.48 ± 12.37	79.65 ± 12.21

6.1 Accuracy Analysis

Tables 2 and 3 present the results obtained by each method used in this analysis. The difference between these two tables is that the first table shows the results of three different classifiers and compares the performance of FlexCon-CE using a homogeneous classifier ensemble (FlexCon-CE(Hom)) with all three FlexCon-C versions (FlexCon-C1(v), FlexCon-C1(s), and FlexCon-C2). For the second table, the best results of Table 2 are selected, for each classification algorithm (Naive Bayes, Decision Tree and k-NN), as well as the FlexCon-CE result using a heterogeneous classifier ensemble (FlexCon-CE(Het)).

Both tables have the same configuration: the first column presents the name of the used classifier while the second column indicates the semi-supervised learning method. Columns 3–7 display the average accuracy and standard deviation obtained by each semi-supervised method over all 20 used datasets, according to the settings of 5%, 10%, 15%, 20%, and 25% of initially labeled instances. Additionally, in both tables, the results that showed the most significant accuracy values are highlighted in bold.

When analyzing results obtained with the Naive Bayes classifier in Table 2, it is possible to observe that the FlexCon-CE method using the homogeneous ensemble - FlexCon-CE(Hom) obtained the best results in 80% of the analysed cases (4 out of the 5% of initially labeled instances). In this same table, considering the results obtained by the Tree classifier, FlexCon-CE(Hom) also achieved the highest accuracy in 100% of cases (5 out of the 5% of initially labeled data). Based on these results, it is possible to state that the FlexCon-CE(Hom) method obtained significantly better performance results in two out of three analyzed classifiers.

Still exploring Table 2, in general, it is possible to observe that the FlexCon-CE(Hom) method performed better than all FlexCon-C versions in 60% of the analyzed cases (9 out of the 15 best results). This is a promising result showing that the use of a more efficient selection and labeling technique leads to more robust SSL methods.

As previously explained, Table 3 presents the performance of the proposed method using the heterogeneous ensemble - FlexCon-CE(Het), and the best results obtained in Table 2, for comparison purposes. When observing the aforementioned table, it is possible to identify that the FlexCon-CE(Hom) method obtained the best performance in 80% of the cases (4 out of 5% of initially labeled instances) when the number of initially labeled instances was higher than or equal to 10%. On the other hand, the other proposed method of this paper, FlexCon-CE(Het), obtained the best performance when the number of initially labeled instances was limited to 5%.

In summary, it is possible to state that one proposed version, FlexCon-CE(Hom) or FlexCon-CE(Het), achieved better accuracy results than all FlexCon-C versions, regardless of the type of classifier used. The next subsection presents the statistical analysis of the obtained results in order to demonstrate that the proposed semi-supervised machine learning methods are promising.

6.2   Statistical Analysis

The statistical analysis performed in this article uses critical difference (CD) diagrams. Figures 1 and 2 display the results obtained using the aforementioned diagrams. The post-hoc Friedman test is used to analyze the results based on their rankings. Additionally, in these diagrams, the methods located further to the left exhibit better results, while the ones on the right presented the poorest performances. Finally, a method is considered statistically different from another when both are not covered by the critical difference line (bold horizontal line in the diagram). Otherwise, when this line covers two or more approaches, it means that the null hypothesis of the Friedman test cannot be rejected.

Figure 1 presents the critical difference diagram with the results of the statistical test of the results obtained in Table 2. The applied classification algorithm is indicated in front of the method name. When analyzing the aforementioned diagram and observing the ranking of methods delivered by each classifier, it can be seen that FlexCon-CE(Hom)-TREE was the method that reached the highest ranking, which confirms the analysis carried out in the previous section regarding the accuracy of these methods. Additionally, according to the statistical analysis, this method is similar to the following four methods, in which two of them applied a decision tree (FlexCon-C2-TREE and FlexCOn-C1(v)) and the remaining two methods used a k-NN method (FlexCon-C2 and FlexCOn-C1(v)). On the other hand, the winning method is statistically superior to the remaining seven methods.

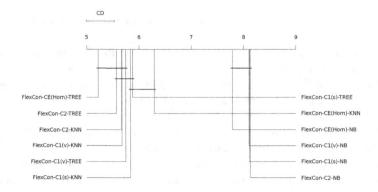

Fig. 1. Critical Difference Diagram using all methods

Figure 2 illustrates the critical difference diagram with the results of the statistical test of the results obtained in Table 3. When analyzing the rankings of each method, it can be seen that the FlexCon-CE(Hom)-TREE and FlexCon-CE(Het) methods are statistically similar and both achieved the best results. This observation confirms what was reported in the previous section, in which the proposed methods outperformed the methods that do not use classifier ensembles.

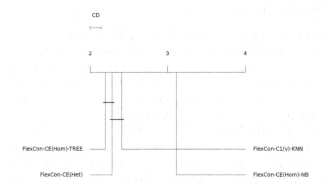

Fig. 2. Statistics performance of the best

In summary, the results obtained in this paper showed significant improvements of the proposed method, in comparison to the FlexCon-C versions. These results corroborate with the results obtained in the previous section, demonstrating an enhancement in the performance of the semi-supervised methods of the proposed methods. Moreover, the proposed methods are validated from two different perspectives, both in terms of performance and statistical analysis.

7   Conclusion and Future Works

This paper proposed a new SSL method, called FlexCon-CE, which was designed based on introducing classifier ensembles in the FlexCon-C functioning. The main difference between FlexCon-C and FlexCon-CE is that the latter uses classifier ensembles to define the adaptive confidence while the predictions of a classifier is used by FlexCon-C. Additionally, this work explores the use of two versions of the proposed method: one that uses an homogeneous ensemble structure and the other one used an heterogeneous ensemble structure.

In order to validate the proposed method, an empirical analysis was conducted, in which a comparative analysis of the performance of FlexCon-CE with FlexCon-C was carried out. In general, the performance of the analyzed methods was assessed in terms of accuracy and standard deviation and they were validated statistically using the Friedman test. After analyzing the results, it was concluded that the FlexCon-CE method obtained better results in most cases, with emphasis on the version that used the homogeneous ensemble with the Tree classifier, followed by the heterogeneous FlexCon-CE version.

For future work, we suggest to explore other classification algorithms and use new databases to address experimental scenarios. Furthermore, it is recommended to make the size of the classifier ensembles more flexible in relation to the number of base classifiers.

References

1. Faceli, K., Lorena, A.C., Gama, J., de Leon Ferreira de Carvalho, A.C.P.: An approach of machine learning. Artif. Intell. (2011)
2. Chapelle, O., Scholkopf, B., Zien, A.: Semi-supervised Learning, vol. 2. The MIT Press, Cambridge, MA (2006)
3. Ovidio Vale, K.M., et al.: Automatic adjustment of confidence values in self-training semi-supervised method. In: 2018 International Joint Conference on Neural Networks (IJCNN), pp. 1–8. IEEE (2018)
4. Wei, W., Jiang, F., Yu, X., Du, J.: An ensemble learning algorithm based on resampling and hybrid feature selection, with an application to software defect prediction. In: 2022 7th International Conference on Information and Network Technologies (ICINT), pp. 52–56. IEEE (2022)
5. Safiya Parvin, A., Saleena, B.: An ensemble classifier model to predict credit scoring-comparative analysis. In: 2020 IEEE International Symposium on Smart Electronic Systems (iSES) (Formerly iNiS), pp. 27–30. IEEE (2020)
6. Zohaib Jan, M., Verma, B.: A novel diversity measure and classifier selection approach for generating ensemble classifiers. IEEE Access 7, 156360–156373 (2019)
7. Lochter, J.V., Zanetti, R.F., Reller, D., Almeida, T.A.: Short text opinion detection using ensemble of classifiers and semantic indexing. Exp. Syst. Appl. 62:243–249 (2016)
8. Li, X., Shi, T., Li, P., Zhou, W.: Application of bagging ensemble classifier based on genetic algorithm in the text classification of railway fault hazards. In: 2019 2nd International Conference on Artificial Intelligence and Big Data (ICAIBD), pp. 286–290. IEEE (2019)

9. Cichosz, P.: Data Mining Algorithms: Explained using R. John Wiley & Sons (2014)
10. Albalate, A., Minker, W.: Semi-supervised and Unsupervised Machine Learning: Novel Strategies. John Wiley & Sons (2013)
11. Zhu, X., Goldberg, A.B.: Introduction to semi-supervised learning. Synth. Lect. Artif. Intell. Mach. Learn. 3(1), 1–130 (2009)
12. Rodrigues, F.M., de M. Santos, A., Canuto, A.M.P.: Using confidence values in multi-label classification problems with semi-supervised learning. In: The 2013 International Joint Conference on Neural Networks (IJCNN), pp. 1–8. IEEE (2013)
13. Nascimento, D.S.C., Coelho, A.L.V., Canuto, A.M.P.: Integrating complementary techniques for promoting diversity in classifier ensembles: a systematic study. Neurocomputing 138, 347–357 (2014)
14. Gharroudi, O.: Ensemble multi-label learning in supervised and semi-supervised settings. Ph.D. thesis, Université de Lyon (2017)
15. Rodrigues, F.M., Câmara, C.J., Canuto, A.M.P., Santos, A.M.: Confidence factor and feature selection for semi-supervised multi-label classification methods. In: 2014 International Joint Conference on Neural Networks (IJCNN), pp. 864–871. IEEE (2014)
16. Gorgônio, A.C., et al.: Análise da variação do limiar para o algoritmo de aprendizado semissupervisionado flexcon-c/threshold variation analysis for flexcon-c semisupervised learning algorithm. Brazil. J. Develop. 5(11), 26654–26669 (2019)
17. Vale, K.M.O., Gorgônio, A.C., Da Luz, E.G.F., De Paula Canuto, A.M.: An efficient approach to select instances in self-training and co-training semi-supervised methods. IEEE Access 10, 7254–7276 (2021)
18. Gorgônio, A.C., Alves, C.T., Lucena, A.J.F., Gorgônio, F.L., Vale, K.M.O., Canuto, A.M.P.: Analysis of the threshold variation of the flexcon-c algorithm for semi-supervised learning. In: Anais do XV Encontro Nacional de Inteligência Artificial e Computacional, pp. 775–786. SBC (2018)
19. Vale, K.M.O., et al.: A data stratification process for instances selection in semi-supervised learning. In: 2019 International Joint Conference on Neural Networks (IJCNN), pp. 1–8. IEEE (2019)
20. Breiman, L.: Bias, variance, and arcing classifiers. Technical report, Tech. Rep. 460, Statistics Department, University of California, Berkeley (1996)
21. Dheeru, D., Karra Taniskidou, E.: UCI machine learning repository (2017)
22. Smith, J.W., Everhart, J.E., Dickson, W.C., Knowler, W.C., Johannes, R.S.: Using the adap learning algorithm to forecast the onset of diabetes mellitus. In: Proceedings of the Annual Symposium on Computer Application in Medical Care, p. 261. American Medical Informatics Association (1988)
23. Pölsterl, S.: scikit-survival: a library for time-to-event analysis built on top of scikit-learn. J. Mach. Learn. Res. 21(1), 8747–8752 (2020)
24. Bisong, E.: Building Machine Learning and Deep Learning Models on Google Cloud Platform. Apress, Berkeley (2019). https://doi.org/10.1007/978-1-4842-4470-8
25. Harris, C.R., et al.: Array programming with numpy. Nature 585(7825), 357–362 (2020)
26. Nelli, F.: Python data analytics with pandas, numpy, and matplotlib (2018)
27. Araújo, Y.N., et al.: A data stratification process for instances selection applied to co-training semi-supervised learning algorithm. In: 2021 International Joint Conference on Neural Networks (IJCNN), pp. 1–8. IEEE (2021)
28. Vale, K.M.O., Gorgônio, F.L., Araújo, Y.N., Gorgônio, A.C., de P Canuto, A.M.: A co-training-based algorithm using confidence values to select instances. In: 2020 International Joint Conference on Neural Networks (IJCNN), pp. 1–7. IEEE (2020)

29. Kuncheva, L.I.: Combining Pattern Classifiers: Methods and Algorithms. John Wiley & Sons (2014)

30. Theodorsson-Norheim, E.: Friedman and quade tests: basic computer program to perform nonparametric two-way analysis of variance and multiple comparisons on ranks of several related samples. Comput. Biol. Med. **17**(2), 85–99 (1987)

Computer Vision

Single Image Super-Resolution Based on Capsule Neural Networks

George Corrêa de Araújo[1], Artur Jordão[2(✉)], and Helio Pedrini[1]

[1] Institute of Computing, University of Campinas, Campinas, SP 13083-852, Brazil
helio@ic.unicamp.br
[2] Polytechnic School of the University of São Paulo, University of São Paulo,
São Paulo, SP 05508-010, Brazil
arturjordao@usp.br

Abstract. Single image super-resolution (SISR) consists of obtaining one high-resolution version of a low-resolution image by increasing the number of pixels per unit area. This method has been actively investigated by the research community, due to the wide variety of problems ranging from real-world surveillance to aerial and satellite imaging. Most of the improvements in SISR come from convolutional networks, in which approaches often focus on the deeper and wider architectural paradigm. In this work, we decided to step up from the traditional convolutions and adopt the concept of capsules. Since their overwhelming results in image classification and segmentation problems, we question how suitable they are for SISR. We also verify that different solutions share similar configurations, and argue that this trend leads to fewer explorations of network designs. Throughout our experiments, we check various strategies to improve results, ranging from new and different loss functions to changes in the capsule layers. Our network achieved positive and promising results with fewer convolutional-based layers, showing that capsules might be a concept worth applying to the image super-resolution problem. In particular, we observe that the proposed method recreates the connection between the different characters more precisely, thus demonstrating the potential of capsules in super-resolution problems.

Keywords: Neural Network · Capsule Networks · Image Super-Resolution

1 Introduction

State-of-the-art methods in image super-resolution are based on artificial intelligence concepts, more specifically on deep neural networks, and have achieved visually striking results [6,22,40,61,63] Most recent models are composed of

---

Artur Jordáo: This work was done when Artur Jordao was a post-doctoral researcher at the University of Campinas.

traditional convolutional layers that exhibit limitations, although widely studied and optimized. For example, they prevent a better understanding of the data by the network, not absorbing valuable information such as the interrelationship between its components.

As an alternative to convolutional networks, a large body of work has demonstrated the abstraction power of the capsules [14,42,60]. For example, Sabour et al. [41] presented an implementation of the capsule concept, a group of neurons whose activation vector represents the parameters that describe a specific entity, such as an object or part of it, and its presence. Using a three-layer capsule network, Sabour et al. achieved results comparable to those of deeper convolutional neural networks for the digit identification problem. The authors also obtained a 5% error rate on the segmentation of two different digits with an 80% overlap – a rate previously achieved only in much simpler cases (with less than 4% overlap between images). Hinton et al. [15] reduced by 45% the error rate on the problem of identifying objects from different perspectives while making the network more resistant to adversarial attacks. LaLonde and Bagci [28] have created a network able to process images tens of times larger while reducing the number of parameters by 38.4% and increasing the quality of medical image segmentation.

A few works proposed the usage of capsule-based networks to solve problems that involve low-resolution images. Singh et al. [46] proposed a Dual Directed Capsule Network, named DirectCapsNet, which employs a combination of capsules and convolutional layers for addressing very low resolution (VLR) digit and face recognition problems. Majdabadi and Ko [35] implemented a Generative Adversarial Network (GAN) that uses a CapsNet as the discriminator for facial image super-resolution, surpassing strong baselines in all metrics. Hsu et al. [18] developed two frameworks to incorporate capsules into image SR convolutional networks: Capsule Image Restoration Neural Network (CIRNN) and Capsule Attention and Reconstruction Neural Network (CARNN). The results outperformed traditional CNN methods with a similar number of parameters. Despite the positive results, most of these works rely on plain CapsNet [41], failing to explore novel capsule concepts. Our method bridges this gap.

This work focuses on the single image super-resolution problem using capsules. We implement a model based on newer concepts of capsules for single image super-resolution (SISR) problems. Throughout our evaluation, we used publicly available and standard benchmarks such as Set5 [7], Set14 [59], B100 [36], Urban100 [19] and DIV2K [2]. Our model yields fine-grained super-resolution images and achieves positive results with fewer convolutional-based layers than baselines.

2 Background

Super-Resolution. Super-resolution (SR) is the process of obtaining one or more plausible high-resolution images (HR) from one or more low-resolution images (LR) [4,37,56]. It is an area that has been studied for decades [21] and has a wide variety of application fields such as smart surveillance, aerial imaging, medical image processing, and traffic sign reading [4,37,44,56]. The relationship between LR and HR images may vary depending on the situation. Many studies assume that the LR image is a reduced version of the HR image by bicubic interpolation. However, other degradation factors can be considered in real examples, such as quantization errors, acquisition sensor limitations, presence of noise, blurring, and even the use of different interpolation operators aiming resolution reduction for storage [50].

The first successful usage of neural networks for SISR problems was developed by Dong et al. [11]. With the Super-Resolution Convolutional Neural Network (SRCNN) model, they created a complete solution that maps LR images to SR versions with little pre/post-processing, yielding superior results. After their achievement, several other works have advanced the state-of-the-art in the SISR problem [10,24,25,54], but they have strong limitations. These models receive as input an enlarged version of the LR image, usually through bicubic interpolation, and seek to improve the quality of the image. This means that the operations performed by the neural network are all done in high-resolution space, which is inefficient and incurs high processing cost. The computational complexity of the convolution grows quadratically with the size of the input image, whose generation of an SR image with a scaling factor n would result in a cost n^2 compared to the processing in the low-resolution space [12].

Looking for a way of postponing the resolution increase in the network, Shi et al. [44] developed a new layer, called the subpixel convolution (or PixelShuffle). Such a layer works equivalent to deconvolution with kernel size divisible by spacing, but it is $\log_2 r^2$ times faster than deconvolution. Their network, named Efficient Sub-pixel Convolutional Neural Network (ESPCN), achieved speed improvements of over $10\times$ compared to SRCNN [11] while having a higher number of parameters and achieving better results for an upscaling factor of $4\times$.

The concept of subpixel convolution is currently a common choice to perform upscaling in neural networks. It has been used by several solutions that have reached the best results [6,30,33,40,53,61,63,64] and participated in several editions of the SISR competition that took place during the New Trends in Image Restoration and Enhancement workshop [8,50,51,62].

Capsules. Initially introduced by Hinton et al. [16], the concept of capsule proposes to solve some of the main flaws found in traditional convolutional networks: inability to identify spatial hierarchy between elements and lack of rotation invariance. Hinton et al. conclude that, after several stages of subsampling, these networks lose information that makes them possible to identify the spatial relationship between the elements of an image. The authors argue that, contrary to looking for a point of view invariance of the neurons' activities that use a single output value, neural networks should use local "capsules" which learn to recognize a visual entity implicitly under a limited domain of viewing conditions and deformations. Capsule structures encode complex calculations into a small, highly informative output vector. This vector contains information such as the probability of that entity being present in a compact domain. In addition, it comprises a set of instantiation parameters, which would include deformation, pose (position, size, orientation), hue, texture, and illumination condition of the visual entity relative to the version learned by the capsule [41].

Although idealized by Hinton et al., the first successful implementation of the capsule concept was made by Sabour et al. [41]. In their work, the authors created a three-layer capsule network that achieved comparable results with the best results in the MNIST [29] digit classification problem – previously achieved by deeper networks only. For this, Sabour et al. [41] developed two innovative concepts: dynamic routing and a new activation function.

Leveraged by Sabour et al. [41], many other authors have enhanced the concept of capsules. Hinton et al. [15] proposed a new type of capsule composed of a logistic unit that indicates the probability of the presence of an entity and a pose matrix of 4×4 representing the pose of that entity. The authors also introduced a new routing algorithm, which allows the outputs of the capsules to be routed to those of the next layer so that the active capsules receive a group of votes from similar poses. Hinton et al. showed that their model surpasses the best result in the smallNORB dataset, reducing the number of errors by more than 40% while being significantly more resistant to white-box adversarial attacks.

A remarkable work, which made possible the development of our solution, was developed by LaLonde and Bagci [28]. The authors expanded the use of capsules for the problem of object segmentation and made innovations that allowed, among other gains, to increase the data processing capacity of the capsule network, increasing inputs from 32×32 to 512×512 pixels. Most significantly, they advanced the state-of-the-art in the problem of segmentation of lung pathologies from computed tomography, while reducing the number of parameters by approximately 38%. In particular, the authors modified the capsule routing algorithm and the reconstruction part, and modified the concept of convolutional capsules.

Recently, a few authors have employed capsules in their solutions for problems involving LR images [18,35,46]. It is worth noting that most of these solutions only made small changes to the first capsule networks introduced by Sabour et al. [41] and Hinton et al. [15]. For example, Majdabadi and Ko [35] employed a two-layered capsule network with dynamic routing as the discriminator for its Multi-Scale Gradient capsule GAN. Leveraged by the matrix capsules by Hinton et al. [15], Hsu et al. [18] created two different approaches: (i) capsules as its main component in the network and reconstructing HR images directly from it (CIRNN) and (ii) capsules for the channel attention mechanism (CARNN).

3   Proposed Method

The proposed model, named Super-Resolution Capsules (SRCaps), is shown in Fig. 1. It consists of four main parts: an initial convolutional layer, followed by B sequentially connected residual dense capsule blocks, a new convolutional layer and, finally, a neural network to increase resolution. All the convolution-based layers use the weight normalization technique [43], as it accelerates the training convergence and has a lower computational cost if compared to batch normalization, without introducing dependencies between the examples of the batch [61].

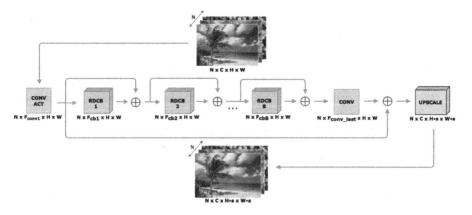

Fig. 1. Diagram of SRCaps model.

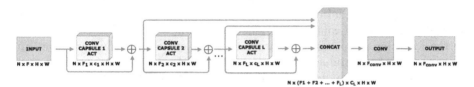

Fig. 2. RDCB diagram.

The first convolutional layer, $CONV\ ACT$ in Fig. 1, generates F filters from convolutional kernels of size $k \times k$ with stride st and padding p, followed by an activation function act. This layer is responsible for converting pixel intensities to local resource detector activations that are used as inputs to the next step in the capsule blocks.

The residual dense capsule blocks, RDCBs, are composed of L convolutional capsule layers followed by an activation function act, with residual connection to their inputs, sequentially connected. The outputs of these layers are concatenated, forming a dense connection, followed by a convolutional layer, as shown in Fig. 2. This convolutional layer, with kernels 1×1, stride 1 and padding 0, acts as a weighted sum between the various filters, allowing the network to learn which filters are more important, thus reducing dimensionality more efficiently [34,45,49]. The output of the RDCB is weighted by a residual scale constant.

All capsules layers within an RDCB module have the same parameters: the number of capsules per layer c, amount of filters F, kernel size k, stride st and padding p.

Our capsules employ the routing algorithm suggested by LaLonde and Bagci [28], because it provides an efficient version of the original capsule definition, as we mentioned before. This algorithm differs from the routing-by-agreement implementation by Sabour et al. [41] in two ways. First, we route capsules from the previous layer to capsules in the next layer within a specific spatial window. The original algorithm, on the other hand, directs the output of all previous layer capsules to all capsules in the next layer, varying only the routing weight. Second, we share the transformation matrices among all capsules of the same type. In a later step, we decided to replace the initial routing algorithm for the no-routing introduced by Gu et al. [13]. The authors argue that the routing procedure contributes neither to the generalization ability nor to the affine robustness of the CapsNets; therefore, distinct ways to approximate the coupling coefficients do not make a significant difference since they will be learned implicitly. In the no-routing approach, the iterative routing procedure is removed by setting all coupling coefficients as a constant $\frac{1}{M}$, where M is the number of capsules in the next layer. We also tried different values for the squashing constant sq used in the squashing function, as done by Huang and Zhou [20].

The RDCBs are sequentially connected, each having a residual connection with the block input, followed by a new convolutional layer, and the output of that layer has a residual connection with the output of the first convolutional layer. We use residual connections, identified by the symbol $\oplus$ in Figs. 1 and 2, for the following purposes: they avoid the problem of vanishing gradients (it becomes zero) by introducing shorter paths, which can take the gradient over the entire length of very deep networks, as demonstrated by Veit et al. [52]; and the use of residual connections seems to greatly improve training speed [48].

At the end of the model, there is a network used to upscale, called upsampling network (UPNet) (see Fig. 3). Following previous works [33,64] and several participants of the NTIRE 2017 [50] and NTIRE 2018 [51] competitions, the UPNet is composed of subpixel convolutions [44]. The UPNet allows the network to implicitly learn the process required to generate the larger version by adding the LR space feature maps and creating the SR image in a single step, saving memory and processing. We prefer this method over deconvolution since it naturally avoids checkerboard artifacts, which with deconvolution must be done using a kernel size that is divisible by stride to avoid the overlapping problem as demonstrated by Odena et al. [38]. Besides, UPNet has a considerably lower computational cost, becoming $\log_2 r^2$ times faster during training [44].

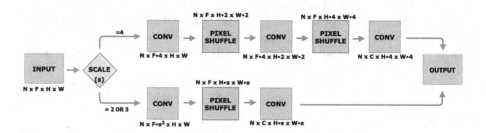

Fig. 3. UPNet diagram.

Loss Functions. During training, we evaluated loss functions commonly used in super-resolution problems [56]. Due to its simplicity and effectiveness, the first loss function we assess is the L1. Previous studies [33,65] showed that a network trained with L1 has achieved superior results compared to the same network trained with L2.

Still in the work of Zhao et al. [65], the idea of using indices based on the Structural Similarity Index (SSIM) [57] is introduced for training neural networks. As previously noted by Dong et al. [11], if a metric based on visual perception is used during training, the network can adapt to it. The SSIM is calculated as:

$$\text{SSIM}(p) = \frac{2\mu_x\mu_y + C_1}{\mu_x^2 + \mu_y^2 + C_1} \cdot \frac{2\sigma_{xy} + C_2}{\sigma_x^2 + \sigma_y^2 + C_2} = l(p) \cdot cs(p) \tag{1}$$

where μ_x and μ_y are the average pixel values in the SR and HR patches, respectively, σ_x and σ_y are the standard deviations of the same patches, σ_{xy} is the covariance between them, and C_1 and C_2 are constants added to avoid instabilities when the values of $\mu_x^2 + \mu_y^2$ and $\sigma_x^2 + \sigma_y^2$ are very close to 0. The $l(p)$ part of the equation calculates the comparison between clipping luminances, while the comparison between their contrasts and structures is calculated by $cs(p)$. Since the highest possible value for SSIM is 1, and because training a neural network usually aims to minimize the loss function, we can define the $\mathcal{L}^{\text{SSIM}}$ function as:

$$\mathcal{L}^{\text{SSIM}}(P) = 1 - \text{SSIM}(\tilde{p}), \tag{2}$$

in which $\tilde{p}$ is the central pixel of patch p.

The best performance in the work of Zhao et al. [65] was obtained by combining L1 and the Multi-Scale Structural Similarity Index (MS-SSIM) shown in Eq. 4 weighing 0.16 and 0.84, respectively. The authors argue that MS-SSIM preserves contrast in high-frequency regions, while the L1 preserves color and brightness regardless of the local structure. The MS-SSIM value is obtained by combining measurements at different scales using the Equation:

$$\text{MS-SSIM}(p) = l_M^{\alpha}(p) \cdot \prod_{j=1}^{M} cs_j^{\beta_j}(p), \tag{3}$$

where scales are used ranging from 1 (original image) to M (the largest scale used), reducing the image by a factor of 2 every iteration; l_M and cs_j are the same terms as defined in Eq. 3 at M and j scales, respectively, while the α and β_j exponents are used to adjust the relative significance of different components. It is worth noting that the luminance comparison (l_M) is calculated only at M scale, while contrast and structure comparisons (cs_j) at each scale. As with SSIM, the largest possible value for MS-SSIM is 1, so we can use it in a loss function in the form:

$$\mathcal{L}^{\text{MS-SSIM}}(p) = 1 - \text{MS-SSIM}(\tilde{p}). \tag{4}$$

We also explored the combination of functions employing several different layers of the network, as suggested by Xu et al. [58], in which the weighted sum between the calculation of the L1 function after two, three and four residual blocks are used, with weights of 0.5, 0.5 and 1, respectively. After each residual block composing the loss, we add a network based on subpixel convolutions to perform upscaling. Besides the above settings, we investigate the benefits of edge maps in the loss function, since L1 may smooth the edges. Similarly to Pandey et al. [39], we evaluated a combination of the L1 using the SR and HR images and the L1 between its edge maps. However, although their work uses the Canny operator [9] to generate the edge map, our work investigated the usage of the Sobel operator [47].

We also consider the three-component weighted PSNR (3-PSNR) and SSIM (3-SSIM) loss functions [31]. Importantly, such metrics can measure the quality of images and videos. This approach breaks up an image into three parts: edges, textures, and more homogeneous regions. To do this, the Sobel operator is applied to the luminance channel of the image and, from the highest calculated value and some pre-established values, the thresholds that delimit each region are calculated.

The value of each metric is calculated by applying different weights for each region. Li and Bovik [31] showed that the weights that achieved the best results were 0.7, 0.15 and 0.15 for 3-PSNR, and 1, 0 and 0 for 3-SSIM, considering edges, textures, and homogeneous regions, respectively. These values are consistent with the observation that perturbations at the edges of an object are perceptually more significant than in other areas. Based on recent solutions available in the literature, Barron [5] presented a loss function that is a superset of Cauchy/Lorentzian, Geman-McClure, Welsch/Leclerc, generalized Charbonnier, Charbonnier/pseudo-Huber/L1-L2, and L2. This function has two hyperparameters: robustness (α) and scale (c), with the variation of which is possible to reach all previous functions as specific cases. The general loss function is calculated as follows:

$$\mathcal{L}^{\text{general}}(x, \alpha, c) = \begin{cases} \frac{1}{2}(x/c)^2 & \text{if } \alpha = 2 \\ \log\left(\frac{1}{2}(x/c)^2 + 1\right) & \text{if } \alpha = 0 \\ 1 - \exp\left(-\frac{1}{2}(x/c)^2\right) & \text{if } \alpha = -\infty \\ \frac{|2-\alpha|}{\alpha}\left(\left(\frac{(x/c)^2}{|2-\alpha|} + 1\right)^{(\alpha/2)} - 1\right) & \text{otherwise} \end{cases}$$

where x is the difference between the HR and SR pixel values. Barron also showed that it is possible to modify its function so that the network learns optimal values for the α and c parameters, thus providing an appropriate exploration by the network of different error functions. Due to its unique features, we use the adaptive loss function for the SRCaps training.

4   Experimental Results

Datasets. In this work, we employed datasets widely used in the literature [4, 56]. Currently, the DIV2K training set is used in the training of neural networks

Table 1. Model metrics for each evaluated dataset. For the PSNR, SSIM, and MS-SSIM metrics the higher the better, while for the ℱLIP metric the lower the better.

Datasets	Metrics	Bicubic	EDSR	RCAN	RDN	**SRCaps**	SRCNN	WDSR
B100	PSNR	25.42985	27.04472	27.42409	27.37293	27.11657	25.99553	27.32658
	SSIM	0.84128	0.78966	0.79804	0.79646	0.78034	0.76109	0.79549
	MS-SSIM	0.94109	0.95965	0.96149	0.96108	0.95999	0.95198	0.96083
	ℱLIP	0.08295	0.06551	0.06198	0.06306	0.06914	0.0771	0.06349
DIV2K(validation)	PSNR	26.82531	28.5879	29.06375	28.92058	28.25471	27.48115	28.84742
	SSIM	0.85062	0.80914	0.82104	0.81771	0.79927	0.7788	0.8156
	MS-SSIM	0.94505	0.9641	0.96706	0.9662	0.96184	0.95525	0.96561
	ℱLIP	0.09288	0.07262	0.06759	0.07001	0.07622	0.08599	0.06868
Set5	PSNR	26.88569	29.6565	30.21205	30.04888	29.64867	27.94313	29.97913
	SSIM	0.86689	0.8554	0.86464	0.86269	0.85474	0.8119	0.86149
	MS-SSIM	0.95716	0.97831	0.97988	0.97954	0.97827	0.96853	0.97929
	ℱLIP	0.11142	0.06874	0.06404	0.06523	0.06966	0.09493	0.06467
Set14	PSNR	24.49373	26.33897	26.77688	26.61797	26.30902	25.26719	26.579
	SSIM	0.80909	0.74944	0.75981	0.75649	0.74529	0.71373	0.75594
	MS-SSIM	0.93117	0.95337	0.95576	0.95498	0.95318	0.94356	0.95467
	ℱLIP	0.10949	0.08205	0.07684	0.07872	0.08311	0.10072	0.07823
Urban100	PSNR	22.36726	24.43614	25.27712	25.03441	24.35517	23.01649	24.90592
	SSIM	0.80426	0.77287	0.79915	0.7915	0.76232	0.71431	0.78676
	MS-SSIM	0.92388	0.95592	0.96239	0.96041	0.95458	0.93913	0.95901
	ℱLIP	0.11929	0.0845	0.07557	0.0788	0.08941	0.10886	0.07929

for the super-resolution problem, while the validation set of DIV2K, B100, Set5, Set14, and Urban100 are used to validate the results. All datasets are composed of original versions of the images (HR) and their reduced versions by bicubic interpolation algorithm for 2×, 3×, and 4×. In this work, we focus on the 4× scale factor.

Metrics. For the validation process of the results obtained, we employed metrics commonly used in the literature. More specifically, Peak Signal-to-Noise Ratio (PSNR) [17], Structural Similarity Index (SSIM) [57] and Multi-Scale Structural Similarity Index (MS-SSIM) [55]. Due to space limitations, we refer interested readers to the work developed by Wang et al. [56] for a detailed formulation.

Algorithms that measure the differences between two images often assume that the images are shown side by side or alternated with an empty image displayed in between for a short period before the next image is shown. In contrast, flipping (or alternating) between two similar images reveals their differences to an observer much more effectively than showing the images next to each other. Aiming to better approximate human evaluators' methods, Andersson et. al. [3] developed a full-reference image difference algorithm, namely ℱLIP, which carefully evaluates differences inspired by models of the human vision.

ℱLIP is designed to have both low complexity and ease of use. It not only evaluates differences in colors and edges, but also pays attention to discrepancies in isolated pixels with colors that greatly differ from their surroundings. ꟻLIP outputs a new image indicating the magnitude of the perceived difference between two images at every pixel. The algorithm can also pool the per-pixel differences down to a weighted histogram, or generate a single value, which is the approach we will use during our analysis. This value is zero when both images are the same, and it increases as the more noticeable are the differences.

Computational Resources. The Super-Resolution Capsules (SRCaps) model was implemented using the PyTorch open-source platform on the PyTorch Lightning wrapper. The implementation made available by Uchida[1] was used as a foundation, and also metrics from the PyTorch Image Quality collection [23] and the official implementation of the ꟻLIP metric [3]. For optimizers implementations, we used the torch-optimizer package[2] and for comparative visualization of the metrics and generation of graphs we employed Tensorboard [1] and Comet.ml[3] tools.

Parameter Search. To find the best set of parameters for our SRCaps model, we used the Ray [32] open-source framework (Tune), which is a scalable hyperparameter tuning library built on top of Ray Core[4]. We employed the Async Successive Halving (ASHA) scheduler during the search, as it decides at each iteration which trials are likely to perform badly, and stops these trials, avoiding wasting resources on poor hyperparameter configurations. We train the models for a maximum of 100 epochs in the DIV2K training dataset and evaluate them using the MS-SSIM performance on the first 10 images from the DIV2K validation dataset. We select these values to allow fast experimentation of different sets of parameters, as it was observed during numerous training processes that the model tends to start stabilizing performance at around 100 epochs.

Experimental Setup. During the training process of the different models used for comparison, the entries have $N = 16$ (batch size) pairs of image clippings from the dataset. For all models, during validation, the LR and HR images are used entirely one by one ($N = 1$). All models discussed here were evaluated for $4\times$ super-resolution for 2000 epochs, where each epoch involves only one iteration through the training dataset, and is trained with two different loss functions: L1 and adaptive. Other functions have been evaluated, and their results will be briefly discussed throughout this section.

The final SRCaps model has convolutional kernels with $k = 3$ in its first layer, followed by 7 RDCBs ($B = 7$). The value used for the hyperparameters L and c are the same for all blocks: 3 and 4, respectively. In the last convolutional layer, we chose $k = 3$, as well as for the convolutional layers internal to the UPNet. Throughout the neural network, we used the values of $act = ReLU$, $k = 3$, $F = 128$, $st = 1$ and $p =$ 'same'. Setting the padding to the 'same' mode means using its value so that the input (H and W) dimensions are preserved

---

[1] github.com/S-aiueo32/sr-pytorch-lightning.

[2] https://github.com/jettify/pytorch-optimizer.

[3] https://www.comet.ml/.

[4] https://docs.ray.io/en/master/tune/index.html.

in the output, that is, $p = \frac{k-1}{2}$. Although having a smaller number of layers and only seven residual blocks in its composition, the SRCaps network has a considerable number of parameters - 15M. Such a value represents 13.5M, 2.4 and 10.2 more parameters than EDSR, RCAN and WDSR, respectively. This is due to the vectorial nature of the capsule, which adds an extra dimension to its composition.

HR image slices (*patch_size*) of size 128×128 and its corresponding 32×32 LR slice were used during their training. The updating of the weights of the networks is done by the Adam optimizer [27], with $\beta_1 = 0.9$, $\beta_2 = 0.999$, and $\epsilon = 10^{-8}$, being the networks trained with an initial learning rate of $lr = 10^{-4}$ which decays to half of the current value every 500 epochs.

When used, the adaptive loss function was initialized with the default values from the official implementation [5]. Employing these values is equivalent to starting the training with the Charbonnier/Pseudo-Huber function and letting the network learn from it what values for its parameters and, consequently, which function of the subset of the general function is more appropriate.

Baselines and Results. The EDSR model used is the base model defined by Lim et al. [33]. We chose the simplest version of the model because it has a smaller number of parameters than the SRCaps model. It is composed of 16 residual blocks without residual scale application since only 64 feature maps (filters) are used per convolutional layer. All convolutions, including the ones internal to the upscale network, have a kernel size of 3×3. During its execution, the input images are subtracted from the mean RGB values of the training images of the DIV2K dataset, which are 0.4488, 0.4371 and 0.4040. These values range from 0 to 1 and are multiplied by the maximum pixel value, 255. The RDN [64] model, as well as the EDSR, generates 64 filters as the output of its convolutional layers, and uses $k = 3$ for all convolutional kernels, except for those used in the fusion layers of LFF and GFF, which have $k = 1$. In this model, 16 RDB blocks were used, with 8 convolutional layers each.

For the WDSR model, we used the original implementation by Yu et al. [61]. The chosen version was `wdsr-b`, with 16 large residual blocks that generate 128 filters, but which internally generate convolutional layers with 6× more filters. This model, such as EDSR, also subtracts the mean RGB values from the DIV2K images. The RCAN model we used is also the original implementation [63] and is available in the same repository used as a baseline. It is composed of 10 residual groups (RG) that form the Residual in Residual (RiR) network structure, in which each RG is composed of 16 Residual Channel Attention Blocks (RCAB). It has $k = 3$ kernel sizes which generate $C = 64$ filters in all convolutional layers, except for those in the channel reduction and amplification mechanism, which have $k = 1$, and $\frac{C}{r} = 4$ and $C = 64$ respectively, with reduction factor $r = 16$.

Table 1 summarizes the results obtained for all models and datasets after the learning phase. From this table, we highlight the following points. First, the SRCaps model obtained comparable results to the EDSR model, sometimes surpassing it for some metrics, particularly in the B100 dataset. Second, as shown in Figs. 4 and 5, the proposed model manages to recreate the connection between

Fig. 4. Model results for "0891" image from DIV2K dataset.

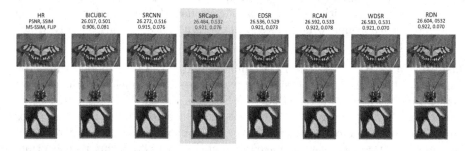

Fig. 5. Model results for "0829" image from DIV2K dataset.

the different characters more precisely, while models with better metrics such as RCAN and RDN tend to thin the connection, as they do for the leftmost part of the symbol on top. Finally, despite being able to reconstruct with quality rounded edges, a deficiency of the SRCaps model is in the reconstruction of linear edges, usually diagonal.

It is remarkable the results obtained with the RCAN model, reaching the highest value in all metrics. We highlight that our goal is not to push the state-of the art but to bring insights into capsules applied to super-resolution problems. We believe that future research on capsule networks could benefit from our findings.

5   Conclusions

The purpose of this work was to evaluate the use of the capsule concept in the solution of single image super-resolution problems, as well as to verify new forms of training and validate the results of neural networks for this purpose. It was evidenced that, despite the inferior result, a trained network with a smaller number of layers obtained a relevant result, indicating that networks that use capsules can have applications in super-resolution. Hypotheses have been raised that the nonlinearity function applied together with the capsules may be a limiting factor, given the different nature of the problem as to its initial usage (super-resolution × classification).

Throughout our work, we investigate the contribution of many hyperparameters such as the activation function, learning rate, optimizer and loss function. Regarding the latter, we evaluate loss functions that take the human visual system into account, as suggested by previous works [55,57]. Additionally, we study different architectural designs to compose our capsule network. The fact that the adaptive function [5] is a superset of several others and that it is possible to make the network learn, along with the other weights, the optimal values for its two main parameters (α and c), allow the network to experiment which loss function best fits the problem. Thus, it is possible to train the network starting from a function similar to L1, while modifying it at each iteration to extract as much useful information as possible from the data. The current limitations of the most used metrics in the literature [31] were also emphasized in this work, showing that visual evaluation of the results is still essential. Existing metrics have been suggested such as MS-SSIM [55] and ⅎLIP [3], encouraging discussion of new metrics.

Several points of possible improvements that could not be deeply evaluated were identified. As a future line of research, we intend to replace the composition of the UPNet network, which is used in much the same way by several networks that have reached the state of the art. One can again verify the usage of the concept of reverse convolutions, or deconvolutions, as used by Dong et al. [12], and also of deconvolutional capsules created by LaLonde and Bagci [28], or use more recent methods from the literature. Kim and Lee [26] recently proposed the enhanced upscaling module (EUM), which achieves better results through nonlinearities and residual connections.

References

1. Abadi, M., et al.: TensorFlow: Large-Scale Machine Learning on Heterogeneous Systems (2015). Software available from tensorflow.org
2. Agustsson, E., Timofte, R.: NTIRE 2017 challenge on single image super-resolution: dataset and study. In: CVPR Workshops, pp. 1–8 (2017)
3. Andersson, P., Nilsson, J., Akenine-Möller, T., Oskarsson, M., Åström, K., Fairchild, M.D.: FLIP: a difference evaluator for alternating images. In: ACM on Computer Graphics and Interactive Techniques (2020)
4. Anwar, S., Khan, S.H., Barnes, N.: A deep journey into super-resolution: a survey. ACM Comput. Surv. **53**(3), 60:1–60:34 (2020). https://doi.org/10.1145/3390462
5. Barron, J.T.: A More General Robust Loss Function. arXiv preprint arXiv:1701.03077 (2017)
6. Behjati, P., Rodriguez, P., Mehri, A., Hupont, I., Tena, C.F., Gonzalez, J.: OverNet: lightweight multi-scale super-resolution with overscaling network. In: WACV, pp. 1–11 (2021)
7. Bevilacqua, M., Roumy, A., Guillemot, C., Alberi-Morel, M.L.: Low-complexity single-image super-resolution based on nonnegative neighbor embedding. In: BMVC (2012)

8. Cai, J., Gu, S., Timofte, R., Zhang, L.: NTIRE 2019 challenge on real image super-resolution: methods and results. In: CVPR Workshops, pp. 1–8 (2019)
9. Canny, J.: A computational approach to edge detection. Trans. Pattern Anal. Mach. Intell. **8**(6), 679–698 (1986)
10. Dong, C., Loy, C.C., He, K., Tang, X.: Image super-resolution using deep convolutional networks. Trans. Pattern Anal. Mach. Intell. **38**(2), 295–307 (2016)
11. Dong, C., Loy, C.C., He, K., Tang, X.: Learning a deep convolutional network for image super-resolution. In: Fleet, D., Pajdla, T., Schiele, B., Tuytelaars, T. (eds.) ECCV 2014. LNCS, vol. 8692, pp. 184–199. Springer, Cham (2014). https://doi.org/10.1007/978-3-319-10593-2_13
12. Dong, C., Loy, C.C., Tang, X.: Accelerating the super-resolution convolutional neural network. In: Leibe, B., Matas, J., Sebe, N., Welling, M. (eds.) ECCV 2016. LNCS, vol. 9906, pp. 391–407. Springer, Cham (2016). https://doi.org/10.1007/978-3-319-46475-6_25
13. Gu, J., Tresp, V.: Improving the robustness of capsule networks to image affine transformations. In: CVPR, pp. 1–15 (2020)
14. Gu, J., Wu, B., Tresp, V.: Effective and efficient vote atack on capsule networks. In: ICLR (2021)
15. Hinton, G., Sabour, S., Frosst, N.: Matrix capsules with EM routing. In: ICLR, pp. 1–10 (2018)
16. Hinton, G.E., Krizhevsky, A., Wang, S.D.: Transforming auto-encoders. In: Artificial Neural Networks and Machine Learning, pp. 44–51 (2011)
17. Hore, A., Ziou, D.: Image quality metrics: PSNR vs. SSIM. In: ICPR, pp. 2366–2369 (2010)
18. Hsu, J., Kuo, C., Chen, D.: Image super-resolution using capsule neural networks. IEEE Access (2020)
19. Huang, J.B., Singh, A., Ahuja, N.: Single image super-resolution from transformed self-exemplars. In: CVPR, pp. 1–9 (2015)
20. Huang, W., Zhou, F.: DA-CapsNet: dual attention mechanism capsule network. Sci. Rep. (2020)
21. Irani, M., Peleg, S.: Improving resolution by image registration. In: CVGIP: Graph. Model. Image Process. **53**(3), 231–239 (1991)
22. Ji, X., Cao, Y., Tai, Y., Wang, C., Li, J., Huang, F.: Real-world super-resolution via Kernel estimation and noise injection. In: CVPR Workshops, pp. 1–8 (2020)
23. Kastryulin, S., Zakirov, D., Prokopenko, D.: PyTorch image quality: metrics and measure for image quality assessment (2019). https://github.com/photosynthesis-team/piq
24. Kim, J., Kwon Lee, J., Mu Lee, K.: Accurate image super-resolution using very deep convolutional networks. In: CVPR, pp. 1–8 (2016)
25. Kim, J., Kwon Lee, J., Mu Lee, K.: Deeply-recursive convolutional network for image super-resolution. In: CVPR, pp. 1–13 (2016)
26. Kim, J.H., Lee, J.S.: Deep residual network with enhanced upscaling module for super-resolution. In: CVPR Workshops, pp. 1–15 (2018)
27. Kingma, D.P., Ba, J.: Adam: a method for stochastic optimization. In: Bengio, Y., LeCun, Y. (eds.) ICLR (2015)
28. LaLonde, R., Bagci, U.: Capsules for Object Segmentation. arXiv preprint arXiv:1804.04241 (2018)
29. Lecun, Y., Bottou, L., Bengio, Y., Haffner, P.: Gradient-based learning applied to document recognition. Proc. IEEE **86**(11), 2278–2324 (1998)
30. Ledig, C., et al.: Photo-realistic single image super-resolution using a generative adversarial network. In: CVPR, pp. 1–8 (2017)

31. Li, C., Bovik, A.C.: Content-weighted video quality assessment using a three-component image model. J. Electron. Imag. **19**, 19 (2010)
32. Liaw, R., Liang, E., Nishihara, R., Moritz, P., Gonzalez, J.E., Stoica, I.: Tune: a research platform for distributed model selection and training. arXiv preprint arXiv:1807.05118 (2018)
33. Lim, B., Son, S., Kim, H., Nah, S., Lee, K.M.: Enhanced deep residual networks for single image super-resolution. In: CVPR Workshops, pp. 1–8 (2017)
34. Lin, M., Chen, Q., Yan, S.: Network in Network. arXiv preprint arXiv:1312.4400 (2013)
35. Majdabadi, M.M., Ko, S.B.: Capsule GAN for Robust Face Super-Resolution. Multim. Tools Appl. (2020)
36. Martin, D., Fowlkes, C., Tal, D., Malik, J.: A database of human segmented natural images and its application to evaluating segmentation algorithms and measuring ecological statistics. In: ICCV, vol. 2, pp. 416–423 (2001)
37. Nasrollahi, K., Moeslund, T.B.: Super-resolution: a comprehensive survey. Mach. Vis. Appl. **25**(6), 1423–1468 (2014)
38. Odena, A., Dumoulin, V., Olah, C.: Deconvolution and checkerboard artifacts. Distill (2016). http://distill.pub/2016/deconv-checkerboard
39. Pandey, R.K., Saha, N., Karmakar, S., Ramakrishnan, A.G.: MSCE: an edge preserving robust loss function for improving super-resolution algorithms. arXiv preprint arXiv:1809.00961 (2018)
40. Ren, H., Kheradmand, A., El-Khamy, M., Wang, S., Bai, D., Lee, J.: Real-world super-resolution using generative adversarial networks. In: CVPR Workshops, pp. 1–8 (2020)
41. Sabour, S., Frosst, N., Hinton, G.E.: Dynamic routing between capsules. In: NeurIPS, pp. 3856–3866 (2017)
42. Sabour, S., Tagliasacchi, A., Yazdani, S., Hinton, G.E., Fleet, D.J.: Unsupervised part representation by flow capsules. In: Meila, M., Zhang, T. (eds.) ICML, vol. 139, pp. 9213–9223 (2021)
43. Salimans, T., Kingma, D.P.: Weight normalization: a simple reparameterization to accelerate training of deep neural networks. In: NeurIPS, pp. 901–909. Curran Associates, Inc. (2016)
44. Shi, W., et al.: Real-time single image and video super-resolution using an efficient sub-pixel convolutional neural network. In: CVPR, pp. 1–8 (2016)
45. Simonyan, K., Zisserman, A.: Very deep convolutional networks for large-scale image recognition. In: ICLR (2015)
46. Singh, M., Nagpal, S., Singh, R., Vatsa, M.: Dual directed capsule network for very low resolution image recognition. In: ICCV, pp. 1–8 (2019)
47. Sobel, I., Feldman, G.: A 3 × 3 Isotropic Gradient Operator for Image Processing (1968). Talk at the Stanford Artificial Project
48. Szegedy, C., Ioffe, S., Vanhoucke, V., Alemi, A.A.: Inception-v4, inception-resnet and the impact of residual connections on learning. In: AAAI, pp. 4278–4284 (2017)
49. Szegedy, C., et al.: Going deeper with convolutions. In: CVPR, pp. 1–8 (2015)
50. Timofte, R., et al.: NTIRE 2017 challenge on single image super-resolution: methods and results. In: CVPR Workshops, pp. 1110–1121 (2017)
51. Timofte, R., Gu, S., Wu, J., Van Gool, L.: NTIRE 2018 challenge on single image super-resolution: methods and results. In: CVPR Workshops, pp. 1–17 (2018)
52. Veit, A., Wilber, M.J., Belongie, S.: Residual networks behave like ensembles of relatively shallow networks. In: NeurIPS, pp. 550–558 (2016)
53. Wang, X., et al.: ESRGAN: enhanced super-resolution generative adversarial networks. In: ECCV, pp. 63–79 (2019)

54. Wang, Z., Liu, D., Yang, J., Han, W., Huang, T.: Deep networks for image super-resolution with sparse prior. In: ICCV, pp. 370–378 (2015)
55. Wang, Z., Simoncelli, E.P., Bovik, A.C.: Multiscale structural similarity for image quality assessment. In: The Thirty-Seventh Asilomar Conference on Signals, Systems Computers, vol. 2, pp. 1398–1402 (2003)
56. Wang, Z., Chen, J., Hoi, S.C.H.: Deep learning for image super-resolution: a survey. Trans. Pattern Anal. Mach. Intell. **43**(10), 3365–3387 (2021)
57. Wang, Z., Bovik, A.C., Sheikh, H.R., Simoncelli, E.P.: Image quality assessment: from error visibility to structural similarity. Trans. Image Process. **13**(4), 600–612 (2004)
58. Xu, J., Zhao, Y., Dong, Y., Bai, H.: Fast and accurate image super-resolution using a combined loss. In: CVPR Workshops, pp. 1093–1099 (2017)
59. Yang, J., Wright, J., Huang, T.S., Ma, Y.: Image super-resolution via sparse representation. Trans. Image Process. **19**(11), 2861–2873 (2010)
60. Yu, C., Zhu, X., Zhang, X., Wang, Z., Zhang, Z., Lei, Z.: HP-capsule: unsupervised face part discovery by hierarchical parsing capsule network. In: CVPR, pp. 4022–4031 (2022)
61. Yu, J., Fan, Y., Yang, J., Xu, N., Wang, X., Huang, T.S.: Wide Activation for Efficient and Accurate Image Super-Resolution. arXiv preprint arXiv:1808.08718 (2018)
62. Zhang, K., Gu, S., Timofte, R.: NTIRE 2020 challenge on perceptual extreme super-resolution: methods and results. In: CVPR Workshops, pp. 1–10 (2020)
63. Zhang, Y., Li, K., Li, K., Wang, L., Zhong, B., Fu, Y.: Image super-resolution using very deep residual channel attention networks. In: ECCV, pp. 1–8 (2018)
64. Zhang, Y., Tian, Y., Kong, Y., Zhong, B., Fu, Y.: Residual dense network for image super-resolution. In: CVPR, pp. 2472–2481 (2018)
65. Zhao, H., Gallo, O., Frosio, I., Kautz, J.: Loss functions for image restoration with neural networks. IEEE Trans. Comput. Imaging **3**(1), 47–57 (2017)

Development of a Deep Learning Model for the Classification of Mosquito Larvae Images

Ramon Mayor Martins[1(✉)], Bruno Manarin Espíndola[1], Pedro Philippi Araujo[1],
Christiane Gresse von Wangenheim[1], Carlos José de Carvalho Pinto[2],
and Gisele Caminha[3]

[1] Department of Informatics and Statistics, Federal University of Santa Catarina, Florianópolis,
SC 88040-900, Brazil
ramon.mayor@posgrad.ufsc.br, {bruno.manarin,
pedro.pa}@grad.ufsc.br, c.wangenheim@ufsc.br
[2] Department of Microbiology, Immunology and Parasitology/Center for Biological Sciences,
Federal University of Santa Catarina, Florianópolis, SC 88040-900, Brazil
carlos.pinto@ufsc.br
[3] Coordination of the Laboratory Reference Network, Santa Catarina, Florianópolis,
SC 88010-001, Brazil

Abstract. Dengue is a disease that is endemic to certain regions, and in 2022,
it was responsible for more than three million cases in the Americas. One of the
most effective ways to prevent Dengue is by preventing the formation of breeding
sites for the *Aedes aegypti* mosquito, which is the primary vector of the disease.
Unfortunately, identifying these breeding sites remains a challenge as citizens
lack knowledge to distinguish *Ae. Aegypti* larvae from other species. A solution
can be the development of a Deep Learning model, to be deployed in a mobile
application that classifies mosquito species using photos of larvae. Currently only
a few models are available that mostly differentiate only between genera (Aedes
versus non-Aedes), or present very low accuracy. Therefore, the objective of this
research is to develop an image classification model that can differentiate between
Ae. Aegypti, *Aedes albopictus*, and *Culex* sp. Larvae using pictures taken with
a cellphone camera by comparing various Deep Learning models (Mobilenetv2,
ResNet18, ResNet34, EfficientNet_B0 and EfficientNet_Lite0). Best results were
obtained with EfficientNet_Lite0 with an accuracy of 97.5% during validation and
90% during testing, an acceptable result considering the risks related to a misclassification in this context. These results demonstrate the viability of a classification
of mosquito larvae differentiating even between Aedes species and thus providing
a contribution to the prevention of dengue.

Keywords: *Aedes aegypti* · Dengue · Deep Learning · Image classification

1 Introduction

Dengue is an acute febrile disease caused by a virus, which is transmitted mainly by
the *Aedes aegypti*, a mosquito that also transmits chikungunya fever and the zika virus
[20]. It is an endemic disease in the Americas, Africa, Asia and Oceania. The World

M. C. Naldi and R. A. C. Bianchi (Eds.): BRACIS 2023, LNAI 14197, pp. 129–145, 2023.
https://doi.org/10.1007/978-3-031-45392-2_9

Health Organization estimates that, annually, between 100 and 400 million cases occur each year, with 500,000 to 1 million developing the severe form, which can be fatal [33]. In Brazil, dengue is a public health problem. According to the [24], approximately 1.5 million probable cases were registered in Brazil in 2022, an increase of 206% compared to 2021. This shows the importance of disease prevention for the well-being of the population. One of the most effective ways to prevent Dengue is to eliminate breeding sites for the *Aedes aegypti* mosquito, the primary vector of the disease. Larvae foci can be found commonly in artificial deposits and containers with stagnant water (e.g., flower pots or tires), where they allow the development of the mosquito larvae. Therefore, an important actor in controlling this disease control is the population, by preventing the formation of breeding sites in their homes. Unfortunately, identifying these breeding sites remains a challenge, as the general population lacks sufficient knowledge to distinguish *Ae. Aegypti* larvae from other species, such as *Aedes albopictus* or *Culex* sp. Also to be found in urban regions in Santa Catarina [12].

Mosquitoes is a holometabolous insect which goes through the complete metamorphosis process from eggs, larva, pupa, to the adult stage. The most favorable stage for collecting mosquito samples may be the larval stage, since it is contained in a place with water. Morphological characteristics of the larva, color, bristles, length of breathing tube, its positioning in the water are factors taken into consideration for the characterization of the insect, such as the differentiation of *Culex* and *Aedes* mosquitoes. Usually the classification of mosquito larvae is based on a microscopic analysis by biologists [14].

An alternative solution in order to enable citizens to identify the presence of *Ae. Aegypti* larvae in their homes could be to develop a Deep Learning (DL) model for the automatic classification of mosquito larvaes deployed in a mobile application that enables the classification using a photograph of a mosquito larva found by the user taken with the cellphone camera. This would make the diagnosis quick and accessible, not requiring specialist knowledge of mosquito morphology.

There is already research on the development of DL models aiming at the classification of mosquito larvae images. Yet, most of the existing models do not classify between *Aedes* species, thus omitting important information when to be used for dengue prevention. The only model developed to distinguish between *Aedes* species [5] reported a very low accuracy of only 64.58% that is inappropriate considering the risk of misclassification. Another limitation of existing research is the use of either microscopic images or pictures taken with a special zoom lens [17, 25], not available to a broader audience.

Therefore, aiming at the development of a DL model to be deployed in a mobile application to be used by citizens in the urban region of Florianopolis/SC, we analyze the following research question: Is it possible to develop a DL model to classify larvae of *Ae. Aegypti*, *Ae. Albopictus* and *Culex* sp. Mosquitoes based on photos taken with a cellphone camera with an accuracy of at least 90%, considering the risk related to erroneous classification in the context of an intelligent mobile application to be used by a wide target audience.

2   Background: Morphology of Mosquito Larvae

There are fundamental differences in the morphology of mosquito larvae between differ-ent genera and species as shown in Fig. 1. Mosquito larvae have three body regions: head, thorax, and abdomen. The head of mosquito larvae is large and sclerotized (composed of a hardened exoskeleton). The shape of the head of *Aedes* and *Culex* larvae is broad. The head has two eyes, two antennae, and a brush-like or comb-like mouth apparatus. The eyes are usually small, simple (not compound) and are found on both sides of the head [8]. The head of the *Aedes* is generally shorter and wider than that of the Culex. Aedes antennae have small scales on the surface, while *Culex* antennae are smooth. *Aedes* jaws are wider and shorter than *Culex* jaws. In general, the thorax of larvae is wider than the head and has three rows of bristles. The thorax of the *Aedes* has white spots. *Culex*'s thorax has no distinctive spots. In addition, the bristles on the surface of the thorax of *Aedes* are longer and more sparse than those of *Culex*. Commonly, the abdomen is elon-gated, composed of ten segments, and its eighth segment has the respiratory siphon that is used to distinguish between species [7, 8]. The abdomen of the *Aedes* is shorter and wider than that of the *Culex*, while the scales on the abdomen of the *Aedes* are rounder and darker than those of the *Culex*.

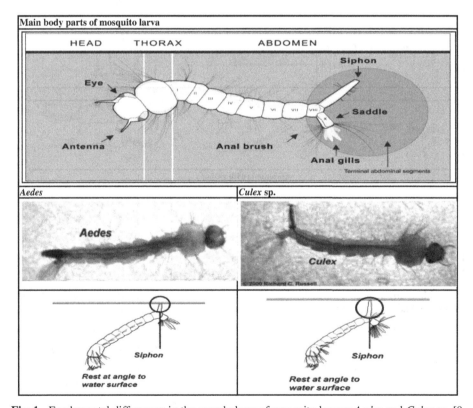

Fig. 1. Fundamental differences in the morphology of mosquito larvae: *Aedes* and *Culex* sp. [9, 14].

The siphon in the abdomen allows the larvae to breathe oxygen from the air while remaining submerged in water. The siphon is one of the main characteristics used to identify mosquito species. Its shape, size and length can vary greatly between species, as well as the presence or absence of a specialized structure called a "pecten". The pecten is a row of spines located at the base of the siphon. There are some notable differences in the siphon between *Aedes* and *Culex*. The *Culex* siphon is longer and narrower than the one of *Aedes* [7, 8, 14, 32].

The genus Aedes has two main disease vector species, *Ae. Aegypti* and *Ae. Albopictus*. Some morphological characteristics distinguish the larvae of these species [7]. The head of the *Ae. Aegypti* is more rounded, while the head of the *Ae. Albopictus* is more elongated. There are also subtle differences in the bristles of the antennae. The bristles on the thorax of the *Ae. Aegypti* are longer, while in the *Ae. Albopictus* they are shorter. And on the abdomen, the bristles of the *Aedes albopictus* are simpler and without many branches (Fig. 2) [7, 8, 14, 32].

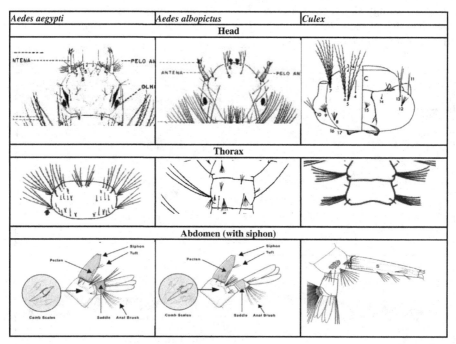

Fig. 2. Main differences in the head, thorax and abdomen of *Aedes* species (*Ae. Aegypti* and *Ae. Albopictus*) and *Culex* [8, 14].

All these characteristics are commonly distinguishable only by biologists and specialists, typically with the aid of microscopes.

3 Related Work

To summarize research adopting Deep Learning for the automatic classification of images of mosquito larvae during the last ten years, a systematic mapping following the procedure proposed by [26] was conducted. As a result a total of 12 research articles were found published mostly during the last five years (Table 1).

Table 1. Overview on research adopting DL for the classification of mosquito larvae images.

Reference	Classified Species	Country	Type of image	Quantity of images	DL models(s)	Performance			
						Accuracy	Precision	Recall	F1-Score
(Arista-Jalife,2018) [2]	Aedes and Non-Aedes	Mexico		570		91%	NI	NI	NI
(Arista-Jalife,2020) [3]	Aedes and Non-Aedes	Mexico	Microscope (60x zoom)	916	VGG-16 VGG-19	88.50% 88.50%	NI	NI	NI
(Asmai, 2019) [4]	Aedes and Non-Aedes	Malaysia	From platforms like Flickr.com and Shutterstock.com under the microscope	NI	VGG16, VGG19, ResNet50, InceptionV3	81.29% 87.25% 86.38% 83.50%	NI	NI	NI
(Azman, 2020) [5]	Aedes Aegypti, Aedes Albopictus, Anopheles and Culex	Malaysia	Captured with the micro blips lens of a Samsung Smartphone	NI	MobileNetV2	64.58%	88%	NI	NI
(De Silva, and Jayalal 2020) [9]	Aedes and Non-Aedes	Sri Lanka	Microscope (60x zoom) and zoomed digital	160 (microscope) 238 (zoom)	ResNet50	86.65%	NI	NI	NI
(Fuad et al., 2018) [11]	Aedes Aegypti and Non-Aedes	Malaysia							
(Hossain, 2022) [17]	Aedes and Non-Aedes	Swiss	Online sources and taken with a photo camera (65mm f/2.8 1–5 × microlens)	900	VGG16, VGG19, ResNet50, ResNet152, InceptionV3	95% 95% 96% 94% 88%	NI	NI	NI
(Munoz, 2018) [25]	Aedes, Culex, Anopheles or Unknown	U.S	Cell phone camera with attached microscope	NI	Caffenet AlexNet	NI	100%	47.4%	NI
(Rajasekhar, 2021) [27]	Anopheles and Non-Anopheles	U.S							
(Sanchez-Ortiz, 2017) [29]	Aedes and Non-Aedes	Mexico	Microscope	300	Alexnet	96.80%	30%	NI	NI
(Surya, 2022) [30]	Aedes, Culex or Unknown	NI	Microscope from database: GLOBE Mosquito Habitat Mapper	10.000	ViT-Base CvT-13, ConvNeXT, ResNet-18	63.74% 64% 65.63% 59.67%	60.61% 62.92% 63.86% 60.34%	63.74% 64% 65.63% 59.67%	58.68% 62.09% 63.55% 57.56%
(Garcia et al., 2019) [13]	Aedes and Non-Aedes	Spain	Microscope from database: GLOBE Mosquito Habitat Mapper	155	DenseNet	97%	97%	NI	NI

NI - Not informed/not identified.

Most of the research focuses on the classification of *Aedes* vs. Non-*Aedes* mosquito larvae, omitting detailed information when using the model for dengue prevention, while few aim at distinguishing between Aedes species. Few also consider the *Culex* species [5, 30], another vector of diseases (such as encephalitis, lymphatic filariasis and Nile fever). Focusing on the development of such a model to be used by citizens for dengue prevention, especially in the urban region of Florianópolis/SC, it is imperative to consider all relevant species, including also a differentiation between *Ae. Aegypti* and *Ae. Albopictus* in order to help the users to take adequate actions to prevent proliferation. Yet, so far no research in Brazil with such a focus has been encountered.

Only [5] presents a research that addresses the primary mosquito species of concern, including *Ae. Aegypti, Ae. Albopictus, Anopheles*, and *Culex* sp. However, the results achieved were not significant, with a reported accuracy of only 64%. This low level of accuracy poses a significant risk for users and could put human lives in danger in the context of the usage scenario.

Another limitation regarding most existing research is that the DL models have been trained with microscopic images, different from photos that would be taken by a citizen with a cellphone camera. Only [5, 17] and [25] use images taken in with a cellphone, yet using a special zoom lens, which again will not be available to a broader audience.

The current research mostly used Convolutional Neural Networks (CNNs), including ResNets as well as older models such as Alexnet. No research with more recent CNNs, such as EfficientNets have been encountered, although high performances have been reported in other application domains comparing it with other models [31].

The performance of the trained models varies largely, with an accuracy ranging from 59.67% [30] to about 97% [13, 29], while most performance results are rather low especially considering the risk related to a misclassification in this context. The highest performances reported are also related to classifications distinguishing only *Aedes* vs. Non-*Aedes*, but not among *Aedes* species. Although DenseNet achieved a high accuracy (97%), as reported by [13], it was not designed specifically for mobile applications, and generally has a higher number of parameters and computational complexity. The only research reporting performance with respect to the classification between *Aedes* species [5] indicates a very low accuracy of only 64.58%.

These results show that there is currently a lack of DL models for the automatic classification of mosquito larvae that are able to distinguish between *Aedes* species based on pictures taken with a cellphone camera with minimal acceptable accuracy.

4   Research Methodology

This research is characterized as applied research [16] by identifying and characterizing a need and aiming to contribute a practical solution adopting DL. Based on the results of a previous systematic mapping as presented in Sect. 3, we follow a systematic process for the human-centric interactive development of Deep Learning models [1, 22, 28]. As presented in Table 2 this includes the analysis of requirements, data preparation and an experiment on training and evaluating different DL models, as well as testing the trained models with new unseen images. The evaluation results are compared between the different DL models as well as in comparison with models encountered in literature.

Table 2. Overview on phases of the research methodology.

Elicitation of the state of the art	Systematic mapping following the procedure proposed by [26]
Development of DL models following [1, 22, 28]	
Requirements analysis	During this stage, the main objective of the DL model and its target features are specified
Data preparation	This phase includes the collection of images, including also the cleaning and standardized formatting of the images. The data is labeled in cooperation with domain experts. Regarding the management of data quality we followed IEEE Std 2801 [18]
Model training and evaluation & comparison	An experiment is conducted training different CNNs and evaluating and comparing their performance
Model prediction/testing & comparison	Using a set of previous unseen images, a test is run and the performance of the trained models is evaluated and compared following ISO/IEC 4213 [19]
Model export	The trained models are exported in order to enable deployment in a mobile application

5 Development of the Image Classification Model

5.1 Requirements Analysis

Adopting [23] requirements notation, the goal is to develop a DL model that learns from experience E with respect to some class of tasks T and with performance measure P, if its performance on tasks in T, measured by P, improves with experience E. Here, Task (T) is to classify mosquito larvae (single label) from a photo taken with a cellphone camera. Experience (E) is a corpus of labeled images of mosquito larvae of the genus Aedes including the species *Ae. Aegypti*, *Ae. Albopictus*, genus *Culex* and a non-mosquito object. In terms of performance (P), considering the risk of misclassification in this context, an accuracy performance level greater than 90% is expected in order to ensure that the model can effectively identify mosquito larvae and contribute to dengue prevention efforts. This level of accuracy would provide a sufficient degree of confidence in the model's ability to identify mosquito larvae correctly and minimize the risk of false positives or false negatives.

5.2 Data Preparation

Due to the unavailability of public datasets of cellphone photos of larvae of the relevant mosquito species, we collected a set of 1.999 images, including 748 images of *Ae. Aegypti* larvae, 464 images of *Ae. Albopictus* larvae, 447 images of *Culex* sp. Larvae and 340 images of Non-Mosquito objects[1].

The images were collected and labeled by researchers from the initiative Computação na Escola/INCoD/INE/UFSC in cooperation with the Laboratory of Hematozoan Transmitters/UFSC and the Coordination of the Laboratory Reference Network/Central Public Health Laboratory of SC. The images were taken with cell phone cameras (Samsung Galaxy S10, Xiaomi Redmi Note 9 and Xiaomi Mi 11 Pro) without any additional lenses (Table 3).

Table 3. Camera specifications.

Cellphone	Camera specifications
Samsung Galaxy S10	12 MP main camera with variable aperture, 16 MP ultra-wide camera, and 12 MP telephoto camera with 2× optical zoom
Xiaomi Redmi Note 9	48 MP main camera, 8 MP ultra-wide camera, 2 MP macro camera, and 2 MP depth camera
Xiaomi Mi 11 Pro	50 MP main camera, 8 MP telephoto camera with 5× optical zoom, and 13 MP ultra-wide camera

The images were collected with varying backgrounds, angles, and resolutions and saved in.jpg format (Fig. 3).

Sample image of *Aedes aegypti*   Sample image of *Aedes albopictus*

Sample image of *Culex* sp.   Sample image of Non-mosquito

Fig. 3. Examples of images from the data set.

In order to assure data quality the following data quality characteristics were considered in accordance with IEEE Std 2801 [18] as presented in Table 4.

The data set was divided into a training (79.2%), validation (19.8%), and test (2%) set.

---

[1] The dataset and notebook for the model are being prepared for availability.

Table 4. Data quality characteristics.

Quality characteristic	Dataset characteristic
Accuracy	The images were labeled by experts (biologists) with an accuracy of no less than 99%
Completeness	The dataset covers all main mosquito species within the context of this research
Uniqueness	The dataset does not contain duplicated images
Authenticity	The mosquito larvae samples were collected by biologists directly from field studies
Accessibility	The dataset can be previewed by authorized researchers of the project
Understandability	The dataset is labeled by species, numbered, stored in training/validation and test folders
Efficiency	The images are of varying resolution, taken with current cell phone cameras with a resolution between 2 MP and 108 MP
Portability	The images are stored in.jpg format that can be accessed by all major image viewers and operating systems. They are stored in the cloud (Google Drive), and can be accessed from any connected device by authorized users
Representativeness	The dataset includes larvae of all relevant mosquito species (*Ae. Aegypti*, *Ae. Albopictus*, and *Culex* sp.) collected in the Santa Catarina region. To mitigate interpretation out of interest, images that do not contain mosquitos are also included (class nonmosquito)

5.3 Model Training, Evaluation and Comparison

In accordance with literature five different DL architectures for image classification that are also indicated for deployment in mobile applications have been trained: MobileNetv2, ResNet18, ResNet34, EfficientNet_B0, and EfficientNet_Lite0. MobileNetv2 is an architecture designed specifically for mobile device applications, with a focus on efficiency and low resource consumption [15]. ResNets 18 and 34 were chosen because of their efficiency and being the least complex of the ResNet architecture family considering a deployment in a mobile app [6]. EfficientNet_B0 is the most basic version of the EfficientNet family, and is designed to be efficient in terms of both computational resources and performance [31] and the lightest version, EfficientNet_Lite0, designed specifically for mobile device applications [21].

The models were developed in Python using Jupyter Notebook/Google Colab. For the development we used the Fast.ai library [10], an open source deep learning library that provides a high-level API to simplify the process of training and validating DL models as well as its support for transfer learning. In order to run EfficientNet and Mobilenet architectures we also used the TIMM library, which provides a collection of pre-trained SOTA (State-of-the-Art) computer vision models [10].

The images were resized to 224×224 pixels and subjected to random data augmentation. An overview of the training parameters and results is shown in Table 5.

Table 5. Overview on training parameters and results.

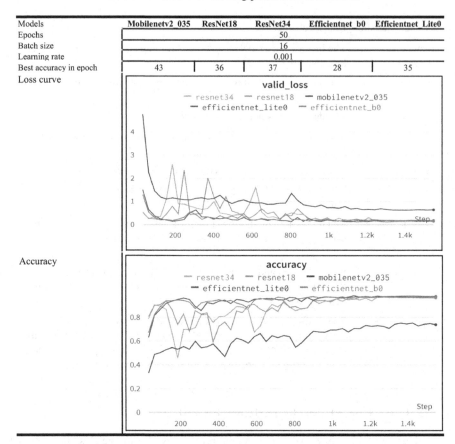

Models	Mobilenetv2_035	ResNet18	ResNet34	Efficientnet_b0	Efficientnet_Lite0
Epochs			50		
Batch size			16		
Learning rate			0.001		
Best accuracy in epoch	43	36	37	28	35

In an exploratory analysis of the results of the models' validation, EfficientNet_Lite0 stands out by showing the best overall performance in terms of accuracy, precision, recall, and F1-score. This result is particularly notable considering that this architecture was designed to be efficient in terms of computational resources according to [21]. The ResNet18, ResNet34, and EfficientNet_B0 models performed almost similarly, with all metrics above 96%. Although there are subtle differences. ResNet34 had slightly higher accuracy, while EfficientNet_B0 had slightly higher precision, recall, and F1-score. MobileNetv2 demonstrated the worst performance among the models, with all metrics significantly below the others. Despite being designed to be fast and lightweight, the model compromises performance in exchange for efficiency. The results of the evaluation of each model are presented in Table 6.

Table 6. Results of evaluation metrics.

Model	Accuracy (%)	Precision (%)	Recall (%)	F1 score (%)
MobileNetv2	73.84	75.0	75.0	75.0
ResNet18	96.17	96.0	96.0	96.0
ResNet34	97.18	96.0	96.0	96.0
EfficientNet_B0	97.16	97.0	97.0	97.0
EfficientNet_Lite0	97.58	98.0	98.0	98.0

6   Prediction Test

Following ISO/IEC 4213 [19] we performed a test with the trained models predicting the classification of previously unseen images.

6.1   Test Preparation

Test Dataset. The test set is composed of a total of 40 images with the data quality characteristics as described in Table 7.

Metrics. To evaluate the performance of the models, the following metrics were employed: Accuracy, to evaluate overall performance; precision and recall to support an understanding how the model handles false positives and false negatives, and F1 score, the harmonic mean of the accuracy and recall results; specificity was measured for correctly identifying negative examples and minimizing false positives.

Execution Environment. The test was run using Jupyter Notebooks in Google Colab.

6.2   Test Results

The test results are presented in Table 8.

The results show that the ResNet18 model obtained the highest accuracy, precision, recall, and F1 score during the test, while the MobileNetv2 model obtained the lowest overall performance. EfficientNet_B0 and EfficientNet_Lite0 perform similarly to ResNet34 in terms of accuracy, precision, and recall. Due to the higher accuracy, ResNet18 performs better in correctly classifying both positive and negative examples, minimizing classification errors, including false positives and false negatives. Due to higher precision, ResNet18 and EfficientNet_Lite0 perform better in correctly identifying positive examples and minimizing false positives.

The specificity of 97% for the ResNet and EfficientNet models families indicates that they are very effective at identifying true negatives while avoiding false positives. This high specificity value suggests that these models can correctly discern the negative classes, reducing the misclassification rate and improving the reliability of the predictions.

The lowest performance results were observed with regard to MobileNetv2, which indicates that the model has a lower performance in correctly classifying examples in

Table 7. Test data set characteristics

Characteristic	Specification
Total no. of images in test set	40 images
Distribution of test set	10 images for each class. Class 'nonmosquito' includes also images of objects/insects visually similar to mosquito larvae to evaluate specificity
Usage of images	These images have not been used during model training
Source	The images were collected by researchers from the initiative Computação na Escola/ INCoD/ INE/UFSC in cooperation with biologists from the LTH/UFSC and CRLAB/LACEN/SC
Ground truth	The images have been labeled by researchers from the initiative Computação na Escola/ INCoD/ INE/UFSC in cooperation with biologists from the LTH/UFSC and CRLAB/LACEN/SC
Preprocessing/data augmentation	No preprocessing/data augmentation has been applied to the test set
Testset accuracy	The accuracy of the labels of the images is 100%
Testset completeness	The images in the dataset cover all classes of interest with a variety of examples
Testset uniqueness	No duplicates of images are in the test set
Testset consistency	Uniformity of images information separating classes in folders. Images are named by class and number
Testset authenticity	The images were collected locally taking pictures of the species encountered in the region of interest (Florianópolis/SC)
	The images match the images to be collected by the target audience as defined in the use case
Testset robustness	Images have been collected varying the background, angles, lighting etc. in order to prevent channeling effects
Testset confidentiality	Not applicable

general. Despite these results, and also having the lowest specificity among the models, the specificity of 87%, still indicates that MobileNetv2 can handle the identification of true negatives and avoid false positives with some efficiency.

Comparing ResNet18, ResNet34 and EfficientNet_B0 only small differences between precision and recall (1%) were observed, suggesting that the models have a balanced performance between avoiding false positives (high precision) and identifying true positives (high recall). This is also confirmed by the F1 score of > 90% of ResNet18, ResNet34 and EfficienttNet_B0, EfficientNet_Lite0, which indicates a robust performance of the models for correct identification of examples (Table 9).

Table 8. Test results.

Model	Accuracy(%)	Precision (%)	Recall (%)	F1 score (%)	Specificity
MobileNetv2	60%	64%	60%	60%	87%
ResNet18	92.5%	94%	93%	93%	97%
ResNet34	90%	91%	90%	90%	97%
EfficientNet_B0	90%	91%	90%	90%	97%
EfficientNet_Lite0	90%	92%	90%	90%	97%

Table 9. Confusion matrices

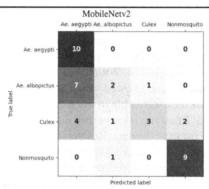

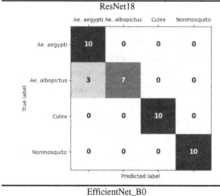

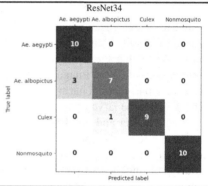

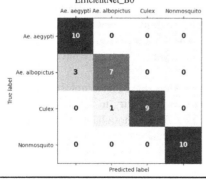

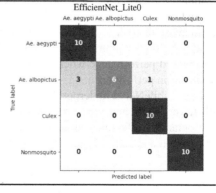

The confusion matrices indicate that all models have accurately classified the images of *Ae. Aegypti*. In our usage scenario, any misclassification would have severe consequences, as it could mistakenly lead the user to believe that the larva found in their home is not a primary vector for dengue. ResNet18, ResNet34, EfficientNet_B0 and EfficientNet_Lite0 demonstrate a similar misclassification of *Ae. Albopictus* as *Ae. Aegypti*, still being a classification error, yet with less risk to the user. ResNet34, EfficientNet_B0 also misclassified in one case an image of a *Culex* sp. as *Ae. Albopictus*. Again, MobileNetv2 demonstrated the worst performance misclassifying *Ae. Albopictus*, *Culex* sp. And even non-mosquito objects. However, even this model did not misclassify any of the images of *Ae. Aegypti*.

7 Discussion

In this research we studied the viability of classifying images of mosquito larvae (*Ae. Aegypti*, *Ae. Albopictus*, *Culex*, and nonmosquito) taken with a cellphone camera comparing the performance of different DL models (MobileNetv2, ResNet18, ResNet34, EfficientNet_B0 and EfficientNet_Lite0). With the exception of MobileNetv2, the models reached convergence quickly and showed satisfactory validation results, suggesting that in a few training epochs they were able to learn the main features relevant for classification. The EfficientNet_Lite0 model achieved convergence the fastest, while EfficientNet_B0 obtained the best fit to the data. ResNet18, ResNet34, EfficientNet_B0 and EfficientNet_Lite0 achieved similar accuracy during validation ranging from 96.17% to 97.58%. Similar results were also achieved during testing on previously unseen data, during which ResNet34, EfficientNet_B0 and EfficientNet_Lite0 demonstrated an accuracy of 90% and ResNet 92.5%. These results provide a first indication that these models were able to learn the generalization of the classification of this kind of image, achieving performance results above the required minimum of 90% in the context of our usage scenario. Analyzing the confusion matrices it can also be observed that none of the models misclassified the images of *Ae. Aegypti*, which would be the worst case in our usage scenario as it could harm the user due to leading him/her to not taking actions by not recognizing the larva found as a potential dengue vector.

During training and validation the EfficientNet families converged and fitted the dataset better due to the optimization techniques for automatic model size adjustment by EfficientNet_Lite0, and the resolution scaling adjustment technique of EfficientNet_B0. Yet, during testing the ResNet18 model performed best with regard to all metrics due to a combination of factors. And, although ResNet34 has a deeper architecture, which may allow it to learn more complex representations of the data, the increased complexity may also increase the risk of overfitting and make generalization to new data more difficult. In addition, ResNet34 may be more susceptible to performance degradation issues, such as decreased performance when adding additional layers, which may have affected its classification performance.

The MobileNetv2 underperformed the other models due to its simpler architecture, with fewer layers and fewer parameters. This limits its ability to learn complex data representations, affecting classification performance. And, although these characteristics make it light and fast, they also result in lower performance.

Threats to Validity. As any empirical study there are several threats to validity. Concerning selection bias, we aimed at preparing a diverse and representative sample of images that covers the range of variability in the population being studied. We considered various cameras, backgrounds, angles, and lighting for external validity, but our dataset may not cover all contexts, potentially limiting generalizability. Considering the size of our dataset containing 1.999 images we consider the risk of a sample size bias minimal especially when also compared to existing approaches reported in the literature. In order to prevent data preprocessing bias we only used standardized preprocessing techniques that are applicable to all models being compared. We also maintained consistent training parameters for comparability between models. With regard to evaluation metric bias, we used multiple standard evaluation metrics that cover different aspects of model performance and to report the results of all metrics used.

8 Conclusion

The results of our research demonstrate that it is possible to classify images of mosquito larvae even distinguishing between *Aedes* species with sufficient performance of testing accuracy > 90% and without misclassifying images of *Ae. Aegypti*. Our results are also much better than the ones reported by the only other study aiming at the differentiation between *Aedes* species reporting an accuracy of only 64.58%.

Both of the the models (ResNet18 and EfficientNet_Lite0), which demonstrated the best results during training/validation or testing, are also small architectures specifically indicated for the deployment in mobile applications, which thus will allow the implementation of such an application in order to allow the operationalization of the intended usage scenario of citizens using the automated classification for dengue prevention in their homes.

Acknowledgments. This work was supported by the CNPq (National Council for Scientific and Technological Development), a Brazilian government entity focused on scientific and technological development.

References

1. Amershi, S., et al.: Software engineering for machine learning: a case study. In: Proceedings of the 41st International Conference on Software Engineering, Montreal, Canada (2019)
2. Arista-Jalife, A., et al.: Deep learning employed in the recognition of the vector that spreads dengue, chikungunya and zika viruses. In: Frontiers in Artificial Intelligence and Applications, vol. 303, pp. 108–120. IOS Press, Amsterdam (2018)
3. Arista-Jalife, A., et al.: Aedes mosquito detection in its larval stage using deep neural networks. Knowl.-Based Syst. **189**, 104841 (2020)
4. Asmai, S.A., et al.: Mosquito larvae detection using deep learning. Int. J. Innov. Technol. Explor. Eng. **8**(12), 804–809 (2019)
5. Azman, M.I.A.B.Z., Sarlan, A.B.: Aedes larvae classification and detection (ALCD) system by using deep learning. In: 2020 International Conference on Computational Intelligence (ICCI), pp. 179–184 (2020)

6. Canziani, A., Paszke, A., Culurciello, E. An analysis of deep neural network models for practical applications. arXiv preprint arXiv:1605.07678 (2016)
7. Clements, A.N.: The Biology of Mosquitoes. Volume 2: Sensory Reception and Behavior. CABI Publishing (1999)
8. Consoli, R., Oliveira, R.L.: Principais mosquitos de importância sanitária no Brasil. Editora FIOCRUZ, Rio de Janeiro, Brasil (1994)
9. De Silva, W.D.M., Jayalal, S.: Dengue mosquito larvae identification using digital images. In: 2020 International Research Conference on Smart Computing and Systems Engineering (SCSE), pp. 31–36 (2020)
10. Fast.ai (2023). https://www.fast.ai/
11. Fuad, M., et al.: Training of convolutional neural networks using transfer learning for aedes aegypti larvae. Telkomnika (Telecommun. Comput. Electron. Control) **16**, 1894–1900 (2018)
12. FUNASA. Instruções para Pessoal de combate ao Vetor - Manual de Normas Técnicas (2001). https://bvsms.saude.gov.br/bvs/publicacoes/funasa/man_dengue.pdf
13. García, Z., et al.: Mosquito larvae image classification based on DenseNet and Guided Grad-CAM. In: Morales, A., Fierrez, J., Sánchez, J.S., Ribeiro, B. (eds.) IbPRIA 2019. LNCS, vol. 11868, pp. 239–246. Springer, Cham (2019). https://doi.org/10.1007/978-3-030-31321-0_21
14. GLOBE. GLOBE Mission Mosquito Mapper (2022)
15. Goh, Y.H., Lee, Y.B., Lum, K.Y.: American sign language recognition based on MobileNetV2. Adv. Sci. Technol **5**(6), 481–488 (2020)
16. Hedrick, T.E., Bickman, L., Rog, D.J.: Applied Research Design: A Practical Guide. Sage Publications, Thousand Oaks (1993)
17. Hossain, M.S., et al.: Aedes larva detection using ensemble learning to prevent dengue endemic. BioMedInformatics **2**(3), 405–423 (2022)
18. IEEE Std 2801–2022. IEEE Recommended Practice for the Quality Management of Datasets for Medical Artificial Intelligence. IEEE (2022)
19. ISO/IEC 4213. Information technology—Artificial intelligence—Assessment of machine learning classification performance. ISO/IEC (2022)
20. Kularatne, S.A.: Dengue fever. BMJ (Clin. Res. ed.) **351**, h4661 (2015)
21. Liu, R.: Higher accuracy on vision models with EfficientNet-Lite. https://blog.tensorflow.org/2020/03/higher-accuracy-on-vision-models-with-efficientnet-lite.html. Accessed 27 Apr 2023
22. Mathewson, K.W.: A Human-Centered Approach to Interactive Machine Learning. arXiv: 1905.06289v1 [cs.HC] (2019)
23. Mitchell, T.M.: Machine Learning. McGraw-Hill, New York (1997)
24. Ministério da Saúde. Boletim Epidemiológico (2023)
25. Munoz, J., et al.: Image recognition of disease-carrying insects: a system for combating infectious diseases using image classification techniques and citizen science. In: 51st Hawaii International Conference on System Sciences (HICSS), pp. 3594–3603 (2018)
26. Petersen, K., et al.: Systematic mapping studies in software engineering. In: Proceedings of the 12th International Conference on Evaluation and Assessment in Software Engineering, Bari, Italy, pp. 68–77 (2008)
27. Rajasekhar, V., et al.: Identifying anopheles/non-anopheles larvae with AI implications. In: NASA SEES Mosquito Mapper Virtual Internship Science Fair, Student Research Reports (2021). https://www.globe.gov/do-globe/research-resources/student-research-reports/-/projectdetail/10157/identifying-anopheles-non-anopheles-larvae-with-ai-implications
28. Ramos, G., Meek, C., Simard, P., Suh, J., Ghorashi, S.: Interactive machine teaching: a human-centered approach to building machine-learned models. Hum.-Comput. Interact. **35**(5-6), 413–451 (2020). https://www.microsoft.com/en-us/research/uploads/prod/2020/05/Interactive_Machine_Teaching__Free_access_.pdf

29. Sanchez-Ortiz, A., et al.: Mosquito larva classification method based on convolutional neural networks. In: 2017 International Conference on Electronics, Communications and Computers, pp. 1–6 (2017)

30. Surya, A., Peral, D., VanLoon, A., Rajesh, A.: A mosquito is worth 16×16 larvae: evaluation of deep learning architectures for mosquito larvae classification. arXiv:2209.07718 (2022)

31. Tan, M., Le, Q.V.: EfficientNet: rethinking model scaling for convolutional neural networks. In: Proceedings of the 36th International Conference on Machine Learning, pp. 6105–6114. PMLR (2019)

32. UTEP. University of Texas at El Paso, Laboratory for Environmental Biology - Centennial Museum - Team Mosquito: Larvae Identification Guide (2004). https://www.utep.edu/leb/mosquito/larvaeID.pdf

33. World Health Organization: Geographical Expansion of Cases of Dengue and Chikungunya Beyond the Historical Areas of Transmission in the Region of the Americas. In: Disease Outbreak News (2023). https://www.who.int/emergencies/disease-outbreak-news

A Simple and Low-Cost Method for Leaf Surface Dimension Estimation Based on Digital Images

Karla Gabriele Florentino da Silva[1,2] ⓘ, Jonas Magalhães Moreira[2] ⓘ,
Gabriel Barreto Calixto[1] ⓘ, Luiz Maurílio da Silva Maciel[1,2(✉)] ⓘ,
Márcio Assis Miranda[2] ⓘ, and Leandro Elias Morais[3] ⓘ

[1] Department of Computer Science, Federal University of Juiz de Fora,
Juiz de Fora, Minas Gerais, Brazil
{karla.florentino,gabriel.calixto}@estudante.ufjf.br,
luiz.maciel@ufjf.br
[2] Department of Computing, Federal Institute of Education,
Science and Technology of Minas Gerais, Ouro Branco, Minas Gerais, Brazil
marcio.assis@ifmg.edu.br
[3] Department of Natural Sciences, Federal Institute of Education, Science and
Technology of Minas Gerais, Ouro Branco, Minas Gerais, Brazil
leandro.morais@ifmg.edu.br

Abstract. The leaf is the organ of the plant body that performs photosynthesis and its area is one of the morphological parameters that most respond to droughts, climate changes, and attack of pathogens, associated with the accumulation of biomass and agricultural productivity. In addition, leaf area and other surface data (for example, width and length) are widely used in studies of plant anatomy and physiology. The methods of measuring these leaf surface parameters are often complicated and costly. In this context, this work aims to develop a simple and low-cost method capable of accurately measuring the leaf surface size of plant species with significant agricultural interest. Our method extract the information through images of leaves accompanied by a scale pattern whose real area is known, captured by a simple camera. To evaluate our method, we performed experiments with images of 118 leaves of 6 species. We compared the results to the ImageJ software, which is widely used to estimate leaf dimensions from images. The results showed our method present performance similar to ImageJ. However, unlike ImageJ, our method does not require user interaction during the dimensions estimation.

Keywords: Computer vision · Image processing · Leaf phenotyping

1 Introduction

The leaves constitute the organ of the vegetable body capable of converting solar energy into chemical energy through photosynthesis [7]. Vegetable growth

M. C. Naldi and R. A. C. Bianchi (Eds.): BRACIS 2023, LNAI 14197, pp. 146–161, 2023.
https://doi.org/10.1007/978-3-031-45392-2_10

and development are determined by the leaf morphology, especially its area, the largest component of biomass accumulation and productivity [17,34]. The leaf size and morphology undergo an intraspecific and interspecific variation, and many physiological parameters (e.g., nitrogen content, photosynthesis and leaf respiration) are normalized by leaf area in order to study the relationship between them [23,41].

The leaf morphology (e.g. area, perimeter, width and length) is strongly used to investigate vegetable responses to different environmental conditions, such as the soil fertility [4,13], soil water/dry availability [29,38], light availability [25], effect of pesticides application [20], climate changes [15,27,31], phytopathogens [18], herbivory [9], screening of genotypes for breeding [39], pollution [12]. In fact, measuring the area, length, width, perimeter and their derivations (e.g. specific leaf area, perimeter/leaf area ratio, length/ width ratio) is a powerful tool used by professionals of several areas of knowledge such as vegetable improvers, botanists, phytopathologists, ecologists and agronomists [14,36,37,40].

There is equipment that measures the leaf dimensions with great accuracy, but it is quite expensive, quotes around US$3,500.00 (quotation performed in January 2021). In addition, it is possible to estimate leaf area using allometric models (equations) using leaf morphology variables (e.g. length and width), adjusted for each species and/or crop. However, a big leaves sample is required to develop these models. For example, Antunes *et al.* [1] analyzed 1563 coffee leaves. It is important to note that these models estimate the leaf area of a single leaf at a time. In tests with many sample units, estimating this parameter will be a time consuming and laborious process. Therefore, it is necessary to develop practical and low-cost methodologies, in order to guarantee the execution of works that require a verification of the leaf area and other dimensions with celerity and accuracy.

In this paper, we present a simple, intuitive and low-cost method to calculate leaf dimensions. The method can be implemented in a desktop software or a mobile application, allowing researchers to use the mobile version to perform field measurements and the desktop version if they wish to perform measurements in the laboratory. For the leaf dimensions determination, we need only capture images of the leaves accompanied of a scale pattern printed on paper. The images can be taken by a simple camera. There is no need to purchase extra equipment. Personal devices can be used, at no additional cost.

Our method presents the following characteristics, which are the main contributions of this work:

- Specific to perform leaf sizing (area, width, length and perimeter);
- It does not require pre-processing;
- Automatically detects the scale pattern.

2   Related Work

Several methods for leaf area estimation based on digital images were proposed in last years. Some works use the color information of the images to extract the interest objects and estimate leaf dimensions [6,16]. Extracting color information may require user intervention at filtering step and even the need of using other software. In our method, we use grayscale images, because the contrast between the objects and the background is enough to separate them, even without the color information.

As in our method, some works used an object of known dimensions as a scale pattern. Jadon [11] chose a coin, but this shows strong reflection when illuminated. Thus, that proposal demonstrates a greater sensitivity to lighting conditions. Easlon et al. [6] and Tech et al. [35] opted for a square. Easlon et al. [6] detect the square automatically using color information. However, as previously mentioned, the color needs to be calibrated by the user. The method of Tech et al. [35] requires the square to be positioned at a specific position in the image. In our method, the square is automatically detected through its shape. Moreover, it can be placed anywhere in the image.

ImageJ[1] software [28] has been used to calculate the leaf area [8,19]. However, it is a general-purpose image processing software, which requires the user to perform several steps until obtaining the measurement, such as selecting a measurement line known in the image. We propose to create an intuitive computational tool so that the user can open their images in the software and through a few simple steps perform the leaf measurements calculation. In addition, we intend to create a specific tool for leaf morphology analysis, allowing the insertion of new functionalities related to plant physiology. As it is an application widely used by researchers, ImageJ will be used to validate our method results.

There are also some mobile applications for leaf area calculation. The Petiole application [24] uses a very controlled scenario. The calculation starts with a calibration step and the device position should be kept for the area estimation. Moreover, it requires user interaction to identify the leaves position on the image. The application Easy Leaf Area Free [6] needs a color calibration step, which makes the application less intuitive for the user. We propose a simple method that does not require user calibration.

Besides methods based only on computer vision techniques, methods for leaf area measurement with Artificial Neural Networks (ANNs) are appearing recently. Siswantoro and Artadana [30] used a fixed Webcam to capture digital images of the individual leaves. They applied a thresholding step for image segmentation and extract the leaf length, width, area, and perimeter in pixels. Finally, ANN proposed uses the extracted features to estimate the real leaf area. Sa-bouri et al. [26] compared three methods to predict leaf area in cereals: regression models, ANN, and ANN combining fuzzy logic. The neural network developed by the authors is similar to the previously cited work and uses leaf width and length as inputs. Two fluorescent light sources were fixed in specific positions to keep the illumination conditions. The proposal based on neuro-fuzzy inference obtained better performance concerning the work experiments. The advantages of our proposal compared to these networks are automatic detection of the interest objects and no need for training. In addition, while these works need a fixed capture scenario, our method can deal with images captured from different positions and less controlled lighting conditions.

---

[1] https://imagej.nih.gov/ij/.

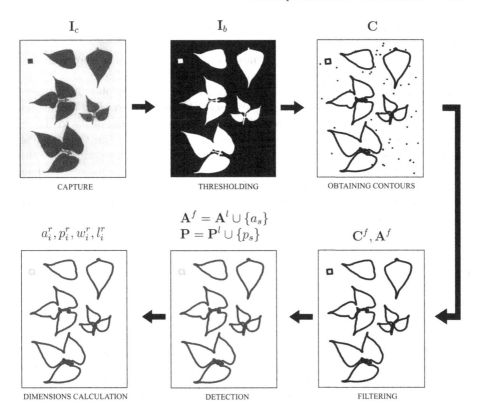

Fig. 1. Steps of the proposed method. Above each image, we have the notation referring to the information obtained in the respective step.

3 Proposed Method

Most existing methods for leaf dimensions estimation depends on user intervention. This user interaction can introduce subjectivity and lead to inaccuracies in results. Other proposals require a specific positioning of scene objects and calibration steps. These constraints make the process difficult. In this work we propose a method to estimate the leaves dimensions in few steps, which do not depend on the user interference. Our proposed method for leaf dimensions calculation is composed of six steps: capture, thresholding, obtaining contours, filtering, detection and dimensions calculation. The Fig. 1 shows an overview of the proposed method. We present images and information extracted in each step.

3.1 Capture

In the first step, an image of the leaf (or leaves) is captured so that its dimensions can be estimated. The leaf must be accompanied by a scale pattern whose real size is known. We chose a square pattern printed on paper. This pattern can be placed anywhere on the image and is automatically detected to estimate the leaf dimensions, as described in Sect. 3.5.

The leaf is not a plane surface. It must be planned to avoid underestimating the measures. For this purpose, a glass or acrylic plate may be used to compress the leaf and keep it stretched. We used a commercial clear glass plate in the tests to keep the projection area of the leaf parallel to the surface, avoiding curvatures and folds. Furthermore, the background should be a light color in order to ensure contrast with the interesting objects. Reflection is critical to the thresholding process described in Sect. 3.2 and should be avoided. It is recommended the image be captured at a perpendicular angle to avoid distortion problems. Various capture distances and scale pattern sizes were tested and the results will be discussed in Sect. 4. It is important to mention that our method allows more than one leaf to be captured in the same image. The dimensions of each leaf are individually estimated.

3.2   Thresholding

In the capture step, a color image I_c is obtained, containing the leaf (or leaves) and the scale pattern. The image is converted to grayscale, resulting in a new image I_g. This conversion and most of the image processing functions were implemented using OpenCV library [2] in C++ programming language.

We applied an Otsu's thresholding [21] on the I_g image. This algorithm is based on the histogram image. It finds a threshold t which lies in between two peaks such that variances to both classes are minimal. Based on this threshold t we achieve a binary image I_b where:

$$I_b(x, y) = \begin{cases} 0, & \text{if } I_g(x, y) > t \\ 1, & \text{if } I_g(x, y) \le t \end{cases}.$$

3.3   Obtaining Contours

In the thresholding step, a binary image I_b is obtained, where the leaf and the scale pattern are white and the background is black. It is necessary to extract the contours of the interest objects, that is, the leaves and the scale pattern.

In this step, we apply Suzuki *et al.* [33] algorithm in the I_b image. This method extracts a set **C**:

$$\mathbf{C} = \{\mathbf{C}_1, \mathbf{C}_2, ..., \mathbf{C}_n\},$$

where each $\mathbf{C}_i$ is a contour, represented by the set of points that compose it:

$$\mathbf{C}_i = \{(x_1, y_1), (x_2, y_2), ..., (x_m, y_m)\},$$

where each (x_j, y_j) is the position of a point (pixel) in the image.

3.4   Filtering

In the previous step, a set $\mathbf{C}$ of n contours is obtained. This set can contain noise, i.e., contours that do not represent interest objects. These contours should be filtered. In general, the bad contours are much smaller (image noise, for example) or much larger (a table contour, for example) than the interest objects. To overcome this problem, a filtering by contour area is performed. The contour areas in pixels are estimated by Green formula [32]. Thus, from $\mathbf{C}$, we obtain the set of areas:

$$\mathbf{A} = \{a_1, a_2, ..., a_n\},$$

where each $a_i \in \mathbf{A}$ is the area of the respective contour $\mathbf{C}_i \in \mathbf{C}$.

The areas $a_k \in \mathbf{A}$ that are not in the range $[a_{min}, a_{max}]$ of expected areas, are discarded, as well as their respective contour $\mathbf{C}_k$. The sets $C^f \subseteq C$ and $A^f \subseteq A$ are the filtered sets of contours and areas, respectively. Experimentally, we defined the values $a_{min} = 10^3$ and $a_{max} = 10^{10}$ which proved to be efficient. However, some noise, such as shadows, may not have been completely removed. It is possible to manually remove these artifacts with a simple tool available in the software. We observed the noise presence in 30% of the analyzed images.

3.5   Detection

In the filtering step, the sets $\mathbf{C}^f$ and $\mathbf{A}^f$, are obtained, representing the contours and their respective areas after filtering. In order to automate the process, we need to separate the scale pattern from the leaves. As the scale pattern is a square and it presents a different geometry from the leaves, we can perform a shape-based detection of the square in the contours set.

First, we apply a polygonal approximation on each contour $\mathbf{C}_i^f \in \mathbf{C}^f$ using the Douglas-Peucker algorithm [5]. This function reduces each contour $\mathbf{C}_i^f$ to a simplified contour $\mathbf{C}_i^p \subseteq \mathbf{C}_i^f$. Thus, we have the set:

$$\mathbf{C}^p = \{\mathbf{C}_1^p, \mathbf{C}_2^p, ..., \mathbf{C}_n^p\},$$

where $\mathbf{C}_i^p$ is the polygonal approximation of the contour $\mathbf{C}_i^f \in \mathbf{C}^f$.

To determine the square contour $\mathbf{C}_s^p \in \mathbf{C}^p$, we analyze some conditions. The number of points in $\mathbf{C}_s^p$ must be equal to 4, i.e., $|\mathbf{C}_s^p| = 4$. In addition, the cosine of the angles θ formed by the contour sides are calculated. If $|cos(\theta)| < 0.3$ for all 4 angles, this contour will be $\mathbf{C}_s^p$. This method was efficient in extracting the square for 98% of the tested images. Thus, we obtain the contour $\mathbf{C}_s^p \in \mathbf{C}^p$ and its corresponding $\mathbf{C}_s \in \mathbf{C}^f$ in the set of non approximated contours. In terms of area, we have the respective area of the square in pixels $a_s \in \mathbf{A}^f$ and the set of leaf areas $\mathbf{A}^l = \mathbf{A}^f - \{a_s\}$.

3.6   Dimensions Calculation

At this point, all the interest objects are properly separated and contoured. Therefore, our method is ready to calculate the dimensions of each leaf in the image. To facilitate the calculation of real leaf length and width, the leaves are aligned according to the Principal Component Analysis (PCA) [3]. PCA is applied to each original leaf contour $\mathbf{C}_i \in \mathbf{C}^f - \{\mathbf{C}_s\}$ to find the main leaf direction. PCA returns the centroid $K_i = (x_k, y_k)$ of each leaf contour and the angle of the main direction θ_i. We associated this direction with the length of the leaf. Each contour is aligned so that the main direction coincides with the horizontal axis of the image. Thus, each contour point $(x_j, y_j) \in \mathbf{C}_i$ is rotated by the equation:

$$x_j = (x_j - x_k) \, cos(\theta_i) - (y_j - y_k) \, sin(\theta_i) + x_k,$$
$$y_j = (x_j - x_k) \, sin(\theta_i) + (y_j - y_k) \, cos(\theta_i) + x_k.$$

The real area a_s^r of the scale pattern in cm^2 is known. We have the scale pattern area a_s in pixels and each leaf area $a_i \in A^l$ in pixels. Thus, the real area of each leaf a_i^r in cm^2 can be obtained by the following equation:

$$a_i^r = \frac{a_i a_s^r}{a_s}.$$

The real perimeter p_s^r of the scale pattern in cm is known. We calculated the perimeter in pixels of each leaf and the scale pattern. Thus, we achieved the set:

$$\mathbf{P} = \{p_1, p_2, ..., p_n\} \cup p_s,$$

where, each $p_i \in \mathbf{P}$ is the perimeter of the respective leaf contour $\mathbf{C}_i \in \mathbf{C}^f$. And p_s is the scale pattern.

To obtain the real perimeter p_i^r of the leaf, we use the following equation:

$$p_i^r = \frac{p_i p_s^r}{p_s}. \tag{1}$$

We know the real side d_s^r of the scale pattern in cm. The pixel measurement on each side d_s of the pattern can be obtained by dividing the perimeter p_s by four. The length l_i and width w_i, in pixels, of each leaf can be obtained by the rectangular bounding box of the rotated contour.

In this way, the base and height of this rectangle will correspond to leaf length l_i and width w_i in pixels, respectively. Thus, the real dimensions of each leaf can be obtained by the following equations:

$$w_i^r = \frac{w_i d_s^r}{d_s}, \qquad l_i^r = \frac{l_i d_s^r}{d_s}.$$

As a restriction of the proposed method, the largest between the two dimensions found is considered to be the length of the leaf. Leaves that do not have this characteristic will have their dimensions reversed, that is, the researcher will have to analyze and correct the results manually before performing any analysis.

4   Results and Discussion

In this section, we describe the experiments performed to evaluate our proposed method. We performed controlled tests with known objects and tests with leaves. For capture the images, we used five devices: three smartphones (Samsung Galaxy J2 Core, Motorola Moto C and Motorola One Vision) and two scanners (HP Deskjet 2050 and MultiXpress M5360RX).

Table 1. Pearson's correlation coefficient (r), RER average (μ), RER standard deviation (σ), R^2 and regression equation by dimension for the ellipses pattern.

Metric	Area	Width	Length	Perimeter
r	98.57	98.24	98.17	98.26
RER (μ)	2.37	1.75	1.69	3.39
RER (σ)	2.03	1.44	1.39	1.28
R^2	0.97	0.97	0.96	0.97
Equation	1.02× − 0.03	1.00× + 0.04	1.04× − 0.27	1.00× + 0.55

4.1   Metrics

In this subsection, we present the metrics used to evaluate quantitatively our method results. The relative error rate (RER) represents the percentage of the difference between the value estimated v_{est} and achieved with the standard method v_{std}. It can be calculated as:

$$RER(\%) = \frac{|v_{est} - v_{std}|}{v_{std}} \cdot 100.$$

The RER measure the accuracy of the proposed method in relation to the standard method.

Given a set of values estimated by our method V_{est} and the respective values achieved with the standard method V_{std}, we analyzed the correlation between these values. We estimated the linear regression equation $y = ax + b$, the determination coefficient R^2 and the Pearson's correlation coefficient r, using the Google Sheets software.

4.2   Controlled Tests

In order to evaluate our method accuracy, we performed preliminary experiments with a controlled pattern. For these tests, we used a pattern containing four ellipses and a scale pattern of 4 cm side. We used a glass plate to plan the pattern as explained in the Sect. 3.1 and captured perpendicular images from distances between 20 and 50 cm. In a general analysis, it is possible to see that the method was able to measure the dimensions accurately, achieving a good performance using simple smartphone cameras.

Table 1 presents the results of the ellipse tests. Analyzing the regression equation and the correlation coefficient, it is possible to verify a strong correlation between the values estimated by the proposed method and the expected values. In addition, we achieved $R^2 > 0.95$ for all dimensions. We achieve RER mean below 3.5% for all the dimensions, which shows the accuracy of our method. Furthermore, we note the regression equations present slope close to 1 and small intercept values.

4.3   Leaf Tests

In this subsection, we introduce the dataset collected for measuring leaf dimensions, and present the results of the tests performed to evaluate the proposed method.

Table 2. R^2 values and regression equations by species and dimensions in relation to the standard method adopted.

Species	Area Equation	R^2	Width Equation	R^2	Length Equation	R^2	Perimeter Equation	R^2
Bean	1.00× − 0.01	1.00	0.95× + 0.14	0.99	0.94× + 0.49	0.98	1.00× − 0.07	1.00
Coffee	1.07× − 0.51	1.00	0.98× + 0.02	0.99	1.01× − 0.17	0.98	0.98× + 1.43	0.97
Cotton	0.98× + 0.93	0.99	0.92× − 0.19	0.74	0.91× + 0.37	0.72	1.02× − 1.08	0.98
Eucalyptus	1.05× + 0.09	0.99	0.94× + 0.10	0.95	0.96× + 0.48	0.98	0.98× + 0.76	0.98
Orange	1.06× − 0.11	1.00	0.99× − 0.04	0.99	0.99× − 0.05	0.99	1.03× − 0.21	0.99
Soybean	1.06× − 0.16	1.00	0.99× − 0.06	1.00	1.00× − 0.08	1.00	1.01× + 0.31	1.00

Data Acquisition. For this work, annual and perennial species were chosen, whose cultures have significant participation in the Brazilian and international Gross Domestic Product (GDP) [10]. All leaves were collected in the morning (before 9:00 am, to guarantee high turgidity), with no apparent/visual symptoms of nutritional deficiency, pathogen attacks, and diseases. The leaves were collected and analyzed on the same day, at the Natural Sciences laboratory of the Instituto Federal de Educação Ciência e Tecnologia de Minas Gerais - Ouro Branco Campus.

The sample is composed of 118 leaves of 6 species: 18 of the bean, 18 of coffee, 11 of cotton, 21 of eucalyptus, 20 of orange, and 30 of soybean. Images were captured between 12 and 50 cm of one or multiple leaves, accompanied by a scale pattern of sides measuring 1, 2, 4, or 6 cm. We used Motorola Moto C and Motorola One Vision smartphones. In addition, the leaves also were scanned accompanied by the 1 cm scale pattern using the printer scanner MultiXpress. Overall, our dataset is composed of 529 leaves images. The images are available at the following link https://bit.ly/42DbyNV.

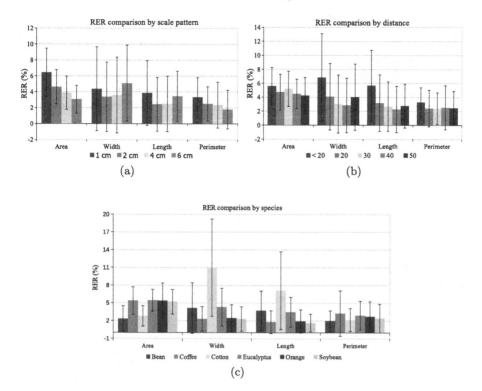

Fig. 2. Average RER in relation to standard method. (a) By scale pattern. (b) By capture distance. (c) By species. The error bar represents the standard deviation

Method Evaluation. To verify the results accuracy, we compared with two methods, the ImageJ software [28] and a simple, but very widespread manual method. The ImageJ software was used as the standard method. As previously stated in Sect. 2, this application is widely used by researchers in the literature.

The steps used to estimate the leaf dimensions with ImageJ software were: 1) select image file; 2) convert image to grayscale; 3) perform a thresholding; 4) draw a line indicating the side of the scale pattern; 5) set the scale known distance; 6) select the leaf and measure area and perimeter; 7) draw a line indicating the width of the leaf and measure the length; 8) draw a line indicating the length and measure. It is important to note that those steps depends on the user interaction and, consequently, are subject to errors.

The manual weight-based method for area estimation [22] starts by tracing the outline of the leaf on a paper, whose grammage g is known. The contour is cut and its weight w is measured by a precision balance. The area a is estimated by the formula $a = \frac{w}{g}$. The same contour used in the weight-based method is measured with a simple rule to estimate the expected width and length. In addition, a line or string is passed around the contour and then measured with a rule to estimate the expected perimeter.

Table 2 presents the coefficient of determination (R^2) and the regression equations by species and dimensions in relation to the standard method (ImageJ software). In general, the performance for area and perimeter was satisfactory. For these dimensions, the R^2 values were above 0.96 and the equations presented angular coefficients close to one for all species. Regarding width and length calculation, we note some problems in the cotton results with R^2 values below 0.75. The morphology of the leaves of this species makes it difficult to estimate these dimensions. However, for the other species, our method achieved R^2 values above 0.94 for these dimensions. In general, the results were promising and indicate our method perform similarly to the standard method.

Figure 2c shows the average RER for each species and dimension in relation to the standard method. The errors for perimeter were the smallest, between 1.98 and 3.25%. For area, we achieved errors between 2.32 and 5.40%. Cotton had the highest errors for length and width. As previously mentioned, the leaves morphology makes it difficult these dimensions estimation for cotton. However, for the other species, the average RER for length and width were between 1.65 and 4.31%. Overall, the errors achieved showed the method applicability for measuring leaves dimensions, specially those with simpler morphologies.

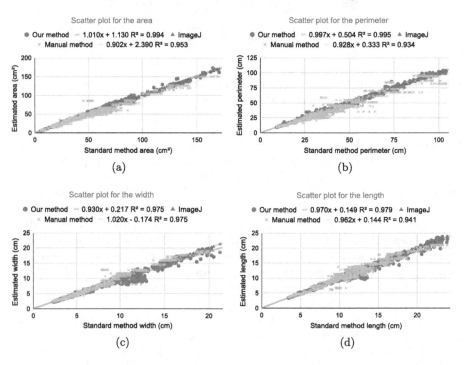

Fig. 3. Scatter plot for three methods and all images. (a) Area. (b) Perimeter. (c) Width. (d) Length.

Figure 2a presents the average RER for each scale pattern size used. We observe that our method has a worse performance for the 1 cm pattern. The hypothesis for this variation is the manual line selection in the ImageJ procedure. Depending on the direction of this line selection, the real dimension can present a significant difference, especially for images with distortion and the smallest pattern. On the other hand, our method estimates the square measures automatically, without user interaction. For the other scale sizes, the results for the methods was quite similar. In general, the results indicate that the size of the pattern does not significantly influence the method's performance.

The average RER for each capture distance is presented in Fig. 2b. To facilitate the comparison, we grouped the tests with capture distances shorter than 20 cm. The tests for this group achieved errors slightly higher than the others. As previously explained, the ImageJ method needs a manual line selection in the image. Since images captured more closely can present distortion, the user's choice when selecting the line can affect the results. For the other distances, the methods performed similarly. This indicates that the capture distance does not have a strong influence, when the capture conditions are respected.

Table 3. Pearson's correlation coefficient (r), RER average (μ) and standard deviation (σ) for Our method $\times$ Manual methods.

Metric	Our method				Manual methods			
	Area	Width	Length	Perimeter	Area	Width	Length	Perimeter
r	99.71	98.73	98.96	99.77	97.63	98.76	97.02	96.64
RER (μ)	4.53	3.95	2.94	2.53	9.36	4.82	7.12	10.32
RER (σ)	2.55	4.79	3.58	2.54	7.96	4.75	7.78	8.93

Figure 3a, 3c, 3d and 3b present the scatter plots considering all the images used for tests. These plots also include the results of tests performed using the manual methods previously described. Measurements by manual methods are performed per leaf. Thus, the yellow dots on the same line of the graph represents the same leaf. Our method achieved R^2 values above 0.975 for all dimensions. These values were greater than those of manual method. Furthermore, the straight lines of our method presented a slope closer to one, except in the length graph. For this dimension, as previously discussed, our method presents limitation to deal with certain leaves morphology.

Table 3 presents the Pearson's correlation coefficient r, RER average (μ) and RER standard deviation (σ) for our method and manual method in relation to the standard method. Pearson's coefficients indicate a strong correlation for both methods. In relation to the RER metric, we observe the manual method present higher values. Due to the several manual steps, manual method is more difficult to control and more susceptible to errors.

Discussion. We verify that the proposed method was able to estimate leaf dimensions satisfactorily in comparison to ImageJ, which is widely applied in the literature. In terms of time to estimate the measures, our method is more efficient than the ImageJ software, which requires less user-side intervention. In a comparison made with nine images, the proposed method estimated all dimensions with an average time of 15 s, while ImageJ took 78 s.

However, our method presents some limitations. It is sensitive to the capture angle and distance, illumination, and the noises that may appear in the image. Images taken from an inclined angle suffer from perspective distortion. Depending on the lighting location may cause shadows or reflections in the image. All these factors compromise the calculation of leaf dimensions. These are limitations typical of applications involving image processing, and we intend to propose improvements to minimize such problems.

Leaf morphology is also a limitation of our method, especially on the width and length calculation. These dimensions still need some adjustments for species with more complex leaf morphologies, such as trifoliate or lobed leaves. These morphologies make difficult the PCA alignment, due to the irregular shape. In the ImageJ software, the width and length direction are manually defined. On the other hand, the area and the perimeter have already proved to be quite efficient. Except for the trifoliate bean leaf that overestimated the area in the petiole region, where the leaflets are inserted. This leaf presented problems for both methods, especially for higher capture distances. Finally, the low contrast between the leaf and the background is another limitation of our proposal. When the leaf presents a light hue, the thresholding process does not work correctly. An alternative for these light leaves is to use a light scale pattern and a dark background to increase the contrast.

5   Conclusions

We proposed a simple and low-cost method for leaf dimensions calculation based on image processing. Our method estimates area, perimeter, length, and width for one or more leaves directly from the images. The software automatically detects each leaf in the image and allows estimate statistics such as average and standard deviation. The only information provided by the user is the size of the scale pattern. No additional user interaction is required. The method estimates automatically the location of leaves and scale pattern.

In general, the measurement of leaf morphological parameters (e.g. area, perimeter, length, and width) by this method is an excellent tool to explain plant responses in several areas: evolution, ecophysiology, improvement, management. The results demonstrated the accuracy of the method proposed for this purpose, in addition to its practicality and speed. Moreover, these measurements can be employed in derivations and other parameters, such as the leaf area/ perimeter ratio, the leaf area index, the specific leaf area, and the length/width ratio. To conclude, as future works, we intend to improve the method, especially length and width estimation. Furthermore, we can perform an improvement in the segmentation step, avoiding the detection of noise in the detected contours. Finally, we are developing free mobile and desktop applications from this method that will be made available to any professional and student who wants to perform leaf morphological phenotyping. The source code of these applications can be found in the links https://github.com/KarlaFlorentino/LIMA-desktop (Desktop) and https://github.com/LuizMaurilio/LIMA (Mobile).

Acknowledgments. This work received financial support from the FAPEMIG process number APQ-00603-21. We thank the agencies CNPq and CAPES for their financial support in this research. And all the people who collaborated directly or indirectly on this work.

References

1. Antunes, W.C., Pompelli, M.F., Carretero, D.M., DaMatta, F.: Allometric models for non-destructive leaf area estimation in coffee (coffea arabica and coffea canephora). Annal. Appl. Biol. **153**(1), 33–40 (2008)
2. Bradski, G.: The opencv library. Dr Dobb's J. Softw. Tools **25**, 120–125 (2000)
3. Cohen-Or, D., et al.: A Sampler of Useful Computational Tools for Applied Geometry, Computer Graphics, and Image Processing. CRC Press (2015)
4. Dornbusch, T., et al.: Plasticity of winter wheat modulated by sowing date, plant population density and nitrogen fertilisation: dimensions and size of leaf blades, sheaths and internodes in relation to their position on a stem. Field. Crop. Res. **121**(1), 116–124 (2011)
5. Douglas, D.H., Peucker, T.K.: Algorithms for the reduction of the number of points required to represent a digitized line or its caricature. Cartographica Int. J. Geograph. Inf. Geovisual. **10**(2), 112–122 (1973)
6. Easlon, H.M., Bloom, A.J.: Easy leaf area: automated digital image analysis for rapid and accurate measurement of leaf area. Appl. Plant Sci. **2**(7), 1400033 (2014)
7. Evert, R.F., Eichhorn, S.E.: Raven: biology of plants. No. 581 RAV (2013)
8. Gao, J., et al.: Measuring plant leaf area by scanner and imagej software. China Vegetables **2**, 73–77 (2011)
9. Gely, C., Laurance, S.G., Stork, N.E.: How do herbivorous insects respond to drought stress in trees? Biol. Rev. **95**(2), 434–448 (2020)
10. IBGE. Agricultura, pecuária e outros — ibge (2023). https://www.ibge.gov.br/estatisticas/economicas/agricultura-e-pecuaria.html. Accessed 17 May 2023
11. Jadon, M.: A novel method for leaf area estimation based on hough transform. JMPT **9**(2), 33–44 (2018)
12. Janhäll, S.: Review on urban vegetation and particle air pollution-deposition and dispersion. Atmos. Environ. **105**, 130–137 (2015)
13. Laughlin, D.C.: Nitrification is linked to dominant leaf traits rather than functional diversity. J. Ecol. **99**(5), 1091–1099 (2011)
14. Li, Y., et al.: Spatiotemporal variation in leaf size and shape in response to climate. J. Plant Ecol. **13**(1), 87–96 (2020)
15. Liancourt, P., et al.: Leaf-trait plasticity and species vulnerability to climate change in a mongolian steppe. Glob. Change Biol. **21**(9), 3489–3498 (2015)
16. Liang, W.Z., Kirk, K.R., Greene, J.K.: Estimation of soybean leaf area, edge, and defoliation using color image analysis. Comput. Electron. Agricult. **150**, 41–51 (2018)
17. Long, S.P., Zhu, X.G., Naidu, S.L., Ort, D.R.: Can improvement in photosynthesis increase crop yields? Plant Cell Environ. **29**(3), 315–330 (2006)

18. Lu, J., et al.: Detection of multi-tomato leaf diseases (late blight, target and bacterial spots) in different stages by using a spectral-based sensor. Sci. Rep. **8**(1), 2793 (2018)
19. Maloof, J.N., Nozue, K., Mumbach, M.R., Palmer, C.M.: Leafj: an imagej plugin for semi-automated leaf shape measurement. JoVE (J. Visual. Exp.) (71), e50028 (2013)
20. Marek, J., et al.: Photoynthetic and productive increase in tomato plants treated with strobilurins and carboxamides for the control of alternaria solani. Sci. Hortic. **242**, 76–89 (2018)
21. Otsu, N.: A threshold selection method from gray-level histograms. IEEE Trans. Syst. Man Cybern. **9**(1), 62–66 (1979)
22. Pandey, S., Singh, H.: A simple, cost-effective method for leaf area estimation. J. Bot. **2011**(2011), 1–6 (2011)
23. Peterson, A.G.: Reconciling the apparent difference between mass-and area-based expressions of the photosynthesis-nitrogen relationship. Oecologia **118**(2), 144–150 (1999)
24. Polunina, O.V., Maiboroda, V.P., Seleznov, A.Y.: Evaluation methods of estimation of young apple trees leaf area. Bullet. Uman Natl. Univ. Horticult. **2**, 80–82 (2018)
25. Poorter, H., et al.: A meta-analysis of plant responses to light intensity for 70 traits ranging from molecules to whole plant performance. New Phytol. **223**(3), 1073–1105 (2019)
26. Sabouri, H., et al.: Image processing and prediction of leaf area in cereals: a comparison of artificial neural networks, an adaptive neuro-fuzzy inference system, and regression methods. Crop Sci. **61**(2), 1013–1029 (2021)
27. Sanz-Sáez, Á., et al.: Leaf and canopy scale drivers of genotypic variation in soybean response to elevated carbon dioxide concentration. Glob. Change Biol. **23**(9), 3908–3920 (2017)
28. Schneider, C.A., Rasband, W.S., Eliceiri, K.W.: Nih image to imagej: 25 years of image analysis. Nat. Methods **9**(7), 671–675 (2012)
29. Shahnazari, A., et al.: Effects of partial root-zone drying on yield, tuber size and water use efficiency in potato under field conditions. Field Crop. Res. **100**(1), 117–124 (2007)
30. Siswantoro, J., Artadana, I.B.M.: Image based leaf area measurement method using artificial neural network. In: 2019 International Conference of Artificial Intelligence and Information Technology (ICAIIT), pp. 288–292. IEEE (2019)
31. Srinivasan, V., Kumar, P., Long, S.P.: Decreasing, not increasing, leaf area will raise crop yields under global atmospheric change. Glob. Change Biol. **23**(4), 1626–1635 (2017)
32. Stewart, J.: Calculus: concepts and contexts. In: Cengage Learning (2009)
33. Suzuki, S., Be, K.: Topological structural analysis of digitized binary images by border following. Comput. Vis. Graph. Image Process. **30**(1), 32–46 (1985). https://doi.org/10.1016/0734-189X(85)90016-7
34. Taiz, L., Zeiger, E.: Auxin: the first discovered plant growth hormone. In: Plant Physiology, 5th edn, pp. 545–582. Sinauer Associates Inc., Publishers, Sunderland (2010)
35. Tech, A.R.B., et al.: Methods of image acquisition and software development for leaf area measurements in pastures. Comput. Electron. Agric. **153**, 278–284 (2018)
36. Villar, R., et al.: Applying the economic concept of profitability to leaves. Sci. Rep. **11**(1), 1–10 (2021)

37. Wang, L., et al.: QTL fine-mapping of soybean (glycine max l.) leaf type associated traits in two rils populations. BMC Genomics **20**(1), 1–15 (2019)
38. Wellstein, C., et al.: Effects of extreme drought on specific leaf area of grassland species: a meta-analysis of experimental studies in temperate and sub-mediterranean systems. Glob. Change Biol. **23**(6), 2473–2481 (2017)
39. Weraduwage, S.M., et al.: The relationship between leaf area growth and biomass accumulation in arabidopsis thaliana. Front. Plant Sci. **6**, 167 (2015)
40. Wright, I.J., et al.: Assessing the generality of global leaf trait relationships. New Phytol. **166**(2), 485–496 (2005)
41. Wright, I.J., et al.: The worldwide leaf economics spectrum. Nature **428**(6985), 821–827 (2004)

Crop Row Line Detection with Auxiliary Segmentation Task

Igor Ferreira da Costa$^{(\boxtimes)}$ and Wouter Caarls$^{(\boxtimes)}$

Pontifícia Universidade Católica do Rio de Janeiro, Rio de Janeiro, RJ, Brazil
igorcosta@aluno.puc-rio.br, wouter@puc-rio.br

Abstract. Autonomous robots for agricultural tasks have been researched to great extent in the past years as they could result in a great improvement of field efficiency. Navigating an open crop field still is a great challenge; RTK-GNSS is a excellent tool to track the robot's position, but it needs precise mapping and planning while also being expensive and signal dependent. As such, onboard systems that can sense the field directly to guide the robot are a good alternative. Those systems detect the rows with adequate image techniques and estimate the position by applying algorithms to the obtained mask, such as the Hough transform or linear regression. In this paper, a direct approach is presented by training a neural network model to obtain the position of crop lines directly from an RGB image. While, usually, the camera in such systems are looking down to the field, a camera near the ground is proposed to take advantage of tunnels formed between rows. A simulation environment for evaluating both the model's performance and camera placement was developed and made available in Github, and two datasets to train the models are proposed. The results are shown across different resolutions and stages of plant growth, indicating the system's capabilities and limitations.

Keywords: Autonomous Robot · Deep Learning · Agricultural Robot

1 Introduction

The use of robotics in agriculture presents the possibility of improving the efficiency of the field, and at the same time increasing the production and quality of the crop. Autonomous robots are a good tool to deal with simple but long lasting tasks, such as monitoring the state of plants growth [1,2], health and presence of plagues or invasive species [3,4].

This study was financed in part by the Coordenação de Aperfeiçoamento de Pessoal de Nível Superior - Brasil (CAPES) - Finance Code 001; The National Council for Scientific and Technological Development - CNPq under project number 314121/2021-8; and Fundação de Apoio a Pesquisa do Rio de Janeiro (FAPERJ) - APQ1 Program - E-26/010.001551/2019.

M. C. Naldi and R. A. C. Bianchi (Eds.): BRACIS 2023, LNAI 14197, pp. 162–175, 2023.
https://doi.org/10.1007/978-3-031-45392-2_11

Real-time kinematics with Global Navigation Satellite System (RTK-GNSS) can provide, in open fields, an accurate position of the robot which can be used to navigate [4,5]. While precise, this system does not describe the crop field itself, thus requiring precise mapping and route planning. RTK-GNSS can also be expensive to deploy [5,6] in scale and be vulnerable to an unreliable or weak signal [1].

Crops are not always straight and plain, varying widely in topology due to the place's geography, which makes navigation even harder [5], see Fig. 1. As such, onboard solutions for navigation gained traction and different techniques and sensors have been tested and developed. Onboard systems, such as LIDAR [4,7], depth cameras and RGB cameras [6,8], make the robot sense the field directly, allowing it to navigate between plantation rows accordingly.

In order to correctly navigate the field, an algorithm to detect the rows and spaces inbetween must be used. Convolutional Neural Networks (CNNs) have already been successful at distinguishing crops from lanes [9,10]. In a similar way, this work employs a CNN, which is trained in images extracted from simulations and previous tests of the robot.

Given a segmented image, techniques such as Linear regression [5,11], can be applied to extract steering commands to guide the robot. This work aims to extract directly from the image, using a CNN, the steering directions, leveraging the model's capacity of feature extraction to guide the robot more directly.

The previously developed robot is a differential drive design, developed to work across soybean and cotton fields, Fig. 2. The original hardware has three Logitech C270 cameras for navigation, one up top looking down and one in each front wheel, two Intel RealSense D435i RGB-D cameras used for plant analysis and two GPS modules, all controlled by an Intel NUC7i5BNK running Ubuntu 18.04 LTS and ROS Melodic.

For this work, simulations and robot control are performed by the Nvidia Jetson AGX Orin Developer Kit. This new single board computer (SBC) is replacing the robot's onboard computer, improving the available computational power, now running Ubuntu 20.04 LTS with ROS Noetic.

2   Related Work

Deep learning methods for crop field navigation have already been researched and tested. Ponnambalam et al. [5] propose a ResNet50 to generate a mask of the field using an RGB camera. This work addresses the changing aspects of the field across the season, such as shadows, changes of colours, bare patches or grass across the field by a multi-ROI approach with the generated mask.

Xaud, Leite and From [10], applied a FLIR thermal camera to navigate within sugarcane fields. The thermal camera yielded images with more visibility and better discernible features, especially in poor lighting and weather conditions. The authors tested a simple CNN model with both cameras to compare the performance of the RGB image against the thermal camera with good results in favor of the latter.

Fig. 1. Uneven and curved field.

Fig. 2. The robot roaming in a test field

Silva et al. [9] compared the same training dataset with and without the depth information of a RGB-D camera applying the U-net deep learning model to identify the rows of plantation. RGB only images end up being selected due to RGB-D not increasing the results seemingly. The obtained masks were post processed by a triangular scanning algorithm to retrieve the line used for control.

Similarly to Ponnambalam [5], Martins et al. [11] applied a multi-ROI approach with sliding windows on a mask generated by the excess of green method. This is also the current robot's method of row navigation and one possible direct comparison which shows good results with smaller plant size scenarios, but limitations are observed in testing.

This work attempts to bridge the last gap in all previous examples by obtaining directly the row line or the row path, that way commands are sent straight to the controller and not derived from a mask by a post process algorithm. The mask, in this instance, will be leveraged as an auxiliary task [12] to improve the training model and allow the use of further techniques on the mask itself.

One of the biggest challenges faced previously were situations where the soil was occluded by leaves. With a bottom camera close to the ground, a similar navigation system will be tested to face this challenge using the developed tools.

3 Approach

The controller, which actuates the robot wheels, receives two parameters, an X value that describes the vertical offset of a given line and θ, the angle between the line and the vertical, see Fig. 3. For each frame captured by the robot's RGB-camera, the aim is to obtain a given line and pass both X and θ values to the controller. During the testing phase, the effects of using a segmentation job as an auxiliary task will also be evaluated.

A second approach will also test a bottom camera near the ground which creates a "tunnel" and "walls" effect by the camera's positioning amongst the rows, see Fig. 3.

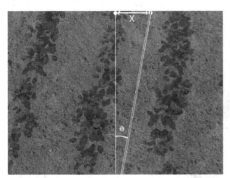

Fig. 3. Top and bottom camera line values given to the controller

3.1   Data Annotation and Preprocessing

The image data previously obtained was separated in groups by weather condition and plant growth. These are RGB images with 640×480 resolution and 8bpc of color depth. One first Python script was made to annotate line coordinates for each image with openCV and another script built a mask for segmentation using the line and various color filters. After the process, this base mask was loaded into JS-Segment Annotator Tool [13] for cleaning and small corrections, see Fig. 4.

The line was defined with two variable X values and two fixed Y values, the line will always be defined by coordinates in the form of $(X_0, 0)$ and $(X_1, 480)$, even if X falls outside of the image size. This special case is caused by lines that start or finish in the sides of the image.

The Top View Dataset was separated in three sections: training, validation and testing, with an split of 70%, 20% and 10%, respectively. Each section contains 50% of images derived from the simulation and also samples of every type from the original image groups.

The Botton View Dataset was only composed by simulated data due to lack of good images for training among the original data.

3.2   Network Model

A modified Network Model based on the DeepLabV3+ [14] was extracted from Keras code examples. The DeepLabV3+ employs an encoder-decoder structure and this variation has ResNet50 [15] as backbone for early feature extraction by the encoder. The ResNet50 was obtained in the Keras Applications Library pretrained with Imagenet weights.

A second head, known here as Line Head, was added alongside DeepLabV3+ responsible for learning the line directly. Its composition is made of a flatten layer, two fully interconnected layers with 25 and 20 neurons each and a output layer. The input is taken from the same tensor that DeepLab uses for the Atrous convolution section of its encoder, see Fig. 5. The interconnected layers apply ReLU as activation function while the output layer uses a sigmoid.

A custom loss function (1) was established due the multitask nature of the model, it combine the mask's Sparse Categorical Cross-entropy and the Root

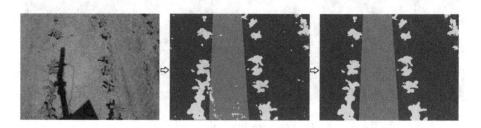

Fig. 4. Masks are generated by algorithm and then cleaned to improve quality

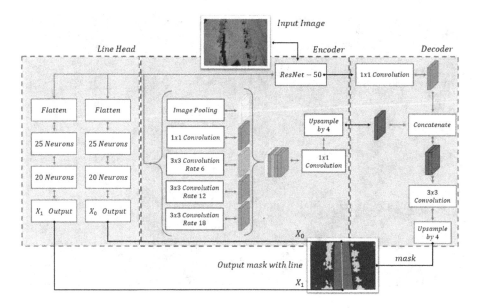

Fig. 5. Multitask proposed model for mask and line generation

Mean Squared Error (RMSE) used for the two values, X_0 and X_1, needed to define the line.

$$Loss = CE_{\text{mask}} + rmse_{X_0} + rmse_{X_1} \tag{1}$$

Training is performed in up to 500 epochs with early stopping configured. The Adam optimizer was chosen with a learning rate of 10^{-5} and training stopped if bad values (NAN) were detected. The best and the final weights are kept to predict the test images and be used in the simulation.

Two variants of the model were tested, the full proposed model and without the DeepLabV3+ decoder alongside the Line Head, evaluating the impact of the auxiliary task. Each variant was tested with 3 image resolutions: 128×128 for faster inference times, 512×512 for better image quality and detail, and 256×256 as a middle ground. All listed configurations were tested for the top and the bottom cameras.

4   Simulation and Test Setup

A simulation for testing was elaborated using the Gazebo-Sim package from ROS, and made avaliable at https://github.com/ifcosta/cnn_line. Plant models were designed in the Blender software based on real image data and a simplified version of the robot was modeled with two cameras, one at a height of 1,8 m looking downward in front of the robot and one just 0,15 m above the ground looking in front slightly upward. Images obtained with the simulations can be automatically annotated using simulation tools and filters (Fig. 6).

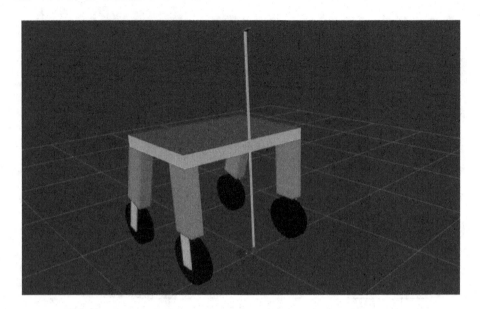

Fig. 6. Simplified robot model built for the simulations

The simulated field is built with five stages of plant growth, each stage has six plant rows that are 50 cm apart and 10 m long, with the plants 25 cm apart. There are multiple models of plants for each stage, each plant in a given row is selected randomly in the stage of growth pool. There is also a random component to the plant placement; 15% of given positions have a missing plant and every plant in the row has a small variance in position, see Fig. 7.

Each model was tested by crossing the field in both directions across all growth stages and the error between the theoretical perfect line and also the mean absolute error of the robot position were both recorded. The robot also starts every run with a small deviation from the center, randomly pointing inward or outward, and has a small error component for the applied velocity as defined by

$$error_{t+1} = (1 - \theta) \times error_t + normal\left(\mu, \sigma\right), \tag{2}$$

with $\theta = 0.25$, $\mu = 0$ and $\sigma = 0.01$.

The tests were designed this way so the model has to be able to compensate for actuation errors while correcting the starting deviation correctly during the run.

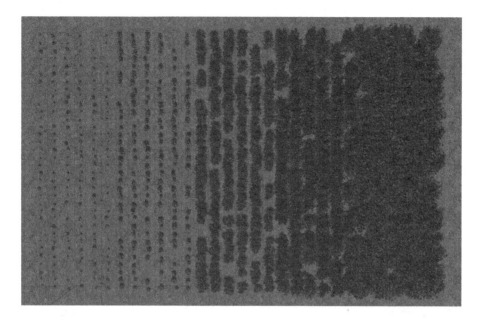

Fig. 7. Field modelled for testing with five growth stages

5   Results

5.1   Training Phase

Concerning the models trained for the bottom camera, both line-only and the variant with an auxiliary task achieved similar results with regards to the line's combined loss function, see Fig. 8. The larger resolution with auxiliary task is the worst performing model by the final loss value and had its training stopped due to NAN values found. Both smallest resolution models were the most unstable, while the best model, in general, was the medium resolution one. Also, the auxiliary tasks had little to no discernible impact except for the high resolution which had a negative one.

Meanwhile, the top camera, in Fig. 9, has better training results than the bottom one. This camera position also exhibits a small but positive effect to all resolutions when using an auxiliary task. The lower resolution had the worst performance, like with the bottom camera, but the high and medium resolution had comparable training performance with this camera placement. The top camera was considerably better during the training phase for all resolutions in regards to the model's loss.

The bottom camera data set is smaller and composed only by simulation derived images, which could be impacting its results. The way that the line was modeled implied that it had a higher variability at the top of the screen with the bottom camera, i.e. X_0, and that could lead to the higher loss observed. This volatility does not happen with the top camera and to the X_1 value of the

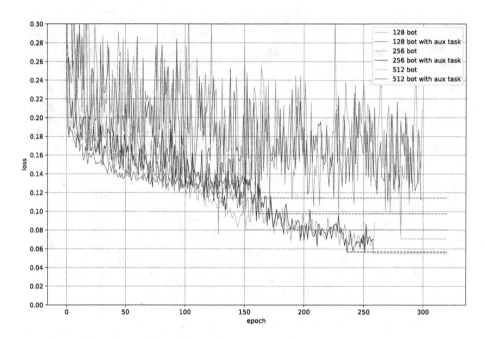

Fig. 8. Bottom camera line loss - dotted line mark the minimum value obtained

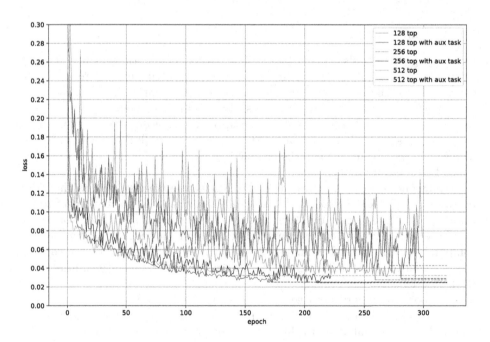

Fig. 9. Top camera line loss - dotted line mark the minimum value obtained

bottom one. As it will be shown later, the bottom camera model can still work with this behavior, but a different training approach for the bottom line might bring better results.

5.2   Simulation Runs

Alongside the model's tests, a control run was executed with ROS transforms to calculate the theoretical perfect line and evaluate the controller performance. These runs have a perfect line, but still have a processing delay comparable to the model's inference times.

For the top camera, there is another available comparison, which is the multi-ROI with excess of green mask [11]. This algorithm is not designed to be used on the bottom camera and its processing time depends of the number of masked pixels inside the sliding window. As such, the larger growth stage, having a greater amount of segmented and processed pixels due to the size of and density of the plants, takes longer to process. It also depends on the CPU's single thread processing power, being less suitable for the Jetson's hardware, see Fig. 10, so this algorithm can be slower depending in plant size and shape while the CNN based algorithms does not get influenced by it.

The growth stage 1 and 2, which has the smaller plants, had similar results shown by Fig. 11 growth stage 1. Also the growth stage 3 and 4 with the medium sized plants, Fig. 12, displayed similar traits. Evaluating the robot's position inside the plantation row can assess the entire performance of the system, i.e. model and controller. Figures 11 and 12 show that only the low resolution without auxiliary task model got consistently bad results while the high resolution with auxiliary task got them only in the growth stage 1 for the top camera.

Figures 12 and 13 also display the weaker result of the Multi-ROI approach from growth stage 3 onward, possibly due to the high processing time needed and the lack of soil visibility. The fifth growth stage marks failures for all top camera models, while the bottom still has good results, in line with the previous scenarios, see Fig. 13.

As such we can conclude that the bottom camera is better suited to this dense environment, with the medium and high resolution displaying the best results.

6   Discussion

The proposed approaches displayed good results in the simulation, in special the medium resolution, which was the most consistent. Meanwhile, the auxiliary task showed potential, improving results for the small and medium resolution, but with problems at the highest.

Overall, with those datasets and this simulated environment, the medium resolution with auxiliary task had the best performance, followed closely by the medium and high resolutions without the auxiliary task.

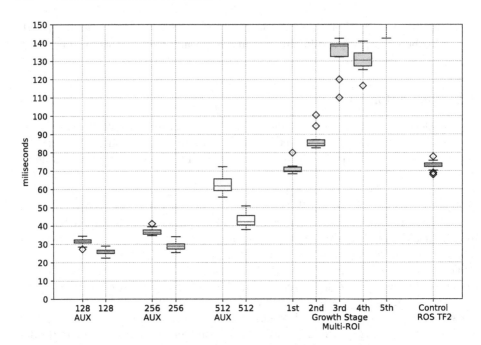

Fig. 10. Processing time for all algorithms, from image reception to command sent. The multi-ROI is split for each growth stage to show the speed decrease observed with bigger plants

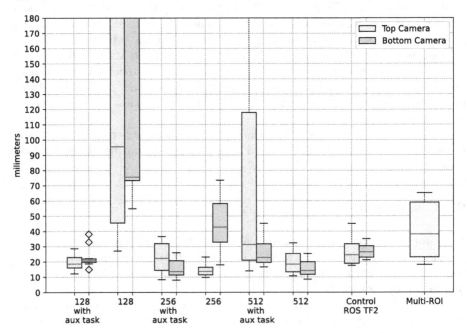

Fig. 11. Robot position deviation from the center of row's path - Growing Stage 1

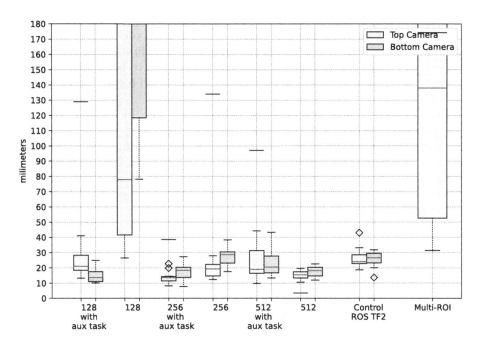

Fig. 12. Robot position deviation from the center of row's path - Growing Stage 3

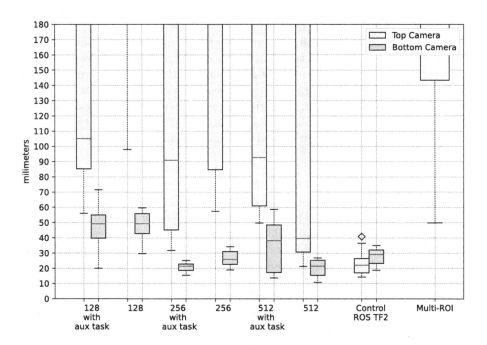

Fig. 13. Robot position deviation from the center of row's path - Growing Stage 5

The camera near the ground delivered comparable results to the top camera in most scenarios and excellent ones when the ground was occluded by the plant leaves. The tunnel formed was enough to guide the robot during the simulations, but tests with objects and grass on the robot's path need to be evaluated. Also, tests in a real environment would improve the dataset and test the model's true capabilities.

It is important to note that the controllers used in this work were not modified from those already in place in the robot. However, with the developed simulation environment other techniques could be applied to better actuate the robot with the given line from the network. Jesus et al. [16] proposed a method leveraging reinforcement learning to navigate an indoor area, reaching goals without hitting obstacles. The same kind of controller could be used in our work, with the line being used as input to the controller agent.

7 Conclusion

We presented a neural network model to extract crop lines from RGB images. The network input can be taken from either a top or a bottom camera, and the loss function was augmented with an auxiliary segmentation task.

With the simulated environment created, the top and the bottom cameras were evaluated in three image resolutions with and without the auxiliary task proposed. All networks configurations trained successfully and were able to guide the robot in the simulation with varying degrees of success while using the data set created with the built environment.

When assessing the overall performance of the entire system, both the top and bottom cameras achieved less than 50 mm of deviation from the center of the path in all but one configuration in the first four growth stages. However, the fifth growth stage resulted in failures for all top configurations, with errors exceeding 90 mm for most configurations, while the bottom camera remained below 50 mm. The processing time of all configurations kept below 70 ms, notably the medium resolution which kept between 30 and 40 ms, comparable to the low resolution model.

The auxiliary task had a positive impact on medium and low resolutions but negatively impacted the high resolution configuration. The bottom camera approach is beneficial to scenarios were the ground is mostly occluded by dense foliage, but can have issues in some configurations containing smaller plants.

With the GPU computing performance available, a fully integrated model could be feasible. This way, the best camera for each situation would not need to be chosen, as both cameras could feed directly into the model, combining channel wise or just as a larger image. Other types of cameras, such as thermal or RGB-D, can be evaluated to take advantage of ground's temperature as it is occluded by bigger plants or the depth perception possible due to the tunnel seen by the bottom camera. Other types of network models could also be evaluated, both for the line head and for the auxiliary task used.

References

1. Ahmadi, A., Nardi, L., Chebrolu, N., Stachniss, C.: Visual servoing-based navigation for monitoring row-crop fields. In: 2020 IEEE International Conference on Robotics and Automation (ICRA) (2020)
2. Nakarmi, A.D., Tang, L.: Within-row spacing sensing of maize plants using 3d computer vision. Biosys. Eng. **125**, 54–64 (2014)
3. McCool, C.S., et al.: Efficacy of mechanical weeding tools: a study into alternative weed management strategies enabled by robotics. IEEE Robot. Automat. Lett. 1 (2018)
4. Barbosa, G.B.P.: Robust vision-based autonomous crop row navigation for wheeled mobile robots in sloped and rough terrains. Dissertação de mestrado em engenharia elétrica, Pontifícia Universidade Católica do Rio de Janeiro, Rio de Janeiro (2022)
5. Ponnambalam, V.R., Bakken, M., Moore, R.J.D., Gjevestad, J.G.O., From, P.J.: Autonomous crop row guidance using adaptive multi-ROI in strawberry fields. Sensors **20**(18), 5249 (2020)
6. Ahmadi, A., Halstead, M., McCool, C.: Towards autonomous visual navigation in arable fields (2021)
7. Shalal, N., Low, T., McCarthy, C., Hancock, N.: Orchard mapping and mobile robot localisation using on-board camera and laser scanner data fusion - part a: tree detection. Comput. Electron. Agric. **119**, 254–266 (2015)
8. English, A., Ross, P., Ball, D., Upcroft, B., Corke, P.: Learning crop models for vision-based guidance of agricultural robots. In: 2015 International Conference on Intelligent Robots and Systems (IROS) (2015)
9. De Silva, R., Cielniak, G., Wang, G., Gao, J.: Deep learning-based crop row following for infield navigation of agri-robots (2022)
10. Xaud, M.F.S., Leite, A.C., From, P.J.: Thermal image based navigation system for skid-steering mobile robots in sugarcane crops. In: 2019 International Conference on Robotics and Automation (ICRA). IEEE (2019)
11. Martins, F.F., et al.: Sistema de navegação autônoma para o robô agrícola soybot. In: Procedings do XV Simpósio Brasileiro de Automação Inteligente. SBA Sociedade Brasileira de Automática (2021)
12. Liebel, L., Körner, M.: Auxiliary tasks in multi-task learning (2018)
13. Tangseng, P., Wu, Z., Yamaguchi, K.: Looking at outfit to parse clothing (2017)
14. Chen, L.-C., Zhu, Y., Papandreou, G., Schroof, F., Adam, H.: Encoder-decoder with atrous separable convolution for semantic image segmentation (2018)
15. He, K., Zhang, X., Ren, S., Sun, J.: Deep residual learning for image recognition. In: IEEE Conference on Computer Vision and Pattern Recognition (CVPR), pp. 382–386 (2016)
16. Jesus, J.C., Bottega, J.A., Cuadros, M.A., Gamarra, D.F.: Deep deterministic policy gradient for navigation of mobile robots in simulated environments. In: 2019 19th International Conference on Advanced Robotics (ICAR), pp. 362–367 (2019)

Multiple Object Tracking in Native Bee Hives: A Case Study with Jataí in the Field

Rodolfo R. V. Leocádio[1]([✉]) [iD], Alan Kardek Rêgo Segundo[1,2] [iD],
and Gustavo Pessin[1,2] [iD]

[1] Programa de Pós-Graduação em Ciência da Computação – Universidade Federal de Ouro Preto (UFOP), Ouro Preto, MG, Brazil
rodolfo.leocadio@ufop.edu.br
[2] Instituto Tecnológico Vale – Mineração, Ouro Preto, MG, Brazil

Abstract. Artificial intelligence approaches, such as computer vision, can help better understand the behavior of bees and management. However, the accurate detection and tracking of bee species in the field remain challenging for traditional methods. In this study, we compared YOLOv7 and YOLOv8, two state-of-the-art object detection models, aiming to detect and classify Jataí Brazilian native bees using a custom dataset. Also, we integrated two tracking algorithms (Tracking based on Euclidean distance and ByteTrack) with YOLOv8, yielding a mean average precision (mAP_{50}) of 0.969 and mAP_{50-95} of 0.682. Additionally, we introduced an optical flow algorithm to monitor beehive entries and exits. We evaluated our approach by comparing it to human performance benchmarks for the same task with and without the aid of technology. Our findings highlight occlusions and outliers (anomalies) as the primary sources of errors in the system. We must consider a coupling of both systems in practical applications because ByteTrack counts bees with an average relative error of 11%, EuclidianTrack monitors incoming bees with 9% (21% if there are outliers), both monitor bees that leave, ByteTrack with 18% if there are outliers, and EuclidianTrack with 33% otherwise. In this way, it is possible to reduce errors of human origin.

Keywords: Computer Vision · Tracking · Native Bees · Optical Flow

1 Introduction

Insects are fundamental in ecosystems as pollinators, herbivores, detritivores, nutrient cyclers, and food sources for other species [1]. Native bees are pollinators related to various plant foods consumed [2]. Bees are fundamental in world agriculture to improve crop quality and yields [3]. Furthermore, they actively contribute to the recovery of degraded ecosystems [4], can be used in biovectorization - technology that uses insects as biocontrol agents [5], and can also play a crucial role in sustainable development projects and programs [4].

Studies report biodiversity losses among native bees in tropical regions [6, 7]. The main factors responsible for your decline are infections disseminated by parasites and

M. C. Naldi and R. A. C. Bianchi (Eds.): BRACIS 2023, LNAI 14197, pp. 176–191, 2023.
https://doi.org/10.1007/978-3-031-45392-2_12

pathogens, lack of genetic variability, stress due to the seasonal movement of hives to pollinate fruits and vegetables, toxic pesticide residues found in pollen, nectar, and hives (mite control), the low nutritional value of agro-landscapes dominated by monocultures (such as corn, soybeans, and cotton), and the most adverse weather conditions in recent decades [1, 6, 8–10].

The decline of insects results in adverse effects on ecosystems. Thus, preserving its abundance and diversity is fundamental for ecology [1]. Individuals with little entomological knowledge are incapable differentiates categories of insects, their stages of maturity, and their behavior. Therefore, it is necessary to develop faster and more effective approaches that solve these problems [11]. The issues addressed in the insect recognition and classification process are to quickly detect the animal of interest positioned on a complex background, accurately distinguish insect species with high similarity, and effectively identify patterns of interest in the behavior of the classes. Artificial Intelligence (AI) tools can have a mutualist relationship with methods applied in ecology. The hives monitoring can help combat the decline of bees related to foraging, pollen characteristics, hive behavior, and invasive species.

Attempting to help monitor hives, we used an object detector to detect and classify *Tetragonisca angustula* bees. We incorporated a simple tracking system based on the object detector to describe the movement of insects, generating information about count and their direction concerning the hive entrance. We also tested a more robust tracking system, in an attempt to treat occlusions, compared to the previous one.

To next section contains recent work on artificial intelligence assisting insect management and conservation. Section 3 presents the methods of detection, classification, tracking, and optical flow of the species. Section 4 contains the results obtained by the steps highlighted in anterior with your corresponding discussions. Finally, we present the conclusions and future perspectives in Sect. 5.

2 Related Works

Some works developed in the last six years stand out in an attempt to automate processes applied to Biological Sciences involving the Class Insecta of the Phylum Arthropoda (Kingdom Animalia). AI techniques used to monitor insects from images can provide an alternative to reduce manual activities and human errors [12]. In this sense, there is a tendency to use computer vision approaches for insect detection and recognition.

Object detection techniques such as Faster R-CNN [13], VGG19 [11], YOLO + SVM [14], and ResNet (mAP around 90%) [15] proved adequate. Such approaches became possible to identify, classify and count insects in plants, leaves, photographs, traps, and grains [13, 16]. The insects include wasps and moths [11].

Liu and his team [17] developed a system called PestNet for field pest detection. Monitoring the number of pest species prevents the indiscriminate use of pesticides that result in crops that are harmful to health. The acquisition equipment is a multispectral light trap with an HD camera to capture pest specimens in a tray. Its best performance achieved an mAP of 75.46%.

Abiotic and biotic factors - such as social interactions - are used to understand social insects. Individuals in a colony engage in various tasks, such as foraging and hive care, depending on social contexts and interactions [18]. Challenges arise when monitoring these colonies because individuals are small and numerous [19]. Tracking systems have been used for this function, as insect movements can provide information about their cognition and decision-making [20].

Most tracking uses classical machine learning techniques [21]; however, computer vision can automate tracking, featuring velocity, straightness, direction, and exploration of trajectories [22]. Markers make the tracking system based on images more robust, making it possible to measure locomotor activities [18] of members of an ant colony - in the laboratory - to study their circadian activity [19]. Some authors, such as [23], try to integrate real-time data analytics and monitor them under natural conditions, such as in a tropical forest [22] without markers. Bioluminescence and behavior studies of insects such as fireflies [24] and pest counting in automatic traps with live insects also use tracking [25].

Concomitantly emerged were works involving the Apidae Family of the Hymenoptera Order. Bee foraging studies using RFID tags [2] merged with temperature, barometric pressure, and solar irradiance data define a Recurrent Neural Networks (RNN) architecture [8] that performs best in predicting bee behavior.

There are also systems with deep learning for automated monitoring of beehives (DeepBees) [10]. In this case, the footage of the bees takes place at the entrance to the hive. DeepBees monitors health and hygiene, attacks and invasions (wasps), pollen, bees dead, and drones. However, there are still challenges in the detection, direct mapping of activities, and with diseases or mite infestation; caused by lack of data.

Tracking for pollination studies targets bees. Monitoring natural habitats are essential for behavioral studies, preferably with non-invasive techniques. Hybrid Detection and Tracking (HyDaT) map interactions between bees and wildflowers; kNN segments the image background and foreground, and YOLO locates bees [20, 26]. Perez-Cham [27], Sun [28], and their teams use 3D tracking to analyze and reconstruct the bees' flight paths.

This work presents the following contributions: It uses non-invasive techniques, without sensors in the hive and markers in the insects; Uses a one-stage object detector to classify and detect Jataí; It only uses tracking based on detections; Compares the performance of two tracking algorithms to perform the task; Contains specific implementation for the optical flow of bees and; Compares the performance of computer vision with humans. In this way, we believe it is possible to contribute to the ecological preservation of the species, combating the decline of insects and facilitating their management.

3 Methods

3.1 Custom Dataset – *Tetragonisca angustula* (Jataí)

The Jataí (Fig. 1) is a bee native to Brazil, and its geographic distribution occurs throughout the Brazilian territory. Considering its natural environment, this species has the habit of nesting in tree hollows and rock cavities. Its adaptability promotes nesting in hollow walls, light boxes, letter boxes, stone walls, and other unusual places, considering an urbanized environment.

The colony has a queen mother and an average of 5000 (from 2000 to 8000) workers. Guards are a caste with larger sizes than other workers. During the day, the workers remain positioned over the inlet tube or hover close to it. On average measures 4 mm and can forage 600 m from the hive.

Fig. 1. Tetragonisca angustula. Adapted from A.B.E.L.H.A.

It is a significant pollinator of crops such as acapu, avocado, coffee, carrot, cupuaçu, guava, orange, mango, watermelon, strawberry, moressuma, pepper, cucumber, tangerine, umbu, and annatto. Therefore, proper management of hives ensures the productivity of such crops and protects the colony from natural enemies.

The custom dataset aims to monitor automatically of native bee hives using non-invasive techniques. It consists of videos of the entry of several colonies for tracking identified as XXXY-Y, where XXX identifies the hive location and Y-Y the recording interval. We identified images taken from the videos for object detectors as the complement of the previous XXXY-YZ..., where Z... is an id associated with each image. Each video and image contains an information file with the metadata.

Knowledge about the Jataís showed that monitoring on cold, rainy days and at night is unnecessary. Also, we avoid windy days. In this way, the data represent reality and are valid only for Jataís.

We used the SONY HDR-AS20 camera, with a resolution of 1920 × 1080p and a recording speed of 60 fps. In an attempt to represent the natural environment of the hives, the recordings took place at different times of movement at the hive's entrance. We withdraw videos of a beehive from the BeeKeep CS database, with a resolution of 960 × 540p and a recording speed of 30 fps. We acquired videos of natural and artificial hives in urban and rural environments relatively close, but climatic conditions are different.

We positioned the video camera, trying to optimize the capture of movements around each hive. Therefore, there is only variation in the capture angle for different beehives. The dynamic in the hive's entrance guarantees the variability of the poses of the bees.

The custom dataset is structured as required by the family of object detectors known as YOLO (test, valid, and train). These directories have a separation between images, labels, and information. The info folder contains the metadata. Incomplete files are due to data not provided by BeeKeep CS (16.7%). The custom dataset has 26487 instances in 2100 images, with 19000 in 1500 images for training, 3719 in 300 for validation, and 3768 in 300 for testing.

Climate variations, preferably days with the sun between clouds (83.2% of the images), periods of the days (16.7% obtained in the morning and 66.7% in the afternoon), and seasons (16.7% obtained in winter and 66.7% in spring), during the recordings provided differences in lighting. We noticed differences in lighting in 45.6% of the images.

The correlogram between labels contains a bimodal distribution that indicates a mixture of two trends with different labels, with a predominance of the right trend with the highest number of observations; A Chi-square distribution inferring that the labels are homogeneous; and two Poisson distributions indicating that the label sizes are random but repeat at a defined average rate. This way, the labels have adequate precision and cover all the bees in the images.

The custom dataset has no duplicates and leaks, considering the SSIM [29] of 0.97. As the bees occupy a small part of the image, for being small, the differences between images are subtle.

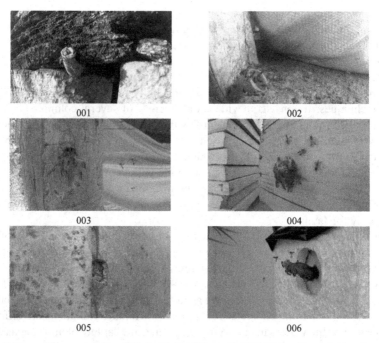

Fig. 2. Custom dataset samples. 001 (-20.381099, -43.506333)* - Natural hive in an urban environment. 002 (-20.502591, -43.520515)*, 003 (-20.502507, -43.520330)* and 005 (-20.345640, -43.608437)* - Natural hive in a rural environment. 004 (-20.345266, -43.608529)* - Artificial hive in a rural environment. 006 - BeeKeep CS artificial beehive. *Geographical coordinates obtained from Google.

There is a background for each beehive present (Fig. 2), totaling 0.29% of the images. When we do not obtain images without bees, use data augmentation to remove them (0.24% of them). We fill in the resulting space of an unlit bee with the texture of its nearest neighbors.

1737 images have occlusions, that is, 82.7% of them. We observed occlusions due to: the overlapping of bees, by background item, per entrance pipe, by light beams, bees that touch each other, and bees cut by frames.

The perceived anomalies are shadows and distortions during bee flight. The incidence of sunlight causes shadows that can be confused with objects by computer vision models. Bee shadows are monitored but have only been ignored so far (361 images have bee shadows, i.e., 17.2%). Distortions during bees' flight occur due to a higher flight speed than the capture capacity of the camera.

Not possible to infer over class distribution because the custom dataset only has one class so far. We did not observe intraclass variations, but they need monitoring during updates.

3.2 Detection and Classification of the Native Bee

YOLO [30] (You Only Look Once) is an object detection model used in computer vision. The model treats object detection as a regression problem using a single Convolutional Neural Network (CNN). It does this by splitting the image into a grid, making multiple predictions for each grid cell, filtering these predictions, and removing overlapping boxes to produce its final output. Its evolution includes YOLOv2 [31], YOLOv3 [32], YOLOv4 [33], YOLOv5 [34], and YOLOv6 [35]; each incorporates contributions to the model.

The ResNeXt backbone and the dual head of YOLOv7 [36], combined with the Label Assigner engine that assigns flexible labels, allow the model to learn from data more effectively. It presents a new multiscale training strategy and a technique called Focal Loss, designed to solve the class imbalance problem.

The latest release, YOLOv8 [37], surpasses previous by incorporating advances such as a new backbone, a head split without anchor, and new loss functions.

3.3 Tracking and Optical Flow of Jataís

Tracking is obtaining an initial set of objects, creating a unique Id for each one, and accompanying each object as they move in frames, maintaining the assignment of Id's. An ideal tracking algorithm should detect objects only once, be fast, control when the tracked object disappears or moves out of frame, be robust with occlusions, and recover lost objects between frames.

This work uses a detector-based tracking similar to [23]; it depends on the Euclidean distance between existing and new centroids. Bounding boxes are the basis for calculating centroids with Id's assigned to the initial set of centroids. After that, it computes the Euclidean distance between the existing and new centroids, associating new centroids with the old ones. Then, minimize distances and update your coordinates. Finally, there is the registration of new centroids and the cancellation of the registration of objects that leave the video. This a simple and fast method that controls when the tracked object disappears or leaves the frame and recovers lost objects between frames.

We attempt to improve the tracking with the algorithm ByteTrack [38] (BT) to it does the same as the previous one, solve occlusions, and detect objects only once. The algorithm performs associations between the detections of the current frame concerning the previous frame. For that, BT uses similarity calculations consider the distances between the resources of appearance (Re-ID) and IoU resources. Then the Hungarian algorithm finalizes the match based on similarity.

Optical flow is used in computer vision to characterize and quantify the movement of objects in a video stream, often for motion-based object detection and tracking systems. When estimating optical flow between frames is possible to measure an objects speed in the video. It is also possible to follow the movement of Jataís, count the bees that leave and enter the hive, and your total. This approach can generate information about hive nutrition.

For this, we propose using the centroids of the detections containing the same Id and a displacement limit from the current frame to the previous frame. We define a circle containing the entrance of the hive as a limit. A Jataí has entered the beehive if the Euclidean distance between the circle center and detection centroid is greater than your radius in the previous frame and smaller than the current frame. Otherwise, the Jataí leaves the hive.

4 Results and Discussion

4.1 Detection and Classification of Jataí

Results from YOLOv8 [37] trained on the custom dataset are in Fig. 3. In just 300 epochs of training, the YOLOv8x model converged to a precision of 0.97, recall of 0.957, mAP_{50} of 0.984, and $mAP_{50\text{-}95}$ of 0.744 (Fig. 3a). Considering the test data, the confusion matrix indicates that: every time the model predicts a class and is background, the reason is Jataí; 98% of Jataí predictions are Jataí and; 2% of background predictions are Jataí. The precision (0.935) – recall (0.937) curve for mAP_{50} of 0.969 and $mAP_{50\text{-}95}$ of 0.682 (Fig. 3b) shows that most of the classifications that the model made are correct, and most of the predicted bounding boxes are as expected.

The YOLOv7 [36] model converged to a precision of 0.952, recall of 0.948, mAP_{50} of 0.979, and $mAP_{50\text{-}95}$ of 0.634 (Fig. 3c). Considering the test data, the confusion matrix is identical to the previous one. The precision (0.922) – recall (0.901) curve for mAP_{50} of 0.955 and $mAP_{50\text{-}95}$ of 0.579 (Fig. 3d) also shows the same as the previous one.

We noticed that both present acceptable results for the application, but the YOLOv8 metrics reach better scores than the YOLOv7 ones. We believe this is due to differences in their backbones. In this way, we use YOLOv8 for the tracking implementations.

4.2 Tracking and Optical Flow of Detected Bees

In the tracking algorithm called EuTrack (ET), which depends on Euclidean distance, a parameter can be changed to control when the tracked object disappears or leaves the frame and its ability to recover lost objects between frames. We set the parameter as 20

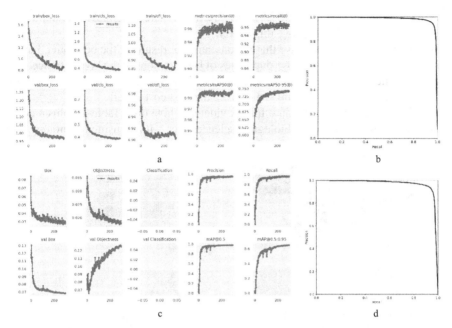

Fig. 3. Training results and Precision-Recall curve with YOLOv8 (a, b) and YOLOv7 (c, d), respectively.

for all experiments, i.e., an object is only considered for a new Id if it is not detected for 20 consecutive frames.

The BT [38] is a robust system in treating occlusions because it performs similarity analysis, adopts the Kalman filter to predict the object's movement, and uses additional Re-ID models to improve the long-range association. We set its parameters for the whole experiment as track_thresh equal to 0.05, track_buffer equal to 20, match_thresh equal to 1.0, aspect_ratio_thresh equal to 3.0, min_box_area equal to 1.0 and mot20 equal to True.

In Fig. 4, one can view the bounding box, detected class, and probability. The Tracking algorithms generate the Id and mark the centroid of each bounding box. The optical flow, implemented from the centroids and Id's, informs about movements in and out of the hive (delimited by the red circle). The bee trails, with a displacement limit equal to 120, were hidden to facilitate the visualization of the other considerations. Also, observe that Jataí with Id 7 (Fig. 4 - 001) entered and the one with Id 14 (Fig. 4 - 003) left their respective hives. Timestamps can collaborate with this information to generate possible monitoring of the hive entrance.

Results for a video of each hive are in Table 1, along with their absolute error and mean relative error. We use a process similar to pseudo-labeling (Pseudo) to count bees in the videos (we call it a procedure with the aid of technology). We used ET to aid in counting. Note in Fig. 4, specifically at 003 (Id 2 and 5), 005 (Id 4), and 006 (Id 0 and 9), the presence of soldiers hovering around the hives entrances. These soldiers generate errors in counting the bees that enter and leave the hive. Therefore, we remove duplicates from these counts (RD) to improve this aspect.

Summarizing Table 1, we have that ET has superior results in the bee count, but we can also use BT. It is best to remove duplicates of bees to monitor the hive entrance. The error is less when BT counts the bees that get out of the hive and when ET counts those that enter. In this way, there is a tie considering the best cost benefit.

We repeated the same procedure for six videos of hive 001 (Table 2), plus a count performed by a person from the biological sciences area without using technology.

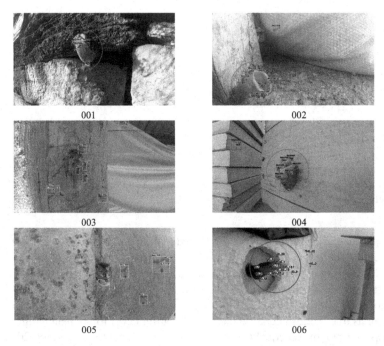

Fig. 4. Results of the tracking and optical flow. Detailed information in Fig. 2.

Summarizing Table 2, we have that BT has superior results in bee count. It is best to remove duplicates of bees to monitor the hive entrance. ET performs better in accounting for incoming and outgoing bees. In this way, ET obtains a better cost-benefit concerning the movement at the hive entrance and BT to count bees. We also noticed that using the system to assist human activity reduces errors caused by fatigue and distractions.

The errors in Tables 1 and 2 are caused by some limitations. ET does not resolve Id's swaps caused by crossing their positions and occlusions caused by soldiers and the hive tube. The movement in the hive tube is responsible for most of the exchanges of Id's, and the model presents difficulties for bees that touch each other (Example in Fig. 4 - 005 Id 5 and 6).

Table 1. Results of the trackings and humans with the aid of technology.

Videos	Parameters	Total	Out	RD*	In	RD*
0010-1	Pseudo	96	14		52	
	BT	84	33	9	62	27
	|BT – P|	12	19	5	10	25
	ET	150	57	19	89	53
	|ET – P|	54	43	5	37	1
0020-1	Pseudo	42	7		11	
	BT	70	0	0	2	2
	|BT – P|	28	7	7	9	9
	ET	93	2	2	10	9
	|ET – P|	51	5	5	1	2
0031-2	Pseudo	118	21		32	
	BT	86	73	18	76	23
	|BT – P|	32	52	3	44	9
	ET	140	100	31	121	47
	|ET – P|	22	79	10	89	15
0040-1	Pseudo	156	85		42	
	BT	118	29	8	45	24
	|BT – P|	38	56	77	3	18
	ET	169	21	5	47	32
	|ET – P|	13	64	80	5	10
00517-18	Pseudo	133	16		30	
	BT	70	17	14	31	24
	|BT – P|	63	1	2	1	6
	ET	119	20	16	34	26
	|ET – P|	14	4	0	4	4
0062-2M	Pseudo	53	15		17	
	BT	67	64	13	71	28
	|BT – P|	14	49	2	54	11
	ET	66	6	3	13	10
	|ET – P|	13	9	12	4	7
WO*	BT	-	-	18.2	83.2	38.9
	ET	-	-	40.9	-	20.6

(*continued*)

Table 1. (*continued*)

Videos	Parameters	Total	Out	RD*	In	RD*
ARE*	BT	31.3	-	60.8	65.8	42.4
	ET	27.9	-	70.9	76.1	21.2

* ARE – Average Relative Error; WO – Average Relative Error Without Outliers; RD – Removed Duplicates

Table 2. Results of the trackings and humans with and without the aid of technology.

Videos	Parameters	Total	Out	RD*	In	RD*
0010-1	Humans	74	14		51	
	Pseudo	96	14		52	
	|H – P|	22	0		1	
	BT	84	33	9	62	27
	|BT – P|	12	19	5	10	25
	ET	150	57	19	89	53
	|ET – P|	54	24	5	37	1
0011-2	Humans	75	8		41	
	Pseudo	81	9		45	
	|H – P|	6	1		4	
	BT	92	25	7	49	28
	|BT – P|	11	16	2	8	17
	ET	152	31	13	71	50
	|ET – P|	71	6	4	26	5
0012-3	Humans	67	15		40	
	Pseudo	74	14		40	
	|H – P|	7	1		0	
	BT	80	27	9	47	26
	|BT – P|	6	13	5	7	14
	ET	128	27	12	58	40
	|ET – P|	54	13	2	18	0
0013-4	Humans	74	16		36	
	Pseudo	77	17		38	
	|H – P|	3	1		2	

(*continued*)

Table 2. *(continued)*

Videos	Parameters	Total	Out	RD*	In	RD*
	BT	82	7	5	18	16
	|BT – P|	5	9	11	20	22
	ET	110	15	10	42	33
	|ET – P|	33	2	7	4	5
0014-5	Humans	101	17		56	
	Pseudo	144	23		72	
	|H – P|	43	6		16	
	BT	157	21	16	37	29
	|BT – P|	13	2	7	35	43
	ET	208	39	22	78	56
	|ET – P|	64	16	1	6	16
0015-6	Humans	80	36		38	
	Pseudo	142	45		49	
	|H – P|	62	9		11	
	BT	121	19	17	34	27
	|BT – P|	21	26	28	15	11
	ET	145	29	24	55	49
	|ET – P|	3	16	21	6	0
ARE*	BT	11.1	69.7	47.5	32.1	44.6
	ET	45.4	63.1	32.8	32.8	9.1

* ARE – Average Relative Error; RD – Removed Duplicates

In agitated hives, agglomerations of bees occur. This fact splits a bounding box containing one bee into others and unifies bounding boxes into one containing several bees. This process contributes to the acquisition of new Ids, along with occlusions.

The transition of Jataís between frames and occlusions causes the resumption of Id by different bees. The camera is not steady in 006, and the systems recognize some anomalies in 002 and 004. We monitored anomalies, but it was not possible to extinguish them in the field.

The speed of the videos, frames per second, cannot be too low because the centroids must be close between the frames. BT relocates Id's in the agglomerations with difficulties, causing the assignment of new Id's and generating duplicates. The same error of dividing and merging bounding boxes also occurs in BT, contributing to new Id's assignment.

There are difficulties in treating occlusions as well. The authors of [38] show that BT is robust for occlusions where one of the objects does not move during occlusion. This type of occlusion is rare with Jataís, as the occluded bee also moves. When the occlusion ends, the occluded object may be in a different position in rotation or translation. The movement causes the bee to have different characteristics from when it was occluded.

The similarity between individuals in the hive also impaired the similarity analysis used in BT. There are no physical differences between workers and castes in the same or different beehives. The bees themselves only differentiate through pheromones.

However, using the system in the field can help to combat the decline of bees related to foraging and hive behavior. Generating reports and doing repetitive work brings advantages to the approach, thus reducing errors of human origin. But the use in practical applications must consider the BT to count the bees with an average relative error of 11%. ET to monitor bees that enter with an average relative error of 9% (21% if there are outliers). Both to observe leaving bees, BT with an average relative error of 18% if there are outliers and ET with an average relative error of 33% otherwise.

5 Conclusion

We believe it is a proposal capable of detecting, classifying, and generating tracking information for Jataí in the field. In this way, it is possible to help combat the decline of bees related to foraging and hive behavior. We must consider a mixture of both systems in practical applications. BT counts bees with an average relative error of 11%. ET monitors incoming bees with an average relative error of 9% (21% if there are outliers). Both monitor bees that leave, BT with an average relative error of 18% if there are outliers, and ET with an average relative error of 33% otherwise.

The biggest challenge in this type of implementation is to obtain conditions and data of the species of interest in the field to compose the custom dataset. Harsh conditions common in natural environments provide difficulty in hardware use and maintenance. The context also introduced issues mainly caused by occlusions and outliers that we must overcome to reduce the relative error.

In future work, we intend to optimize the custom dataset by adding greater diversification to the data - considering invasive species, acquiring videos of different hives, adding data for tracking to the custom dataset presented, and completing the metadata of the videos with pseudo-labeling for counts and the same performed by biologists. To implement the MOT20 format [39] in the tracking custom dataset introducing evaluation metrics. To compare results with other Tracking algorithms in an attempt to reduce the limitations of this work. Finally, we will implement pollen identification in monitoring to generate information about hive nutrition. (Additional information at https://github.com/Rodolfoloc/Native-bees).

Acknowledgments. The authors would like to thank the Universidade de São Paulo (USP BeeKeep CS - https://beekeep.pcs.usp.br), the Empresa Brasileira de Pesquisa Agropecuária (Embrapa), and the Associação Brasileira de Estudo das Abelhas (A.B.E.L.H.A. - https://abelha.org.br) by the data and videos. People who allowed filmings on their properties. To the Laboratório Multiusuário de Práticas Simuladas (LaMPS - https://lamps.medicina.ufop.br) and the Laboratório de Controle e Automação Multiusuário (LABCAM) for the infrastructure and equipment provided.

Google Collaboratory by the technologies that make AI research possible with scarce resources. To Carlos J. Pereira, Eduardo Carvalho, Levi W. R. Filho, and André A. Santos (Instituto Tecnológico Vale) along with Diego M. Alberto (Efí) for their support with the computational methods. This research received financial support from the Coordenação de Aperfeiçoamento de Pessoal de Nível Superior - Brazil (CAPES) - Financing code 001.

References

1. Hallmann, C.A., et al.: More than 75 percent decline over 27 years in total flying insect biomass in protected areas. PLoS ONE **12**, 18–22 (2017). https://doi.org/10.1371/journal.pone.0185809
2. Arruda, H., Imperatriz-Fonseca, V., de Souza, P., Pessin, G.: Identifying bee species by means of the foraging pattern using machine learning. In: 2018 International Joint Conference on Neural Networks (IJCNN), pp. 1–6. IEEE (2019). https://doi.org/10.1109/IJCNN.2018.8489608
3. Kuan, A.C., et al.: Sensitivity analyses for simulating pesticide impacts on honey bee colonies. Ecol. Model. **376**, 15–27 (2018). https://doi.org/10.1016/j.ecolmodel.2018.02.010
4. Giannini, T.C., et al.: Climate change in the Eastern Amazon: crop-pollinator and occurrence-restricted bees are potentially more affected. Reg. Environ. Change **20**(1), 1–12 (2020). https://doi.org/10.1007/s10113-020-01611-y
5. Macharia, J.M., Gikungu, M.W., Karanja, R., Okoth, S.: Managed bees as pollinators and vectors of bio control agent against grey mold disease in strawberry plantations. Afr. J. Agric. **16**(12), 1674–1680 (2020). https://doi.org/10.5897/AJAR2020.15203
6. Sánchez-Bayo, F., Wyckhuys, K.A.G.: Worldwide decline of the entomofauna: a review of its drivers. Biol. Cons. **232**, 8–27 (2019). https://doi.org/10.1016/j.biocon.2019.01.020
7. Borges, R.C., Padovani, K., Imperatriz-Fonseca, V.L., Giannini, T.C.: A dataset of multi-functional ecological traits of Brazilian bees. Sci. Data **7**(1), 1–9 (2020). https://doi.org/10.1038/s41597-020-0461-3
8. Gomes, P.A.B., et al.: An Amazon stingless bee foraging activity predicted using recurrent artificial neural networks and attribute selection. Nat. Res. **10**(1), 1–12 (2020). https://doi.org/10.1038/s41598-019-56352-8
9. Filipiak, M.: A better understanding of bee nutritional ecology is needed to optimize conservation strategies for wild bees - the application of ecological stoichiometry. Insects **9**(3), 1–13 (2018). https://doi.org/10.3390/insects9030085
10. Marstaller, J., Tausch, F., Stock, S.: DeepBees - building and scaling convolutional neuronal nets for fast and large-scale visual monitoring of bee hives. In: 2019 IEEE/CVF International Conference on Computer Vision Workshop (ICCVW), pp. 271–278. (2019). https://doi.org/10.1109/ICCVW.2019.00036
11. Xia, D., Chen, P., Wang, B., Zhang, J., Xie, C.: Insect detection and classification based on an improved convolutional neural network. Sensors **18**(12), 1–12 (2018). https://doi.org/10.3390/s18124169
12. Abreu, V.H.R., Pimentel, A.D.A., Absy, M.L., Rech, A.R.: Pollen sources used by Frieseomelitta Ihering 1912 (Hymenoptera: Apidae: Meliponini) bees along the course of the Rio Negro, Amazonas. Brazil. Acta Botanica Brasilica **24**(2), 371–383 (2020). https://doi.org/10.1590/0102-33062019abb0391
13. Júnior, T.C., Rieder, R.: Automatic identification of insects from digital images: a survey. Comput. Electron. Agric. **178**(5), 105784 (2020). https://doi.org/10.1016/j.compag.2020.105784

14. Zhong, Y., Gao, J., Lei, Q., Zhou, Y.: A vision-based counting and recognition system for flying insects in intelligent agriculture. Sensors **18**(5), 1489 (2018). https://doi.org/10.3390/s18051489

15. Qing, Y., et al.: Development of an automatic monitoring system for rice light-trap pests based on machine vision. J. Integr. Agric. **19**(10), 2500–2513 (2020). https://doi.org/10.1016/S2095-3119(20)63168-9

16. Shen, Y., Zhou, H., Li, J., Jian, F., Jayas, D.S.: Detection of stored-grain insects using deep learning. Comput. Electron. Agric. **145**, 319–325 (2018). https://doi.org/10.1016/j.compag.2017.11.039

17. Liu, L., et al.: PestNet: an end-to-end deep learning approach for large-scale multi-class pest detection and classification. IEEE Acess **7**, 45301–45312 (2019). https://doi.org/10.1109/ACCESS.2019.2909522

18. Fujioka, H., Abe, M.S., Okada, Y.: Ant activity-rest rhythms vary with age and interaction frequencies of workers. Behav. Ecol. Sociobiol. **73**(3), 30 (2019). https://doi.org/10.1007/s00265-019-2641-8

19. Fujioka, H., Abe, M.S., Okada, Y.: Individual ants do not show activity-rest rhythms in nest conditions. J. Biol. Rhythms **36**(3), 297–310 (2021). https://doi.org/10.1177/0748730421102934

20. Ratnayake, M.N., Dyer, A.G., Dorin, A.: Tracking individual honeybees among wildflower clusters with computer vision-facilitated pollinator monitoring. PLoS ONE **16**(2), e0239504 (2021). https://doi.org/10.1371/journal.pone.0239504

21. Lima, M.C.F., Leandro, M.E.D.A., Valero, C., Coronel, L.C.P., Bazzo, C.O.G.: Automatic detection and monitoring of insect pests - a review. Agriculture **10**(5), 161 (2020). https://doi.org/10.3390/agriculture10050161

22. Imirzian, N., et al.: Automated tracking and analysis of ant trajectories shows variation in forager exploration. Sci. Rep. **9**(1), 1 (2019). https://doi.org/10.1038/s41598-019-49655-3

23. Sclocco, A., Ong, S.J.Y., Aung, S.Y.P., Teseo, S.: Integrating real-time data analysis into automatic tracking of social insects. R. Soc. Open Sci. **8**(3), 202033 (2021). https://doi.org/10.1098/rsos.202033

24. Tathawee, T., Wattanachaiyingcharoen, W., Suwannakom, A., Prasarnpun, S.: Flash communication pattern analysis of fireflies based on computer vision. Int. J. Adv. Intell. Inf. **6**(1), 60–71 (2020). https://doi.org/10.26555/ijain.v6i1.367

25. Bjerge, K., Nielsen, J.B., Sepstrup, M.V., Helsing-Nielsen, F., Hoye, T.T.: An automated light trap to monitor moths (Lepidoptera) using computer vision-based tracking and deep learning. Sensors **21**(2), 343 (2021). https://doi.org/10.3390/s21020343

26. Howard, S.R., Ratnayake, M.N., Dyer, A.G., Garcia, J.E., Dorin, A.: Towards precision apiculture: traditional and technological insect monitoring methods in strawberry and raspberry crop polytunnels tell different pollination stories. PLoS ON **16**(5), e0251572 (2021). https://doi.org/10.1371/journal.pone.0251572

27. Perez-Cham, O.E., et al.: Parallelization of the honeybee search algorithm for object tracking. Appl. Sci. **10**(6), 2122 (2020). https://doi.org/10.3390/app10062122

28. Sun, C., Gaydecki, P.: A visual tracking system for honey bee (Hymenoptera: Apidae) 3D flight trajectory reconstruction and analysis. J. Insect Sci. **21**(2), 1–12 (2021). https://doi.org/10.1093/jisesa/ieab023

29. Wang, Z., Bovik, A.C., Sheikh, H.R., Simoncelli, E.P.: Image quality assessment: from error visibility to structural similarity. IEEE Trans. Image Process. **13**(4), 600–612 (2004)

30. Redmon, J., Divvala, S., Girshick, R., Farhadi, A.: You only look once: unified, real-time object detection. In: Proceedings of the IEEE Conference on Computer Vision and Pattern Recognition, pp. 779–788 (2016)

31. Redmon, J., Farhadi, A.: YOLO9000: better, faster, stronger. In: Proceedings of the IEEE Conference on Computer Vision and Pattern Recognition, pp. 7263–7271 (2016)

32. Redmon, J., Farhadi, A.: YOLOv3: an incremental improvement. arXiv preprint: arXiv:1804.02767 (2018)
33. Bochkovskiy, A., Wang, C., Liao, H.M.: YOLOv4: optimal speed and accuracy of object detection. Cornell Univ. arXiv preprint: arXiv:2004.10934 (2020)
34. Jocher, G.: ultralytics/yolov5: v3.1. (2020). https://doi.org/10.5281/zenodo.4154370
35. Li, C., et al.: YOLOv6: a single-stage object detection framework for industrial applications. arXiv preprint: arXiv:2207.02696 (2022)
36. Wang, C., Bochkovskiy, A., Liao, H.M.: Designing network design strategies through gradient path analysis. arXiv preprint: arXiv:2211.04800 (2022)
37. Jocher, G., Chaurasia, A., Qiu, J.: YOLO by Ultralytics v8. (2023)
38. Zhang, Y., et al.: ByteTrack: multi-object tracking by associating every detection box. In: Proceedings of the European Conference on Computer Vision (2022)
39. Dendorfer, P., et al.: MOT20: a benchmark for multi object tracking in crowded scenes. arXiv: 2003.09003 [cs] (2020)

An Open Source Eye Gaze Tracker System to Perform Remote User Testing Evaluations

Marc G. Capdevila$^{(\boxtimes)}$, Karine Aparecida P. Rodrigues ,
Camila F. Jardim , and Renato M. Silva

Facens University, Sorocaba, São Paulo, Brazil
{marc.capdevila,karine.rodrigues,renato.silva}@facens.br

Abstract. Eye gaze trackers are devices designed to identify an individual's gaze fixation in relation to a screen or another reference point. These tools are widely applied in usability testing as they provide various metrics for studying how people interact with applications. In the past, these tools were expensive and required a controlled environment, as well as trained personnel for proper operation. Although nowadays, new implementations do not require physical hardware to perform these tests, they often rely on license-based models instead of being open source. The objective of this work is to create a standalone system that enables any user to implement a low-cost eye gaze tracker using web technologies. The goal is to facilitate its use in remote and face-to-face studies in a simple way, requiring only a computer and a webcam. We evaluated the impact of three different calibration techniques on the performance of a regression-based prediction algorithm in eye-tracking. In our experiments, the best result of linear regression was obtained with a circular calibration system that uses 160 points. Furthermore, we integrated the system with a web interface and an API, enabling users to record their usability sessions and analyze fixation points through heatmaps.

Keywords: eye gaze tracking · user testing · user experience

1 Introduction

Usability and User eXperience (UX) have become used in most of the user-centered design methodologies [4] to understand and later improve the way that users interact with a system [18]. One of the well-known techniques is called "user testing", which is a technique that involves evaluating a system with real users to see how easy it is to use and how well it meets their needs [3].

The versatility of the technique enables the use of different kinds of configurations based on the specific goals of the research [37]. They can be divided between moderated or unmoderated [19], and can be applied in a physical environment like a usability laboratory [37] to perform in-person tests, or in a remote environment [34] using multiple testing tools [46,51]. They can also be applied

© The Author(s), under exclusive license to Springer Nature Switzerland AG 2023
M. C. Naldi and R. A. C. Bianchi (Eds.): BRACIS 2023, LNAI 14197, pp. 192–207, 2023.
https://doi.org/10.1007/978-3-031-45392-2_13

in one-on-one testing procedures, where one moderator works with one user at a time, or in a group where a moderator addresses multiple users simultaneously. To capture information through these different configurations, multiple tools can be used, such as eye gaze tracker systems [11], which are the object of the study of this paper.

Eye gaze trackers and their remote extensions, also known as "remote eye gaze trackers" (REGT), are devices capable of identifying a person's gaze fixation. These devices are widely used in usability testing. Generally, they are accompanied by native software that generates various metrics, such as fixation time, areas of interest, saccade speed, and gaze duration. However, these tools are expensive and require highly controlled spaces, typically found in usability facilities. Consequently, the possibility for regular users to perform remote tests using these tools is almost null.

The general objective of this study is to investigate, validate, and develop a software that can be used in usability studies with the purpose of capturing the eye gaze of a person and saving their fixation points to generate usability reports. The proposed system will be composed only of a computer and a conventional webcam, creating an open-source and low-cost project that can be used by individuals or groups, both in person or remotely.

Finding open-source contributions to this topic is a challenge, as highlighted by Shehy et al. [47]. Although the authors introduced several organizations and software options, we could not find a software that was still in active development, and incorporate both the Open API specifications and make use of web technologies that would allow easy integration. For this reason, we developed our own system to be published under open source license. Moreover, we also prepared the system to work standalone and also as a plugin for an open source organization called Remote User Experience Lab (RUXLAB)[1] with the objective to be integrated as a tool in their user testing system.

The development workflow to be implemented has to be capable of identifying the direction of the eye gaze based on the position of the iris and pupil in relation to the screen. Several machine learning methods have already been proposed in the literature to accomplish this task, such as convolutional neural networks [25], recurrent neural networks [22], adversarial generative networks [28], and siamese neural networks [2]. However, although these models are used with success in gaze estimation and other related learning problems, they require a large amount of training data, and are computationally expensive. For an open-source and low-cost system like the one being proposed, it is important to use a predictive algorithm that is simple and fast. Therefore, this study used a regression algorithm to predict the direction of the eye. The performance of this method or any other employed in eye tracking may be impacted by the calibration system. Therefore, we also investigated three different calibration techniques to analyze their effect on the system's performance.

---

[1] Remote User Experience LAB (RUXLAB). Available at: https://github.com/uramakilab/remote-usability-lab.

The calibration system is used to obtain personalized information from each user, according to the position of their head and also the level of brightness when they perform the calibration process. This information feed the model to improve the performance, giving better insights about the user environment and features. We also developed a graphical interface as a tool for users to easily use the model for analyzing and conducting their studies on their desired web pages.

The remainder of this study is organized as follows. Section 2 presents the context about user testing and usability alongside with the eye tracker systems. In Sect. 3, we explore existing research in the field and examine various eye tracker systems that were tagged as open source. Our proposed prototype is described in Sect. 4, where we segmented it into the gaze tracker system, the web implementation, and the API architecture. Finally, the conclusions and future work are presented in Sect. 5.

2   Context

The following section describes the topics related to conducting a user testing evaluation, focusing on eye tracker tools. We describe the methods and necessary steps to conduct such evaluations, as well as the main features that these systems must contain.

2.1   User Testing and Usability

The term usability was created to define the ease with which people interact with a tool or interface while performing a task [24]. It was defined by the ISO/CD 9241-11 as "the ability of software to be understood, learned, used under specific conditions"[2]. To evaluate usability, different techniques can be applied [36], which can be grouped into three different categories: *inspection methods*, *inquiry methods*, and *test evaluations*. User testing belongs to the *test evaluations* technique and it can be performed during product development to assess the effectiveness, speed, user error tolerance, ease of learning, and user engagement of the product. The goal is to ensure that users experience satisfaction instead of frustration when using the solution [3].

User Testing can be divided into three different steps or sections in which users have to complete different tasks based on provided information from the team of observers or evaluators. Those sections are the *pre-test*, *test phase*, and *post-test* sections. Inside the *pre-test* section, participants are typically presented with a consent form and a pre-test form to get insights, such as demographic information, about the user that will participate in the study. The *test phase* involves different tasks that the user must accomplish, which can be highly customized by the evaluators to gather as much qualitative or quantitative data as possible. Each task can have additional forms attached, known as pre-task or

---

[2] ISO/CD 9241-11: Ergonomics of human-system interaction - Part 11: Guidance on usability (1998). Available at: https://www.iso.org/standard/63500.html. Accessed date: September 15, 2023.

post-task forms, that are typically completed before or after the task is finished. Examples of these forms are the *Single Ease Question (SEQ)* [45] or the NASA Task Load Index (NASA TLX) [17]. Finally, in the *post-test* section, a post-test form such as *The System Usability Scale (SUS)* [7] can be shared.

Recruiting participants is another key challenge for *user testing*. It involves recruiting representations of the target user group for the product or service being tested. When a test is performed in physical facilities, the representative sample of the target group may not be heterogeneous enough, as highlighted by several authors [10,34,51]. After the participants are selected, a set of tasks are presented to them to test the product or service, while the evaluator observes and takes notes.

When the different tasks of a *user testing* are created, the evaluation team must decide what type of data needs to be collected. These decisions can influence the nature of the results, for example, whether they are qualitative or quantitative. They can also determine whether it is possible to measure or calculate the effectiveness, efficiency, or user satisfaction of the system [15]. To achieve this, a variety of configurations can be used, such as recording tools for capturing time, voice, screen activities, and video [10]. In the last case, we can distinguish between face recognition, which enables the application of Artificial Intelligence (AI) for emotion recognition [33], and gaze tracking, which allows for the implementation of eye-tracking procedures combined with heat map technologies [21].

2.2   Eye Tracker

The detection and recognition of human faces are extensively studied tasks in the area of AI, and their applications can be expanded in several use cases, such as security, entertainment, marketing, identification of people in social networks, forensic investigations, participant control in events, and many more [5,9,11,23, 33].

A face has very rich features that are essential for social interaction. While humans can effortlessly detect and interpret these characteristics, this task is not as simple for a computer [57]. Initially, it is necessary for the computer to understand which features are related to a human face and differentiate them from other objects that may have similar visual characteristics [23]. In most face detection algorithms, the first point to be identified are the eyes [57], as they are usually the simplest features to detect. Next, it follows other parts, such as the mouth, nose, eyebrows, and contours to increase confidence in the detected area being a face [31]. For better chances of success and to address a broad spectrum of human physiology and morphology [23], these models are usually trained with very large datasets of positive and negative images [47], since different factors can affect face detection, such as the pose, the presence or absence of structures (glasses and beard for example), facial expression, occlusion, orientation, and image quality [31].

Eye tracking systems aim to discover where the eyes are focused or study how they move when interacting with a graphical interface or environment [5]. This technology enables people with physical or cognitive disabilities [9] to perform activities using only their eyes and opens up a wide range of studies related to the usability of interfaces [44], where eyes can be analyzed to discover how users interact with websites, desktop, and mobile applications [35].

Eye tracking systems can be divided into multiple categories depending on the technology that they are based on, their final application, or their method of operation [47]. Some of them are more intrusive than others, and regarding this configuration, they can be divided into two categories: *(i) head-mounted* and *(ii) remote-based* eye trackers. The first category consists of wearable devices that track the movements through a system mounted on a headset like glasses that can be worn. While these systems are more intrusive compared to the remote option, they offer more freedom of movement to the user. On the other hand, *remote-based* are a non-intrusive method, as they capture eye movements from a distance. In this case, cameras or other sensors that track the movements are normally coupled into a layout or screen. In such cases, the main problems that can arise are related to a lower accuracy because they can be affected by light conditions, as well as a more time-consuming calibration process [9,29].

Some of the most used analysis tools in these types of studies are heatmaps, gaze replays, and gazeplots. However many other options exist, and a combination of them might help to find more accurate information [43].

(a) Heatmap.   (b) Gaze replay.   (c) Gazeplot.

Fig. 1. Examples of different output solutions from eye tracker systems.

- **Heatmaps** (Fig. 1a) are able to show areas where users fixed their gaze the most. They can be generated using information from a single person, or from all participants if the objective is to find the most fixed consensus point among all of them. In addition, different metrics can be chosen to generate hot and cold zones, such as the fixation time in an area or the number of fixations.
- **Gaze Replays** (Fig. 1b) enable researchers to watch the test session and see exactly where the user was looking. Usually, in this method, a cursor is placed on the screen to simulate the position of the user's eyes, while the recorded screen video plays in the background. This helps to evaluate task flow and determine exactly the order of each fixation.
- **Gazeplots** (Fig. 1c) show where the user is fixing its attention, which order was followed, and for how long the fixation occurred at that point. In most representations, the size of two circles represents the fixation time (larger circles indicate longer durations), and their position indicates the location of

the fixation on the canvas. The track represent the direction of the next fixing point, and they are typically numbered in sequential order.

There are different eye tracking devices on the market, such as TobiiPro[3] and PupilLabs Invisible[4]. These tools have multiple light sensors and cameras to record the user's gaze, and they come with native software that generates various metrics such as fixation time, areas of interest, saccade speed, and gaze duration. They also incorporate functionality for locating the eyes in the image and performing calibration processes.

3   Related Work

Usability tests with eye trackers have been studied since early 2000 ss. At the same time, AI started to become more affordable, and in the field of REGT, it contributed to the development of systems that do not need special hardware and are released as open source [38]. In this study, an open source organization called RUXLAB [8][5] has been investigated, which is dedicated to carrying out different usability tests, including user tests. This organization allows to implement different plugins to increase the functionalities that can be used when a user test is performed. For this reason, we studied different open-source REGT systems that can be reused instead of developing a new one.

We used the study performed by Shehu et al. [47] as a starting point as they present a general overview of different configurations in which REGT can be grouped. One of those presented configurations is based on mapping open-source REGT. We have conducted in-depth research to analyze which platforms are still available and also find new options that might have appeared. Table 1 summarises our finds. The first block represents the different organizations that were found by Shehu et al. [47], while the second block presents organizations that were found in another studies. We tried to identify critical aspects that correspond with our objectives. Specifically, we assessed whether the organization provided an open repository, if the code was available, also which kind of open source license they were using, if it contains Open API specifications, and also if the final result was an executable or not. Moreover, we have identified the technologies that were used during the development and the last updates that we were able to find in the code.

4   Prototype Results

We developed our prototype taking into account the main features encountered in the previous section. One of our objectives was to create a system that does

---

[3] Global leader in eye tracking for over 20 years - Tobii. Available at: https://www.tobii.com/. Accessed date: September 15, 2023.

[4] Pupil Invisible - Eye tracking glasses for the real world - Pupil Labs. Available at: https://pupil-labs.com/. Accessed date: September 15, 2023.

[5] Remote User Experience LAB. Available at: https://retlab.web.app/. Accessed date: September 15, 2023.

Table 1. List of Gaze Tracking Software.

Software	Repo.	License	API	Exe.	Technology	Last Update
iTracker [29]	Yes[a]	Custom	No	Yes	MATLAB	2021
RecurrentGaze [40]	Yes[b]	MIT	No	Yes	Python	2019
NNET [20]	-	-	No	Yes	C++, Python, Java	2012
EyeTab [55]	Yes[c]	MIT	No	Yes	Python, C, C++, Objetive-C	2014
Opengazer [58]	Yes[d]	GPLv2	No	Yes	C++	2010
TurkerGaze [16]	Yes[e]	MIT	No	No	JavaScript, HTML	2016
Camgaze	Yes[f]	-	No	Yes	Python	2013
ITU gaze tracker [1]	Yes[g]	GPLv2	No	Yes	C#	2014
CVC ET [14]	Yes[h]	GPLv2	No	Yes	C, C++	2016
xLabs [56]	Yes[i]	-	Yes	No	JavaScript	2018
GazePointer	Yes[j]	MIT	Yes	Yes	C#	2021
MyEye [30]	No	GPLv2	No	Yes	-	2009
NetGazer [13]	No	GPLv2	No	Yes	C++, C#	2009
OpenEyes [32]	Yes[k]	GPLv3	No	Yes	PHP, JavaScript	2022
Ogama [53]	Yes[l]	GPLv3	-	Yes	C++, C#	2022
GazeParser [50]	Yes[m]	-	No	Yes	Python, C++	2022
PyGaze [12]	Yes[n]	GPLv3	No	Yes	Python	2022
EyeTribe [26,39,52]	Yes[o]	BSD	Yes	Yes	Java, C# and C++	2016
Pupil Labs [27]	Yes	BSD	Yes	Yes	Python, Java, C# and C++	2016
RealEye [54]	No[p]	-	Yes	No	-	2023
Camgaze.js [48]	Yes[q]	-	No	No	Javascript	2019

[a] GazeCapture. Available at: https://github.com/CSAILVision/GazeCapture. Accessed date: September 15, 2023.
[b] Available at: https://github.com/crisie/RecurrentGaze. Accessed date: September 15, 2023.
[c] Available at: https://github.com/errollw/EyeTab. Accessed date: September 15, 2023.
[d] Available at: https://github.com/opengazer/OpenGazer. Accessed date: September 15, 2023.
[e] Available at: https://github.com/PrincetonVision/TurkerGaze. Accessed date: September 15, 2023.
[f] Available at: https://github.com/a20r/camgaze. Accessed date: September 15, 2023
[g] GazeTracker. Available at: https://github.com/devinbarry/GazeTracker. Accessed date: September 15, 2023.
[h] OpenGazer. Available at: https://github.com/tiendan/OpenGazer. Accessed date: September 15, 2023.
[i] Available at: https://github.com/xLabsgaze/xlabs-demo. Accessed date: September 15, 2023.
[j] GazePointer. Available at: https://github.com/MSREnable/GazePointer. Accessed date: September 15, 2023.
[k] Openeyes. Available at: https://github.com/AppertaFoundation/openeyes. Accessed date: September 15, 2023.
[l] Ogama. Available at: https://github.com/avosskuehler/ogama. Accessed date: September 15, 2023.
[m] Gazeparser. Available at: https://github.com/hsogo/gazeparser. Accessed date: September 15, 2023.
[n] PyGaze. Available at: https://github.com/esdalmaijer/PyGaze. Accessed date: September 15, 2023.
[o] Available at: https://github.com/eyetribe. Accessed date: September 15, 2023.
[p] Realeye. Available at: https://www.realeye.io/es/. Accessed date: September 15, 2023.
[q] Available at: https://github.com/a20r/camgaze.js/tree/master. Accessed date: September 15, 2023.

not rely on specific hardware, thereby avoiding dependencies. This means that the system should be compatible with conventional webcams that are available in the market at a low cost. Additionally, since we intended for the system to run on web browsers, we aimed to develop it using web technologies. Moreover, it has to comply with open source compliance as the system will be licensed by an open source license and able to be integrated with RUXLAB system. Finally, communication layer must be created using Open API[6] specification [6]. For this

---

[6] OpenAPI Specification - Version 3.0.3. Available at: https://swagger.io/specification/. Accessed date: September 15, 2023.

reason we created three different parts to be the most modular possible: (i) the gaze tracker with the calibration system with the use of an AI model (linear regression), (ii) the web application that allows to run tests and serves as the user interface, and (iii) the API that does all the communication between the parties and manages the created data using the Open API specification.

4.1 Gaze Tracker System

In order for the proposed model to accurately predict the direction in which the eyes focus, it is necessary to carry out a point calibration process [5,9]. Before starting a usability test on the platform, users need to undergo small exercises using their eyes. This step is essential before each session, as each user possesses unique physical characteristics, so it is possible to have a special analysis for each face instead of a pre-trained generic model with other people's faces. Environmental conditions, including factors such as luminosity, can also interfere with the access during eye tracking sessions [11]. Therefore, each session will be trained considering the specific environmental conditions in which the participant is situated.

The calibration process starts with the user positioning correctly within the camera coverage area, so that it is possible to capture the face and eyes without obstructions or poor lighting conditions. Once positioned, the user will go through the exercises, which consist of following instantiated points on the screen with the eyes. The user has to do his best to follow the moving circle that goes around the whole screen. While the user performs them, the X and Y position of the user's pupil and the X and Y position of the moving object are saved. Pupil points are captured using the *Face Landmarks Detection* model[7] from the Tensorflow.js package[8] which has a map of the entire face, including the center of the pupils. This pre-trained model is based on Media Pipe's FaceMesh package[9]. All points captured during this process are saved and stored along with the session. The calibration points are stored in a CSV file containing the entire relationship between the position of the points and the corresponding pupil position. This data is later used to train the machine-learning model.

The method chosen to perform this task was the linear regression [49], as it is ideal for cases where you need to predict a value based on the relationship between two or more variables. Moreover, it is a light and fast model, which suits for the scenario considered in this work, that involves low-cost cameras and multiple browsers. To implement the linear regression method we used the *LinearRegression* function from the scikit-learn library [42] with default

---

[7] TensorFlow.js: Face Landmarks Detection. Available at: https://github.com/tensorflow/tfjs-models/tree/master/face-landmarks-detection. Accessed date: September 15, 2023.

[8] Tensorflow Javascript. Available at: https://www.tensorflow.org/js. Accessed date: September 15, 2023.

[9] MediaPipe. Available at: https://developers.google.com/mediapipe. Accessed date: September 15, 2023.

parameters. We performed the experiments using a holdout validation with 80% of the data in the training set and 20% in the test set.

For the calibration process, we compared the three different techniques presented in Fig. 2. The *(a) 9-point coordinate system* (Fig. 2a) calibration system was implemented with a double forward and backward movement that increases the number of points to 36; the *N-point calibration*(Figure 2b), where the number of points is determined by the width and height relation of the screen, and finally a system based on the design of four concentric circles of varying sizes, presented in descending order from largest to smallest which uses 160 points (Fig. 2c) [9, 11].

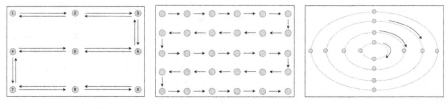

(a) 36 points calibration.   (b) N-points calibration.   (c) 160 points calibration.

Fig. 2. Calibration strategies.

We used the coefficient of determination and also the mean squared error to analyze the performance of the calibration strategies adopted. The formula for the coefficient of determination (R-squared) is given by:

$$R^2 = 1 - \frac{SS_{res}}{SS_{tot}} \tag{1}$$

where SS_{res} is the residual sum of squares and SS_{tot} is the total sum of squares. On the other hand, the formula for the mean squared error (MSE) is given by:

$$MSE = \frac{1}{n} \sum_{i=1}^{n} (y_i - \hat{y}_i)^2 \tag{2}$$

where y_i is the actual value of the dependent variable, $\hat{y}_i$ is the predicted value of the dependent variable, and n is the total number of observations.

Table 2 shows the different results for the three different techniques that were used. It can be observed that the *(c) 160 points calibration* technique obtained higher values for both X R^2 and Y R^2. However, regarding MSE, we obtained different results. For X *MSE*, the lowest value was achieved with the *(c) 160 points calibration* technique. However, for Y *MSE*, the lowest value was obtained with the *(a) 36 points calibration*.

The number of calibration points generated with the 160 points calibration technique is higher than the other two evaluated techniques. Therefore, with this technique, more data was provided for training the linear regression, which may have contributed to this learning model capturing more of the variability in the relationship between eye movements and screen coordinates, which resulted in a better performance in terms of R^2.

Table 2. Comparative table of Coefficient of Determination (R-squared) and Mean Squared Error (MSE) coefficients.

Technique	X MSE	X R^2	Y MSE	Y R^2
(a) 36 points calibration	0.027	0.76	0.11	0.52
(b) N-points calibration	0.120	0.38	0.27	0.49
(c) 160 points calibration	0.018	0.85	0.56	0.67

4.2 Web System

The proposed project has to run first standalone so it must be easily deployable into any environment. We have used Docker to package our build and to facilitate dependencies between users. Regarding dependencies, any browser can run the project as only will ask for webcam and screen recording permissions.

Functionalities implemented for the prototype include a landing page with a social authentication system, and a home page to perform create, read, update and delete operations, also known as CRUD. Additionally, it incorporates a calibration system, screen and webcam recording system, as well as the iris capture model integration.

(a) Calibration screen.   (b) Session evaluation screen.

Fig. 3. Screenshots from the Gaze module implemented.

For the website development, the Vuejs framework[10] was used along with the CSS Framework called Vuetify[11] which provides UI components following

---

[10] Vuejs. Available at: https://vuejs.org/ Accessed date: September 15, 2023.
[11] Vue Component Framework. Available at: https://vuetifyjs.com/. Accessed date: September 15, 2023.

Material Design guidelines as long as their icons[12]. For the recording of the screens were used the Native APIs for Javascript Media Devices[13] and Media Recorder[14]. For the Login feature, Google Sign-in using Firebase[15] is used as the authentication provider.

Therefore, another important part of the proposed Gaze Tracker system is the generation of heatmaps that we have used[16] to represent the focus of interest of a user on the studied page. This system needs to be integrated with an API, so that the calibration process, the usability session and the delivery of the results can be shown in a web application.

4.3   API

The API has a set of CRUD endpoints to manage sessions for each logged-in user. Table 3 contains the list of all routes defined in the API together with their descriptions and HTTP methods correspondents.

Those endpoints are used for the integration into the RUXLAB system but also might be useful for others that might use the present work, as it presents a decoupled layer of interconnectivity between a user testing evaluation and an eye tracker system. The final version of our code can be found in the same organization umbrella where is attached the RUXLAB in Github repository[17].

5   Conclusions and Future Work

In this paper, we have presented a complete architecture and system that enables to perform eye tracking sessions that work standalone. We have used a linear regression along with three different calibration procedures and developed a graphical user interface to display data using heatmaps with an API specification to facilitate communication between both systems.

One of the problems we encountered while analyzing previous solutions was that approximately 70% of the software applications were standalone executable programs with their own interfaces, whereas only 20% had a library or API that could allow integration. This topic has limited our research in terms of reusability, especially when we dive deep into the details of trying to use some

---

[12] Material Design. Available at: https://m2.material.io/design/guidelines-overview Accessed date: September 15, 2023.

[13] MediaDevices - Web APIs — MDN. Available at: https://developer.mozilla.org/es/docs/Web/API/MediaDevices. Accessed date: September 15, 2023.

[14] MediaRecorder - Web APIs — MDN. Available at: https://developer.mozilla.org/en-US/docs/Web/API/MediaRecorder. Accessed date: September 15, 2023.

[15] Firebase. Available at: https://firebase.google.com/. Accessed date: September 15, 2023.

[16] Dynamic heatmap for the web. Available at: https://www.patrick-wied.at/static/heatmapjs/. Accessed date: September 15, 2023.

[17] Available at: https://github.com/uramakilab/web-eye-tracker. Accessed date: September 15, 2023.

Table 3. API endpoints and their descriptions.

Method	Endpoint	Description
POST	/api/session	Saves a new test session. Receives the recorded video files and other additional information about the test in the request body
PATCH	/api/session/:id	Updates a session with new information. It is not possible to edit the videos, only the text information. The request body contains the fields to be updated and the session ID is sent in the route parameters
DELETE	/api/session/:id	Deletes a session by sending the session ID in the route parameters
GET	/api/user/sessions	Returns all sessions for the logged-in user
GET	/api/session/:id	Returns only the session whose ID is passed in the route parameters
GET	/api/session/results	Returns the results of the session, an array containing the prediction of the x and y coordinates
GET	/api/session/results/record	Returns the recorded video of the eye tracking session

of the identified items like *PupilLabs*, which was identified as having an open source code [27], but requiring specific hardware to run. A similar situation occurred with *EyeTribe*, as the project was acquired by Facebook in 2016 and later discontinued.

Moreover, Table 1 shows that around the 70% of the softwares identified as open source, have not received updates in the past three years. Moreover we were not able to find repositories, partly because the cited links in the bibliography encountered were broken [1,14,30,55,56,58] and also we do not find references or keywords to conduct a deep research. Regarding technologies, we found that most of them are based on Java, C, or Python, and around only 10% use web pure technologies like Javascript [41]. It has a direct correlation if the software is executable or not, being most of the applications or softwares developed using Java or C# as an executable. As we could not identify a solution that met our requirements, we implemented our own solution taking into consideration different topics found in the other systems like AI model, calibration process, and data visualization technique.

We have used different calibration systems based on Girshick et al. [16] for their trade-off between simplicity and performance, similar to the choice of the linear regression as an artificial algorithm core due to that kind of unsupervised appearance-based method have demonstrated their high effectiveness in other works [47]. The results obtained from the calibration method suggest that there is room for improvement in achieving a high-accuracy and high-precision system. Therefore, in future work, we intend to explore other techniques for the calibration process and alternative machine-learning models.

Still regarding the calibration method, different authors cited the importance of this step as it can help to greatly improve the final results [9]. We just implemented only three different models but we found gamification processes with better results than ours that also use linear regression [16]. In future work, we plan to study how different strategies can perform on different calibration methods to better analyze which configuration suits better for conventional webcam users. Moreover, as the system was build as a standalone module, we pretend to integrate it with RUXLAB to be used as a complementary tool to perform user testing evaluations.

Finally, regarding the visualization of the results, it was used a heatmap library to display that information. However, it exists other kinds of maps that can be created like the gaze-replay (Fig. 1b) and gazeplot (Fig. 1c), which might be implemented to improve the understanding of the results. Also, we started with the integration of the plugin into RUXLAB and we are working with the release of a new version that will allow the automatic generation of reports.

References

1. Agustin, J.S., et al.: Evaluation of a low-cost open-source gaze tracker. In: Proceedings of the 2010 Symposium on eye-tracking research & applications, pp. 77–80 (2010)
2. Ahn, H., Jeon, J., Ko, D., Gwak, J., Jeon, M.: Contactless real-time eye gaze-mapping system based on simple Siamese networks. Appl. Sci. **13**(9) (2023). https://doi.org/10.3390/app13095374
3. Bastien, J.C.: Usability testing: a review of some methodological and technical aspects of the method. Int. J. Med. Inform. **79**(4), e18–e23 (2010)
4. Bevan, N.: What is the difference between the purpose of usability and user experience evaluation methods. In: Proceedings of the Workshop UXEM. vol. 9, pp. 1–4. Citeseer (2009)
5. Biedert, R., Buscher, G., Dengel, A.: The eye book. Informatik-Spektrum **33**(3), 272–281 (2009)
6. Braunschweig, K., Eberius, J., Thiele, M., Lehner, W.: The state of open data. Limits Curr. Open Data Platforms **1**, 72–72 (2012)
7. Brooke, J., et al.: Sus-a quick and dirty usability scale. Usability Eval. Indust. **189**(194), 4–7 (1996)
8. Capdevila, M.G., Saltiveri, T.G.: Heurísticas de usabilidad utilizando una plataforma abierta y colaborativa. V Congreso Internacional de Ciencias de la Computación y Sistemas de Información 2021 (2022)
9. Carter, B.T., Luke, S.G.: Best practices in eye tracking research. Int. J. Psychophysiol. **155**, 49–62 (2020)
10. Castillo, J.C., Hartson, H.R., Hix, D.: Remote usability evaluation: can users report their own critical incidents? In: CHI 98 Conference Summary on Human Factors In Computing Systems, pp. 253–254 (1998)
11. Chennamma, H., Yuan, X.: A survey on eye-gaze tracking techniques. arXiv preprint arXiv:1312.6410 (2013)

12. Dalmaijer, E.S.: Pygaze: an open-source, cross-platform toolbox for minimal-effort programming of eyetracking experiments. Behav. Res. Methods **46**(4), 913–931 (2014)
13. Ferhat, Onur e Vilariño, F.: Rastreamento ocular de baixo custo: o panorama atual. Inteligência computacional e neurociência 2016 (2016)
14. Ferhat, O., Vilarino, F., Sánchez, F.J.: A cheap portable eye-tracker solution for common setups. J. Eye Movement Res. **7**(3) (2014)
15. Frøkjær, E., Hertzum, M., Hornbæk, K.: Measuring usability: are effectiveness, efficiency, and satisfaction really correlated? In: Proceedings of the SIGCHI conference on Human Factors in Computing Systems, pp. 345–352 (2000)
16. Girshick, R., Donahue, J., Darrell, T., Malik, J.: Rich feature hierarchies for accurate object detection and semantic segmentation. In: Proceedings of the IEEE Conference on Computer Vision and Pattern Recognition, pp. 580–587 (2014)
17. Hart, S.G.: Nasa-task load index (nasa-tlx); 20 years later. In: Proceedings of the human factors and ergonomics society annual meeting. vol. 50, pp. 904–908. Sage publications Sage CA: Los Angeles, CA (2006)
18. Hassan, H.M., Galal-Edeen, G.H.: From usability to user experience. In: 2017 International Conference on Intelligent Informatics and Biomedical Sciences (ICIIBMS), pp. 216–222. IEEE (2017)
19. Hertzum, M., Borlund, P., Kristoffersen, K.B.: What do thinking-aloud participants say? a comparison of moderated and unmoderated usability sessions. Int. J. Human-Comput. Interact. **31**(9), 557–570 (2015)
20. Holanda, Corey e Komogortsev, O.: Eye tracking em tablets comuns não modificados: desafios e soluções. In: Proceedings of the Symposium on Eye Tracking Research and Applications, pp. 277–280 (2012)
21. Holmqvist, K., Nyström, M., Mulvey, F.: Eye tracker data quality: What it is and how to measure it. In: Proceedings of the symposium on eye tracking research and applications, pp. 45–52 (2012)
22. Hwang, B.J., Chen, H.H., Hsieh, C.H., Huang, D.Y.: Gaze tracking based on concatenating spatial-temporal features. Sensors **22**(2), 545 (2022). https://doi.org/10.3390/s22020545
23. Jain, A.K., Li, S.Z.: Handbook of face recognition, vol. 1. Springer (2011)
24. Jordan, P.W.: An introduction to usability. CRC Press (1998)
25. Kanade, P., David, F., Kanade, S.: Convolutional neural networks (CNN) based eye-gaze tracking system using machine learning algorithm. Europ. J. Electr. Eng. Comput. Sci. **5**(2), 36–40 (2021)
26. Karlsson, H., Berglund, E., Larsson, J.: Method and apparatus for eye tracking (2014). https://patents.google.com/patent/US8723875B2/, the Eye Tribe Aps
27. Kassner, M., Patera, W., Bulling, A.: Pupil: An open source platform for pervasive eye tracking and mobile gaze-based interaction. In: Proceedings of the 2014 ACM International Joint Conference on Pervasive and Ubiquitous Computing: Adjunct-publication, pp. 1151–1160 (2014)
28. Kim, J.H., Jeong, J.W.: Gaze estimation in the dark with generative adversarial networks. In: ACM Symposium on Eye Tracking Research and Applications. ETRA '20 Adjunct, Association for Computing Machinery, New York, NY, USA (2020). https://doi.org/10.1145/3379157.3391654
29. Krafka, K., et al.: Eye tracking for everyone. In: Proceedings of the IEEE Conference on Computer Vision and Pattern Recognition, pp. 2176–2184 (2016)
30. Kuling, E., et al.: Myeye: an open-source wearable gaze tracker. In: Proceedings of the Symposium on Eye Tracking Research and Applications, pp. 1–10. ACM (2019)

31. Lee, S.: Understanding face detection with the viola-jones object detection framework. Towards data science (2020)
32. Li, D., Babcock, J., Parkhurst, D.J.: Openeyes: a low-cost head-mounted eye-tracking solution. In: Proceedings of the 2006 symposium on Eye tracking research & applications, pp. 95–100 (2006)
33. Lim, J.Z., Mountstephens, J., Teo, J.: Emotion recognition using eye-tracking: taxonomy, review and current challenges. Sensors **20**(8), 2384 (2020)
34. Madathil, K.C., Greenstein, J.S.: An investigation of the efficacy of collaborative virtual reality systems for moderated remote usability testing. Appl. Ergon. **65**, 501–514 (2017)
35. Manhartsberger, M., Zellhofer, N.: Eye tracking in usability research: What users really see. In: Usability Symposium. vol. 198, pp. 141–152 (2005)
36. Martins, A.I., Queirós, A., Silva, A.G., Rocha, N.P.: Usability evaluation methods: a systematic review. Human Factors Softw. Develop. Design 250–273 (2015)
37. Nielsen, J.: Usability laboratories. Behav. Inform. Technol. **13**(1–2), 3–8 (1994)
38. Nielsen, J., Pernice, K.: Eyetracking web usability. New Riders Publishing (2003)
39. Ooms, K., Dupont, L., Lapon, L., Popelka, S.: Accuracy and precision of fixation locations recorded with the low-cost eye tribe tracker in different experimental setups. J. Eye Movement Res. **8**(1) (2015)
40. Palmero, C., Selva, J., Bagheri, M.A., Escalera, S.: Recurrent CNN for 3D gaze estimation using appearance and shape cues (2018)
41. Papoutsaki, A., Laskey, J., Huang, J.: Searchgazer: Webcam eye tracking for remote studies of web search. In: Proceedings of the 2017 Conference On Conference Human Information Interaction and Retrieval, pp. 17–26 (2017)
42. Pedregosa, F., et al.: Scikit-learn: Machine learning in python. J. Mach. Learn. Res. **12**, 2825–2830 (2011)
43. Pernice, K., Nielsen, J.: How to conduct eyetracking studies. Nielsen Norman Group 945397498 (2009)
44. Poole, A., Ball, L.J.: Eye tracking in HCI and usability research. In: Encyclopedia of human computer interaction, pp. 211–219. IGI global (2006)
45. Sauro, J.: things to know about the single ease question (seq). Measuring U 2012 (2012)
46. Scholtz, J.: Adaptation of traditional usability testing methods for remote testing. In: Proceedings of the 34th Annual Hawaii International Conference on System Sciences, pp. 8-pp. IEEE (2001)
47. Shehu, I.S., Wang, Y., Athuman, A.M., Fu, X.: Remote eye gaze tracking research: a comparative evaluation on past and recent progress. Electronics **10**(24), 3165 (2021)
48. Sjöberg, A., Rominger, M.: Beyond hand-eye coordination: An exploration of eye-tracking and speech recognition as a navigation tool for interactive systems (2015)
49. Skodras, E., Kanas, V.G., Fakotakis, N.: On visual gaze tracking based on a single low cost camera. Signal Process. Image Commun. **36**, 29–42 (2015)
50. Sogo, H.: Gazeparser: an open-source and multiplatform library for low-cost eye tracking and analysis. Behav. Res. Methods **45**, 684–695 (2013)
51. Thompson, K.E., Rozanski, E.P., Haake, A.R.: Here, there, anywhere: remote usability testing that works. In: Proceedings of the 5th Conference on Information Technology Education, pp. 132–137 (2004)
52. Venugopal, D., Amudha, J., Jyotsna, C.: Developing an application using eye tracker. In: 2016 IEEE International Conference on Recent Trends in Electronics, Information & Communication Technology (RTEICT), pp. 1518–1522. IEEE (2016)

53. Voßkühler, A., Nordmeier, V., Kuchinke, L., Jacobs, A.M.: Ogama (open gaze and mouse analyzer): open-source software designed to analyze eye and mouse movements in slideshow study designs. Behav. Res. Methods **40**, 1150–1162 (2008)

54. Wisiecka, K., et al.: Comparison of webcam and remote eye tracking. In: 2022 Symposium on Eye Tracking Research and Applications, pp. 1–7 (2022)

55. Wood, E., Bulling, A.: Eyetab: Model-based gaze estimation on unmodified tablet computers. In: Proceedings of ETRA (2014). http://www.cl.cam.ac.uk/research/rainbow/projects/eyetab/

56. Zhang, M., Bulling, A.: Xlabs: A platform for rapid design, prototyping and evaluation of ubiquitous gaze interfaces. In: Proceedings of the Symposium on Eye Tracking Research and Applications, pp. 69–76. ACM (2018)

57. Zhao, W., Chellappa, R., Phillips, P.J., Rosenfeld, A.: Face recognition: a literature survey. ACM Comput. Surv. (CSUR) **35**(4), 399–458 (2003)

58. Zielinski, P.: Opengazer: open-source gaze tracker for ordinary webcams. Samsung and The Gatsby Charitable Foundation. http://www.inference.phy.cam.ac.uk/opengazer (2007)

Language and Models

Who Killed the Winograd Schema Challenge?

Hugo Neri and Fabio G. Cozman$^{(\boxtimes)}$

Universidade de São Paulo, São Paulo, Brazil
fgcozman@usp.br

Abstract. In which we investigate the technical issues surrounding the defeat, or perhaps the sudden assassination, of the Winograd Schema Challenge. We argue that, while the obvious suspect is the WinoGrande-based solution, the real cause of death was the masked language modeling technique for learning large language models. The Winograd Schema Challenge was, in the end, just a test for masked language closure, and as such it was killed by the use of this technique at scale.

1 Introduction

The Winograd Schema Challenge has been met; in fact, one might say that it has been defeated or, perhaps more dramatically, that it has had a violent death. This is a sad ending for a challenge once hailed as a leading alternative to the Turing test [24–27].

A very detailed discussion of the defeat of the Winograd Schema Challenge (WSC) has been produced by Kocijan et al. [21]. It seems that very little can be added when it comes to describing the tools for the WSC. The "autopsy" by Kocijan et al. also offers valuable insight as to the reasons why the WSC did not resist when attacked, and the reasons why the WSC was perhaps not a very good test for commonsense reasoning and for overall intelligence in the first place.

Still, it is fair to ask: What exactly killed the WSC? Was it too weak that it in fact died of some common disease while still young? Was it really murdered? If so, who did it, and what was the murder weapon? By reading commentary around the WSC, one feels that there is no consensus on the cause of death. How could a leading substitute for the venerable Turing test die suddenly with so little understanding of the event?

Two previous references deserve attention.

The most prominent one is the already mentioned analysis by Kocijan et al. [21], whose authors have produced substantial contributions to the WSC [20,22]. They seem to argue, and cite several colleagues who seem to agree with them, that the problem with the WSC was that the challenge was weaker than expected in a number of ways: schemas were harder to build than expected and were less meaningful than originally thought; schemas were more constrained than planned and subjected to large scale correlations that led to death in the hands of large language models such as BERT and RoBERTa. In short, they argue that the WSC was weak and that any specific challenge may be weak in capturing commonsense reasoning [21]. This is a far-reaching position, in which the large language model RoBERTa killed the challenge in a context where the challenge was so weak that it almost did not deserve to live. Hence, there is no prison time for RoBERTa, but not much credit for it either.

Another analysis can be found in the original paper on the WinoGrande corpus, the corpus that was used to refine RoBERTa so as to obtain human performance [42]. Indeed, one might argue that the pair RoBERTa + WinoGrande was the murderer. Or one might argue that the WinoGrande corpus alone was the murderer, on the grounds that the real defeat came only when researchers realized that more data, even of low quality, could be used to refine a large language model to victory over the WSC. In any case, the authors of the Wino-Grande corpus seem to believe that biases in corpora let language models excel in the WSC. For those authors, the Achilles heel of the WSC was that, when building a corpus with questions and answers, human subjects introduce biases that, once captured by large language models, lead to the correct resolution of Winograd schemas. They seem to argue that, if one could produce a very large unbiased corpus with Winograd schemas, then the WSC with respect to that corpus would require commonsense. So the problem was not with the challenge, but with the difficulty in building the required corpus. (Note that such difficulty was deemed by Kocijan et al. to be a defect of the WSC, so there are divergent opinions even on the meaning of the difficulty.) In any case, the position by the authors of WinoGrande seems to be that the WSC is not dead but it may be in a frozen state to be resuscitated when we find a difficult enough corpus.

We offer a different analysis. We agree with others that the RoBERTa language model, and even the WinoGrande corpus, were key elements in this story; however, we think previous analysis has not been able to explain why is it that the WSC succumbed so easily and what exactly killed it — and we believe we have a credible answer to these questions. To develop our argument, we first describe the main elements of the WSC in Sect. 2 and then we focus on RoBERTa and the WinoGrande corpus in Sect. 3. Our solution to the puzzle surrounding the death of the WSC is elaborated in Sects. 4, 5, and 6. Our closing arguments are given in Sect. 7.

2   The Winograd Schema Challenge

The WSC consists of a set of Winograd Schemas, where each schema is a pronoun disambiguation problem such as:

> The city councilmen refused the demonstrators a permit because they
> FEARED/ADVOCATED FOR violence. Who are "they"?

A Winograd Schema contains a pronoun that refers to one of two entities (in the example, city councilmen/demonstrators); the reference depends on a particular word. Each schema can be instantiated in two ways by selecting this particular word (in the example, FEARED/ADVOCATED). Then the resulting anaphora must be solved for each sentence. The average human performance in the challenge ranges from 92.1% [3] to 94% [43]. Winograd schemas are devised to display no strong statistical correlations between words; hence, knowledge, and more importantly, commonsense knowledge should be needed to solve them [10].

Hector Levesque, who proposed the WSC, submitted to the AI community, at IJCAI-16, when the challenge ran as an actual competition, that: "Doing this [solving the challenge] correctly appears to rely on having a solid base of commonsense knowledge and the ability to reason intelligently with that knowledge" [24]. In fact, in the original paper about the WSC the following ambitious statements were given by Levesque [24]:

> The claim of this paper in its strongest form might be this: with a very high probability, *anything that answers correctly a series of these questions . . . is thinking in the full-bodied sense we usually reserve for people.* To defend this claim, however, we would have to defend a philosophical position that Turing sought to avoid with his original Turing Test. So like Turing, it is best to make a weaker claim: with a very high probability, *anything that answers correctly is engaging in behaviour that we would say shows thinking in people.* Whether or not a subject that passes the test is really and truly thinking is the philosophical question that Turing sidesteps. [Emphases added]

We may thus distinguish two different claims in Levesque's proposal:

Strong (Ontological) claim Anything that solves the WSC is thinking in the full-bodied sense we usually reserve for people.
Weak (Pragmatic) claim Anything that solves the WSC is engaging in behaviour that we would say shows thinking in people.

In any case, the idea behind the WSC is to test the ability to answer commonsense questions as related to sentence comprehension. The research community followed this lead with various degrees of enthusiasm, taking the WSC as a new type of Turing test [2, 15, 32–34, 51].

Machines performed poorly up to 2017. That suddenly changed with language models, esp. GPT-2 (Generative Pre-trainedTransformer) [38], BERT (Bidirectional Encoder Representations from Transformers) [12], and RoBERTa (Robustly optimized BERT approach) [29]. Figure 1 shows the astonishing evolution in accuracy. Today we have (extremely) large language models that can solve a variety of linguistic tasks with high accuracy [1, 46].

We can say that the defeat of the WSC took place by the end of 2019 [44]. Interestingly, high expectations were assigned to the WSC until its sudden defeat. For instance, Marcus and Davis insisted in their book for a non-specialized audience, in 2019, that the WSC is "one of the most challenging tests for machines that is currently available" [30].

The literature on the WSC is relatively small after the WinoGrande solution was widely disseminated in 2020 [42]. Variants of the WSC have been proposed, and various applications have been tried — for instance, to test gender bias or to

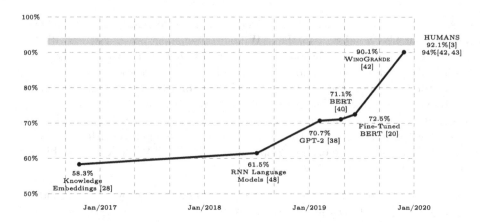

Fig. 1. Accuracy of systems on the Winograd Challenge [6].

require explanations. Such work is reviewed with exceptional acumen by Kocijan et al. [21], so we do not repeat here what they have done in a definite fashion.

3   LLMs, RoBERTa and WinoGrande: Main Suspects

A language model usually yields probabilities over tokens of a language, given other tokens [18]. Several language models resort to Markov chains and similar probabilistic tools; recently, large language models have been built by training deep neural networks with properly designed architectures. There has been a shift up in the performance with the introduction of Large Language Models (LLMs) based on the transformers architecture [50]. In short, a transformer is a neural network model that uses attention as it main mechanism; because this mechanism lends itself to parallelization, learning techniques can rely on larger training sets. Besides parallelization, the self-attention mechanism encodes contextual information.[1] The most popular LLMs, such as BERT, T5, GPT, are transformers. Transformer-based LLMs frequently achieve state-of-the-art results in tasks like question answering and sentiment analysis [17,49].

LLMs rose to center stage in 2018 after Trihn and Le [48] employed RNN-based models. Even though their models' performance was relatively low (63.7%), they optimistically claimed that they had a good indication their model had a "good grasp of commonsense knowledge" [48]. They accepted that the better the performance in the challenge the more confidently one can say the model carries commonsense knowledge.

Significant evolution on the WSC was observed after Thihn and Le's work, as one can see in Fig. 1. An spectacular gain in performance was then reported

---

[1] The *self-attention* mechanism guarantees that long-distance context has "equal opportunity" to show up. When it comes to an anaphora resolution, the self-attention mechanism tackles it at its core.

with the publication of the WinoGrande corpus and associated solution for the WSC (first in arXiv [44] and then in the AAAI-2020 conference [42]). At that point the WSC was defeated.

WinoGrande is a dataset of schemas inspired by the original Winograd schemas. The full version WinoGrande contained 43,972 problems while the debiased version contained 12,282 problems.

The WinoGrande paper reported at the AAAI-2020 attained 90.1% accuracy in the Winograd Schema Challenge [43]. The content of that AAAI-2020 paper was originally uploaded in November 2019 on arXiv [42]. However, there was a first version of this paper uploaded to arXiv on July 2019 which presented a significant lower performance, 77.6%, still the state-of-art at that moment [44]. Basically, the difference in performance was produced by the RoBERTa fine-tuned with WinoGrande in the second version, whereas the first version adopts BERT fine-tuned with WinoGrande. Comparing both versions, the sole reason for the performance difference is the adoption of RoBERTa in the final version in contrast to BERT in the first version.

The Winogrande team argued that the performance of existing systems could be due to the "extent to which spurious effects are prevalent in existing datasets, which run the risk of overestimating the true capabilities of machine intelligence on commonsense reasoning" [44]. The initial motivation for WinoGrande focused on dealing with biases that might exist in the hand-crafted Winograd schema and that might thus unfairly help existing solutions. Such biases might be of two types [47]: (a) language-based and (b) dataset-specific biases. In fact, Trichelair et al. observed that at least 37 sentences in the WSC273 dataset (13.6%) are conceptually rather easy due to language associations [47]. For instance, take

In the storm, the tree fell down and crashed through the roof of my house. Now, I have to get it REPAIRED/REMOVED.

and note that trees are removed more often than repaired, while roofs are repaired more often than removed. On the other hand, 131 sentences (out of 273) were then found to yield meaningful examples even if candidates in the sentence were switched. This was called "swichtability" [47]. As an example, take

Bob collapsed on the sidewalk. Soon he saw Carl coming to help. He was very ILL/CONCERNED.

In this sentence, Bob and Carl can be switched to obtain an equivalent example with the opposite answers. Such schemas were called "switchable". Trichelair et al. [2018] encouraged future researchers to additionally report results on the switchable dataset (when the candidates are switched, and when they are not).

The WinoGrande team also discussed dataset-specific biases, such as annotation artifacts or spurious correlations in crowdsourced datasets.

Their solution was to develop an unbiased dataset (to be robust against both types of biases discussed above [44]), so presenting "problems that are more challenging by reducing such biases, while also scaling to a significantly larger number of problems (273 to 44k) by crowdsourcing" [44].

In the second version of the WinoGrande paper, the team's declared ambitions with the new dataset increased. The dataset would allow a "true estimation of the machine commonsense capabilities ... with 44k problems that are inspired by the original design of WSC, but modified to improve both the scale and hardness of the problems" [42, 43]. Their goal was then to minimize this bias in the original Winograd Schema. The key steps in WinoGrande construction consisted of (1) a carefully designed crowdsourcing procedure, followed by (2) a novel algorithm AFLITE that generalizes human-detectable biases based on word occurrences to machine-detectable biases based on embedding occurrences. The key motivation for this procedure was that it is difficult for humans to write problems without accidentally inserting unwanted biases.

After training BERT [44] and RoBERTa [42] in the WinoGrande-All dataset and achieving good results, the team was somewhat skeptical about performance. They questioned whether neural language models successfully acquired commonsense or it was just an *overestimation* of the true capabilities of machine commonsense:

> the potential overestimation leads to another crucial question regarding potential unwanted biases that the large-scale neural language models might be exploiting, essentially solving the problems right, but for wrong reasons. ... While such biases and annotation artifacts are not apparent for individual instances, they get introduced in the dataset as problem authors subconsciously repeat similar problem-crafting strategies.

To proceed, we must examine one key element in BERT (and RoBERTa) training; that is, masked language modeling. We will then be able to understand the way commonsense is absorbed by LLMs.

4   BERT and Cloze Procedures

The Bidirectional Encoder Representations from Transformers, BERT for short, is an LLM that handles contextual information in the pretraining phase. Notable previous efforts can be cited: semi-supervised sequence learning was developed in 2015 [9],[2] and well designed embeddings (ELMo) appeared in 2017 [36]. The problem with those efforts is that they only used either left context or right context of a word, but language understanding is *bidirectional*, as indicated in decades of psychological studies, particularly in *Gestalt Theory*, and in philosophical discussion dating back to Frege [16]. The main idea there is that meaning lies in a sentence and not in single words. For a century now, experimenters have been reporting findings that may be interpreted as showing that language behavior depends on total context:

> The results indicate that the ability to identify, learn, recognize, remember or produce any language "symbol" (element or pattern) depends heavily

---

[2] As a digression, we note that Quoc V. Le, the second author of this paper, tackled the WSC in one of the first successful approaches using language models [48].

on the variable degrees to which it is associated with everything else by larger and meaningful (familiar) overall combinations. [31]

The authors of BERT proposed a solution for bidirectional language learning, where they masked out k% words during training.

The authors of BERT explicitly mentioned the psychological test called *cloze test*, also known as *cloze procedure*, in their commentary about BERT.[3] In our analysis, this is an essential character in the death of the WSC, so it pays to examine cloze procedures in more detail.

A cloze procedure is a psychological tool for measuring the effectiveness of communication, introduced in 1953 by Wilson Taylor [45]. Such a procedure is based on *cloze units*, in which the word "cloze" derives from the concept of *closure*. Gestalt psychology defines the latter concept as the human tendency to complete a familiar but not-quite-finished pattern — for instance, when seeing a broken circle as a whole one. In other words, "closure" refers to our natural tendency to mentally close gaps. The same tendency applies to language; for instance, given "Chickens cackle and — quack" almost anyone can instantly supply "ducks" [45]. In a cloze procedure, if the word uttered by a human subject is really the same as the word omitted, the person scores one cloze unit for correctly closing the gap in the language pattern.

It should be clear that a cloze procedure is really similar to masked language modeling as performed when learning BERT and similar large language models.

For our investigation to proceed, it makes sense to examine in more detail the definition of linguistic commonsense:

> [T]he sentence pattern is a complex one made up of many subpatterns. *One must know not only the meanings* (i.e., patterns of symbol- meaning relationships) *and forms* (patterns of letters) *of all the five words, but also the meanings of given combinations* of them – plus the fact that the sentence structure seems to demand a term parallel to "cackle" but associated with ducks instead of chickens. In other words, one must guess what the mutilated sentence means as a whole, then complete its pattern to fit that whole meaning. [Emphases added] [45]

Any method that intercepts a message from a "transmitter" (writer or speaker), mutilates it, and administers it to "receivers" (readers or listeners) that then attempt to fill the missing parts, is a method that potentially yields a number of cloze units. As Taylor argued, different persons may express the same meaning in somewhat differing ways, and the same language patterns may have differing meanings for different people.

The total context of a language behavior, such as the one involved in the resolution of a cloze unit, includes everything that tends to motivate, guide, assist or hinder that behavior. As noted by Miller, all such factors are combined into a notion of commonsense:

---

[3] In an interview, Jacob Devlin has commmented on the cloze test as being a fairly known test in psycholinguistics for accessing levels of readership ability.

"I heard a — bark" is likely to elicit "dog" both because that word is habitually associated with "bark" and because it fits in with past experience with noisy dogs. If the verbal context is enlarged to "For the first time, I heard a — bark," the impulse to supply "dog" may be *reduced by commonsense*; the subject may ask himself: "Who is this guy that has never heard a dog? Could he be referring to some other animal?" And if the preceding sentence has mentioned a voyage to the Pribilof Islands, the reader may draw on past knowledge to produce "seal." [31]

We emphasize the expression "reduced by commonsense" as it reflects exactly what researchers supporting the WSC argued for. Habits of expression take over most of the work of translating an individual's meaning into an organized series of language symbols for transmission to others. Likewise, habits of reading or listening cause the reader or listener to anticipate words, almost automatically, when receiving messages.

In many ways, language masked modeling is already discussed approvingly in connection with cloze procedures by Taylor:

How can a random system play fair when some words are easier to replace than others? Obviously, one is more likely to be able to supply "an" in "A quinidine is — alkaloid isomeric ..." than to guess "$6,425" in "The city council voted — for a new swimming pool." Yet the former example is far more difficult reading. The answer is that if enough words are struck out at random, the blanks will come to represent proportionately all kinds of words to the extent that they occur. [45]

As a digression, we note that both BERT's masked language modeling and cloze procedures have to decide on how many words to delete. BERT was trained using 15% of words randomly masked out, a number based on empirical performance. It was then noted that deleting too few words leads to too much training effort, while deleting too many words may eliminate too much context.

In short: BERT's pretraining applies a sort of cloze procedure at scale; we may refer to BERT's pretraining as going through a large number of "masked language closure" steps. In a sense, BERT's training is teaching a machine to solve a linguistic commonsense task. As Winograd schemas are actually similar to cloze units, it turns out that BERT's designed training clearly address Winograd schemas. We emphasize the connection with cloze units in the next section.

5   Textual Entailment, and Schemas as Cloze Units

The explicit linguistic problem to be solved given a Winograd schema is anaphora resolution. However, Winograd schemas are also variants of another task, *Recognizing Textual Entailment* (RTE) [4,8,41]. RTE asks a machine to recognize what is the *Textual Entailment* (TE) from a given sentence. A Textual Entailment is a linguistic phenomenon that appears everywhere in daily life communication; it expresses the relationship between sentences or fragments of texts in which one implies the other. For instance, "John bought a novel yesterday," textually entails, "John bought a book." Dagan et al. have proposed a simple yet well-accepted definition of Textual Entailment: T textually entails H if and only if a human reading T would typically infer that H is most probably true [7].

As we can see here, solving a RTE problem qualifies as a test for implicit causes in both human understanding and perception. Winograd schema inherit such a feature. More accurately, 94.6% of all existing Winograd schemas have this structure according to our own manual analysis.

However, recognizing the correct direction of causation requires a prior understanding of what the entities in the text mean. Consider the following example. The sentence "Ferrous sulfate heptahydrate is green" (T) textually entails "$FeSO_47H_20$ is green" (H) [23], which is a valid entailment because "Ferrous sulfate heptahydrate' and "$FeSO_47H_20$" refer to the same entity in the world in philosophical jargon. Although this entailment looks, at first sight, a synonym problem, a human reader cannot tell whether T entails H by working with words alone. A piece of appropriate background knowledge is required.

This example has the same structure as the Morning Star - Evening Star puzzle that Gottlob Frege posed a century ago with respect to the difference between meaning (*Bedeutung*) and reference (*Sinn*) [16]. Knowing that both stars are the same astronomical object (Venus) required new empirical information. As the philosopher of language W.V. Quine put it, "astronomical observation was needed, and not mere reflection on meanings, to determine the sameness of the entity in question" [37]. This sort of difficulty appears in the " — barks" example discussed before in connection with cloze procedures.

Not surprisingly, research in RTE has arrived at a definition that takes into account appropriate background knowledge:

> A text T textually entails a hypothesis H relative to a group of end-users G just in case, typically, a member of G reading T would be justified in inferring the proposition expressed by H from the proposition expressed by T. [23]

The challenge here lies in the high level of dependency on contextualized information within a community. Even though the intuition behind RTE seems correct as a test for commonsense reasoning, it is hard to come up with a well-delimited test in which knowledge representation is not about general issues.

Indeed, Levesque did not take RTE as the best way of testing commonsense because there is an *indefinite number of valid and correct consequences* that may follow from a sentence within RTE [27]. The problem of inferential causality [35] is unsurmountable as a practical test of machine intelligence. As an alternative, he proposed the WSC as a test that is not only feasible but also measurable.

Levesque's innovation is the addition of a double ambiguity: a pronominal anaphora plus a coreferential ambiguity. For instance:

(information) The city councilmen denied the demonstrators a permit.
(ambiguity) Because they FEARED/ADVOCATED FOR violence.

220   H. Neri and F. G. Cozman

The personal pronoun "they" may refer to either the city councilmen or the demonstrators. If we ask "who feared violence?" the answer is "the city councilmen". Conversely, if we ask "who advocated for violence," the answer is "the demonstrators". Therefore, the coreferential ambiguity is based on a pair of special words that considerably shift the meaning of the sentence.

We can see that the WSC, in an attempt to move away from the complexities of RTE, arrived at a modified form of cloze test. First of all, the blank to be replaced, in a Winograd schema, by either one of the pair of keywords is just a mask with only one correct answer. For instance, we have:

> The city councilmen denied the demonstrators a permit. Because they [MASK] violence.

To make the connection with anaphora resolution even more evident, suppose the pronoum that has to be disambiguated is also masked. We would end up with two masks as follows:

> The city councilmen denied the demonstrators a permit. Because [MASK] [MASK] violence.

In this particular example we have a little bit more than 15% of masked tokens, and we have preserved enough context to a good guessing from the machine. (As a digression, note that, by limiting the flexibility of placeholders, Winograd schemas introduce some sort of bias, a social desirability bias — the requirement that nouns are mapped correctly to keywords so as to infer what the designers of the test take as the correct and obvious alternative.)

We must now examine what is it that RoBERTa brought to the plot, and verify the effect of the WinoGrande corpus, so as to understand the role played by those suspects.

6   A Second Look at RoBERTa and WinoGrande

The large language model RoBERTa is a refined version of BERT. As indicated by the authors of RoBERTa, its improved pretraining, with more corpora and better hyperparameters, led to significance performance gains across many tasks [29]. Besides BookCorpus [52] and the English Wikipedia (the original data used to train BERT), RoBERTa was trained with the English portion of the CC-NEWS dataset,[4] OpenWebText,[5] and, finally, and most notably, STORIES, a dataset introduced by Trinh and Le [48] containing a subset of CommonCrawl data *filtered exactly* to match the style of Winograd schemas.

From the second version of the WinoGrande paper [42], we have that the fine-tuning of RoBERTa with DPR (Definite Pronoun Resolution Dataset) achieved 83.1% accuracy in the WSC. DPR was a fairly common used dataset for fine-tuning models for solving the Winograd Schema that introduced 1886 additional

---

[4] At https://commoncrawl.org/2016/10/news-dataset-available/.
[5] At http://Skylion007.github.io/OpenWebTextCorpus.

Winograd schemas authored by 30 undergraduate students. That result was already a state-of-art performance for the WSC. And then the fine-tuning of RoBERTa with WinoGrande, as we know, achieved 90.1% accuracy in the WSC.

Indeed, RoBERTa + DPR already produced excellent results. For instance, consider the PDP Pronoun Disambiguation Problems) dataset. This dataset consists of 80 pronoun disambiguation problems, all formulated through multiple choice questions, in which a pronoun must be resolved to one of up to 5 (but mostly binary) options. This is clearly related to the WSC, and often used as proxy for the WSC itself. For the PDP, RoBERTa + DPR and RoBERTa + WinoGrande had close results, 86.3% and 87.5% respectively. We can see similar results for other datasets in Table 1.

Table 1. Comparison between DPR and WinoGrande fine-tuned RoBERTa. Note that SuperGLUE-WSC is discussed at https://super.gluebenchmark.com/tasks/.

DATASET	RoBERTa + WinoGrande	RoBERTa + DPR
WSC [24]	90.1%	83.1%
PDP [11]	87.5%	86.3%
SuperGLUE-WSC	85.6%	83.6%
DPR [39]	93.1%	91.7%
KnowRef [14]	85.6%	84.2%
COPA [19]	90.6%	86.4%

The authors of the WinoGrande corpus briefly discussed those points:

Our model is based on RoBERTa fine-tuned with WinoGrande (train and dev sets). To compare different corpora used as a resource, we also fine-tune RoBERTa on DPR (train and test sets). For hyper parameter search, we use the same grid search strategy as in Sect 4. Overall, RoBERTa fine-tuned on WinoGrande helps improve the accuracy on all the related tasks (Table 6), and performs consistently better than when RoBERTa is fine-tuned on DPR. While improvements on some related datasets (particularly WSC, PDP, and DPR) might seem expected, the significant improvement on COPA is not so. [42]

Intuitively, the larger the dataset, the better; the WinoGrande corpus is valuable because it is a large one. One should expect the performance of RoBERTa to increase by adding 40k+ Winograd Schemas.

Our goal through this discussion is to suggest that RoBERTa was a key character in our plot, due to the masked learning modeling that lies beneath it, while WinoGrande was an important element of the plot but one that simply added learning material to the underlying infrastructure.

Before we leave this discussion, we must comment on the analysis by Elazar et al. [13]. They argue that the perceived progress in the WSC is due to flawed evaluation, artifacts, and commonsense knowledge gleaned from a supervised training set rather than advancements in LLMs; clearly their analysis emphasizes features of the WSC that are not aligned with the analysis in this paper. While it is undeniable that flawed evaluation and artifacts have influenced the perceived progress in the WSC, evidence suggests that the contribution of pre-trained LLMs is significant and should not be disregarded. Pre-trained LLMs indeed play a significant role in improving word sense disambiguation and substantial progress has been observed through BERT to RoBERTa (and note that BERT alone got 64.9% accuracy in the WSC, while RoBERTa alone reached 79.1% accuracy). Further evidence supporting this claim is the change in performance across different versions of the WinoGrande paper: despite minimal changes in the WinoGrande dataset, Sakaguchi et al.'s BERT-based model achieved 77.6% in July 2019, and with the availability of RoBERTa in November 2019, performance jumped to 90.1% [42, 44].

To conclude this section, we note that we have focused on the events that took place around 2019/2020, when the WSC was defeated. LLMs have had an explosive evolution since them; masked-modeling training is certainly not the only technique that deserves attention now. While gap-filling (cloze) training aligns with the human reading experience, where we move forward and backward through text, models such as GPT predict the next token using solely previous tokens. The latter strategy is similar to everyday human conversation, where information is exchanged linearly over time, and participants cannot go forward to examine tokens yet to be uttered. Given the surprising success in question answering and dialogue by GPT-like models, one might thus think that there are other mechanisms that may defeat the WSC, perhaps provided that one can use really large amounts of data and really large models. But even an extraordinarily large model such as GPT-3 [5] does not lead to spectacular performance in the WSC: our testing led to mere 67% accuracy. Things improve for the even larger GPT-4 [1], where we then got impressive 85.2% accuracy — but we do not really any access to that LLM, so we do not know whether it has been trained with even larger sets of Winograd-like sentences. More investigation is required, and we leave the matter to future investigation.

7   Closing (Clozing?) Arguments

The Winograd Schema Challenge was killed because the cloze-style training that is used to produce many recent language models is perfectly tailored to solve Winograd schemas. Both RoBERTa and WinoGrande were important in this plot, and RoBERTa in particular was a key character as it carried the critical information needed for the killing.

But in fact the WSC was not murdered by a single language model or a corpus. Rather, the WSC died because it was supposedly designed to capture commonsense but in fact it only offered a task that is in essence a version of cloze procedures, a task that masked language modeling, as employed in large language models such as BERT, is specifically designed to address.

Acknowledgements. The first author was supported by FAPESP through grant 2018/0968-1. The second author was partially supported by CNPq through grant 305753/2022-3. This work was carried out at the Center for Artificial Intelligence (C4AI-USP), with support by FAPESP (grant 2019/07665-4) and by the IBM Corporation.

References

1. Open AI. GPT4 Technical Report. arXiv:2303.08774, 2023
2. Bailey, D., Harrison, A., Lierler, Y., Lifschitz, V., Michael, J.: The winograd schema challenge and reasoning about correlation. In: Working Notes of the Symposium on Logical Formalizations of Commonsense Reasoning (2015)
3. Bender, D.: Establishing a human baseline for the Winograd schema challenge. In: Modern AI and Cognitive Science Conference, pp. 39–45 (2015)
4. Bobrow, D.: Precision-focussed textual inference. In: Proceedings of the Workshop on Textual Entailment and Paraphrasing ACL, Prague (2007)
5. Brown, T.B., et al.: Language models are few-shot learners. arXiv:2005.14165 (2020)
6. Cozman, F.G., Neri, H.: Some thoughts on knowledge-enhanced machine learning. Int. J. Approximate Reasoning **136**, 308–324 (2020)
7. Dagan, I.: Recognizing textual entailment: Rational, evaluation and approaches. Natural Lang. Eng. **15**(4), i-xvii (2009)
8. Dagan, I., Glickman, O., Magnini, B.: The PASCAL Recognising Textual Entailment Challenge. In: Quiñonero-Candela, J., Dagan, I., Magnini, B., d'Alché-Buc, F. (eds.) MLCW 2005. LNCS (LNAI), vol. 3944, pp. 177–190. Springer, Heidelberg (2006). https://doi.org/10.1007/11736790_9
9. Dai, A.M., Le, V.Q.: Semi-supervised sequence learning. In: C. Cortes, N. Lawrence, D. Lee, M. Sugiyama, and R. Garnett, eds, Advances in Neural Information Processing Systems, vol. 28. Curran Associates Inc (2015)
10. Davies, E.: Winograd schemas and machine translation. arXiv:1608.01884 (2016)
11. Davis, E., Morgenstern, L., Ortiz, C.L.: The first Winograd Schema Challenge at IJCAI-16. AI Mag. **38**(3), 97–98 (2017)
12. Devlin, J., Chang, M.-W., Lee, K., Toutanova, K.: BERT: pre-training of deep bidirectional transformers for language understanding. arXiv:1810.04805v2 (2019)
13. Elazar, Y., Zhang, H., Goldberg, Y., Roth, D.: Back to square one: artifact detection, training and commonsense disentanglement in the Winograd schema. In: Proceedings of the 2021 Conference on Empirical Methods in Natural Language Processing, pages 10486–10500. Association for Computational Linguistics (2021)
14. Emami, A., et al.: The KnowRef coreference corpus: removing gender and number cues for difficult pronominal anaphora resolution. In Proceedings of the 57th Annual Meeting of the Association for Computational Linguistics, pp. 3952–3961, Florence, Italy, July 2019. Association for Computational Linguistics (2019)
15. Emami, A., Trischler, A., Suleman, K., Chi, J., Cheung, K.: A generalized knowledge hunting framework for the Winograd Schema Challenge. In: NAACL-HLT 2018: Student Research Workshop, pp. 25–31 (2018)
16. Frege, G.: Sense and reference. Philos. Rev. **57** (1948)
17. Joshi, B., Shah, N., Barbieri, F., Leonardo Neves, L.: The devil is in the details: evaluating limitations of transformer-based methods for granular tasks. In: Proceedings of the 28th International Conference on Computational Linguistics, pp. 3652–3659, Barcelona, Spain (Online). International Committee on Computational Linguistics (2020)

18. Jurafsky, D., Martin, J.H.: Speech and Language Processing (3rd ed. draft) (2023)
19. Kavumba, P., Inoue, N., Heinzerling, B., Singh, K., Reisert, P., Kentaro Inui, K.: When choosing plausible alternatives, Clever Hans can be clever. In: Proceedings of the First Workshop on Commonsense Inference in Natural Language Processing, pp. 33–42, Hong Kong, China. Association for Computational Linguistics (2019)
20. Kocijan, V., Cretu, A.-M., Camburu, O.-M., Yordanov, Y., Lukasiewicz, T.: A surprisingly robust trick for the Winograd scheme challenge. In: Annual Meeting of the Association for Computational Linguistics, pp. 4837–4842 (2019)
21. Kocijan, V., Davis, E., Lukasiewicz, T., Marcus, G., Leora Morgenstern, L.: The defeat of the Winograd Schema Challenge. arXiv:2201.02387 (2023)
22. Kocijan, V., Lukasiewicz, T., Davis, E., Marcus, G.: A review of Winograd Schema Challenge datasets and approaches. arXiv:2004.13831v1 (2020)
23. Korman, D.: Defining textual entailment. J. Assoc. Inf. Sci. Technol. **69** (2018)
24. Levesque, H.: The winograd schema challenge. In: AAAI (2011)
25. Levesque, H.: On our best behaviour. In: IJCAI (2013)
26. Levesque, H.: Common Sense, the Turing Test, and the Quest for Real AI. The MIT Press (2017)
27. Levesque, H., Davis, E., Morgenstern, L.: The Winograd Schema Challenge. Knowledge Representation (2012)
28. Liu, Q., Jiang, H., Ling, H.-Z., Zhu, X,. Wei, S., Hu, Y.: Commonsense knowledge enhanced embeddings for solving pronoun disambiguation problems in Winograd schemes challenge. arXiv:1611.04146 (2016)
29. Yinhan Liu, Y., et al.: RoBERTa: a robustly optimized BERT pretraining approach. arXiv:1907.11692 (2019)
30. Marcus, G., Davis, E.: Rebooting AI: Building Artificial Intelligence We Can Trust. Pantheon (2019)
31. George, A.: Müller. Language and Communication, McGraw-Hili (1951)
32. Nicos, I., Michael Loizos, M.: Tackling the Winograd schema Challenge through machine logical inferences. STAIRS **75** (2016)
33. Nicos, I., Michael Loizos, M.: How the availability of training material affects performance in the Winograd Schema Challenge (2017)
34. Nicos, I., Michael Loizos, M.: A data-driven metric of hardness for WSC sentences. GCAI-2018 (EPiC Series in Computing) **55**, 107–120 (2018)
35. Judea Pearl, J.: Causality: Models, Reasoning, and Inference. Cambridge University Press, 2ª edition (2009)
36. Peters, M.E., Neumann, M., Iyyer, M, Gardner, M., Kenton Lee, K., Luke Zettlemoyer, L.: Deep contextualized word representations, Christopher Clark (2018)
37. Quine. Two dogmas of empiricism. Philos. Rev. **60** (1951)
38. Radford, A., Wu, J., Child, R., Luan, D., Amodei, D., Sutskever, I.: Language models are unsupervised multitask learners. Technical Report 8, OpenAI Blog (2019)
39. Rahman, A., Ng, V.: Resolving complex cases of definite pronouns: The Winograd schema challenge. In: Proceedings of the 2012 Joint Conference on Empirical Methods in Natural Language Processing and Computational Natural Language Learning, pp. 777–789, Jeju Island, Korea (2012). Association for Computational Linguistics
40. Ruan, Y.-P., Zhu, X., Ling, Z.-H., Shi, Z., Liu, Q., Wei, S.: Exploring unsupervised pretraining and sentence structure modelling for Winograd schemes challenge. arXiv:1904.09705 (2019)
41. Rus, V.: A study of textual entailment. Int. J. Art. Intell. Tools **17** (2007)

42. Sakaguchi, K., Bras, R.L., Bhagavatula, C., Choi, Y.: Winogrande: an adversarial Winograd Schema Challenge at scale. arXiv:1907.10641v2 (2019)
43. Sakaguchi, K., Bras, R.L., Bhagavatula, C., Choi, Y.: Winogrande: an adversarial Winograd Schema Challenge at scale. AAAI-20 Technical Tracks **34**(05) (2019)
44. Sakaguchi, K., Bras, R.L., Bhagavatula, C., Yejin Choi, Y.: Winogrande: an adversarial Winograd Schema Challenge at scale. arXiv:1907.10641v1 (2019)
45. Taylor, W.: Cloze procedure: a new tool for measuring readability. J. Quartely Fall (1953)
46. Hugo Touvron, H., et al.: LLaMA: open and efficient foundation language models. Technical report, arXiv:2302.13971 (2023)
47. Trichelair, P., et al.: On the evaluation of common-sense reasoning in natural language understanding. arXiv preprint arXiv:1811.01778 (20180
48. Trinh, T., Quoc Le, Q.: A simple method for commonsense reasoning. arXiv:1806.02847 (2018)
49. van Aken, B., Winter, B., Löser, A., Felix, A.: Gers. How does BERT answer questions? a layer-wise analysis of transformer representations. In: Association for Computing Machinery, editor, Proceedings of the 28th ACM International Conference on Information and Knowledge Management, CIKM '19, pp. 1823–1832 (2019)
50. Vaswani, A., et al.: Attention is all you need. In: 31st Conference on Neural Information Processing Systems (NIPS 2017), Long Beach, CA, USA (2017)
51. Zhang, H., Song, Y.: A distributed solution for Winograd Schema Challenge. In: ICMLC2018 (2018)
52. Zhu, Y, et al.: Aligning books and movies: towards story-like visual explanations by watching movies and reading books. In: The IEEE International Conference on Computer Vision (ICCV) (2015)

Sabiá: Portuguese Large Language Models

Ramon Pires[(✉)] , Hugo Abonizio , Thales Sales Almeida ,
and Rodrigo Nogueira

Maritaca AI, Campinas, Brazil
{ramon,hugo,thales,rodrigo}@maritaca.ai

Abstract. As the capabilities of language models continue to advance, it is conceivable that "one-size-fits-all" model will remain as the main paradigm. For instance, given the vast number of languages worldwide, many of which are low-resource, the prevalent practice is to pretrain a single model on multiple languages. In this paper, we add to the growing body of evidence that challenges this practice, demonstrating that monolingual pretraining on the target language significantly improves models already extensively trained on diverse corpora. More specifically, we further pretrain GPT-J and LLaMA models on Portuguese texts using 3% or less of their original pretraining budget. Few-shot evaluations on *Poeta*, a suite of 14 Portuguese datasets, reveal that our models outperform English-centric and multilingual counterparts by a significant margin. Our best model, Sabiá-65B, performs on par with GPT-3.5-turbo. By evaluating on datasets originally conceived in the target language as well as translated ones, we study the impact of language-specific pretraining in terms of 1) capturing linguistic nuances and structures inherent to the target language, and 2) enriching the model's knowledge about a domain or culture. Our results indicate that most benefits stem from the domain-specific knowledge acquired through monolingual pretraining. Finally, we show that our optimized model for Portuguese demonstrates a reduced performance in English tasks, thereby substantiating the inherent compromise in refining models for specific linguistic domains.

1 Introduction

Language Models have revolutionized the field of natural language processing with their exceptional ability to perform tasks with minimal supervision. Although primarily pretrained on English-centric corpora, the models have shown impressive multilingual capabilities [10]. Given the abundance of languages worldwide, the majority of which are low-resource, it has become a common practice to pretrain single models on multiple languages simultaneously. Models like XLM-R [12], mBART [28], mT5 [70], and BLOOM [54] exemplify this approach.

Despite the success of these multilingual models, we argue that they may not be the optimal approach for capturing the cultural and knowledge richness inherent in individual languages. When a moderately-sized language-specific corpus

M. C. Naldi and R. A. C. Bianchi (Eds.): BRACIS 2023, LNAI 14197, pp. 226–240, 2023.
https://doi.org/10.1007/978-3-031-45392-2_15

is available, continued pretraining could integrate the missing knowledge into the model, enhancing its performance on targeted tasks. To test this hypothesis, we extend the pretraining of English-centric models using Portuguese corpora and evaluate their performance on an extensive range of Portuguese datasets employing a few-shot learning approach. Our results indicate that, even for models trained beyond the recommendations by Hoffmann et al [18], this additional pretraining considerably improves performance compared to multilingual models.

We evaluate our models on datasets comprising texts originally created by native Brazilian Portuguese speakers, as well as datasets translated from English to Portuguese. We observe improvements across all datasets due to the Portuguese pretraining, with the gains being particularly pronounced for datasets created by Brazilian speakers. One of the largest improvements was observed on the ENEM dataset [57], which is derived from entrance exams used by Brazilian universities and requires extensive knowledge of the country's history, geography, and literature. This result provides evidence that the major contribution of our language-specific pretraining is to inject domain-specific knowledge about a particular culture as opposed to solely enhancing language proficiency.

2   Related Work

The success of multilingual pretraining has been well-documented in the literature, with models such as ByT5 [69], mT5 [70], XLM-R [12], XGLM [27] and mGPT [56] paving the way for more inclusive language understanding and generation by leveraging shared knowledge across multiple languages. However, there are limitations to this approach.

BLOOM, a 175B-parameter model pretrained on 46 languages, performs worse on English tasks compared to OPT [74], a similarly sized model pretrained on English-centric corpora using comparable computational resources and data size. We conjecture that BLOOM's underperformance may be attributed to its relatively limited exposure to English tokens during the pretraining phase. Consequently, this observation suggests that monolingual pretraining could offer supplementary advantages.

In support of this hypothesis, models with hundreds of millions of parameters pretrained on monolingual texts have demonstrated gains over multilingual counterparts [2,6–8,21,24,25,32,36,52,59]. Additionally, research has indicated that language adaptation is beneficial even for low-resource languages [4,13,38,72]. However, there is a limited number of published research articles with comprehensive evaluations of the benefits of continued pretraining at the multi-billion-parameter scale [22,50,73]. Through this study, we contribute to the literature by demonstrating the effectiveness of continued language-specific pretraining for Portuguese language models up to the 65B-parameter scale.

The question concerning whether it is advantageous to train models for specific languages is closely associated with the question of whether it is beneficial to train models for particular domains of knowledge. Recent studies, such as

Minerva [26] and Galactica [62], have shown that domain-specific pretraining can lead to significant improvements, even with a smaller pretraining corpus compared to large-scale, general-purpose pretraining corpora. Analogously, Fu et al. [15] demonstrated the feasibility of specializing smaller models to perform multi-step reasoning, a capability typically exclusive to models with at least 50B parameters, at the expense of diminished performance in other, more general tasks.

Pretraining with a combination of general and domain-specific corpora can potentially enhance performance in specialized tasks without compromising effectiveness in general-purpose tasks, albeit at the cost of increased computational demands. For example, BloombergGPT [68], a 50B-parameter model pretrained on heterogeneous corpus in which more than half of texts are from the financial domain, exhibits comparable performance to OPT-66B in general tasks. However, BloombergGPT's pretraining dataset is three times larger, and consequently used more computational resources.

Rather than pursuing a single model that performs well across multiple domains, Gururangan et al. [17] propose an alternative approach: using multiple expert models, each trained on a domain-specific subset within a broader, diverse dataset, to function as a single general-purpose model. Their models outperform dense ones across various domain-specific tasks, at the expense of an increased parameter count, consequently leading to larger memory requirements for efficient inference.[1]

3   Methodology

In this section, we outline the pretraining data and training details used to build our models, including data sources, preprocessing techniques, architectures, hyperparameters, and optimization methods.

3.1   Pretraining Data

The pretraining data is derived from the Portuguese subset of the ClueWeb 2022 dataset [40,41]. To increase the datasets's quality, we apply the quality filters from MassiveText [45], modifying them to accommodate the specific requirements of the Portuguese language. We normalize the text with $ftfy$[2], convert wikitexts into human-readable texts, and exclude documents containing less than 200 unique tokens.

These quality filters are primarily designed for web pages and may not seamlessly transfer to other domains. There is potential for improvement by employing more automated methods; however, this study did not explore such approaches due to the resource-intensive nature of pretraining experiments.

---

[1] To serve their ensemble with a low latency, the weights for each expert must be kept in GPU memory.

[2] $ftfy$ normalization fixes *mojibakes* and remove remnant HTML tags.

Following the cleaning process, all documents are concatenated using an end-of-sequence token as a separator, and then tokenized. The GPT-J tokenizer, which is identical to the GPT-2 tokenizer [44], produces 7.8 billion tokens, while the LLaMA tokenizer produces 7.3 billion tokens. The discrepancy in the total number of tokens is primarily due to the different tokenization strategies each model employs, byte-level BPE and BPE based on sentencepiece [23], respectively along with the variation of the vocabularies used by each tokenizer.

We extended the training of three models - LLaMA 7B and 65B [63] as well as GPT-J [66] - originally trained on English-centric corpora, on Portuguese texts; these further pretrained models from LLaMA are denoted as Sabiá, while the one derived from GPT-J is referred to as Sabiá-J.[3]

3.2   Sabiá Models

The LLaMA 7B and 65B models are decoder-only Transformer models [64] with a similar architecture to PALM's [10]. The models were trained using a causal language modeling objective on a massive dataset sourced from webpages, code, books, and scientific papers. The 7B model was trained on 1 trillion tokens and the 65B model was trained on 1.4 trillion tokens. While the majority of the corpus is in English, it also includes an unspecified amount of Portuguese text.

Starting from the LLaMA weights, we train the Sabiá models on our Portuguese dataset (see Sect. 3.1) using the t5x and seqio frameworks [48]. Adhering closely to the hyperparameters used by PALM, we use the AdaFactor optimizer [55] without factorization, a first-order momentum $\beta_1 = 0.9$, and a second-order momentum $\beta_2 = 1 - k^{-0.8}$, where k represents the step number. We apply global norm clipping at 1.0 and dynamic weight decay of lr^2, with lr denoting the current learning rate.

Besides the standard causal language modeling loss, we use an auxiliary loss of $10^{-4} \log^2(\sum_i e^{z_i})$, where z are the logits, to decrease the likelihood of loss spikes at the 65B-parameter scale. The learning rate is linearly increased from 0 to 1e-3 over the initial 1,000 steps, followed by a constant learning rate of 1e-3 for an additional 9,000 steps.

The models were trained on a TPU v2-512, using batches of 512 sequences, each containing 2048 tokens. We utilized gradient checkpointing, also known as rematerialization, to enable the use of larger batches, thereby increasing TPU utilization. For the 7B model, this configuration results in a throughput of 124,000 tokens/sec, corresponding to a Model FLOPs Utilization (MFU) [10] of 45.2%, excluding the self-attention operations. For the 65B model, we achieve a throughput of 14,000 tokens/sec, resulting in an MFU of 47.4%.

The resulting models were trained on a total of 10.4 billion tokens, or 1.52 epochs of the Portuguese dataset. This equals to 10,000 training steps, which is the same amount used to train Sabiá. We noticed improvements in few-shot tasks

---

[3] Sabiá is a tribute to the eponymous bird, renowned for its diverse and intricate vocalizations.

beyond one epoch, which corroborates results from Taylor et al. [62]. However, due to the high costs of pretraining, we did not continue training.[4]

3.3   Sabiá-J

The GPT-J model is a 6B-parameter decoder-only Transformer model whose architecture and training hyperparameters closely follow GPT-3 6.7B. The main differences reside on computing the MLP and self-attention in parallel, applying attention head with dimension 256 (twice larger than GPT-3 6.7B), and using Rotary Positional Embedding (RoPE) [61]. GPT-J was trained on 400B tokens from The Pile dataset [16], whose 97.4% tokens are in English.

We begin training Sabiá-J from the released GPT-J checkpoint,[5] using the `mesh-transformer-jax` framework [65] and AdamW optimizer [30] with a weight decay of 0.1. We start the pretraining by warming up the learning rate until 1.2e-5 over 13,500 steps, followed by a cosine annealing decay during 135,518 steps until the end learning rate of 2.4e-6, and kept it constant from there on. We train on a TPU v3-8 using an effective batch size of 32 sequences of 2048 tokens. This results in a throughput of 5,200 tokens/sec, corresponding to a MFU of 44.5% without self-attention. The model was trained for 18 d on 7.8B tokens, or one epoch of the Portuguese dataset.[6]

4   Evaluation on Poeta

We evaluate the Sabiá models on the Portuguese Evaluation Tasks (Poeta) benchmark, which comprises 14 downstream NLP datasets in Portuguese: ASSIN 2 RTE and STS [47], ENEM Challenge [57], ENEM 2022 [37], FaQuAD [53], TweetSentBr [5], AG News [75], IMDB [31], MASSIVE [14], MKQA [29], BoolQ [11], SST2 [58], WSC [33], and BLUEX [1]. Half of them (ASSIN 2 RTE and STS, BLUEX, ENEM Challenge, ENEM 2022, FaQuAD, and TweetSentBr) were originally written in Portuguese, and the remaining ones were either manually or automatically translated into Portuguese from their originals in English. We refer to the first group as "Native" datasets and the second group as "Translated" datasets.[7]

The models were evaluated in a few-shot manner using the maximum number of examples that fits into a 2048-token context for each task. We used the *GPT-2* tokenizer as a reference because it results in more tokens. This allowed us to comfortably fit prompts tokenized with other tokenizers.

---

[4] Considering the on-demand pricing of 384 USD per hour for a TPU v2-512, pretraining Sabiá-7B and Sabiá-65B costs approximately 9,000 and 80,000 USD, respectively.

[5] https://huggingface.co/EleutherAI/gpt-j-6b.

[6] Due to constraints in our hardware budget, this model was trained with fewer tokens compared to Sabiá.

[7] The MASSIVE dataset underwent manual translation and localization; however, given that the original text was composed in English, it has been categorized as a translated dataset.

Table 1. A summary of the datasets constituting the Poeta benchmark.

Dataset	Type	Preferred Metric	Rand. Score	Transl.	Avg Len (chars)	Num Train	Num Test	Num Few-shot
AG News	Multiclass classification (4)	Accuracy	25	Yes	282.34	120,000 (110,953)	7,600	12
ASSIN 2 RTE	Binary classification	F1	50	No	139.99	6,500	2448	18
ASSIN 2 STS	Regression	Pearson	0	No	139.99	6,500	2448	15
BLUEX	Multiple choice (4)	Accuracy	25	No	1,228.08	-	178	1
BoolQ	Binary classification	Accuracy	50	Yes	562.30	9,427 (7,015)	3,270	4
ENEM Challenge	Multiple choice (5)	Accuracy	20	No	1,286.68	-	916	1
ENEM 2022	Multiple choice (5)	Accuracy	20	No	1,170.24	-	118	1
FaQuAD	Extractive QA	F1	0	No	1,056.47	-	63	4
IMDB	Binary classification	Accuracy	50	Yes	1,114.56	25,000 (18,613)	25,000	2
MASSIVE	Multiclass classification (18)	F1-macro	0.58	Yes	68.35	11,514	2,974	36
MKQA	Extractive QA	F1	0	Yes	80.32	-	10,000 (6,758)	40
SST2	Binary classification	Accuracy	50	Yes	84.19	67,349	872	34
TweetSentBR	Multiclass classification (3)	F1-macro	32.4	No	93.32	12,990	2010	30
WSC	Binary classification	Accuracy	50	Yes	102.15	-	285	18

To evaluate the models, we manually select a set of few-shot examples for each dataset on Poeta. Depending on the dataset, these examples are balanced by class (except for FaQuAD, BLUEX, ENEM Challenge, ENEM 2022, MKQA, and WSC). For each test example, the prompts are built with the selected few-shot examples in alternating order. Each task on Poeta has a particular instruction that is placed at the beginning of the prompt.

Following Srivastava et al. [60], we adopt the Normalized Preferred Metric (NPM) as our primary evaluation measure:

$$\text{NPM} = \frac{1}{N} \sum_{i=1}^{N} 100 \times \frac{[\texttt{raw preferred metric}]_i - [\texttt{random score}]_i}{[\texttt{high score}]_i - [\texttt{random score}]_i} \quad (1)$$

where N is the number of evaluation datasets, $[\texttt{raw preferred metric}]_i$ is the score obtained by the model on the i-th dataset, $[\texttt{random score}]_i$ is the score of a random model (e.g., 50% for a binary classification task) and $[\texttt{high score}]_i$ is the highest possible score on that dataset, which is either 1 or 100. The preferred metric and random score for each dataset are presented in Table 1. The rationale behind employing NPM rather than a straightforward average across all datasets is to mitigate the undue influence of datasets with inherently high scores, such as binary classification datasets, which could otherwise outweigh datasets characterized by lower scores.

5   Results

The main results can be found in Table 2. Models such as BLOOMZ, XGLM and Bertin-GPT struggled to generate answers in Portuguese. To address this issue, we adopted an approach akin to that used by the XGLM authors: by calculating the likelihood of each candidate answer string based on the input text and subsequently selecting the class with the highest probability. For FaQuAD, the only dataset in the benchmark without predetermined candidate answers, we allowed the models to generate answers in their original format.

We observe that the LLaMA baselines significantly outperform models of equivalent size trained with fewer tokens, such as Galactica and OPT. Furthermore, despite being trained on English-centric corpora, LLaMA-7B surpasses multilingual BLOOM and XGLM of similar sizes. The Sabiá models demonstrate considerable improvement in NPM compared to their respective baseline models. These NPM gains are more substantial for the smaller Sabiá-J and Sabiá-7B models. Notably, Sabiá-65B marginally outperforms OpenAI's GPT-3.5-turbo, which serves as the base model for ChatGPT.

Table 2. Few-shot NPM results on the Poeta benchmark.

	Native	Translated	All
GALACTICA-6.7B	2.2	13.6	7.9
OPT-6.7B	5.3	39.7	22.5
OPT-66B	16.4	47.1	31.7
BERTIN-GPT	5.8	42.5	24.2
BLOOM-7.1B	10.6	44.2	27.4
BLOOMZ-7.1B	18.3	44.7	31.5
XGLM-7.5B	14.0	46.9	30.4
GPT-3.5-turbo	67.9	66.0	67.0
GPT-4	78.8	82.5	80.6
GPT-J	10.2	33.9	22.0
Sabiá-J	25.0	43.1	34.0
LLaMA-7B	20.2	45.8	33.0
Sabiá-7B	43.4	53.6	48.5
LLaMA-65B	59.1	68.4	63.7
Sabiá-65B	69.2	69.6	69.4

Through our Portuguese pretraining, we observed that the improvement in NPM was higher in native datasets than that in translated datasets. For Sabiá-65B, improvements over LLaMA-65B were mostly from the native subset. We hypothesize that this is due to the "mechanistic" nature of translated datasets: since they were translated from English, the baseline model already possesses the knowledge needed to solve them and gains little from learning the linguistic, syntactic, and grammatical knowledge of the target language. For instance, to answer the question *"does p o box come before street address"* (BoolQ dataset), the model gains little from additional pretraining on a Portuguese corpus as it is unlikely that the corpus would provide new information regarding the formatting of US mailing addresses that the model has not already encountered during its initial English-centric pretraining. Conversely, language-specific pretraining introduces the specific knowledge required to solve tasks in the native subset.

Although GPT-J exhibited lower few-shot performance in English tasks relative to LLaMA, we use it in this study to illustrate that not only highly optimized models like LLaMA can benefit from extended pretraining. We chose not to use BLOOM-7.1B as our initial checkpoint for pretraining due to its inferior performance compared to GPT-J in preliminary few-shot experiments on three Portuguese datasets. However, we later discovered that its performance on Poeta surpassed GPT-J's. Nonetheless, BLOOM still exhibits lower performance compared to LLaMA.

Analogous to Sabiá-J, BERTIN-GPT is a model pretrained on Spanish text starting from the GPT-J weights. Since Spanish and Portuguese are similar languages, it is reasonable to expect that BERTIN-GPT would perform better than its baseline model. Nevertheless, the observed NPM for BERTIN-GPT is only slightly higher than GPT-J's.

A noteworthy comparison involves Galactica, a model pretrained on scientific text, predominantly in English, and a similarly-sized OPT model, which utilized comparable pretraining compute but was pretrained on a larger and more diverse English-centric corpus. In their study, the authors demonstrate that Galactica performs on par with OPT on English tasks and largely outperforms OPT on scientific-related tasks. Conversely, OPT significantly outperforms Galactica in Portuguese tasks. This result underscores the trade-offs associated with domain-specific specialization, which often entails diminished performance in other tasks.

BLOOMZ [35], a multilingual instruction-tuned model, demonstrated superior performance compared to its baseline BLOOM model, rivaling LLaMA of equivalent size.[8] Nevertheless, our approach of pretraining in Portuguese appears to yield superior results, as Sabiá-J surpasses BLOOMZ despite originating from a lower-performing baseline model. We envision continued pretraining and instruction tuning as complementary techniques to be combined in future research.

5.1 Results per Dataset

Table 3 presents the results per Poeta dataset for Sabiá models, their baselines, and for the supervised state-of-the-art. The SOTA results reported for the translated datasets were obtained using their original English versions [46,51,71,76]. Since the Poeta benchmark excludes unanswerable examples of the MKQA dataset, we decided not to include the SOTA result for this dataset.

In more challenging datasets, such as ENEM Challenge, ENEM 2022, and BLUEX, which are derived from admission exams to Brazilian universities, we see the most significant gains due to language-specific pretraining. Substantial improvements are also observed in TweetSentBr, a dataset containing tweets with an abundance of slang and references to Brazilian popular culture. We hypothesize that this pretraining imparts specific knowledge about the country's culture, literature, and geography that is less frequently encountered and learned during the original pretraining with more diverse texts.

---

[8] This model was used in the experiments: https://huggingface.co/bigscience/bloomz-7b1-mt.

Table 3. Results per dataset. [1] [49]; [2] [9]; [3] [34]; [4] [3]; [5] [71]; [6] [76]; [7] [46]; [8] [51].

		Native							Translated						
	Avg	ASSIN 2 RTE (F1)	ASSIN 2 STS (Pearson)	BLUEX (Acc)	ENEM (Acc)	ENEM 2022 (Acc)	FaQuAD (F1)	TweetSentBr (F1-macro)	AG News (Acc)	BoolQ (Acc)	IMDB (Acc)	MASSIVE (F1-macro)	MKQA (F1)	SST2 (Acc)	WSC (Acc)
SOTA supervised	-	92.07[1]	86.00[2]	-	-	-	82.40[3]	77.27[4]	95.55[5]	92.40[6]	96.21[6]	-	-	97.50[7]	90.10[8]
GPT-4	64.99	90.96	77.58	76.40	92.00	79.66	64.74	82.40	93.50	86.50	97.00	83.30	55.67	97.50	92.63
GPT-3.5-turbo	76.08	88.28	66.41	60.11	80.57	75.42	78.28	74.59	87.71	71.43	84.86	84.19	44.92	91.71	76.84
Galactica-6.7B	34.11	34.92	11.63	28.65	20.74	22.88	40.16	21.96	38.33	57.13	51.08	36.62	3.01	62.27	49.12
Bertin-GPT-6B	45.18	33.24	6.23	22.47	20.52	23.03	64.00	35.52	82.44	44.25	87.66	55.46	15.56	61.08	62.11
OPT-6.7B	43.35	43.35	21.35	24.16	19.87	20.34	56.45	14.37	55.67	61.31	90.42	51.84	13.64	56.47	47.72
OPT-66B	49.68	65.66	7.88	29.78	20.41	17.80	71.12	32.54	81.87	58.75	92.66	61.64	21.17	57.50	46.67
BLOOM-7.1B	47.01	50.32	12.16	25.84	20.85	17.08	72.67	28.12	79.48	60.43	89.80	56.23	15.72	43.83	48.77
BLOOMZ-7.1B	50.94	33.57	24.50	34.27	25.38	27.12	79.90	50.36	83.82	58.25	93.80	55.31	12.36	86.95	64.56
XGLM-7.5B	48.79	53.75	15.07	24.16	19.10	19.49	44.84	63.23	77.47	49.76	91.46	59.74	13.72	89.11	62.11
GPT-J	43.51	54.88	17.86	24.72	20.85	20.54	59.52	30.98	64.15	48.75	72.68	55.67	10.69	83.94	54.04
Sabiá-J	52.84	35.49	22.97	39.89	39.41	36.44	69.28	64.18	64.30	51.53	90.86	58.92	15.84	87.16	45.51
LLAMA-7B	51.30	56.82	7.59	32.02	29.04	23.73	77.38	44.19	76.94	57.37	86.92	59.90	30.08	88.76	47.72
Sabiá-7B	62.43	64.87	13.63	47.75	60.59	60.17	77.43	67.17	83.28	64.07	92.70	68.95	31.98	90.60	50.88
LLAMA-65B	73.84	74.98	62.85	53.93	75.00	62.71	87.25	68.05	88.01	73.12	94.98	79.71	48.34	94.27	71.58
Sabiá-65B	77.65	88.07	63.29	57.97	90.39	72.03	88.47	72.91	88.34	75.96	92.76	79.41	49.47	93.43	74.74

Certain capabilities only emerge at scale, as evidenced by [67]. For example, 6-7B models perform close to the random baseline in datasets such as ASSIN 2 RTE and STS, and WSC. However, at the 65B scale, we observe substantial improvements, approaching or surpassing state-of-the-art supervised models on the ASSIN 2 RTE and FaQuAD datasets.

GPT-4 [39] results indicate that there is still room for improvement for Sabiá-65B in the majority of the datasets evaluated in this work. Nevertheless, Sabiá-65B performs on par with GPT-4 in datasets such as ASSIN 2 RTE, ENEM Challenge, and FaQuAD.

5.2   Data Contamination

The pretraining data for Sabiá models were collected up until February 2022. Since ENEM 2022 was publicly released in November 2022, the model could not have access to the answers for the questions present within its pretraining data. Consequently, the improvements observed at least for ENEM 2022, which were higher than the average of the datasets, cannot be attributed to data contamination. However, for the other datasets, the possibility of data contamination cannot be ruled out.

5.3   Ablation: English Datasets

In this ablation study, we investigate the potential impact of Portuguese pretraining on the performance of the model in English datasets. We evaluated the LLaMA-7B and the Sabiá-7B models in English multiple-choice tasks. For simplicity, we employed a few-shot evaluation setup with 10 randomly selected examples (dynamic-sampled prompt). Importantly, we did not incorporate any descriptions or include Portuguese keywords to delimit the few-shot examples. We also restricted all the datasets to 350 test examples.

Following LLaMA's [63] approach, given the provided context, we select the answer with the highest likelihood normalized by the number of characters. The results in Table 4 indicate that the Sabiá-7B model exhibits a slightly reduced performance in English tasks compared to the baseline. This result corroborates our premise that model specialization invariably entails a balancing act, where improvements in one domain frequently coincide with degradation in another.

Table 4. Results in English datasets.

	PIQA	HellaSwag	WinoGrande	ARC-e	ARC-c	OBQA	*NPM*
LLaMA-7B	83.43	77.43	74.29	69.43	48.86	43.14	50.10
Sabiá-7B	80.86	75.71	72.29	72.86	50.00	42.29	49.02

6 Limitations

Owing to the financial constraints associated with pretraining and, more significantly, the manual labor involved in collecting and curating evaluation datasets, experiments were conducted exclusively in Portuguese. Given that our models started pretraining from English-pretrained models and that Portuguese and English exhibit relatively close linguistic proximity, we anticipate that other researchers conducting further pretraining on languages closely related to English will observe comparable improvements in their target tasks. However, determining whether the benefits of this method persist for languages more distant from English remains an open research question.

Portuguese is a language with an abundance of high-quality web-based texts. Thus, the gains observed with the proposed method may not necessarily extend to low-resource languages with limited availability of quality texts. In such cases, parameter-efficient methods [19,42,43] could be advantageous, as evidenced by Yong et al. [72]. We did not use these techniques in this study due to the training costs, which are approximately equivalent to training the entire model.[9]

7 Conclusion

In this study, we contributed to the expanding body of scientific evidence that specializing models for individual languages leads to improvements, even when the baseline model is large and extensively trained. We achieved this for the Portuguese language utilizing a near state-of-the-art model with 65 billion parameters. Given the relatively low pretraining cost and significant performance gains observed, we foresee a future landscape consisting of a diverse array of models, each tailored to a specific domain, rather than a single, all-encompassing model.

Acknowledgments. We thank Google Cloud for the generous TPU grant.

---

[9] Although parameter-efficient methods adjust only a fraction of the weights, they use only marginally fewer training FLOPs, as activations and gradients are computed for the entire model. For instance, LoRA [20], a parameter-efficient method, improves training throughput of a GPT-3 175B model by only nearly 32%.

References

1. Almeida, T.S., Laitz, T., Bonás, G.K., Nogueira, R.: Bluex: A benchmark based on Brazilian leading universities entrance exams. To appear (2023)
2. Antoun, W., Baly, F., Hajj, H.: AraBERT: Transformer-based model for Arabic language understanding. In: Proceedings of the 4th Workshop on Open-Source Arabic Corpora and Processing Tools, with a Shared Task on Offensive Language Detection. pp. 9–15. European Language Resource Association, Marseille, France (2020)
3. Barros, T.M.d., et al.: Employing transformers and emoji to perform sentiment classification of social media texts: Utilizando transformers e emoji na classificação de sentimento de textos oriundos de redes sociais (2021)
4. Bhattacharjee, A., et al.: BanglaBERT: Language model pretraining and benchmarks for low-resource language understanding evaluation in Bangla. In: Findings of the Association for Computational Linguistics: NAACL 2022, pp. 1318–1327. Association for Computational Linguistics, Seattle, United States (2022). https://doi.org/10.18653/v1/2022.findings-naacl.98
5. Brum, H., Volpe Nunes, M.d.G.: Building a sentiment corpus of tweets in Brazilian Portuguese. In: Proceedings of the Eleventh International Conference on Language Resources and Evaluation (LREC 2018). European Language Resources Association (ELRA), Miyazaki, Japan (May 2018)
6. Cañete, J., Chaperon, G., Fuentes, R., Ho, J.H., Kang, H., Pérez, J.: Spanish pretrained BERT model and evaluation data. In: PML4DC at ICLR 2020 (2020)
7. Carmo, D., Piau, M., Campiotti, I., Nogueira, R., Lotufo, R.: Ptt5: Pretraining and validating the t5 model on brazilian portuguese data. arXiv preprint arXiv:2008.09144 (2020)
8. Chan, B., Schweter, S., Möller, T.: German's next language model. In: Proceedings of the 28th International Conference on Computational Linguistics, pp. 6788–6796. International Committee on Computational Linguistics, Barcelona, Spain (Online) (2020). https://doi.org/10.18653/v1/2020.coling-main.598
9. Chaves Rodrigues, R., Tanti, M., Agerri, R.: Evaluation of Portuguese Language Models (2023). https://doi.org/10.5281/zenodo.7781848, https://github.com/ruanchaves/eplm
10. Chowdhery, A., et al.: Palm: Scaling language modeling with pathways. arXiv preprint arXiv:2204.02311 (2022)
11. Clark, C., Lee, K., Chang, M.W., Kwiatkowski, T., Collins, M., Toutanova, K.: BoolQ: Exploring the surprising difficulty of natural yes/no questions. In: Proceedings of the 2019 Conference of the North American Chapter of the Association for Computational Linguistics: Human Language Technologies, Volume 1 (Long and Short Papers), pp. 2924–2936. Association for Computational Linguistics, Minneapolis, Minnesota (2019). https://doi.org/10.18653/v1/N19-1300
12. Conneau, A., et al.: Unsupervised cross-lingual representation learning at scale. In: Proceedings of the 58th Annual Meeting of the Association for Computational Linguistics, pp. 8440–8451 (2020)
13. Ebrahimi, A., Kann, K.: How to adapt your pretrained multilingual model to 1600 languages. In: Proceedings of the 59th Annual Meeting of the Association for Computational Linguistics and the 11th International Joint Conference on Natural Language Processing (Volume 1: Long Papers), pp. 4555–4567. Association for Computational Linguistics, Online (Aug 2021). 10.18653/v1/2021.acl-long.351

14. FitzGerald, J., et al.: MASSIVE: A 1m-example multilingual natural language understanding dataset with 51 typologically-diverse languages (2022)
15. Fu, Y., Peng, H., Ou, L., Sabharwal, A., Khot, T.: Specializing smaller language models towards multi-step reasoning. arXiv preprint arXiv:2301.12726 (2023)
16. Gao, L., et al.: The pile: An 800gb dataset of diverse text for language modeling. arXiv preprint arXiv:2101.00027 (2020)
17. Gururangan, S., et al.: Scaling expert language models with unsupervised domain discovery. arXiv preprint arXiv:2303.14177 (2023)
18. Hoffmann, J., et al.: Training compute-optimal large language models. arXiv preprint arXiv:2203.15556 (2022)
19. Houlsby, N., et al.: Parameter-efficient transfer learning for NLP. In: International Conference on Machine Learning, pp. 2790–2799. PMLR (2019)
20. Hu, E.J., et al.: LoRA: Low-rank adaptation of large language models. In: International Conference on Learning Representations (2022). https://openreview.net/forum?id=nZeVKeeFYf9
21. Kalyan, K.S., Rajasekharan, A., Sangeetha, S.: Ammus: a survey of transformer-based pretrained models in natural language processing. arXiv preprint arXiv:2108.05542 (2021)
22. Kim, B., et al.: What changes can large-scale language models bring? intensive study on hyperclova: Billions-scale korean generative pretrained transformers. In: Proceedings of the 2021 Conference on Empirical Methods in Natural Language Processing, pp. 3405–3424 (2021K
23. Kudo, T., Richardson, J.: SentencePiece: A simple and language independent subword tokenizer and detokenizer for neural text processing. In: Proceedings of the 2018 Conference on Empirical Methods in Natural Language Processing: System Demonstrations, pp. 66–71. Association for Computational Linguistics, Brussels, Belgium (2018). https://doi.org/10.18653/v1/D18-2012
24. Le, H., et al.: FlauBERT: Unsupervised language model pre-training for French. In: Proceedings of the Twelfth Language Resources and Evaluation Conference, pp. 2479–2490. European Language Resources Association, Marseille, France (2020)
25. Lee, H., Yoon, J., Hwang, B., Joe, S., Min, S., Gwon, Y.: Korealbert: Pretraining a lite bert model for korean language understanding. In: 2020 25th International Conference on Pattern Recognition (ICPR), pp. 5551–5557. IEEE (2021)
26. Lewkowycz, A., et al.: Solving quantitative reasoning problems with language models. arXiv preprint arXiv:2206.14858 (2022)
27. Lin, X.V., et al.: Few-shot learning with multilingual generative language models. In: Proceedings of the 2022 Conference on Empirical Methods in Natural Language Processing, pp. 9019–9052 (2022)
28. Liu, Y., et al.: Multilingual denoising pre-training for neural machine translation. Trans. Assoc. Comput. Linguist. **8**, 726–742 (2020)
29. Longpre, S., Lu, Y., Daiber, J.: MKQA: a linguistically diverse benchmark for multilingual open domain question answering. Trans. Assoc. Comput. Linguist. **9**, 1389–1406 (2021)
30. Loshchilov, I., Hutter, F.: Decoupled weight decay regularization. In: International Conference on Learning Representations (2019)
31. Maas, A.L., Daly, R.E., Pham, P.T., Huang, D., Ng, A.Y., Potts, C.: Learning word vectors for sentiment analysis. In: Proceedings of the 49th Annual Meeting of the Association for Computational Linguistics: Human Language Technologies, pp. 142–150. Association for Computational Linguistics, Portland, Oregon, USA (2011)

32. Martin, L., et al.: CamemBERT: a tasty French language model. In: Proceedings of the 58th Annual Meeting of the Association for Computational Linguistics. pp. 7203–7219. Association for Computational Linguistics, Online (2020). https://doi.org/10.18653/v1/2020.acl-main.645
33. de Melo, G., Imaizumi, V., Cozman, F.: Winograd schemas in portuguese. In: Anais do XVI Encontro Nacional de Inteligência Artificial e Computacional, pp. 787–798. SBC (2019)
34. Moraes, G., Bonifácio, L.H., Rodrigues de Souza, L., Nogueira, R., Lotufo, R.: A cost-benefit analysis of cross-lingual transfer methods. arXiv preprint arXiv:2105.06813 (2021). https://arxiv.org/abs/2105.06813
35. Muennighoff, N., et al.: Crosslingual generalization through multitask finetuning (2022)
36. Nguyen, D.Q., Tuan Nguyen, A.: PhoBERT: Pre-trained language models for Vietnamese. In: Findings of the Association for Computational Linguistics: EMNLP 2020, pp. 1037–1042. Association for Computational Linguistics, Online (2020). https://doi.org/10.18653/v1/2020.findings-emnlp.92
37. Nunes, D., Primi, R., Pires, R., Lotufo, R., Nogueira, R.: Evaluating gpt-3.5 and gpt-4 models on brazilian university admission exams (2023)
38. Ogueji, K., Zhu, Y., Lin, J.: Small data? no problem! exploring the viability of pre-trained multilingual language models for low-resourced languages. In: Proceedings of the 1st Workshop on Multilingual Representation Learning, pp. 116–126. Association for Computational Linguistics, Punta Cana, Dominican Republic (2021)
39. OpenAI: Gpt-4 technical report (2023)
40. Overwijk, A., Xiong, C., Callan, J.: Clueweb 22: 10 billion web documents with rich information. In: Proceedings of the 45th International ACM SIGIR Conference on Research and Development in Information Retrieval, pp. 3360–3362 (2022)
41. Overwijk, A., Xiong, C., Liu, X., VandenBerg, C., Callan, J.: Clueweb 22: 10 billion web documents with visual and semantic information (2022)
42. Pfeiffer, J., Kamath, A., Rücklé, A., Cho, K., Gurevych, I.: AdapterFusion: Non-destructive task composition for transfer learning. In: Proceedings of the 16th Conference of the European Chapter of the Association for Computational Linguistics: Main Volume, pp. 487–503. Association for Computational Linguistics, Online (2021). https://doi.org/10.18653/v1/2021.eacl-main.39
43. Pfeiffer, J., Vulić, I., Gurevych, I., Ruder, S.: Mad-x: An adapter-based framework for multi-task cross-lingual transfer. arXiv preprint arXiv:2005.00052 (2020)
44. Radford, A., Wu, J., Child, R., Luan, D., Amodei, D., Sutskever, I., et al.: Language models are unsupervised multitask learners. OpenAI blog **1**(8), 9 (2019)
45. Rae, J.W., et al.: Scaling language models: Methods, analysis & insights from training gopher. arXiv preprint arXiv:2112.11446 (2021)
46. Raffel, C., et al.: Exploring the limits of transfer learning with a unified text-to-text transformer. J. Mach. Learn. Res. **21**(1), 5485–5551 (2020)
47. Real, L., Fonseca, E., Gonçalo Oliveira, H.: The ASSIN 2 shared task: a quick overview. In: Quaresma, P., Vieira, R., Aluísio, S., Moniz, H., Batista, F., Gonçalves, T. (eds.) PROPOR 2020. LNCS (LNAI), vol. 12037, pp. 406–412. Springer, Cham (2020). https://doi.org/10.1007/978-3-030-41505-1_39
48. Roberts, A., et al.: Scaling up models and data with t5x and seqio. arXiv preprint arXiv:2203.17189 13 (2022)
49. Rosa, G.M., Bonifacio, L.H., de Souza, L.R., Lotufo, R., Nogueira, R.: A cost-benefit analysis of cross-lingual transfer methods. arXiv preprint arXiv:2105.06813 (2021)

50. la Rosa, J.D., Fernández, A.: Zero-shot reading comprehension and reasoning for spanish with BERTIN GPT-J-6B. In: y Gómez, M.M., (eds.) Proceedings of the Iberian Languages Evaluation Forum (IberLEF 2022). CEUR Workshop Proceedings (2022)
51. Sakaguchi, K., Bras, R.L., Bhagavatula, C., Choi, Y.: Winogrande: an adversarial winograd schema challenge at scale. Commun. ACM **64**(9), 99–106 (2021)
52. Sarti, G., Nissim, M.: It5: Large-scale text-to-text pretraining for italian language understanding and generation. arXiv preprint arXiv:2203.03759 (2022)
53. Sayama, H.F., Araujo, A.V., Fernandes, E.R.: FaQuAD: Reading comprehension dataset in the domain of brazilian higher education. In: 2019 8th Brazilian Conference on Intelligent Systems (BRACIS), pp. 443–448 (2019). https://doi.org/10.1109/BRACIS.2019.00084
54. Scao, T.L., et al.: Bloom: A 176b-parameter open-access multilingual language model. arXiv preprint arXiv:2211.05100 (2022)
55. Shazeer, N., Stern, M.: Adafactor: Adaptive learning rates with sublinear memory cost. In: International Conference on Machine Learning, pp. 4596–4604. PMLR (2018)
56. Shliazhko, O., Fenogenova, A., Tikhonova, M., Mikhailov, V., Kozlova, A., Shavrina, T.: MGPT: Few-shot learners go multilingual. arXiv preprint arXiv:2204.07580 (2022)
57. Silveira, I.C., Maua, D.D.: Advances in automatically solving the enem. In: 2018 7th Brazilian Conference on Intelligent Systems (BRACIS), pp. 43–48. IEEE Computer Society, Los Alamitos, CA, USA (oct 2018). https://doi.org/10.1109/BRACIS.2018.00016
58. Socher, R., et al.: Recursive deep models for semantic compositionality over a sentiment treebank. In: Proceedings of the 2013 Conference on Empirical Methods in Natural Language Processing, pp. 1631–1642. Association for Computational Linguistics, Seattle, Washington, USA (2013)
59. Souza, F., Nogueira, R., Lotufo, R.: BERTimbau: pretrained BERT models for Brazilian Portuguese. In: Cerri, R., Prati, R.C. (eds.) BRACIS 2020. LNCS (LNAI), vol. 12319, pp. 403–417. Springer, Cham (2020). https://doi.org/10.1007/978-3-030-61377-8_28
60. Srivastava, A., et al.: Beyond the imitation game: Quantifying and extrapolating the capabilities of language models. arXiv preprint arXiv:2206.04615 (2022)
61. Su, J., Lu, Y., Pan, S., Wen, B., Liu, Y.: Roformer: Enhanced transformer with rotary position embedding. arXiv preprint arXiv:2104.09864 (2021)
62. Taylor, R. et al.: Galactica: A large language model for science. arXiv preprint arXiv:2211.09085 (2022)
63. Touvron, H., et al.: Llama: Open and efficient foundation language models. arXiv preprint arXiv:2302.13971 (2023)
64. Vaswani, A., et al.: Attention is all you need. In: Advances in Neural Information Processing Systems 30 (2017)
65. Wang, B.: Mesh-Transformer-JAX: Model-Parallel Implementation of Transformer Language Model with JAX. https://github.com/kingoflolz/mesh-transformer-jax (2021)
66. Wang, B., Komatsuzaki, A.: GPT-J-6B: A 6 Billion Parameter Autoregressive Language Model (2021)
67. Wei, J., et al.: Emergent abilities of large language models. Transactions on Machine Learning Research (2022), survey Certification
68. Wu, S., et al.: BloombergGPT: A large language model for finance (2023)

69. Xue, L., et al.: Byt5: towards a token-free future with pre-trained byte-to-byte models. Trans. Assoc. Comput. Linguist. **10**, 291–306 (2022)
70. Xue, L., et al.: mt5: A massively multilingual pre-trained text-to-text transformer. arXiv preprint arXiv:2010.11934 (2020)
71. Yang, Z., Dai, Z., Yang, Y., Carbonell, J., Salakhutdinov, R., Le, Q.V.: XLNet: Generalized Autoregressive Pretraining for Language Understanding. Curran Associates Inc., Red Hook, NY, USA (2019)
72. Yong, Z.X., et al.: Bloom+ 1: Adding language support to bloom for zero-shot prompting. arXiv preprint arXiv:2212.09535 (2022)
73. Zeng, A., et al.: Glm-130b: An open bilingual pre-trained model. arXiv preprint arXiv:2210.02414 (2022)
74. Zhang, S., et al.: Opt: Open pre-trained transformer language models. arXiv preprint arXiv:2205.01068 (2022)
75. Zhang, X., Zhao, J.J., LeCun, Y.: Character-level convolutional networks for text classification. In: NIPS (2015)
76. Zoph, B.: Designing effective sparse expert models. In: 2022 IEEE International Parallel and Distributed Processing Symposium Workshops (IPDPSW), p. 1044. IEEE (2022)

Disambiguation of Universal Dependencies Part-of-Speech Tags of Closed Class Words in Portuguese

Lucelene Lopes[1]([⊠]) [iD], Paulo Fernandes[2] [iD], Marcio L. Inacio[3] [iD],
Magali S. Duran[1] [iD], and Thiago A. S. Pardo[1] [iD]

[1] ICMC, University of São Paulo, São Carlos, Brazil
{lucelene.lopes,taspardo}@icmc.usp.br, magali.duran@uol.com.br
[2] Merrimack College, North Andover, MA, USA
fernandesp@merrimack.edu
[3] CISUC, Universidade de Coimbra, Coimbra, Portugal
mlinacio@dei.uc.pt

Abstract. This paper explores methods to disambiguate Part-of-Speech (PoS) tags for closed class words in Brazilian Portuguese corpora annotated according to the Universal Dependencies annotation model. We evaluate disambiguation methods of different paradigms, namely a Markov-based method, a widely adopted parsing tool, and a BERT-based language modeling method. We compare their performances with two baselines, and observe a significant increase of more than 10% over the baselines for all proposed methods. We also show that while the BERT-based model outperforms the others reaching for the best case a 98% accuracy predicting the correct PoS tag, the use of the three methods as an Ensemble method offers more stable result according to the smaller variance for the numerical results we performed.

Keywords: Part-of-Speech tagging · Universal Dependencies · Disambiguation

1 Introduction

To develop large annotated corpora is a challenging task in the area of Natural Language Processing (NLP). It is a common strategy to rely on automatic analysis tools to pre-annotate the data and then to manually review the annotation, as direct human annotation requires large amounts of time, a strict discipline, and an homogeneous system of annotation rules that must be followed by annotators. In the case of treebanks (i.e., databases of sentences and their corresponding syntactic trees), this must be carried out for both Part-of-Speech (PoS) tagging and syntactic parsing, at least. However, as the annotation of PoS tags can be considered input for syntactic annotation, the better the PoS annotation, the lower the risk of tagging errors leading to parsing errors.

M. C. Naldi and R. A. C. Bianchi (Eds.): BRACIS 2023, LNAI 14197, pp. 241–255, 2023.
https://doi.org/10.1007/978-3-031-45392-2_16

In the pre-annotation of PoS, the hardest part is assigning tags to words from closed classes, as many of them may pertain to different classes, that is, they are ambiguous. These words are among the most frequent in any corpus, regardless of genre and domain. As they constitute marks for syntactic annotation decisions, it is important to annotate them accurately, thus offering safe annotation points to support the effort of producing large annotated treebanks.

Of particular interest to this paper is the initiative to perform PoS tagging for Portuguese language according to the widely adopted Universal Dependencies (UD) model [18,20], which predicts the existence of "universal" PoS tags and syntactic relations that may be applied to all languages in the world. There are currently nearly 200 treebanks in over 100 languages affiliated to UD.

In Portuguese, we highlight the previous work of Rademaker et al. [22], that has produced a UD-annotated version of Bosque treebank [1], and the recent efforts of Silva et al. [25] that have explored PoS-tagging for tweets, Lopes et al. [17] that shows that there are non-ambiguous words and sequences of words that may significantly improve tagging accuracy, and Souza et al. [3] that creates a UD corpus for domain specific documents. Nevertheless, in practically all languages and domains there is a large number of challenging words that are ambiguous and require more efforts to be properly annotated.

We focus here on investigating and proposing methods for solving the problem of PoS tag disambiguation of words of a particular case in Portuguese: the closed class words. Such word classes include prepositions, conjunctions, determiners, numbers, pronouns and primitive adverbs (non-derived adverbs). The motivation for such choice relies on the fact that closed class words may be very ambiguous regarding PoS tagging and, once solved, may serve as solid evidence to help disambiguating words of other word classes, improving tagging accuracy overall. For instance, in PortiLexicon-UD [16], which is a UD-based lexicon for Portuguese, the word *"que"* (in English, it usually corresponds to "what", "that" or "which") may have up to 7 different word classes and represents a challenge for annotation, as discussed in the work of Duran et al. [6]. To exemplify the challenges, we reproduce some examples from this work: the word *"que"* is:

- a **pronoun** in the sentence *"As coisas com **que** sonhamos são pistas para o autoconhecimento"*,
- a **subordinate conjunction** in *"É obvio **que** quero votar"*,
- a **determiner** in *"**Que** maravilha"*, and
- an **adverb** in *"**Que** horrível foi esse acidente"*.

The work of Silva et al. [25] evidences the practical difficulties in PoS tagging. Some of these classes, as pronouns and subordinate conjunctions, achieve some of the lowest accuracy values (below 90%) in the reported experiments.

In our paper we explore three different methods to disambiguate the words of the selected classes:

- a **Markovian model** similar to the one developed by Assunção et al. [2] to disambiguate each ambiguous word, considering the PoS tag sequences of a training set;

- **the tagger/parser UDPipe 2.0** [28] pretrained for Portuguese is used to tag the whole sentences, and we observe the decision took for ambiguous words;
- **a contextual word embedding-based solution** using a pretrained Portuguese model to individually disambiguate the PoS tag for ambiguous words, in a similar approach to Vandenbussche et al. [30] that uses BERT (Bidirectional Encoder Representations from Transformers) [4] by a pretrained language model specific for Brazilian Portuguese, the BERTimbau [26].

We perform a detailed error analysis in comparison with two baselines and also explore an Ensemble solution considering a voting system taking into account the three presented models' outcomes. Our results show accuracy values ranging from 90% to 98% for the three proposed methods, as well as the Ensemble one. While the BERTimbau-based approach delivers the higher accuracy values for the best cases experimented, the Ensemble solution presents more stable results since the accuracy of all experimented cases has a smaller standard deviation than all three methods applied individually. The achieved accuracy is relevant and the gains in terms of annotation effort are considerable, as we correct nearly 25% of all tokens of large Portuguese corpora.

This paper is organized as follows. The next section presents the main related work. The problem definition and the baselines for this study are presented in Sect. 3. Section 4 presents the three experienced models: the Markovian Model, the UDPipe 2.0 application, and our contextual word embedding BERTimbau-based approach, respectively. Section 5 shows our experiments over an example corpus and discusses the achieved effectiveness. Finally, the final section summarizes our contributions and suggests some future work.

2   Related Work

We can find in the literature related works using varied techniques to deal with PoS ambiguities. We comment on the main ones in what follows.

Ehsani et al. [7] perform the disambiguation of PoS tags for Turkish using Conditional Random Fields (CRF) [14]. In such work the authors point out the processing burden associated using CRF to perform the task, but they deliver PoS tag accuracies around 97%. The authors also mention that this high PoS tag accuracy is very beneficial to support reliable lemma and morphological annotations.

A different approach is the work of Assunção et al. [2] that uses a Markovian model to perform PoS tag disambiguation based on Markovian models [13] and a partially annotated corpus. In this work, the disambiguation follows a basic probability computation of PoS tag sequences, and therefore delivers faster computations than more sophisticated approaches.

The work of Hoya Quecedo et al. [10] presents an effort to disambiguate PoS tag and lemma for morphological rich languages, which is the case of Portuguese, using a Neural Network model based on BiLSTM (Bidirectional Long Short-Term Memory) [8]. Hoya Quecedo et al. build word embeddings to submit to the bidirecional model and, thus, estimate the probability of the PoS tag and lemma options of each word. These experiments deliver a precision varying from 80% to 85% to corpora in Finnish, Russian, and Spanish.

Another interesting work published by Muñoz-Valero et al. [19] also employs an LSTM (Long Short-Term Memory) RNN (Recurrent Neural Networks) [5] to disambiguate words in the American National Corpus [11]. Similar to this one, the works of Shen et al. [24] and Zalmout and Habash [32] aim at the disambiguation of morphological features, both using LSTM and RNN.

3   Problem Definition and Baseline Approaches

The problem we are aiming to solve is how to disambiguate closed class words that can be tagged with different PoS tags, for example, the word *"por"* that in Portuguese can be either an ADP (in English, "by") or a VERB (in English, "to put"). Our proposed methods are based on working with a training set of fully annotated sentences using Universal Dependencies (UD) in the CoNLL-U format [20] in order to disambiguate the PoS tag annotation of all ambiguous words of a test set.

The set of UD PoS tags is formed by:

- ADJ - adjectives, as *"bonito"* ("beautiful" in English);
- ADP - adpositions, as *"de"* ("of" in English);
- ADV - adverbs, as *"não"* ("no" in English);
- AUX - auxiliary verbs, as *"foi"* ("was" in English);
- CCONJ - coordinating conjunctions, as *"e"* ("and" in English);
- DET - determiners, as *"cujo"* ("whose" in English);
- INTJ - interjections, as *"tchau"* ("goodbye" in English);
- NOUN - nouns, as *"vida"* ("life" in English);
- NUM - numerals, as *"cinco"* ("five" in English);
- PART - particles, which is not employed in Portuguese;
- PRON - pronouns, as *"ele"* ("he" in English);
- PROPN - proper nouns, as *"Brasil"* ("Brazil" in English);
- PUNCT - punctuations, as *"?"*;
- SCONJ - subordinating conjunctions, as *"porque"* ("because" in English);
- SYM - symbols, as *"$"*;
- VERB - verbs, as *"jogamos"* ("(we) play" in English);
- X - others, as foreign words.

The closed classes considered in this paper are ADP, CCONJ, DET, NUM, PRON and SCONJ, plus subsets of the classes ADV[1], AUX[2], and ADJ[3]. Table 1 shows the considered closed classes, the total number of words (#tot.), the number of ambiguous words (#amb.), and some examples. Considering that many ambiguous tokens belong to several classes, the overall sum of ambiguous tokens is 368 different words. A full list of the considered closed class words is available at https://sites.google.com/icmc.usp.br/poetisa/publications.

Table 1. Closed PoS tags considered in our work.

PoS tag	Definition	#tot.	#amb.	Examples
ADP	Adpositions	30	16	*"até"* ("until" or "even" in English) ADP or ADV
ADJ	Adjectives that are ordinal numbers	92	29	*"segundo"* ("second" or "according to" in English) ADJ, NOUN, or SCONJ
ADV	Adverbs not ending with -*mente*	99	45	*"logo"* ("soon" or "therefore" in English) ADV or CCONJ
AUX	Auxiliary verbs	316	217	*"era"* ("was" or "age" in English) AUX or NOUN
CCONJ	Coordenative conjunctions	23	14	*"nem"* ("nor" or "not" in English) CCONJ or ADV
DET	Determiners	102	94	*"certo"* ("certain" or "right" in English) DET or ADJ
NUM	Numerals	40	4	*"um"* ("one" or "a" in English) NUM, PRON, or DET
PRON	Pronouns	125	96	*"outro"* ("other" or "another" in English) PRON or DET
SCONJ	Subordinate conjunctions	14	12	*"se"* ("if" or "itself" in English) SCONJ or PRON

The first baseline (Baseline 1) that we evaluate takes into account the number of occurrences of each ambiguous word in the training set and uses the most common PoS tag to all occurrences in the test set. For example, if the word *"até"*

---

[1] Similarly to English with the ending -"ly", in Portuguese it is possible to turn adjectives into adverbs by adding -*"mente"* at the end. We disconsider those -ly adverbs, as only the primitive adverbs are a closed class.

[2] The verbs *ser* and *estar* ("to be" in English) are always annotated as AUX, either by being true auxiliary verbs, either by being copula verbs. The verbs *"ir"*, *"haver"*, and *"ter"* ("to go", "to exist", and "to have" in English) are sometimes annotated as VERB, sometimes annotated as AUX (as "going to" and "have" + a past participle in English).

[3] While adjectives are not a closed class, the adjectives that are ordinal numbers are considered belonging to a closed subset of class ADJ.

that can be an ADP ("until" in English) or an ADV ("even" in English) is found 189 times as ADP and 77 times as ADV, we will assume that all occurrences of the word "até" will be considered ADP. However, to prevent decisions based on too sparse data at the training set, we will only consider a prediction when there are at least 3 occurrences of a word with the most frequent PoS tag. Additionally, to prevent decisions based on too close differences, the second more frequent PoS tag has to be less than half frequent than the more frequent one.

The second baseline considered is the implementation following the definitions made by the work of Hoya Quecedo et al. [10] that uses a Neural Network model based on BiLSTM as described before. We employ this baseline using the same hyperparameters defined in [10] and we will refer to this as Baseline 2.

4   Prediction Methods

In this section we present the three methods implemented in these paper experiments. The first one is a traditional Markovian modeling approach inspired in the work of Assunção et al [2] and it is presented in Subsect. 4.1. The second method is the application of the UDPipe 2.0 tool [27] and it is presented at Subsect. 4.2. The third method is an approach based on BERTimbau pretrained neural language model [26] to build a contextual word embedding model and it is presented at Subsect. 4.3.

4.1   Prediction Through Markovian Models

Among the Markovian approaches to stochastically predict language information, we decided to employ the one based on Assunção et al. [2]. This approach is based on developing a Markov chain model to each sentence of the training set describing the probabilities of PoS tag sequences.

For an illustration of the method, consider the sentence "Com a rescisão, as provas apresentadas pelos delatores ainda podem ser usadas.". This sentence will be annotated as described in Fig. 1. This sentence corresponds to the attached Markov chain that represents the sequences of PoS tags. For example, the two words tagged as ADP are followed by DET, therefore the probability of transition from ADP to DET is 100%. Similarly, the three occurrences of NOUN are followed by PUNCT, VERB, and ADV tags, therefore, each existing arc in the chain has a transition probability of 33.3%. Additionally, one additional node representing the boundary of the sentence (⋆) links to the sentence start (before ADP) and the sentence end (after PUNCT).

This Markov chain creation process is repeated to all sentences of the training set. Once all sentences of the training set are used to create a Markov chain, each possible word with ambiguous PoS tags is disambiguated considering the PoS tags preceding and succeeding the ambiguous word.

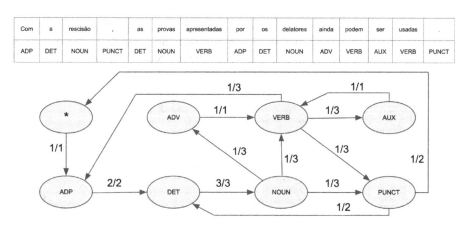

Com	a	rescisão	,	as	provas	apresentadas	por	os	delatores	ainda	podem	ser	usadas	.
ADP	DET	NOUN	PUNCT	DET	NOUN	VERB	ADP	DET	NOUN	ADV	VERB	AUX	VERB	PUNCT

Fig. 1. Annotated sentence and corresponding Markov chain of PoS tag sequences.

4.2 Prediction Through UDPipe 2.0

The second method experimented is the use of the tagger/parser UDPipe 2.0 [28] using one of the available Portuguese training models, which is BOSQUE-PT [29] in our experiment, composed by 9,364 sentences and 210,957 tokens. UDPipe 2.0 uses a RNN based on BiLSTM to perform a full UD annotation based on a previously training set.

Our approach in this paper is to feed the whole sentences of the testing set to UDPipe 2.0 and pinpoint the ambiguous words to observe the accuracy of disambiguation of the target closed class words.

4.3 Prediction Through BERTimbau-Based Model

A more recent approach, which has achieved state of the art results in many NLP tasks [9,15], is based on the usage of pretrained neural Language Models (LM), such as BERT [4]. A LM is capable of providing contextualized word embeddings, which can later be used in a downstream task. Specifically, we use the language model BERTimbau, a BERT model pretrained for Brazilian Portuguese [26], in the smallest version, BERTimbau-base, that, like BERT-base, has 12 layers.

All sentences are previously tokenized using the same BERTimbau model, as it requires the words to be split into subtokens. For words consisting of more than one subtoken, the embedding used during classification represents the first one.

For the PoS disambiguation, our model has two inputs: the sentence duly tokenized and the position of the token to disambiguate. BERTimbau is used to retrieve a sentence vector (which represents the whole context) and also the word embedding for the token to disambiguate (to include specific information about the token), which are later concatenated and passed through a linear layer in order to create a combined representation.

Finally, this vector is fed into another linear layer with Softmax for classification into PoS tags, plus a dropout layer between the combined representation and the final classification layer in order to tackle over fitting issues. Figure 2 depicts the general process for the BERTimbau-based method.

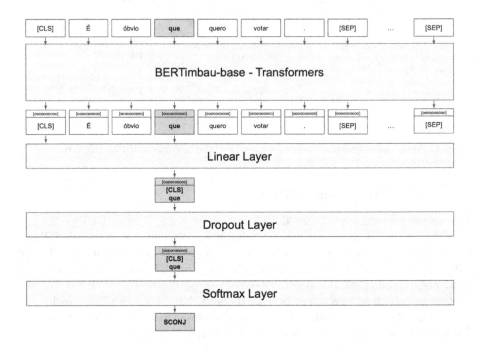

Fig. 2. BERTimbau-based method process.

For the fine tuning of the model, the Cross Entropy loss function was optimized using Adam [12] with learning rate $\alpha = 10^{-5}$ for 2 epochs. The dropout rate used was 0.2 with batch size 8. The whole model was implemented using HuggingFace's Transformers library [31] alongside PyTorch [21] and it is available as supplemental material[4].

5   Experiments

The conducted experiments were performed over an annotated corpus in Brazilian Portuguese language with 8,420 sentences (168,397 tokens) extracted from the first five thousand news from Folha Kaggle data bank [23]. This corpus was manually annotated using UD tags by a group of 10 annotators with redundancy and supervised by a chief linguist. It contains (among its 168,397 tokens) 44,208 occurrences of 368 distinct ambiguous words.

---

[4] https://github.com/nilc-nlp/pos-disambiguation.

To allow performing the experiments and comparing all the methods, we randomly split the corpus into: **train** - a training data set with approximately 80% of the tokens of the full corpus; **dev** - a development data set with approximately 10% of the tokens of the full corpus; **test** - a testing data set with approximately 10% of the tokens of the full corpus. However, to prevent bias in our experiments, we created 10 of such random splits as folds to perform cross-validation in order to establish the accuracy of each method. The application of Baseline 1 and the Markov-based model employs the **train+dev** data sets to training and **test** data set to testing. The application of UDPipe 2.0 method employs **test** data set as testing and ignores the data sets **train** and **dev**. The application of Baseline 2 and BERTimbau-based model employ each split individually.

Table 2 presents the accuracy and F1 obtained for each fold, while Table 3 summarizes the results obtained by each method for each fold in terms of the average[5], standard deviation, minimum, and maximum results of the ten folds individually. In these tables, the methods are abbreviated as Markov, for the Markov-based method (Subsect. 4.1), UDPipe for the UDPipe 2.0 application method (Subsect. 4.2), and BERT for the BERTimbau-based method (Subsect. 4.3). Finally, in these tables we included the accuracy and F1 results of an Ensemble method. The Ensemble is obtained applying the Markov-based, UDPipe 2.0, and BERTimbau-based methods, and deciding on the PoS tag by a simple voting strategy among them.

Table 2. Accuracy and F1 values of each method for each of the ten folds.

Method	Accuracy for each fold									
	0	1	2	3	4	5	6	7	8	9
Baseline 1	77.78%	80.63%	79.15%	80.02%	79.19%	64.11%	79.17%	78.99%	78.98%	79.56%
Baseline 2	79.91%	81.27%	84.00%	83.30%	83.03%	85.32%	81.04%	79.52%	79.45%	80.10%
Markov	90.28%	91.33%	90.48%	91.07%	90.45%	90.37%	90.50%	90.67%	91.10%	90.57%
UDPipe	93.58%	93.96%	93.72%	93.73%	93.52%	93.23%	93.39%	93.71%	92.80%	93.39%
BERT	**97.58%**	**98.23%**	**97.52%**	**98.02%**	**98.35%**	**98.23%**	**97.50%**	**97.70%**	**97.53%**	**97.71%**
Ensemble	97.20%	97.74%	97.20%	97.44%	97.38%	97.30%	97.09%	97.00%	96.93%	97.03%
Method	F1 for each fold									
	0	1	2	3	4	5	6	7	8	9
Baseline 1	87.50%	89.28%	88.36%	88.90%	88.39%	78.13%	88.37%	88.26%	88.26%	88.62%
Baseline 2	88.83%	89.67%	91.30%	90.89%	90.73%	92.08%	89.53%	88.59%	88.55%	88.95%
Markov	94.89%	95.47%	95.00%	95.33%	94.99%	94.94%	95.01%	95.11%	95.34%	95.05%
UDPipe	96.68%	96.89%	96.76%	96.76%	96.65%	96.50%	96.58%	96.75%	96.27%	96.58%
BERT	**98.78%**	**99.11%**	**98.74%**	**99.00%**	**99.17%**	**99.11%**	**98.73%**	**98.84%**	**98.75%**	**98.84%**
Ensemble	98.58%	98.86%	98.58%	98.70%	98.67%	98.63%	98.52%	98.48%	98.44%	98.49%

The first observation from the results in Table 3 is that the three methods are significantly more accurate than the baselines. It is also noticeable the high

---

[5] The values stated as average are the macro average of the values of each fold, but since the folds have about the same size, the values for micro and macro average have are practically the same (less than 0.01% difference).

Table 3. Summarization of accuracy and F1 values considering the ten folds.

Method	Accuracy				F1			
	average	st. dev.	min.	max.	average	st. dev.	min.	max.
Baseline 1	77.76%	0.0485	64.11%	80.63%	87.41%	0.0329	78.13%	89.28%
Baseline 2	81.69%	0.0208	79.45%	85.32%	89.91%	0.0125	88.55%	92.08%
Markov	90.68%	0.0036	90.28%	91.33%	95.11%	0.0020	94.89%	95.47%
UDPipe	93.50%	0.0033	92.80%	93.96%	96.64%	0.0017	96.27%	96.89%
BERT	**97.84%**	0.0034	**97.50%**	**98.35%**	**98.91%**	0.0017	**98.73%**	**99.17%**
Ensemble	97.23%	**0.0024**	96.93%	97.74%	98.60%	**0.0013**	98.44%	98.86%

Table 4. Number of wrong PoS tags by method for each of the ten folds of the PoS tag errors.

	Number of wrongfully predicted PoS tags									
	0	1	2	3	4	5	6	7	8	9
by Markov alone	353	325	361	313	**319**	348	345	335	315	339
by UDPipe alone	178	176	175	171	**187**	190	176	171	208	176
by BERT alone	25	15	23	15	**9**	18	15	13	24	22
by Markov and UDPipe	52	45	43	47	**50**	63	48	49	57	64
by Markov and BERT	18	15	21	19	**18**	15	20	31	26	18
by UDPipe and BERT	12	11	21	9	**16**	8	20	19	17	17
by all three methods	41	27	41	33	**29**	32	39	33	37	32
total by Markov	464	412	466	412	416	458	452	448	435	453
total by UDPipe	283	259	280	260	282	293	283	272	319	289
total by BERT	96	68	106	76	72	73	94	96	104	89
total by Ensemble	123	98	126	108	113	118	127	132	137	131

accuracy achieved by the BERTimbau-based approach that alone provides a better accuracy than the Ensemble method. The second observation is the stability of the Ensemble method delivering smaller standard deviation both for accuracy and F1. The accuracy achieved by the Ensemble method is slightly smaller than the BERTimbau-based one, but since the its variance of results is smaller, it is probably more reliable to employ the Ensemble method in a real annotation where the gold standard is not previously known.

To illustrate the difficulties of the phenomenon, we analyzed the number of mistaken decisions (wrong PoS tags) made by each of the three methods, stated in the first three rows of Table 4. We also present in the middle rows of this table the mistakes shared by the methods, two by two, and all together. The last rows stated the total mistakes by methods and the use of an Ensemble of the three methods application.

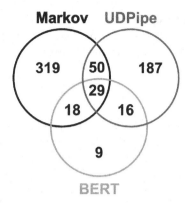

Fig. 3. Number of wrong PoS tags by each method for Fold 4

Figure 3 shows a Venn diagram of the PoS tag errors made by the three methods for Fold 4 (which is a fold with number of errors nearest to the average of all folds, as highlighted in Table 4). For this fold, there were 319 PoS tag errors by the Markov-based method alone, 187 by the UDPipe 2.0 method alone, and 9 by the BERTimbau-based method alone. Similarly, 50 PoS tag errors were made by both Markov-based and UDPipe 2.0 methods, 18 by Markov-based and BERTimbau-based methods, and 16 by the UDPipe 2.0 and BERTimbau-based methods. Finally, 29 errors were made by all three methods. These lead to 416 errors by Markov-based (319+50+18+29), 282 by UDPipe 2.0 (187+50+16+29), and 72 by BERTimbau-based (9+18+16+29). Since the Ensemble method will always decide wrongfully when the three methods are wrong (29 errors), when any two methods are wrong (50+18+16), the errors of the Ensemble method will appear on 113 PoS tags (29+50+18+16).

Analyzing Fold 4 testing set outcome, we observe that it is composed by 842 sentences and a total of 16,938 tokens. From these, 4,354 tokens are ambiguous words from closed classes, and only 72 were incorrect after applying BERTimbau-based method, *i.e.*, the method produced 4,282 correct tokens (25.21% of the total tokens), which correspond to 98.35% of the ambiguous tokens.

To illustrate the errors of the methods, we use two words from the 29 mistakes in the testing set for Fold 4. The token *"segundo"* (*"second"* in English) at the end of the sentence *"Flamengo e São Paulo viraram 12 e 11 pontos de distância, mas não estavam em segundo."* is tagged as ADJ, but the methods delivered a NOUN prediction, because the methods had difficulty to handle the noun elipse, as the full form would be *"segundo lugar"* (*"second place"* in English). The first token *"a"* (*"her"* in English) at the sentence *"Ela recebeu uma mensagem certo dia chamando a a ir a Cariri..."* is tagged as PRON. The three methods delivered ADP prediction because a double token sequence *"a a"* is the usual tokenized form of crasis (*"à"*, in English *"to the"*), a very frequent form in Portuguese that is tagged as ADP DET, respectively. In the training set for Fold 4, the number of crasis occurrences is 276, while other occurrences of double *"a"* were never present, consequently the three methods were unable to learn tags PRON ADP for the sequence, assigning DET instead of PRON for it. One may see that both cases are difficult ones, which somehow explains why the methods failed.

6   Conclusion

We conducted an experiment with three proposed methods that show to be competitive and delivered an accuracy gain over the baselines, reaching average accuracy values ranging from 90.76% to 97.84%. The F1 average values were also impressive, ranging from 94.89% to 99.11%. It is important to mention that, since we experiment using a 10-fold cross-validation, our results are statistically sound, as the standard deviation of the results of each fold was always low[6].

It is particularly noticeable that the BERTimbau-based approach brings a relevant correction power. Recalling Fold 4 testing set correction of 25.21% of the total tokens, we observe that, to a large corpus as the Folha-Kaggle [23] with around 84 million tokens, we can automatically deliver more than 20 million correct tokens for ambiguous words of closed classes.

It is also noticeable that combining the three presented methods in an Ensemble method has shown a small impact in the average accuracy with respect to BERTimbau-based method, respectively 97.23% and 97.84%, but the Ensemble solution offers more stable results, since the standard deviation of the 10 folds dropped from 0.0034 for BERTimbau-based solution to 0.0024 for the Ensemble one.

As future work, we intend to perform a deeper analysis of the Ensemble approach, perhaps adding other methods casting votes together with the three methods already developed. It is also possible to imagine the development of a similar technique to disambiguate other fields of the UD model, as lemma and morphological features, given the promising results achieved by our methods.

Acknowledgements. This work was carried out at the Center for Artificial Intelligence (C4AI-USP), with support by the São Paulo Research Foundation (FAPESP grant number 2019/07665-4) and by the IBM Corporation. The project was also supported by the Ministry of Science, Technology and Innovation, with resources of Law N. 8.248, of October 23, 1991, within the scope of PPI-SOFTEX, coordinated by Softex and published as Residence in TIC 13, DOU 01245.010222/2022-44.

References

1. Afonso, S., Bick, E., Haber, R., Santos, D.: Floresta sintá(c)tica: A treebank for Portuguese. In: Proceedings of the Third International Conference on Language Resources and Evaluation (LREC'02). ELRA, Las Palmas, Canary Islands - Spain (May 2002), http://www.lrec-conf.org/proceedings/lrec2002/pdf/1.pdf

---

[6] For reproducibility purposes, all data (including fold splits) and implementation of all methods are available at https://sites.google.com/icmc.usp.br/poetisa/publications.

2. Assunção, J., Fernandes, P., Lopes, L.: Language independent pos-tagging using automatically generated markov chains. In: Proceedings of the 31st International Conference on Software Engineering & Knowledge Engineering, pp. 1–5. Lisbon, Portugal (2019). https://doi.org/10.18293/SEKE2019-097
3. De Souza, E., Freitas, C.: Polishing the gold-how much revision do we need in treebanks? In: Procedings of the Universal Dependencies Brazilian Festival, pp. 1–11 (2022). https://aclanthology.org/2022.udfestbr-1.2.pdf
4. Devlin, J., Chang, M.W., Lee, K., Toutanova, K.: Bert: Pre-training of deep bidirectional transformers for language understanding (2018). https://doi.org/10.48550/ARXIV.1810.04805, https://arxiv.org/abs/1810.04805
5. DiPietro, R., Hager, G.D.: Chapter 21 - deep learning: RNNs and LSTM. In: Zhou, S.K., Rueckert, D., Fichtinger, G. (eds.) Handbook of Medical Image Computing and Computer Assisted Intervention, pp. 503–519. The Elsevier and MICCAI Society Book Series, Academic Press (2020). https://doi.org/10.1016/B978-0-12-816176-0.00026-0
6. Duran, M., Oliveira, H., Scandarolli, C.: Que simples que nada: a anotação da palavra que em córpus de UD. In: Proceedings of the Universal Dependencies Brazilian Festival, pp. 1–11 (2022). https://aclanthology.org/2022.udfestbr-1.3
7. Ehsani, R., Alper, M.E., Eryiğit, G., Adali, E.: Disambiguating main POS tags for Turkish. In: Proceedings of the 24th Conference on Computational Linguistics and Speech Processing (ROCLING 2012), pp. 202–213. The Association for Computational Linguistics and Chinese Language Processing (ACLCLP), Chung-Li, Taiwan (2012). https://aclanthology.org/O12-1021
8. Gers, F.A., Schmidhuber, J.A., Cummins, F.A.: Learning to forget: continual prediction with LSTM. Neural Comput. **12**(10), 2451–2471 (2000). https://doi.org/10.1162/089976600300015015
9. Hoang, M., Bihorac, O.A., Rouces, J.: Aspect-based sentiment analysis using BERT. In: Proceedings of the 22nd Nordic Conference on Computational Linguistics, pp. 187–196. Linköping University Electronic Press, Turku, Finland (2019). https://aclanthology.org/W19-6120
10. Hoya Quecedo, J.M., Maximilian, K., Yangarber, R.: Neural disambiguation of lemma and part of speech in morphologically rich languages. In: Proceedings of the 12th Language Resources and Evaluation Conference, pp. 3573–3582. European Language Resources Association, Marseille, France (2020). https://aclanthology.org/2020.lrec-1.439
11. Ide, N., Suderman, K.: Integrating linguistic resources: The American national corpus model. In: Proceedings of the Fifth International Conference on Language Resources and Evaluation (LREC'06). ELRA, Genoa, Italy (2006). http://www.lrec-conf.org/proceedings/lrec2006/pdf/560_pdf.pdf
12. Kingma, D.P., Ba, J.: Adam: A method for stochastic optimization. In: Bengio, Y., LeCun, Y. (eds.) Proceedings of the 3rd International Conference on Learning Representations (2015). http://arxiv.org/abs/1412.6980
13. Kupiec, J.: Robust part-of-speech tagging using a hidden markov model. Comput. Speech Lang. **6**(3), 225–242 (1992). https://www.sciencedirect.com/science/article/pii/088523089290019Z
14. Lafferty, J.D., McCallum, A., Pereira, F.C.N.: Conditional random fields: probabilistic models for segmenting and labeling sequence data. In: Proceedings of the Eighteenth International Conference on Machine Learning, pp. 282–289. ICML '01, Morgan Kaufmann Publishers Inc., San Francisco, CA, USA (2001). https://dl.acm.org/doi/10.5555/645530.655813

15. Liu, Y., Lapata, M.: Text summarization with pretrained encoders. In: Proceedings of the 2019 Conference on Empirical Methods in Natural Language Processing and the 9th International Joint Conference on Natural Language Processing (EMNLP-IJCNLP), pp. 3730–3740. Association for Computational Linguistics, Hong Kong, China (2019). https://doi.org/10.18653/v1/D19-1387

16. Lopes, L., Duran, M., Fernandes, P., Pardo, T.: Portilexicon-ud: a Portuguese lexical resource according to universal dependencies model. In: Proceedings of the Language Resources and Evaluation Conference, pp. 6635–6643. European Language Resources Association, Marseille, France (2022). https://aclanthology.org/2022.lrec-1.715

17. Lopes, L., Duran, M.S., Pardo, T.A.S.: Universal dependencies-based pos tagging refinement through linguistic resources. In: Proceedings of the 10th Brazilian Conference on Intelligent System. BRACIS'21 (2021). https://link.springer.com/chapter/10.1007/978-3-030-91699-2_41

18. de Marneffe, M.C., Manning, C.D., Nivre, J., Zeman, D.: Universal Dependencies. Comput. Linguist. **47**(2), 255–308 (2021). https://doi.org/10.1162/coli_a_00402, https://aclanthology.org/2021.cl-2.11

19. Muñoz-Valero, D., Rodriguez-Benitez, L., Jimenez-Linares, L., Moreno-Garcia, J.: Using recurrent neural networks for part-of-speech tagging and subject and predicate classification in a sentence. Int. J. Comput. Intell. Syst. **13**, 706–716 (2020). https://doi.org/10.2991/ijcis.d.200527.005

20. Nivre, J., et al.: Universal Dependencies v1: A multilingual treebank collection. In: Proceedings of the Tenth International Conference on Language Resources and Evaluation (LREC'16), pp. 1659–1666. ELRA, Portorož, Slovenia (2016). https://aclanthology.org/L16-1262

21. Paszke, A., et al.: PyTorch: an imperative style, high-performance deep learning library. In: Wallach, H., Larochelle, H., Beygelzimer, A., d'Alché-Buc, F., Fox, E., Garnett, R. (eds.) Advances in Neural Information Processing Systems 32, pp. 8024–8035. Curran Associates, Inc. (2019). http://papers.neurips.cc/paper/9015-pytorch-an-imperative-style-high-performance-deep-learning-library.pdf

22. Rademaker, A., Chalub, F., Real, L., Cláudia Freitas, Bick, E., De Paiva, V.: Universal dependencies for Portuguese. In: Proceedings of the Fourth International Conference on Dependency Linguistics (Depling), pp. 197–206 (2017)

23. Santana, M.: Kaggle - news of the brazilian newspaper. https://www.kaggle.com/marlesson/news-of-the-site-folhauol, accessed: 2021-06-14

24. Shen, Q., Clothiaux, D., Tagtow, E., Littell, P., Dyer, C.: The role of context in neural morphological disambiguation. In: Proceedings of COLING 2016, the 26th International Conference on Computational Linguistics: Technical Papers, pp. 181–191. Osaka, Japan (2016). https://aclanthology.org/C16-1018

25. Silva, E., Pardo, T., Roman, N., Fellipo, A.: Universal dependencies for tweets in brazilian portuguese: Tokenization and part of speech tagging. In: Anais do XVIII Encontro Nacional de Inteligência Artificial e Computacional. pp. 434–445. SBC, Porto Alegre, RS, Brasil (2021). https://doi.org/10.5753/eniac.2021.18273, https://sol.sbc.org.br/index.php/eniac/article/view/18273

26. Souza, F., Nogueira, R., Lotufo, R.: BERTimbau: pretrained BERT models for Brazilian Portuguese. In: 9th Brazilian Conference on Intelligent Systems, BRACIS, Rio Grande do Sul, Brazil, October 20–23 (2020), https://link.springer.com/chapter/10.1007/978-3-030-61377-8_28

27. Straka, M.: UDPipe 2.0 prototype at CoNLL 2018 UD shared task. In: Proceedings of the CoNLL 2018 Shared Task: Multilingual Parsing from Raw Text to Universal Dependencies, pp. 197–207 (2018). https://aclanthology.org/K18-2020

28. Straka, M., Straková, J.: Tokenizing, POS tagging, lemmatizing and parsing UD 2.0 with UDPipe. In: Proceedings of the CoNLL 2017 Shared Task: Multilingual Parsing from Raw Text to Universal Dependencies, pp. 88–99. Association for Computational Linguistics, Vancouver, Canada (2017), https://aclanthology.org/K17-3009

29. Universal Dependencies: UD Portuguese Bosque - UD version 2. https://universaldependencies.org/treebanks/pt_bosque/index.html. Accessed 14 Jun 2021

30. Vandenbussche, P.Y., Scerri, T., Jr., R.D.: Word sense disambiguation with transformer models. In: Proceedings of the 6th Workshop on Semantic Deep Learning (SemDeep-6), pp. 7–12. Association for Computational Linguistics, Online (2021) https://aclanthology.org/2021.semdeep-1.2

31. Wolf, T., et al.: Transformers: State-of-the-art natural language processing. In: Proceedings of the 2020 Conference on Empirical Methods in Natural Language Processing: System Demonstrations, pp. 38–45. Association for Computational Linguistics, Online (2020). https://www.aclweb.org/anthology/2020.emnlp-demos.6

32. Zalmout, N., Habash, N.: Don't throw those morphological analyzers away just yet: Neural morphological disambiguation for Arabic. In: Proceedings of the 2017 Conference on Empirical Methods in Natural Language Processing, pp. 704–713. Association for Computational Linguistics, Copenhagen, Denmark (2017). https://aclanthology.org/D17-1073

Bete: A Brazilian Portuguese Dataset for Named Entity Recognition and Relation Extraction in the Diabetes Healthcare Domain

Lucas Pavanelli[1(✉)], Yohan Bonescki Gumiel[2], Thiago Ferreira[1],
Adriana Pagano[2], and Eduardo Laber[3]

[1] aiXplain Inc., Los Gatos, USA
{lucas.pavanelli,thiago}@aixplain.com
[2] Universidade Federal de Minas Gerais, Belo Horizonte, Brazil
yohan.gumiel@gmail.com, apagano@ufmg.br
[3] Pontifícia Universidade Católica do Rio de Janeiro (PUC-RJ), Rio de Janeiro,
Brazil
laber@inf.puc-rio.br

Abstract. The biomedical NLP community has seen great advances in dataset development mostly for the English language, which has hindered progress in the field, as other languages are still underrepresented. This study introduces a dataset of Brazilian Portuguese annotated for named entity recognition and relation extraction in the healthcare domain. We compiled and annotated a corpus of health professionals' responses to frequently asked questions in online healthcare forums on diabetes. We measured inter-annotator agreement and conducted initial experiments using up-to-date methods to recognize entities and extract relations, such as BERT-based ones. Data, models, and results are publicly available at https://github.com/pavalucas/Bete.

Keywords: Dataset · Named Entity Recognition · Relation Extraction

1 Introduction

Named entity recognition (NER) identifies named entities in texts; these entities can be proper names, locations, and organizations. Relation extraction (RE) consists in predicting whether and what predefined relation exists between two mentioned entities or if they are unrelated. In the healthcare domain, named entities include mentions relevant to the clinical context, such as treatment plans, drugs, and symptoms while relations can be, for instance, if a certain drug treats a symptom. Both tasks are prevalent in machine learning methods [2] and are considered the primary step for several other Natural Language Processing (NLP) tasks, such as question answering [20], document classification [6], and search engines [12].

M. C. Naldi and R. A. C. Bianchi (Eds.): BRACIS 2023, LNAI 14197, pp. 256–267, 2023.
https://doi.org/10.1007/978-3-031-45392-2_17

Consultation time is not always sufficient for health professionals to answer all the patient and their families questions, and patients may not have easy access to primary care centers [8]. Hence, patients often engage in dedicated public forums to search for answers to their queries. Among chronic diseases, diabetes stands out as a condition much in need of attention because it is an increasingly prevalent and severe long-term problem quickly growing worldwide, particularly in developing countries, as is the case of Brazil. In 2019, it was estimated that almost half a billion of the world's population (9.3% of the adults between 20–79 years) had diabetes [18]. Further, one in every two (50.1%) persons with diabetes is unaware of or has not been diagnosed with this condition. Hence, question answering (QA) systems could help relieve some of the burdens of health care. In QA, an answer to a question is found by querying a corpus of text documents. A QA system incorporates NER and RE components so that entities and their relations can be detected and answers can be more efficiently found.

In this study, the main goal was to annotate a corpus of texts in order to develop a framework to automatically identify healthcare-related entities and relations. To that end, medical and nutrition science students produced texts as answers to queries posted by users in diabetes-related public forums. Our work is thus relevant for studies in the healthcare domain targeting specifically the general public and focusing on Brazilian Portuguese. The obtained dataset is expected to be used to build BeteQA, a community question answering system that provides fast and precise answers to questions about Diabetes Mellitus posed by the lay public [5].

To the best of our knowledge, this is the first study focusing on NER and relation extraction drawing on a novel corpus of real-life questions answered by professional healthcare specialists in Brazilian Portuguese.

1.1 Related Work

Among named entity recognition methods, contextual word representations are employed due to their capacity for modeling complex word usage characteristics and variations across linguistic contexts [17]. A method worth highlighting is BERT [7], a masked language model that benefits from transformers. These pre-trained language models are becoming the new NER paradigm because of their contextualized embeddings. Furthermore, they are fine-tuned for several NLP tasks by adding an additional output layer [13].

As far as relation extraction methods, BERT-based models are also dominant. Soares et al. [21] proposed a model that learns relation representations directly from the text using BERT-based models.

Considering the available BERT models for Brazilian Portuguese, there is a multilingual version [7]; a Brazilian Portuguese version focused on the general domain called BERTimbau [22]; and a Brazilian Portuguese version focused on the clinical domain named BioBERTpt [19].

258   L. Pavanelli et al.

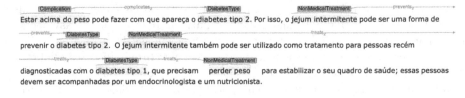

Fig. 1. Annotation example from our corpus.

2   Methodology

2.1   Corpus

We searched for questions posed by users regarding health issues in several online forums. In such forums, users answer each other's questions drawing on their beliefs and understanding of the issues, with no help or supervision of any healthcare professional. To avoid this problem, we designed a study in which answers were produced by medical and nutrition science students relying on their domain knowledge. Answers were curated by expert professionals in our project who supervised students. This way, we ensured that our set of QA is reliable and can be used in a prospective QA system to be queried through a conversational agent. Our corpus comprises two sets of documents made up of diabetes-related general questions with their respective answers: a first set contains 304 real online forum questions answered by medical students under the supervision of medical professionals and a second one contains 201 real online forum questions answered by nutrition science students supervised by professionals. This way, we can have a QA system to provide accurate answers about health and nutrition issues, authored by professionals in both fields.

Annotation Setup. As an annotation tool, we used Webanno [4], an open-source and intuitive software/platform. After comparing several available entity tagging tools, we found Webanno to be the easiest and most efficient tool for our purposes. The system was set up on a web server, text data was uploaded, and entity/relation types were defined within the system.

Annotation Guidelines. The annotation guidelines were created in an iterative process. A first draft was created, containing general guidelines as well as specific examples of types of entities and relations. Specialists were consulted regarding annotators' queries and their answer was used to update the guidelines, then they were tested again. Besides, during the annotation process, whenever one of the annotators ran into a dubious case, this was added to the guidelines. Figure 1 shows an example of a text annotated in Webanno following our guidelines[1]. The annotation guidelines are publicly available for download as part of the dataset.

---

[1] Translation into English: "Being overweight can lead to type 2 diabetes. Therefore, intermittent fasting may be a way to prevent type 2 diabetes. Intermittent fasting

Annotation Process. We recruited undergraduate students pursuing their BA degree to complete the annotation task. The students are part of Empoder@, a multidisciplinary project engaging health sciences, statistics, computer science, and applied linguistics students. The project aims to empower researchers, professionals, and users of health services.

Annotators took part in a training session and were requested to read the guidelines and resort to the project coordinator whenever they encountered problems during annotation. Two students annotated each document, and a third one performed the adjudication so that a gold-standard was obtained upon completion. Annotators were presented with each text on a simple interface (see Fig. 1) and using a mouse or a track pad they selected entities and dragged relationships between them.

2.2   Entity and Relation Extraction

We considered relation extraction as a multi-label classification problem in which, given a pair of entities, a label is assigned out of the relation types available. Also, we decoupled relation extraction from entity recognition, so we performed RE on the gold entities.

The set of entity and relation types devised for our annotation was built drawing on an ontology proposed in [1]. The ontology labels diabetes mentions as "DiabetesType", and diabetes-related diseases are classed with the "Complication" entity type. Diabetes-related temporal expressions, clinical tests, and treatments are also addressed in the ontology. Table 1 lists the 14 entity types, providing a brief explanation of each label with some examples. The 5 relations types are verbalized by the following verbs: "causes", "diagnoses", "has", "prevents", and "treats".

Reliability. To measure dataset reliability, we computed inter-annotator agreement (IAA) considering exact matches. Following the work of [3], we computed pairwise F1 score and Cohen's Kappa. The former is more reliable according to various studies [9,11]. Because of the vast amount of unannotated tokens (labeled "O"), we calculated the scores without the O label, for both annotated entities and relations. Table 2 shows the obtained agreement. Annotated entities can be said to be fully reliable, achieving an IAA of 0.93. As regards relations, moderate agreement (0.58) was found.

3   Dataset Information

Table 3 shows overall statistics of the whole dataset. Table 4 shows the number of annotations per entity type, while Table 5 covers the annotated relations.

---

can also be used as a treatment for people newly diagnosed with type 1 diabetes who need to lose weight to achieve a more stable health condition; these people should be advised and monitored by an endocrinologist and a nutritionist.".

Table 1. Entities description and examples.

Category	Description	Examples of annotated entities
Diabetes Type	subclass of diabetes	type 2 - type 1
Complication	diseases and health conditions causing or caused by diabetes	being overweight - wounds - depression - neuropathy
Symptom	physical or mental condition experienced by the patient regarded as indicating diabetes	low blood sugar
Glucose Value	measurement of blood sugar level	250 - 100 - 80
Insulin	insulin type	NPH - Aspart
Medication	prescribed drugs or medicine	Metformin - Tetracaine hydrochloride
Non Medical Treatment	healthcare activities or behavior other than prescribed medication	intermittent fasting - physical exercise
Food	source of nutritional support for organisms	peanut butter, candies, bread
Dose	amount, quantity or size	150 ml - 200 g - 1 glass
Test	medical exams	blood test - glycosylated hemoglobin test
Date	calendar dates	17/01/2021
Time	point in time	at night - at bedtime - at midday
Duration	length of time for occurrence	half an hour - twenty minutes
Set	frequency of occurrence	twice a week - every day

Table 2. Inter-annotator agreement for entity recognition and relation extraction.

Annotation type	F1 score	Cohen's Kappa
Entity	0.93	0.91
Relation	0.58	0.32

Table 3. Dataset information.

Documents	Sentences	Tokens	Entities	Relations
505	2340	55530	2396	1223

Table 4. Entities: Number of occurrences and percentage per entity type sorted in decreasing order.

Entity	Count	%
Food	631	26.35
Complication	410	17.12
NonMedicalTreatment	405	16.91
Symptom	309	12.90
GlucoseValue	308	12.85
Time	71	2.96
Test	67	2.80
Medication	52	2.17
DiabetesType	48	2.00
Set	29	1.21
Insulin	25	1.04
Dose	23	0.96
Duration	18	0.75

Table 5. Relations: Number of occurrences and percentage per relation type sorted in decreasing order.

Relation	Count	%
has	833	68.11
treats	202	16.52
causes	79	7.03
diagnoses	52	4.25
prevents	50	4.09

4   Experiments Setup

We conducted initial experiments using methods to recognize entities and extract relations. Since the dataset is in Brazilian Portuguese, we chose deep learning models trained on multilingual and Brazilian Portuguese data. These models are multilingual BERT (mBERT), BERTimbau, and the three different versions of BioBERTpt: BioBERTpt-bio, trained on Portuguese biomedical texts, BioBERTpt-clin, trained on clinical narratives from electronic health records from Brazilian Hospitals, and BioBERTpt-all, trained in both biomedical texts and clinical narratives.

Regarding the training setup, we randomly divided the 505 documents into train/dev/test using the split 0.8/0.1/0.1, respectively, tuning the hyperparameters on the development set and reporting the results on the test set. To run the experiments, we used one NVIDIA GeForce RTX 3090 GPU.

As for models, we trained a baseline Conditional Random Field (CRF) for the NER task. We used a Portuguese model from the Python library called Spacy [10] to extract a set of features. Table 6 shows the used features.

Table 6. Used CRF features.

Feature
Part of speech tagging, extracted using Spacy [10]
If the word is in uppercase
If the word is a digit
If the word is a title, i.e. start with an uppercase letter and the rest is in lowercase
The previous word and the above features
The next word and the above features

For BERT models for the NER task, we used the Adam [14] optimizer with a learning rate of 1e-5 and a maximum length of 512. Moreover, we trained for 50 epochs with early stopping of 15 epochs and a batch size of 64.

Regarding the relation extraction task, we experimented with a baseline Support Vector Machine (SVM) model, using the one-vs-the-rest (OvR) multiclass strategy from scikit-learn [16] Python library.

Considering BERT for relation extraction, we used 7e-5 as the learning rate with Adam optimizer and max length of 512, training for 11 epochs, and a batch size of 32.

We considered the following metrics for evaluation: precision, recall, and F1 score. We reported the weighted average F1 score, averaging the support weighted mean per label. All metrics are considering exact matches. In addition, we computed results considering all classes and, for the best-performing model, we reported metrics for each label.

5   Results

Table 7 shows the results for NER models. We trained baseline CRF and BERT-based models. The best one concerning F1 score is the BioBERTpt-clin model, outperforming BioBERTpt-all by 0.9 points. Also, BERTimbau did not perform well with an F1 score 5.3 points lower than the second-lowest.

To evaluate the relation extraction task, we experimented using the two models: SVM, as the baseline, and BERT for relation extraction (BERT-RE) from [21], using mBERT as the BERT encoder.

Table 8 shows the result for both relation extraction models. We can observe that, although SVM has higher precision than BERT-RE, the latter performs better overall with 17 points of F1 score difference.

We also provide an in-depth analysis of the best-performing model in NER and RE. Table 9 shows detailed results for the entity recognition BioBERTpt-clin model. The method produced good scores (>80%) for the three entities with the largest number of examples: Food, Complication, and Symptom. However, for some entities with few examples, the model was not able to perform good predictions (<60%): Test, Time, and Set.

Table 10 describes an in-depth analysis of BERT-RE performance for each relation type. Similarly to NER results, the model performs well for the relation with most examples (has) and falls short in relations with few examples: causes and prevents.

Table 7. Experiments for entity recognition models. The best scores are highlighted in bold.

Model	Precision	Recall	F1
CRF	**80.3**	72.9	76.1
BioBERTpt-bio	73.1	80.5	76.6
BioBERTpt-clin	77.5	81.8	**79.4**
BioBERTpt-all	74.5	**83.2**	78.5
BERTimbau	72.2	70.2	70.8
mBERT	74.3	81.6	77.6

Table 8. Experiments for relation extraction models. The best scores are highlighted in bold.

Model	Precision	Recall	F1
SVM	**80.0**	46.2	58.1
BERT-RE (mBERT)	73.6	**77.8**	**75.1**

Table 9. Detailed results for the best entity recognition model (BioBERTpt-clin).

Entities	Precision	Recall	F1	N° of examples
Complication	84.8	91.8	88.2	61
DiabetesType	100.0	100.0	100.0	2
Dose	50.0	100.0	66.7	2
Duration	100.0	100.0	100.0	1
Food	79.1	81.5	80.3	135
GlucoseValue	60.4	78.4	68.2	37
Insulin	100.0	100.0	100.0	4
Medication	68.8	91.67	78.6	12
NonMedicalTreatment	72.7	64.9	68.6	37
Set	66.7	50.0	57.1	4
Symptom	90.0	94.7	92.3	57
Test	50.0	33.3	40.0	9
Time	44.4	50.0	47.1	8
Weighted Average	77.5	81.8	79.4	369

Table 10. Detailed results for the best relation extraction model (BERT-RE).

Entities	Precision	Recall	F1	N° of examples
causes	33.33	33.33	33.33	6
prevents	66.67	50.00	57.14	4
treats	52.94	6429	58.06	14
has	77.89	86.05	81.77	86
diagnoses	100.00	57.14	72.73	7
Weighted Average	73.56	77.78	75.06	117

6   Discussion

In this study, we introduced a novel dataset and models for NER and RE tasks, gathering answers written by health professionals about Diabetes Mellitus to online forum users.

Corpus. We found that the entities Food and Complication were the most annotated in the corpus. Also, nutrition is a prevalent topic in online forums, as people with diabetes frequently inquire whether they can or cannot consume specific foods, such as chocolate, or alcoholic drinks. Similarly, people are generally keen on finding further information on diabetes-related complications because they feel specific symptoms or want to know about frequent diabetes-related complications.

Regarding the main difficulties found in the process of annotating relations, despite annotators' prior training and availability of guidelines, some entities were annotated as having a different extent in terms of words making up those entities. Thus, "type 2 Diabetes" and "Diabetes" are two different annotations that refer to the same entity type, "DiabetesType". This ends up impacting the relation since a relation is defined by a pair of entities and a relation type, which can justify our lower agreement for relation annotation. So if two annotators annotated the same relation type holding between entities spanning different extents in words, the resulting relations did not match. Further, as evidenced in [15], temporal relation extraction, which is a specific type of relation extraction, usually has lower agreement than span annotation.

Models. For the NER task, among the BERT models, we found that BioBERTpt-clin, which leveraged clinical data, had superior performance compared to the other models. Both BioBERTpt-clin and BioBERTpt-all were trained on clinical data and were initialized with the weights from Multilingual BERT, so as expected, these models achieved similar results. It is also worth noting that, although trained on Brazilian Portuguese data, BERTimbau did not perform well. An explanation is that BERTimbau was trained on brWAC [23], a large corpus extracted from the Web, that differs in vocabulary and context from Bete medical data.

Considering relation extraction, the BERT-based model outperforms the baseline model, showing the dominance of context-aware embeddings compared to kernel-based methods that were dominant in the past.

7 Conclusion

To the best of our knowledge, this is the first study yielding an annotated corpus of NER in Brazilian Portuguese made up of diabetes-related answers authored by domain specialists in response to questions posed by lay users. Moreover, it contributes to the research field by introducing resources for a sensitive context, such as diabetes, and creating models for a low-resource language, such as Brazilian Portuguese.

The fine-tuning of models leveraging clinical data was found to improve the results. Hence, the vocabulary and the context from the clinical context boosted the model's ability to predict entities.

We plan to expand our corpus in future work, especially for the categories of entities and relations that have a small number of instances.

Limitations

Among the limitations of our study is the size of our dataset, which is due to the fact that it was obtained through manual annotation thereby demanding human effort and more time to accomplish this task. Another limitation is that some entity types have few instances, making our dataset slightly imbalanced. For instance, only 1/3 of our entity types have percentages of occurrence higher than 3%. Hence, adding more training examples and making the dataset less imbalanced will certainly enhance our results.

Our dataset targets a particular domain - diabetes; future experiments targeting other chronic diseases, such as cardiovascular disease, will boost our potential. Additionally, in the aftermath of COVID-19, there is growing uncertainty about procedures, symptoms, and treatments, with several questions being asked over social media. Hence, addressing COVID-19-related frequently asked questions would be a valuable contribution.

Ethical Statement. Our study fully complies with ethical standards and did not require any submission to ethical boards, since no data collection with human subjects was carried out. Our dataset was created by our team and contains texts drafted by medical students under the supervision of healthcare professionals, all of whom are research members in our project.

References

1. Ben Abacha, A., Zweigenbaum, P.: MEANS: a medical question-answering system combining NLP techniques and semantic web technologies. Inf. Process. Manag. **51**(5), 570–594 (2015). https://doi.org/10.1016/j.ipm.2015.04.006, https://www.sciencedirect.com/science/article/pii/S0306457315000515

2. Bose, P., Srinivasan, S., Sleeman, W.C., Palta, J., Kapoor, R., Ghosh, P.: A survey on recent named entity recognition and relationship extraction techniques on clinical texts. Appl. Sci. **11**(18) (2021). https://doi.org/10.3390/app11188319, https://www.mdpi.com/2076-3417/11/18/8319

3. Brandsen, A., Verberne, S., Wansleeben, M., Lambers, K.: Creating a dataset for named entity recognition in the archaeology domain. In: Proceedings of the Twelfth Language Resources and Evaluation Conference, pp. 4573–4577. European Language Resources Association, Marseille, France (2020), https://aclanthology.org/2020.lrec-1.562

4. Eckart de Castilho, R., et al.: A web-based tool for the integrated annotation of semantic and syntactic structures. In: Proceedings of the Workshop on Language Technology Resources and Tools for Digital Humanities (LT4DH), pp. 76–84. The COLING 2016 Organizing Committee, Osaka, Japan (2016), https://www.aclweb.org/anthology/W16-4011

5. Castro Ferreira, T., et al.: Evaluating recognizing question entailment methods for a Portuguese community question-answering system about diabetes mellitus. In: Proceedings of the International Conference on Recent Advances in Natural Language Processing (RANLP 2021), pp. 234–243. INCOMA Ltd., Held Online (2021), https://aclanthology.org/2021.ranlp-main.28

6. Choudhary, A., Arora, A.: Linguistic feature based learning model for fake news detection and classification. Expert Syst. Appl. **169**, 114171 (2021)

7. Devlin, J., Chang, M.W., Lee, K., Toutanova, K.: BERT: pre-training of deep bidirectional transformers for language understanding. In: Proceedings of the 2019 Conference of the North American Chapter of the Association for Computational Linguistics: Human Language Technologies, Volume 1 (Long and Short Papers), pp. 4171–4186 (2019). https://doi.org/10.18653/v1/N19-1423, https://www.aclweb.org/anthology/N19-1423

8. Gabarron, E., et al.: Social media for health promotion in diabetes: study protocol for a participatory public health intervention design. BMC Health Serv. Res. **18**(1), 414 (2018). https://doi.org/10.1186/s12913-018-3178-7

9. Grouin, C., Rosset, S., Zweigenbaum, P., Fort, K., Galibert, O., Quintard, L.: Proposal for an extension of traditional named entities: from guidelines to evaluation, an overview. In: Proceedings of the 5th linguistic annotation workshop, pp. 92–100 (2011)

10. Honnibal, M., Montani, I.: spaCy 2: Natural language understanding with Bloom embeddings, convolutional neural networks and incremental parsing. To Appear **7**(1), 411–420 (2017)

11. Hripcsak, G., Rothschild, A.S.: Agreement, the f-measure, and reliability in information retrieval. J. Am. Med. Inform. Assoc. **12**(3), 296–298 (2005)

12. Lahav, D., et al.: A search engine for discovery of scientific challenges and directions. In: AAAI (2022)

13. Li, J., Sun, A., Han, J., Li, C.: A survey on deep learning for named entity recognition. IEEE Trans. Knowl. Data Eng. **34**(1), 50–70 (2020)

14. Loshchilov, I., Hutter, F.: Decoupled weight decay regularization. In: International Conference on Learning Representations (2019). https://openreview.net/forum?id=Bkg6RiCqY7

15. Nikfarjam, A., Emadzadeh, E., Gonzalez, G.: Towards generating a patient's timeline: Extracting temporal relationships from clinical notes. J. Biomed. Inform. **46**, S40–S47 (2013). https://doi.org/10.1016/j.jbi.2013.11.001, supplement: 2012 i2b2 NLP Challenge on Temporal Relations in Clinical Data

16. Pedregosa, F., et al.: Scikit-learn: machine learning in Python. J. Mach. Learn. Res. **12**, 2825–2830 (2011)

17. Peters, M.E., et al.: Deep contextualized word representations. In: Proceedings of the 2018 Conference of the North American Chapter of the Association for Computational Linguistics: Human Language Technologies, Volume 1 (Long Papers), pp. 2227–2237. Association for Computational Linguistics, New Orleans, Louisiana (2018). https://doi.org/10.18653/v1/N18-1202, https://aclanthology.org/N18-1202

18. Saeedi, P., et al.: Global and regional diabetes prevalence estimates for 2019 and projections for 2030 and 2045: Results from the international diabetes federation diabetes atlas, 9th edition. Diabetes Res. Clin. Pract. **157**, 107843 (2019). https://doi.org/10.1016/j.diabres.2019.107843

19. Schneider, E.T.R., et al.: BioBERTpt - a Portuguese neural language model for clinical named entity recognition. In: Proceedings of the 3rd Clinical Natural Language Processing Workshop, pp. 65–72. Association for Computational Linguistics (2020). https://doi.org/10.18653/v1/2020.clinicalnlp-1.7

20. Sharma, V., Kulkarni, N., Pranavi, S., Bayomi, G., Nyberg, E., Mitamura, T.: BioAMA: towards an end to end biomedical question answering system. In: Proceedings of the BioNLP 2018 workshop, pp. 109–117. Association for Computational Linguistics, Melbourne, Australia (2018). https://doi.org/10.18653/v1/W18-2312, https://aclanthology.org/W18-2312

21. Soares, L.B., FitzGerald, N., Ling, J., Kwiatkowski, T.: Matching the blanks: distributional similarity for relation learning. In: ACL 2019–57th Annual Meeting of the Association for Computational Linguistics, Proceedings of the Conference, pp. 2895–2905 (2020). https://doi.org/10.18653/v1/p19-1279

22. Souza, F., Nogueira, R., Lotufo, R.: BERTimbau: pretrained BERT models for Brazilian Portuguese. In: Cerri, R., Prati, R.C. (eds.) BRACIS 2020. LNCS (LNAI), vol. 12319, pp. 403–417. Springer, Cham (2020). https://doi.org/10.1007/978-3-030-61377-8_28

23. Wagner, J., Wilkens, R., Idiart, M., Villavicencio, A.: The brWaC corpus: a new open resource for Brazilian Portuguese (2018)

LegalBert-pt: A Pretrained Language Model for the Brazilian Portuguese Legal Domain

Raquel Silveira[1]([✉])(ID), Caio Ponte[2](ID), Vitor Almeida[2](ID), Vládia Pinheiro[2](ID), and Vasco Furtado[2](ID)

[1] Federal Institute of Ceará (IFCE), Fortaleza, Brazil
`raquel_silveira@ifce.edu.br`
[2] University of Fortaleza (UNIFOR), Fortaleza, Brazil
`{caioponte,vasco}@unifor.br`

Abstract. Language models trained with Bidirectional Encoder Representations from Transformers (BERT) have demonstrated remarkable results in various Natural Language Processing (NLP) tasks. However, the legal domain poses specific challenges for NLP due to its highly specialized language, which includes technical vocabulary, formal style, frequent use of law citations and semantics based on vast knowledge. Therefore, pretrained language models on a generic corpus may not be suitable for performing specific legal domain tasks. They lack the necessary expertise to understand the nuances of legal language, leading to inaccuracies and inconsistencies. This work describes the development of a specialized language model, LegalBert-pt, for the legal domain in Portuguese. The model was pretrained on a large and diverse corpus of Brazilian legal texts and is now open-source and customizable for specific tasks. Experiments were conducted to evaluate the pretrained model's effectiveness in the legal domain, both intrinsically and in two specific tasks: named-entity recognition and text classification. The results indicate that using LegalBert-pt outperforms the generic language model in all tasks, emphasizing the importance of specialization in achieving effective results for specific tasks in the legal domain.

Keywords: Language Models · Legal Texts · BERT · BERTimbau

1 Introduction

As legal documents become increasingly digitized, Natural Language Processing (NLP) has gained importance for automating tasks in the legal field. NLP tools are now commonly used to address real-world legal problems, including the identification of participants in legal proceedings [24], the classification of legal documents [4], named-entity recognition [3], and legal text summarization [13]. These solutions employ traditional machine learning paradigms (e.g., TF-IDF and classifiers) or deep learning techniques to achieve their objectives [28].

Language models structured in the Transformers architecture, such as BERT [11] and its variants [7,33], have achieved promising results in several downstream NLP tasks on generic reference datasets. In recent years, there have been some efforts to pretrain linguistic models for Brazilian Portuguese against more traditional word embeddings (Word2Vec, FastText, etc.), as well as structured models in BERT.

However, these models were trained with general documents and were not designed to represent the Brazilian legal language [28]. The effectiveness of language models on domain-specific tasks can be limited by the model's lack of specialization for that domain. The development of language models for legal texts is justified since the language used for the preparation of legal documents has its own vocabulary, formal style, semantics based on a wide spectrum of knowledge, and frequent use of citations to laws. Evidence shows that using pretrained language models with domain-specific corpus can significantly improve performance on domain-specific tasks [31, 37].

This work describes the production process of a language model for the legal domain in the Portuguese language. The model was pretrained to acquire specialization for the domain, and later it could be adjusted for use in specific tasks. Two versions of the model were created: one as a complement to the BERTimbau model [33], and the other from scratch. The effectiveness of the model based on BERTimbau was evident when analyzing the perplexity measure of the models. Experiments were also carried out in the tasks of identifying legal entities and classifying legal petitions. The results show that the use of specific language models outperforms those obtained using the generic language model in the tasks studied, suggesting that the specialization of the language model for the legal domain is an important factor for improving the accuracy of learning algorithms.

2 Related Work

Progress in the field of NLP is closely linked to advancements in Machine Learning models, primarily due to the emergence of Word Embeddings. However, these models possess a drawback in that they generate static representations of words, meaning the same word always produces a fixed embedding, even when found in sentences with varying contexts. The ELMo model [27] and subsequently, Transformers [34], began to generate contextualized embeddings. As a result, the same word will have different representations in diverse situations, depending on the context of the entire sentence.

By employing self-attention mechanisms, Transformers are able to capture long-range relationships [19]. Models built upon the Transformers architecture, such as BERT [11] and GPT [6], have emerged as the state of the art in NLP tasks. The core principle behind using these pretrained language models lies in the transfer learning and self-supervised learning approach. In other words, the model is initially trained on a sizable unlabeled corpus to learn the universal representation of the language [14]. Subsequently, the knowledge gained during pretraining can be applied to fine-tune downstream tasks, reducing the reliance on labeled data and enhancing the performance of the models [25].

Although pretrained language models, such as BERT, exhibit strong performance in generic texts, they may yield inferior results in domain-specific texts

[7]. Therefore, implementing strategies to enhance training by incorporating data with texts from a specific domain is a widely-used technique in various fields. Examples include: (i) BioBERT [16], which was trained on biomedical texts and outperformed BERT as well as other state-of-the-art models; (ii) FinBERT [38], which utilized a large corpus of financial communication comprising 4.9 billion tokens and demonstrated superior performance to BERT in sentiment classification tasks; (iii) SciBERT [5], which employed full texts from 1.14 million Semantic Scholar articles and showcased improved performance compared to BERT-Base in NLP tasks within the scientific domain; (iv) CodeBERT [12], which made use of open-source code from public GitHub repositories across six programming languages, achieving state-of-the-art results in natural language code search and code-to-documentation generation tasks.

The legal field serves as an excellent example of a domain that could benefit from generating pretrained language models, given the vast amounts of data produced daily in courts and legal digital platforms. LEGAL-BERT [7] was among the pioneers in developing legal language models, utilizing a corpus of approximately 12 GB with texts from European and North American legislation and cases. The strategies employed in LEGAL-BERT's pretraining included: (i) LEGAL-BERT-FP, which considered additional training from BERT, and (ii) LEGAL-BERT-SC, which focused on training from scratch exclusively within the legal corpora. Both strategies outperformed BERT and achieved state-of-the-art results in three end-tasks.

However, pretraining must also take into account not only the specific domain but also the language in which the downstream tasks need to be addressed. A model trained on multiple languages may yield inferior results in languages that lack adequate representation in the dataset [10]. Consequently, efforts have been made to create monolingual pretrained language models that tackle tasks in the legal domain, such as Lawformer [37], ITALIAN-LEGAL-BERT [18], InLegal-BERT [26], and JurisBERT [35].

3   LegalBert-pt: A BERT Model for the Brazilian Legal Domain

This section provides a detailed account of the steps taken to pretrain LegalBert-pt, a language model for the Portuguese legal domain.

3.1   Pretraining Data

To pretrain various versions of the LegalBert-pt language model, we collected a total of 1.5 million legal documents in Portuguese from ten Brazilian courts. These documents consisted of four types: initial petitions, petitions, decisions, and sentences. Table 1 shows the distribution of these documents.

The data were obtained from the Codex system of the Brazilian National Council of Justice (CNJ), which maintains the largest and most diverse set of legal texts in Brazilian Portuguese. As part of an agreement established with the researchers who authored this article, the CNJ provided these data for our research.

Our use of this corpus allowed us to pretrain variations of the LegalBert-pt model that are well-suited to handling the nuances and complexities of legal language in the Brazilian context. We drew upon previous research [3,4,35] that demonstrated the importance of using large and diverse datasets for training language models, particularly in domain-specific contexts.

Table 1. Statistical Legal Documents by Data Source.

Data source	Number of documents	%
Court of Justice of the State of Ceará	80,504	5.37%
Court of Justice of the State of Piauí	90,514	6.03
Court of Justice of the State of Rio de Janeiro	33,320	2.22
Court of Justice of the State of Rondônia	971,615	64.77
Federal Regional Court of the 3rd Region	70,196	4.68
Federal Regional Court of the 5th Region	6,767	0.45
Regional Labor Court of the 9th Region	16,133	1.08
Regional Labor Court of the 11th Region	5,351	0.36
Regional Labor Court of the 13th Region	155,567	10.37
Regional Labor Court of the 23th Region	70,033	4.67
Total	1,500,000	100.00%

To minimize errors in the texts of the documents, we employed a two-stage preparation process consisting of pre-processing and cleaning. In the pre-processing stage, we removed documents with less than 50 words and less than 80% valid words to ensure that the remaining corpus was of high quality. In the cleaning step, we removed special characters and extra spacing to further improve the corpus. Table 2 provides statistics on the types of legal documents that remained after the preparation steps. These documents were then used to train various versions of the LegalBert-pt language model.

After the preparation steps, the documents were divided into sentences with a maximum size of 512 tokens, generating a total of about 12,000,000 sentences.

3.2 Vocabulary Generation

Since BERTimbau [33] is trained on data from the general domain, we believe that training a language model for a specific domain can improve its performance on domain-specific tasks with a specific vocabulary. To this end, we generated a vocabulary consisting of 30,000 subword units using the SentencePiece library [15] and the Byte-Pair Encoding (BPE) algorithm [30]. We used 2 million random sentences from 1 million Wikipedia articles in Portuguese and 2 million random

Table 2. Details of the training corpus used to pretrain the different variations of LegalBert-pt.

Documento type	Number of documents	%
Initial petition	338,648	22.58%
Sentence	199,433	13.30%
Petition	503,310	33.55%
Decision	200,076	13.34%
Other documents	258,533	17.23%
Total	1,500,000	100.00%

sentences from 1.5 million legal documents in our pretraining dataset described in Sect. 3.1.

To ensure compatibility with the original BERT code, we converted the resulting vocabulary to the WordPiece format. To do this, we followed the BERT tokenization rules. First, we added all special BERT tokens ([CLS], [MASK], [SEP], and [UNK]) and punctuation characters to the English vocabulary. Then, we split SentencePiece tokens that contain punctuation characters, removed the punctuation, and added the resulting subword units to the vocabulary. Finally, we prefixed subword units that do not begin with the SentencePiece metacharacter "_" with "##" and removed the "_" symbol from the remaining tokens.

Additionally, we included 5,977 identifiers of Brazilian legislation in the vocabulary. The resulting LegalBert-pt vocabulary consists of 36,345 subwords.

Figure 1 compares the number of subwords in the LegalBert-pt and BERTimbau vocabularies. We found that 16,885 subwords are common between the two vocabularies, while 19,460 subwords are specific to LegalBert-pt.

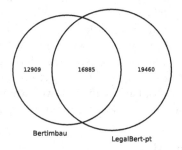

Fig. 1. Number of units of subwords of the BERTimbau and LegalBert-pt vocabularies.

3.3   Variations of the LegalBert-pt Model

The use of BERT in downstream tasks involves two stages: pretraining and model fine-tuning. In the pretraining stage, the model is trained from scratch or with additional steps from an existing model to learn bidirectional context between tokens. This step is computationally intensive and should only be performed once. In the fine-tuning stage, a pretrained model is further trained on a specific task of interest, such as text classification or named entity recognition.

For our study, we developed two variations of the pretraining of legal domain language models in Brazilian Portuguese: (i) pretraining from scratch using a specific domain corpus (LegalBert-pt SC) and (ii) an adaptation of BERTimbau with pretraining using a specific domain corpus (LegalBert-pt FP). Both models were pretrained as case-sensitive, as we focused on developing general-purpose models and capitalization is relevant for tasks such as named-entity recognition.

We pretrained the models using the Masked Language Model (MLM) task. We only used the MLM task in pretraining, as recent research [20] has suggested that the Next Sentence Prediction (NSP) task is not effective. The MLM objective allows the representation to learn left and right context, which allows us to pretrain a deep two-way transformer. The masked language model randomly chooses some tokens from the input and replaces them with either a special [MASK] token with 80% probability, a random vocabulary token with 10% probability, or the original token with 10% probability. The goal is to predict the ID of the masked word in the original vocabulary based on its context.

During training, we used the ADAMW optimizer [21] with the following parameters: $\beta_1 = 0.9$, $\beta_2 = 0.999$, $\epsilon = 1e - 6$, and a learning rate of $1e - 4$.

The LegalBert-pt SC model has the same architecture as BERTimbau-Base, with 12 layers, 768 hidden units, and 12 attention heads (a total of 110 million parameters). We used this architecture in all of our experiments. For this model, we used the specialized vocabulary generated for the legal domain as described in Sect. 3.2. We pretrained this model for 7.5 million steps on a legal domain corpus, as described in Sect. 3.3.

For the LegalBert-pt FP model, we followed the approach outlined in [11], initializing the weights from the pretrained BERTimbau-Base checkpoint [33], and then performing additional pretraining steps using a legal domain corpus, as described in Sect. 3.3. [11] suggests performing additional pretraining steps up to 1,000,000 steps. In our case, we pretrained the model for up to 2.4 million steps to evaluate the prolonged effect of pretraining on downstream tasks. BERTimbau-Base has been significantly pretrained on generic domains such as health, sport, technology and computing, laws and policies, among others, using a vocabulary of 30,000 subwords that is best suited for these generic domains. Therefore, we expect that domain-specific pretraining will result in better accuracy for specific tasks.

4   Evaluation

To evaluate the effectiveness of the pretrained LegalBert-pt language models, LegalBert-pt SC and LegalBert-pt FP, we conducted intrinsic and extrinsic evaluations, comparing them with a generic model, Bertimbau-Base [33], and with a language model of the legal domain of the Portuguese language, Legal-BERTimbau-base [17]. Intrinsic evaluation measures the quality of a model independent of any application, while extrinsic evaluation measures the usefulness of the model in a specific task.

We evaluated the models using two specific NLP tasks: Named Entity Recognition (NER) and Text Classification. These tasks were chosen to evaluate the application of the model at both the token level (NER) and the sentence level (text classification), and due to the availability of labeled datasets for these tasks. For each specific task, we fine-tuned the pretrained model. In sentence-level tasks, classification was performed using the coded representation of the special token [CLS], while in token-level tasks, the coded representation of each token was used.

We measured the performance of the pretrained models using perplexity and F1-score. Perplexity measures how well a language model predicts a sample of text and reflects the model's ability to generate coherent and natural-sounding sentences. The F1-score measures the accuracy of the model in identifying named entities in text or classifying text into predefined categories.

4.1   Perplexity

Perplexity is an important intrinsic evaluation metric used to assess language model performance by quantifying the degree of uncertainty a model has about the predictions it makes. Low perplexity indicates that a model is reliable, but it does not guarantee accuracy. Perplexity is also often correlated with a model's ultimate performance on specific tasks. Therefore, in addition to standard evaluation metrics, NLP researchers have started looking at perplexity to test how well language models capture language [23].

To analyze the perplexity of our language models, we built a corpus consisting of 750 legal documents obtained from various sources. This corpus included 250 legal documents representing initial petitions and complaints from the Court of Justice of the State of Ceará (TJ-CE) in Brazil, 250 legal documents of various types from the Public Ministry of the State of Ceará (MP-CE) in Brazil, and 250 legal documents from Extraordinary Appeals of the STF obtained randomly from the VICTOR dataset [4]. By evaluating perplexity on this corpus, we were able to assess how well our pretrained models captured the language used in legal documents in Brazilian Portuguese.

4.2   Named Entity Recognition

Named Entity Recognition (NER) is the task of identifying snippets of text that mention named entities (NEs) and classifying them into predefined categories such as person, organization, and location. Given a sequence of tokens, the NER model has to output the entity of each token. The NER model was designed as a token labeling task and performs entity identification and classification using the IOB tagging scheme [29].

To date, there are few gold standard datasets for named entities in the legal domain in Portuguese. To evaluate the NER model, we used two datasets separately, LENER-BR [3] and CDJUR [22], with annotated entities in legal documents. The LENER-BR [3] dataset was built by manually annotating 66 legal documents from several Brazilian courts. Additionally, four legislative documents were included, totaling 70 annotated documents. The entities were categorized as "ORGANIZATION", "PERSON", "TIME", "LOCATION", "LEGISLATION", and "LEGAL CASES", resulting in a total of 12,248 entity annotations. The CDJUR [22] dataset contains 1,074 manually annotated legal documents, with a total of 44,526 labeled entities. This dataset provides a detailed annotation of entities specific to the legal domain. For example, the "PERSON" category was specified in 9 entities that are typically present in a judicial process, such as plaintiff, lawyer, defendant, victim, witness, judge, prosecutor, police authority, and others. "ADDRESSES" were specified in 6 entities to identify different addresses present in a lawsuit. The LAWS category was specified in three entities: Main Law, Accessory Law, and Jurisprudence. Similarly, specifications were made for "EVIDENCE", "PENALTY", and "SENTENCE".

We trained a NER model for each of these datasets, and a linear classifier layer was attached on top of each model to predict each token's tag independently. The models' performance was evaluated in terms of F1-score at the entity level, taking into account the partial correspondence between the predicted entity and the actual entity based on the Partial metric defined by MUC [9]. Partial correspondence is considered correct when the entity type of the prediction given by the model corresponds to the same entity type of the golden annotation, but not necessarily in the same position limits. For example, if the annotation is { "entity_type": "MAIN-LAW", "text": "Law n° 8.112/90, of 12/11/1990"} and the model prediction is { "entity_type": "MAIN-LAW", "text": "Law n° 8.112/90"}, it is considered correct.

4.3 Text Classification

Text classification is a widely researched task in NLP and text mining, involving the assignment of one or more categories to a document from a set of options. There are different text classification variants such as binary classification, multi-class classification, and multi-label classification. Language models from the field of NLP and learning models from the field of AI can be developed to automate this task, which can be trained from a gold collection of documents.

In the legal domain, text classification has a crucial application in the initial stages of the judicial process when a petitioner presents a petition to the Justice [1,2]. At this stage, the petitioner is required to specify the matter to which the claim pertains. In Brazil, the petitioner has to choose the topic from a hierarchy of over 4,000 subjects as part of the Unified Procedural Boards (TPU system)

[32], maintained by the CNJ. Making the correct association with the hierarchy theme is not trivial and often done incorrectly, causing delays in the judicial process and negative financial and societal impacts by generating rework and a sense of impunity.

We evaluated the performance of the language model classification using a gold collection of 64,000 textual petitions obtained from various Brazilian courts maintained by the CNJ's Codex system. Each complaint is associated with a legal issue that the case addresses under the TPU. Each judicial process is associated with a hierarchy of matters, represented by three levels. More specifically, the gold collection used in the experiment contains 213 legal matters associated with the initial petitions, 9 matters in the first level of the hierarchy, 41 matters in the second level and 163 matters in the third level.

The evaluation of the models in the classification task was carried out in three scenarios. In the first scenario, the classification task was modeled as Hierarchical Text Classification (HTC), which categorizes text into a set of labels organized in a hierarchical structure. We use a Contrastive Learning Approach to Hierarchical Text Classification [36] with the pre-trained language model to train the classification model. In the second scenario, the classification task was modeled with multiclass classification, in which each process is associated only with the subject of the third level, therefore, contemplating 163 classes. In both scenarios, the text is truncated to use the initial 512 tokens of the court case text. Complementarily, the evaluation was carried out in a third scenario, in which the classification task was modeled with multiclass classification, and using 8,192 initial tokens from the text of the judicial process (according [8]). In both scenarios, we evaluated the models in terms of the F1-score for the third-level subject of the hierarchy.

5   Experimental Results

The initial evaluation of the language models involved intrinsic evaluation using perplexity as a metric. A lower perplexity score indicates a better model. Essentially, if a model assigns a high probability to the test set, it means that it is not surprised to see it, indicating a good understanding of how language works. The results of the language models regarding perplexity are in Table 3. In the tables, the values in bold are the best values among the experiments with the models.

The LegalBert-pt FP language model performed the best with a perplexity of 3,700, followed by the LegalBert-pt SC language model with a perplexity of 3,822, indicating that pretraining with a domain-specific corpus that includes diverse legal documents leads to a better understanding of the specific language. Models with lower perplexity are also expected to perform better on specific tasks. Therefore, the language models were fine-tuned for specific tasks, namely NER and Text Classification, according to [33].

The results of the application of language models in the Named Entity Recognition task, using corpora LENER-BR and CDJUR, are presented in Tables 4 and 5, respectively, as described in Sect. 4.2. Both experiments were run 5 times

Table 3. Results of language models in terms of perplexity.

Language Model	Perplexity
BERTimbau-Base [33]	3.903
Legal-BERTimbau-base [17]	3.949
BERTiKal [28]	6.01
LegalBert-pt SC	3.822
LegalBert-pt FP	**3.700**

in 10 epochs to avoid bias, and the F1-score results are displayed in terms of the mean and the Standard Error of the Mean (SEM) of the runs. Values marked in bold indicate the best results, considering the mean and SEM.

Table 4. Results of the language models in terms of the mean + standard error of the F1-score in Named Entity Recognition in the LENER-BR corpus.

Language Model	F1-micro	F1-macro	F1-weighted
BERTimbau-Base	0.930 ± 0.000	0.916 ± 0.002	0.930 ± 0.000
Legal-BERTimbau-base	0.930 ± 0.000	0.920 ± 0.003	0.930 ± 0.000
LegalBert-pt SC	**0.932 ± 0.002**	**0.932 ± 0.002**	**0.932 ± 0.002**
LegalBert-pt FP	**0.936 ± 0.002**	**0.932 ± 0.002**	**0.936 ± 0.002**

Table 5. Results of the language models in terms of the mean ± standard error of the F1-score in Named Entity Recognition in the CDJUR corpus.

Language Model	F1-micro	F1-macro	F1-weighted
BERTimbau-Base	0.674 ± 0.002	0.584 ± 0.005	0.670 ± 0.004
Legal-BERTimbau-base	0.670 ± 0.003	0.580 ± 0.004	0.664 ± 0.002
LegalBert-pt SC	0.668 ± 0.002	**0.604 ± 0.002**	0.662 ± 0.002
LegalBert-pt FP	**0.680 ± 0.000**	**0.606 ± 0.002**	**0.680 ± 0.000**

The LegalBert-pt FP model demonstrated superior results, outperforming the BERTimbau-Base by 0.65% in F1-score micro, 1.75% in F1-score macro, and 0.65% in F1-score weighted for NER in the LENER-BR dataset, and by 0.89% in F1-score micro, 3.78% in F1-score macro, and 1.49% in F1-score weighted for NER in the CDJUR dataset. This reinforces the benefits of using domain-specific pretrained models compared to generic ones, as LegalBert-pt FP can understand both the generic context and specific legal language.

Moreover, when compared with Legal-BERTimbau-base, the LegalBert-pt FP model outperformed by 0.65% in F1-score micro, 1.31% in F1-score macro, and 0.65% in F1-score weighted for NER in the LENER-BR dataset, and by 1.49% in F1-score micro, 4.48% in F1-score macro, and 2.41% in F1-score weighted for NER in the CDJUR dataset. These results reinforce the effectiveness of pretraining a legal domain model with a diverse set of legal documents, enabling a wider coverage of specific terms and acquiring greater domain specialization.

The superiority of the language models in NER with the LENER-BR dataset compared to the CDJUR dataset is also notable. This difference can be attributed to the specificity and number of entities present in each dataset. The LENER-BR dataset has only 6 specific entities of the legal domain (organization, person, time, place, legislation, and jurisprudence), while the CDJUR dataset has 21 specific entities of the legal domain (author, lawyer, defendant, victim, witness, judge, prosecutor, police authority and other persons, author's address, offense address, defendant's address, witness address, victim's address, other addresses, main law, accessory law, jurisprudence, evidence, penalty, and sentence). Thus, it is more challenging for the model to recognize the more specific entities present in the CDJUR dataset.

Table 6, 7 and 8 presents the results in terms of F1-score for the application of language models in the Text Classification task, in different scenarios, as described in Sect. 4.3. To avoid biasing the results, both experiments were run 5 times and the F1-score results are displayed in terms of the mean and SEM of the runs. Both models were run for 30 epochs, with early stopping and a 5 epoch patience criterion.

Table 6. Results of the language models in terms of the mean ± standard error of the F1-score in hierarchical text classification.

Language Model	F1-micro	F1-macro	F1-weighted
BERTimbau-Base	0.506 ± 0.002	**0.470 ± 0.000**	0.500 ± 0.000
Legal-BERTimbau-base	0.504 + 0.002	0.466 ± 0.002	0.498 ± 0.002
LegalBert-pt SC	**0.510 ± 0.000**	**0.470 ± 0.000**	**0.504 ± 0.002**
LegalBert-pt FP	**0.512 ± 0.002**	0.468 ± 0.003	**0.502 ± 0.002**

In the text classification task, in all evaluated scenarios, the LegalBert-pt models demonstrate superior results compared to other models. In the scenario that evaluates the hierarchical text classification, LegalBert-pt FP outperforms the generic BERTimbau-base model by 1.19% in F1-micro and 0.4% in F1-weighted. When compared with Legal-BERTimbau-base, LegalBert-pt FP also outperforms F1-micro by 1.59% and F1-weighted by 0.8%.

It is important to note, however, that the relatively close results between the generic and specific models in the text classification task suggest that the use of a generic textual structure in the composition of the initial petitions may not require in-depth domain-specific knowledge. Thus, relevant linguistic features for classification may be more universal and captured by a generic language model. It is also important to recognize that the performance of a model may vary depending on the task and dataset, and a specific model may be better suited for a particular task.

Table 7. Results of the language models in terms of the mean ± standard error of the F1-score in text classification with 512 tokens.

Language Model	F1-micro	F1-macro	F1-weighted
BERTimbau-Base	0.538 ± 0.001	0.474 ± 0.002	0.526 ± 0.001
Legal-BERTimbau-base	0.533 + 0.001	0.467 ± 0.001	0.520 ± 0.001
LegalBert-pt SC	**0.551 ± 0.001**	0.487 ± 0.001	**0.541 ± 0.001**
LegalBert-pt FP	**0.552 ± 0.001**	**0.496 ± 0.002**	**0.542 ± 0.001**

Table 8. Results of the language models in terms of the mean ± standard error of the F1-score in text classification with 8,192 tokens.

Language Model	F1-micro	F1-macro	F1-weighted
BERTimbau-Base	0.570 ± 0.000	0.511 ± 0.000	0.558 ± 0.000
Legal-BERTimbau-base	0.567 + 0.000	0.504 ± 0.000	0.554 ± 0.000
LegalBert-pt SC	**0.589 ± 0.001**	**0.531 ± 0.001**	**0.577 ± 0.001**
LegalBert-pt FP	**0.587 ± 0.001**	**0.530 ± 0.003**	**0.576 ± 0.002**

In the multi-class classification approach, the LegalBert-pt models outperform the Bertimbau-base results in all experiments. When analyzing the classification with the representation of the text by 512 tokens, the result LegalBert-pt FP exceeds by 2.6% in the F1-micro, 4.6% in the F1-macro and 3.0% in the F1-weighted. When analyzing the classification with the representation of the text by 8,192 tokens, the LegalBert-pt FP result exceeds 3.0% in the F1-micro, 3.7% in the F1-macro and 3.2% in the F1-weighted.

One possible reason for LegalBert-pt FP's better performance in text classification is that it was pretrained on a dataset with language similar to the language of the texts used in the classification task. This could have allowed the model to learn relevant linguistic features that contribute to improved performance.

The results of our experiments demonstrate that domain-specific language models outperform domain-general language models on domain-specific tasks such as NER and text classification. This suggests that pretraining with domain-specific texts allows the language model to learn richer and more specific representations of the domain-specific language compared to domain-general language models. However, it is important to note that specialization in a single domain limits the model's applicability to tasks in that domain. Therefore, in scenarios where the application needs to handle multiple domains, a broader approach may be necessary, such as using generic language models.

6   Conclusion and Future Works

This article outlined the process of training, using, and evaluating a language model specifically designed for the legal domain. It demonstrated that the language model enhances accuracy in particular tasks within the legal field.

Experiments were conducted to determine the optimal strategy for generating a language model for a new domain, weighing the options of additional pretraining from a pre-existing generic domain model or starting from scratch. Consequently, two versions of the language model pretrained on legal domain documents were developed: LegalBert-pt SC (trained from scratch) and LegalBert-pt FP (pretrained using the BERTimbau-Base model). For the LegalBert-pt SC model, legal documents and articles from the Portuguese Wikipedia were utilized to create a vocabulary that includes both generic and legal domain-specific terms in Portuguese.

The legal domain language models were compared to the BERTimbau-Base (a generic domain language model for the Portuguese language) using intrinsic evaluation (measured through perplexity) and extrinsic evaluation (by fine-tuning the language model for specific tasks such as Named Entity Recognition and Text Classification). In both evaluations, the language models pretrained on legal domain documents yielded superior results compared to the BERTimbau-Base. Notably, the most significant performance gains were observed in the most challenging final tasks. The language models developed in this study are available under the OpenRAIL license and can be accessed at http://huggingface.co/raquelsilveira/legalbertpt_sc and http://huggingface.co/raquelsilveira/legalbertpt_fp. The models can be further fine-tuned using other types of legal documents, which presents an opportunity for ongoing improvement. This contribution is significant for both the scientific and practical communities, as it advances our understanding of how language models can enhance legal tasks.

In future research, we aim to examine the performance of LegalBert-pt on additional datasets and explore the model's application in other tasks within the legal domain. This will help assess the extent to which understanding the specific language of these models impacts the accuracy of domain-specific tasks. Comparisons with GPT models and the utilization of the pretrained model to distill LLMs are also planned for future endeavors.

References

1. Aguiar, A., Silveira, R., Pinheiro, V., Furtado, V., Neto, J.A.: Text classification in legal documents extracted from lawsuits in Brazilian courts. In: Anais da X Brazilian Conference on Intelligent Systems, SBC, Porto Alegre, RS, Brasil (2021). https://sol.sbc.org.br/index.php/bracis/article/view/19093
2. Aguiar, A., Silveira, R., Furtado, V., Pinheiro, V., Neto, J.A.M.: Using topic modeling in classification of Brazilian lawsuits. In: Pinheiro, V., et al. (eds.) PROPOR 2022. LNCS (LNAI), vol. 13208, pp. 233–242. Springer, Cham (2022). https://doi.org/10.1007/978-3-030-98305-5_22
3. Luz de Araujo, P.H., de Campos, T.E., de Oliveira, R.R.R., Stauffer, M., Couto, S., Bermejo, P.: LeNER-Br: a dataset for named entity recognition in Brazilian legal text. In: Villavicencio, A., et al. (eds.) PROPOR 2018. LNCS (LNAI), vol. 11122, pp. 313–323. Springer, Cham (2018). https://doi.org/10.1007/978-3-319-99722-3_32
4. Luz de Araujo, P.H., de Campos, T.E., Ataides Braz, F., Correia da Silva, N.: VICTOR: a dataset for Brazilian legal documents classification. In: Proceedings of the Twelfth Language Resources and Evaluation Conference, pp. 1449–1458. European Language Resources Association, Marseille (2020). https://aclanthology.org/2020.lrec-1.181
5. Beltagy, I., Lo, K., Cohan, A.: Scibert: a pretrained language model for scientific text. arXiv preprint arXiv:1903.10676 (2019)
6. Brown, T., et al.: Language models are few-shot learners. Adv. Neural. Inf. Process. Syst. **33**, 1877–1901 (2020)
7. Chalkidis, I., Fergadiotis, M., Malakasiotis, P., Aletras, N., Androutsopoulos, I.: Legal-bert: the muppets straight out of law school. arXiv preprint arXiv:2010.02559 (2020)
8. Chalkidis, I., et al.: Lexglue: a benchmark dataset for legal language understanding in english (2022)
9. Chinchor, N., Sundheim, B.M.: Muc-5 evaluation metrics. In: Fifth Message Understanding Conference (MUC-5): Proceedings of a Conference Held in Baltimore, Maryland, 25–27 August 1993 (1993)
10. Conneau, A., et al.: Unsupervised cross-lingual representation learning at scale. arXiv preprint arXiv:1911.02116 (2019)
11. Devlin, J., Chang, M.W., Lee, K., Toutanova, K.: Bert: pre-training of deep bidirectional transformers for language understanding. arXiv preprint arXiv:1810.04805 (2018)
12. Feng, Z., et al.: Codebert: a pre-trained model for programming and natural languages. arXiv preprint arXiv:2002.08155 (2020)
13. Jain, D., Borah, M.D., Biswas, A.: Summarization of legal documents: where are we now and the way forward. Comput. Sci. Rev. **40**, 100388 (2021)
14. Kalyan, K.S., Rajasekharan, A., Sangeetha, S.: Ammus: a survey of transformer-based pretrained models in natural language processing. arXiv preprint arXiv:2108.05542 (2021)
15. Kudo, T., Richardson, J.: Sentencepiece: a simple and language independent subword tokenizer and detokenizer for neural text processing. arXiv preprint arXiv:1808.06226 (2018)
16. Lee, J.: Biobert: a pre-trained biomedical language representation model for biomedical text mining. Bioinformatics **36**(4), 1234–1240 (2020)
17. Legal-bertimbau-base. https://huggingface.co/rufimelo/Legal-BERTimbau-base

18. Licari, D., Comandè, G.: Italian-legal-bert: a pre-trained transformer language model for Italian law (2022)
19. Lin, T., Wang, Y., Liu, X., Qiu, X.: A survey of transformers. AI Open **3**, 111–132 (2022). https://doi.org/10.1016/j.aiopen.2022.10.001
20. Liu, Y., et al.: Roberta: a robustly optimized bert pretraining approach. arXiv preprint arXiv:1907.11692 (2019)
21. Loshchilov, I., Hutter, F.: Decoupled weight decay regularization. arXiv preprint arXiv:1711.05101 (2017)
22. Brito, M., et al.: Cdjur-br - a golden collection of legal document from Brazilian justice with fine-grained named entities. arXiv preprint arXiv:2023.49053 (2023)
23. Meister, C., Cotterell, R.: Language model evaluation beyond perplexity. arXiv preprint arXiv:2106.00085 (2021)
24. Nguyen, T.S., Nguyen, L.M., Tojo, S., Satoh, K., Shimazu, A.: Recurrent neural network-based models for recognizing requisite and effectuation parts in legal texts. Artif. Intell. Law **26**, 169–199 (2018)
25. Pan, S.J., Yang, Q.: A survey on transfer learning. IEEE Trans. Knowl. Data Eng. **22**(10), 1345–1359 (2010)
26. Paul, S., Mandal, A., Goyal, P., Ghosh, S.: Pre-training transformers on indian legal text. arXiv preprint arXiv:2209.06049 (2022)
27. Peters, M.E., et al.: Deep contextualized word representations (2018)
28. Polo, F., et al.: Legalnlp - natural language processing methods for the Brazilian legal language. In: Anais do XVIII Encontro Nacional de Inteligência Artificial e Computacional, pp. 763–774. SBC, Porto Alegre (2021). https://doi.org/10.5753/eniac.2021.18301. https://sol.sbc.org.br/index.php/eniac/article/view/18301
29. Sang, E.F., Veenstra, J.: Representing text chunks. arXiv preprint arXiv:cs/9907006 (1999)
30. Sennrich, R., Haddow, B., Birch, A.: Neural machine translation of rare words with subword units. arXiv preprint arXiv:1508.07909 (2015)
31. Shao, Y., et al.: Bert-pli: modeling paragraph-level interactions for legal case retrieval. In: IJCAI, pp. 3501–3507 (2020)
32. Sistema de gestão de tabelas processuais unificadas. https://www.cnj.jus.br/sgt/consulta_publica_assuntos.php. Accessed 09 Aug 2022
33. Souza, F., Nogueira, R., Lotufo, R.: BERTimbau: pretrained BERT models for Brazilian Portuguese. In: Cerri, R., Prati, R.C. (eds.) BRACIS 2020. LNCS (LNAI), vol. 12319, pp. 403–417. Springer, Cham (2020). https://doi.org/10.1007/978-3-030-61377-8_28
34. Vaswani, A., et al.: Attention is all you need. Adv. Neural Inf. Process. Syst. **30**, 1–11 (2017)
35. Viegas, C.F.O.: Jurisbert: transformer-based model for embedding legal texts (2022)
36. Wang, Z., Wang, P., Huang, L., Sun, X., Wang, H.: Incorporating hierarchy into text encoder: a contrastive learning approach for hierarchical text classification. arXiv preprint arXiv:2203.03825 (2022)
37. Xiao, C., Hu, X., Liu, Z., Tu, C., Sun, M.: Lawformer: a pre-trained language model for Chinese legal long documents. AI Open **2**, 79–84 (2021)
38. Yang, Y., Uy, M.C.S., Huang, A.: Finbert: a pretrained language model for financial communications. arXiv preprint arXiv:2006.08097 (2020)

A Framework for Controversial Political Topics Identification Using Twitter Data

Kenzo Sakiyama[1]([✉])[iD], Lucas de Souza Rodrigues[2][iD],
Bruno Magalhães Nogueira[2][iD], Edson Takashi Matsubara[2][iD],
and Roseli A. F. Romero[1][iD]

[1] Universidade de São Paulo - ICMC, São Carlos, Brazil
kenzosakiyama@usp.br, rafrance@icmc.usp.br
[2] Universidade Federal de Mato Grosso do Sul - FACOM, Campo Grande, Brazil
lucas.rodrigues@ifms.edu.br, bruno@facom.ufms.br, edsontm@facom.ufms.br

Abstract. Social networks have become the main stage for discussion on various current topics. In particular, electoral processes tend to bring many publications with polarized opinions on political issues addressed by candidates. A comprehensive analysis of social media publications on high-impact controversial topics and the opinions expressed in them could contribute to a clearer understanding of the dynamics of political discussion, providing valuable insights for society. In this context, we investigate how to apply a clustering-based topic modeling approach to produce public evaluation information on different current issues, in particular controversial political topics. We propose a framework that enriches text representations, combining state-of-the-art unsupervised (HDBSCAN) and supervised (BERTimbau) techniques to identify controversial political topics in social media publications in Brazilian Portuguese. To this end, weekly collections were carried out on the social network Twitter, making it possible to identify controversial events for each analyzed date. We compare the controversial topics uncovered with real-world news to validate the results and compare our method with a traditional method described in the literature.

Keywords: Topic Modelling · Clustering · Sentiment Analysis

1 Introduction

Political topics generate heated discussions and opinion polarization in social networks between supporters and opponents about a given topic. This dynamic is especially evident during election times. When a candidate gives a speech, social networks are flooded with publications supporting or opposing the topics addressed in that speech. In Brazil, a survey presented by the Data Senado Institute [26] shows that the main communication channels for seeking information about politics are: TV (37%), social networks (24%), and Internet sites (23%). Young people most often use social networks and journalistic portals. TV attracts the more advanced age group and, from social network users, at least 20% confirm that they use them to talk about politics.

Therefore, the period before and after the 2022 Brazilian elections indicates an excellent opportunity to analyze social media data on political discussions. As shown in Fig. 1, using social media posts, it would be possible to better understand the political discussion by using a topic model to identify and study the different topics in the data. Furthermore, using sentiment analysis, it is possible to estimate user ratings on different topics. In this article, we expose a framework based on topic modeling, using Twitter publications about politics in Brazil. However, this approach can be used in various current scenarios and topics. Our goal is to automatically identify potential high-impact policy-related topics by combining two NLP tasks: topic modeling and sentiment analysis. The proposal is presented in Fig. 2.

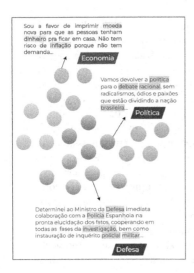

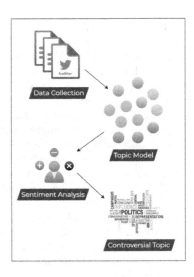

Fig. 1. Topic Modelling Techniques in NLP.

Fig. 2. Controversial Topic Identification.

In the literature, there are several techniques for clustering and identifying topics. A traditional method for topic discovery and semantic mining of unsorted documents is Latent Dirichlet Allocation (LDA). However, the method has some drawbacks. It does not consider the temporal aspect of the data, when creating topics in data with different epochs. It is very computationally intensive to train a model with millions of examples [18]. Also, LDA does not consider sequential or semantic words as a prerequisite for creating topics [4].

As an alternative to the standard approach, the work of Top2Vec and BERTopic [4,14] explored a topic modeling methodology based on the use of state-of-the-art dense representations, followed by dimensionality reduction and clustering, to extract meaningful topics from document collections. In this sense, a series of studies emerged to evaluate the use of correlation of the aforementioned techniques, as described in [11], in which the performance of the different algorithms is evaluated in terms of their strengths and weaknesses in a social science context. Based on the results of this study, the authors indicate that BERTopic supports more embedding models than Top2Vec, allows multilingual analysis, and automatically determines the number of topics. The advantages of Top2Vec are the support for hierarchical topic reduction, the ability to work with very large datasets, and its use of embeddings so that no preprocessing of the original data is required.

Focusing on Portuguese texts, BERTopic has already been successfully applied to Brazilian Portuguese documents by [1]. In their work, the authors applied the methodology to automatically classify legal documents into the six most representative document classes of the Brazilian National Council of Justice (CNJ), achieving 89% of the macro F1 score. Similarly, the authors of [28] examined the social impact of law changes on Twitter social media.

On the other hand, some papers have also conducted public opinion experiments on the political aspects of social networks. A recent study mapped a large-scale cross-party sentiment analysis in Greece, Spain, and the United Kingdom on tweets [5]. The study showed a preponderance of negatively-tweeted tweets from politicians and examined trends in popularity and sentiment. Another study in Spain used Twitter to analyze the impact of elite discourse on effective citizen polarization [16]. According to this study, users' contact with candidates does not affect polarization.

From this perspective, the aim of this article is to develop a framework that uses a topic modeling approach based on clustering techniques proposed by [1, 14, 28] for publications in Brazilian Portuguese on Twitter about political topics. Furthermore, we hypothesize that combining the extracted topics with state-of-the-art sentiment analysis can add more information and identify potentially controversial topics. Thus, we can identify controversial issues and assess public appreciation. The main contributions of this work are:

1. A new framework for identifying controversial topics in Twitter data, combining clustering and sentiment analysis based on state-of-the-art Transformer representations;
2. Evaluation of discovered topics. The results are evaluated by associating real-world events with the topics and by using quantitative metrics;
3. Comparison between our methodology and a simpler approach;

This work is organized as follows. In Sect. 2, we briefly describe the related works. Section 3 describes the methodology, Sect. 4 shows our experimental results. Finally, in Sect. 5 are presented the conclusions and future works.

2 Related Works

Many works deal with information extraction using social networks, opinion mining, and general linguistic representations. In this section, we briefly survey examples of applications.

Text mining is an important area in Knowledge Discovery in Databases (KDD). It focuses on discovering interesting patterns in structured and unstructured data [13]. As a result, the applications in this area are diverse. In addition, it fosters strong connections to natural language processing, data mining, machine learning, information retrieval, and knowledge management [23]. In the following, we present some examples of recent studies on topic modeling and text clustering related to this proposal.

In the work proposed in [3], a machine learning-based approach is presented to improve cognitive distortions classification of the Arabic content over Twitter, enriching text representation by defining latent topics in tweets. Another study on data clustering tools is the use of topic modeling for customer service chats [15]. The mentioned work focuses on finding new intents from user messages, that are not yet included in any previous intents and reorganizing existing intents by analyzing the topic model generated.

In relation to text clustering, HDBSCAN [19] was applied to investigate how to link popular social media topics and news stories using Transformer models and neural networks [2]. Other works highlight the use of HDBSCAN to compare latent semantic analysis and latent Dirichlet allocation, analyzing the topic COVID-19 [27]. This study checks which most frequent words from each cluster will be displayed and compared with factual data about the outbreak to find out if there are any correlations. The authors state how well HDBSCAN clusters its data in comparison to K-Means [22]. [17] investigates the most effective way of performing text classification and clustering of duplicate texts in technical documentation written in Simplified Technical English. Vector representations from pre-trained Transformers and LSTM models were tested against TF-IDF using the density-based clustering algorithms DBSCAN and HDBSCAN.

Monitoring Social Media Research has become essential for government entities, large corporations, and global companies. Several data mining tools currently assess public reaction to measures taken by a government, company and famous personalities. For example, some studies use social media data to predict a country's elections based on public sentiment [8]. Other recent work examines social media to study the public awareness of COVID-19 pandemic trends and uncover meaningful themes of concern posted by Twitter users [7].

Focusing on data collection, there are several applications for extracting and analyzing data from social networks. Tweepy is an advanced Twitter scraping tool written in Python allowing to scrape tweets from Twitter profiles without using Twitter's official API [10]. Another variant, Snscrape, is a scraper for social networking services (e.g., Instagram, Facebook, and Twitter) [6]. It scrapes data like user profiles, hashtags, or searches and returns the discovered items.

3   Methodology

In this Section, we will describe the steps we took to identify controversial political themes using Twitter data. In Fig. 3 is shown an overview of our methodology, presenting its main components: Data Collection (Sect. 3.1), Tweet Preprocessing (Sect. 3.2), Calibration and Clustering (Sects. 3.3 and 3.4), and the

Cluster Analysis performed (Sect. 3.7). At the end, in Sect. 3.8 is discussed the evaluation of the controversial topics.

| Data Collection | Pre-processing | Calibration and Clustering | Sentiment Analysis | Controversial Topic Identification | Cluster Analysis |

Fig. 3. Overview of the controversial topic identification and analysis.

3.1 Data Collection

The 38th former Brazilian president Jair Messias Bolsonaro used to perform weekly live streams on his channel on the YouTube video sharing platform. During the live streams, which last an average of one hour, he discussed events of the week that relate to his government.

We assumed that even if the live streams take place on a specific platform (YouTube), supporters and opponents end up generating publications about the content of the broadcast on other social networks such as Twitter. Twitter is a social network in which users are capable of publishing and interacting with others through small text messages called tweets. Given this context, we considered Twitter posts, published after the start of the ex-president's live streams, a good source of controversial topics.

Given the start time of the weakly live streams, we collected Twitter posts related to the former Brazilian president, and published up to 3 h after the start of the broadcasts. We focused on the lives performed in May 2022 and collected tweets using the snscrape[1] social media scrapper. Our tweet data consists of tweets mentioning the ex-president Twitter profile (using ' @jairbolsonaro') or his name ('Bolsonaro'). This way, we expect that the collected tweets at least mention Jair Bolsonaro.

Table 1. Examples collected for each May live stream. The last live stream was delayed due to the schedule of Bolsonaro.

Date	Collected tweets	After pre-processing
May 05th	9.243	7.987
May 12th	9.257	8.221
May 19th	8.827	6.883
May 27th	8.041	7.029

---

[1] https://github.com/JustAnotherArchivist/snscrape.

In Table 1 is shown the number of tweets collected for each live stream of May. In total, 30.120 unique tweets were analyzed with an average of 20 tokens (space-separated tokens) per tweet. In addition, it's worth mentioning that the amount of collected tweets is superior to the amount of posts published in the YouTube comments section for each live stream.

3.2   Tweet Pre-processing

As a text pre-processing step for the following analyses, we removed mentions (usernames), URLs, empty texts and duplicated texts using the spacy[2] Python package. We preserved hashtags as they can be good indicators of themes or subjects. In Table 1, we present the number of examples remaining after the text pre-processing.

Following the text pre-processing, we converted the resulting texts to state of the art contextualized dense representations based on Transformers [31]. We used efficient multilingual representations based on the MiniLM [32] language model, trained using the Sentence Transformers framework [24], and publicly available to use and research[3].

We limited the token sequence length to be 128 WordPiece tokens (smaller sequences are padded and larger sequences are truncated). The resulting representations for each tweet are vectors of 384 dimensions, generated by the mean of the contextualized token representations and well suited to be compared using similarity metrics such as cosine distance [24]. Considering representations that use context, we expect publications semantically similar (or related to similar subjects) to be close in the representation space.

Finally, we used the UMAP (Uniform Manifold Approximation and Projection) [20] to reduce the dimension of the MiniLM vectors. UMAP is able to reduce dimensions and preserve the global distribution of the original data on the lower dimension space with competitive execution time when compared to other dimension reduction techniques (ex: t-SNE). Since our goal is to use clustering as an intermediate step to identify controversial topics, the dimension reduction is used to improve the efficiency of the clustering method. We observed that performing clustering without dimension reduction led to worse results.

3.3   Clustering Tweets with HDBSCAN

We used the HDBSCAN (Hierarchical Density-Based Spatial Clustering of Applications with Noise) [19] algorithm to cluster tweets and extract common topics, inspired by the success of previous works [4,14]. The method consists in a hierarchical variation of DBSCAN clustering algorithm [12] that is able to identify clusters of different densities and be more robust in relations to

---

[2] https://spacy.io/.

[3] https://huggingface.co/sentence-transformers/paraphrase-multilingual-MiniLM-L12-v2.

its parameters. HDBSCAN identifies regions of large concentration of examples (high density) as clusters, and it's able to ignore noisy examples (label as noise).

When representing Twitter publications as multidimensional vectors, we expect that regions of common subjects can be of any shape. Hence, we chose to use HDBSCAN to identify these regions of common subjects as clusters. The motivation for using HDBSCAN is related to its ability to identify clusters of different densities, handle outliers, automatically discover the number of clusters and provide a hierarchical structure of clusters. Furthermore, when dealing with social media posts, not every publication is associated to a common theme. Assuming such examples are not close to high density areas, HDBSCAN can easily identify them as noise.

3.4   Parameter Calibration

Since clustering is used to identify topics, the parameters of the both methods, UMAP[4] and HDBSCAN[5] were chosen adequately. To find a good set of parameters, we used Random Search with the objective of maximizing the DBCV (Density-Based Clustering Validation index) metric [21]. DBCV is a metric created with the evaluation of density based clustering as its main goal. The metric evaluates the density of the obtained clusters (considering their shape properties) and consider the number of noise examples in its definition. It evaluates both the density sparseness of the examples of a cluster and the density separation of different clusters, using the core distance and mutual reachability distances presented by the authors. Thus, we choose the DBCV metric, to evaluate the quality of the clustering using a metric that is suited to evaluate density-based clustering, instead of relying on metrics such as silhouette and DB index, which may not capture the specific characteristics of density-based clustering algorithms.

Considering the UMAP dimension reduction pre-processing step, two parameters were adjusted: the number of components and the number of neighbors. The number of components determines the number of dimensions in which we want to embed our representations. The number of neighbors is a parameter that controls the balance between the local (favored by lower values) and the global structure (favored by larger values) of the data. Furthermore, the cosine metric was used to determine distances in the original space and we fixed the minimum distance in the lower dimensions space as zero.

For HDBSCAN, we chose to tune the following parameters: minimum cluster size and minimum samples. Minimum cluster size defines the smallest number of examples that can be considered as a cluster. Minimum samples specify the number of neighbors used to estimate the probability density function. For other parameters, we preserved the default parameters of HDBSCAN Python package.

Although HDBSCAN determines the number of clusters automatically, we observed empirically that a minimum number of clusters is beneficial for the

---

[4] https://umap-learn.readthedocs.io/en/latest/.
[5] https://hdbscan.readthedocs.io/en/latest/.

analysis. This way, parameter sets that generate less than 4 clusters were discarted. In Table 2 is presented the obtained UMAP and HDBSCAN parameter sets for each day of analysis.

Table 2. Parameters obtained for each May tweet collection and value ranges used.

Date	UMAP		HDBSCAN	
	num. components	num. neighbours	min. cluster size	min. samples
May 05th	195	40	55	105
May 12th	105	20	15	145
May 19th	35	10	60	35
May 27th	125	30	15	75
Ranges	[5, 10, ..., 200]	[10, 20, ..., 70]	[10, 15, ..., 100]	[5, 10, ..., 200]

3.5 Sentiment Analysis

Sentiment analysis is used as an intermediate step of the controversial topic identification. Contextualized dense representations generated by Transformers [31] are predominant in different NLP tasks. Souza et al. [30] investigated different usages of Transformer representations in the sentiment analysis of reviews in Brazilian Portuguese. As shown by the authors results, fine-tuning a language model led to better performance in all the datasets used. Therefore, to perform sentiment analysis on the collected tweets, we used the BERTimbau-base [29] language model fine-tuned in the TweetSentBR corpus [9].

The corpus consists of tweets published during popular Brazilian TV shows. A total of 12.312 examples were recovered from the original corpus. The examples are divided in three classes: positive (45%), neutral (25%) and negative (29%). We preserved the class imbalance, since the minority class has an expressive amount of examples (25%), and used the same text pre-processing described in the start of the Sect. 3.2.

We fine-tuned BERTimbau for sentiment analysis using the popular Hugging-Face library[6]. We limited the token sequence length to a maximum of 128 tokens (padding the smaller sequences and truncating the larger ones). We trained the model for 5 epochs using: batch size of 128, learning rate of 2×10^{-5} and a cosine scheduler with warm-up (10% of the train data used for warm-up). Evaluating using a stratified holdout (80% train and 20% test), the model was able to achieve 69.15% macro-F1. We used the fine-tunned model to perform sentiment analysis in the collected tweets described in Sect. 3.1.

3.6 Controversial Topic Identification

The controversial topic detection is done by combining the clustering and sentiment analysis of the collected tweets. For each date of analysis, HDBSCAN is

---

[6] https://huggingface.co/.

used for generating the cluster labels. Then, the fine-tuned BERTimbau extracts the sentiments of each tweet in the clusters. As the next step, the clusters are sorted by the percentage of positive, neutral and negative examples in the clusters. Finally, assuming that each cluster corresponds to a certain subject, we observed that potential controversial topics are usually located in clusters with large amounts of negative publications. Therefore, a controversial cluster is defined as being a cluster with a negative percentage above a threshold C ($C \in \mathbb{R}$, $0 \leq C \leq 1$), where C is a configurable parameter. It was observed that a threshold $C = 0.7$ was enough to identify controversial clusters in the collected data. Following this methodology, by identifying a controversial cluster, we can also quantify how bad the topic was received by the Twitter users by analysing the C value.

3.7 Cluster Analysis

The same cluster based TF-IDF used by BERTopic [14], is used to discriminate and analyse the tweet clusters. The procedure consists in a generalization of TF-IDF score. By modifying the definition of term frequency and inverse document frequency, a cluster based TF-IDF allow us to generalize TF-IDF representation from documents to collections (clusters) of documents. The modification of TF-IDF is defined as $W_{t,c} = tf_{t,c} \cdot \log\left(1 + \frac{A}{tf_t}\right)$, where $(tf_{t,c})$ represents the frequency of a token t in a cluster c, (tf_t) represents the frequency of t in all clusters and (A) average number of tokens per cluster. TF-IDF aims to associate high scores to the most discriminative tokens in a text example. Treating all examples of a cluster as a single example, it is expected that the most discriminative tokens of a cluster have the highest TF-IDF scores. Thus, we summarized each cluster by the tokens with the highest TF-IDF scores. This way, by analysing the filtered tokens, we can easily identify the topics present in the clusters.

To generate the cluster TF-IDF score, we added lemmatisation and punctuation and stop-word removal to the text pre-processing step (Sect. 3.2), treated the cluster examples as a single document (by concatenation) and generated the scores using unigrams.

3.8 Evaluation

In addition to identifying a controversial cluster, it is interesting to associate a real world event to its content as a form of validation. In order to associate news or presidential declarations to the controversial clusters, we searched manually over the news published during the week period analyzed. We also analyzed the topics discussed in the Bolsonaro's live streams, in search of subjects that could also appear in the clusters.

To quantitatively assess the discoveries, we computed the coherence measurement C_V [25] for each controversial topic identified. Based on the distributional hypothesis of words, the coherence measurement C_V aims to quantify how much the topic is highlighted by the documents analysed. We evaluated the C_V, summarizing the controversial topic by it's top-10 tokens with the highest cluster TF-IDFs (using the same pre-processing described in the Sect. 3.7).

4 Results and Discussions

In this section, we will present the results of exploring the HDBSCAN clusters with the sentiment analysis data.

4.1 Clustering and Sentiment Analysis

Table 3 shows the results of applying the parameters obtained by the calibration step (Sect. 3.4) in the Twitter data collected. We obtained DBCV values above 0.45 for all analyzed dates, and the collection at May the 19th generated the largest number of clusters and the lowest DBCV.

Table 3. Clustering and sentiment analysis results obtained.

Date	Clustering			Sent. Analysis		
	Clusters	Outliers	DBCV	Positive	Neutral	Negative
May 05th	6	2.008	0,498	1486 (18.6%)	2345 (29,3%)	4156 (52,0%)
May 12th	7	1.864	0,523	1037 (12.6%)	2462 (29,9%)	4722 (57,4%)
May 19th	17	1.428	0,479	1365 (19.8%)	2252 (32,7%)	3266 (47,4%)
May 27th	8	1.357	0,483	849 (12.1%)	2176 (30,9%)	4004 (56,9%)

In relation to sentiment analysis, Table 3 presents the quantities of positive, neutral and negative tweets in the collected data. In general, we observed similar percentages of neutral tweets. On the other hand, the percentages of positive and negative tweets varied the most, with the highest percentages being associated with negative tweets. The predominance of negative comments, in the brazillian scenario, corroborates the previous mentioned work [5].

4.2 Controversial Topic Discussion

Table 4. Controversial clusters detected and their respective sizes (number of examples), negative percentage and C_V.

Date	Size	Negative(%)	C_V
May 05th	588	72.2	0.429
May 12th	310	77.1	0.681
May 19th i)	137	83.5	0.402
May 19th ii)	62	80.6	0.666
May 27th	96	86.4	0.972

(a) May 05th.

(b) May 12th.

(c) May 19th i).

(d) May 19th ii).

(e) May 27th.

Fig. 4. Word cloud visualizations for the 50 tokens with the highest cluster-TF-IDF in the cluster TF-IDF. The bigger the token, the higher its TF-IDF score.

In Table 4 is shown information about the controversial clusters identified. We observed a large variation in the number of examples of the controversial clusters and in the C_V metric between the dates of analysis. In addition, May 19th was the only date with more than one controversial cluster identified.

To visualize the controversial clusters identified by HDBSCAN, we generated the word clouds shown in the Fig. 4. We analyzed the tweets in each controversial cluster and will discuss the discovered controversial events below.

- **May 05th (Fig. 4a):** a large quantity of tweets may have been motivated by a declaration made by Bolsonaro during his broadcast, in which he expressed his concern about frauds in the upcoming election. To add more context, in the same week of the live stream, the Brazilian military forces questioned the entity responsible for the electoral process about possible vulnerabilities in the electronic voting machine.[7]
- **May 12th (Fig. 4b):** during his live stream, Bolsonaro talked about a possible deal with Petrobras (important Brazilian petroleum corporation) about reduction in the price of fuel. In the same week, the Brazilian Senate approved a bill that changes taxes residing on fuel. We observed tweets related to both these events in the cluster.[8]

---

[7] May 05th - Links: www.uol.com.br | www.em.com.br.

[8] May 12th - Links: www.uol.com.br | www.senado.leg.br.

- **May** 19th **(Figs.** **4c and 4d):** on this day, we identified two clusters with negative percentages over 70%. The first one (Fig. 4c) may be related to a speech made by the Bolsonaro during an event in the capital of Rio de Janeiro in the same day. In his speech, Bolsonaro repeated that in his government there is no corruption and criticized the predecessor government. We observed the lowest C_V (0.402) for this cluster. The second cluster (Fig. 4d) seems to contain tweets related to news published during the week, which discloses the ex-president's expenses using his corporate card.[9]
- **May** 27th **(Fig.** **4e):** on the same day of the live stream, the Brazilian government announced a cut in spending destined for the Brazilian Ministry of Education. We observed a large amount of tweets related to this event in the cluster, and it was the worst reception by users (86% negative) and the topic with the highest C_V.[10]

4.3   Evaluating a Simpler Approach

Supposing that a simpler clustering approach could obtain similar results, we repeated the methodology using the K-Means algorithm and no dimension reduction. We chose K-Means, instead of sparse topic modelling methods such as LDA, since it could benefit from using the dense MiniLM representations. We fixed the K-Means parameters (200 random initialisations and 500 maximum iterations), and determined the number of clusters using the elbow rule.

Focusing only on the clusters with the highest percentage of negative tweets, Table 5 compares the top-10 tokens based on their cluster TF-IDF for HDBSCAN and K-Means. As we can see, even tough we observed similar results for the first date of analysis (May 05th), the use of HDBSCAN in conjunction with

Table 5. Top-10 tokens and C_V, based on the cluster TF-IDF for each date of data collection. HDBSCAN top tokens on the left, and K-means top tokens on the right.

Date	HDBSCAN		K-Means	
	Top-10	Cv	Top-10	Cv
May 05th	votar, eleição, urna, eleitoral, ser, voto, governar, fraudar, tse, eleito	0,429	votar, eleição, ser, eleitoral, urna, governar, país, campanha, ano, auditoria	0,307
May 12th	preço, imposto, impor, aumentar, redução, combustível, lucrar, governar, reduzir, mercar	0,681	votar, eleição, ser, governar, ciro, ter, dizer, pt, falar, ano,	0,328
May 19th	governar, político, corrupção, poder, presidência, corrupto, ano, ser, público, apoiar	0,402	governar, votar, ser, ter, ano, político, eleição, stf, dinheiro, país	0,327
May 27th	educação, cortar, universidade, bilião, orçamentar, governar, 3.2, federar, bloquear, mec	0,972	pesquisar, governar, votar, ser, ver, eleição, ter, perder, ir, ruir	0,271

---

[9] May 19th - Links: www.uol.com.br | www.correiobraziliense.com.br.
[10] May 27th - Links: www.brasil247.com.

the UMAP dimension reduction led to clusters with more discriminative and meaningful tokens in the other days. This result is also indicated in the Cv, in which, although the C_V of the HDBSCAN topics varied the most, their values were always superior to the topics extracted by K-Means.

4.4 Other Discoveries

(a) May 05th, Positive cluster (49.6% positive, 23.4% neutral and 26% negative).

(b) May 27th, 'Laugh' cluster (8% positive, 41% neutral and 49% negative).

Fig. 5. Examples of clusters with high percentage of positive and neutral tweets. Word Clouds generated the same way used in the Sect. 4.2

To extract controversial topics, we focused only on the negative tweets. Although the percentages of positive and neutral tweets were not as representative as the negatives, we choose to investigate their examples as well. Then, by analyzing the clusters with high percentages of these classes, we also discovered interesting patterns. Considering all dates of analysis and all identified clusters, we will briefly discuss the findings in the following.

- **Most positive clusters (Fig. 5a):** analyzing the most positive clusters obtained for each date, we mainly identified tweets from supporters of Bolsonaro. In specific, at the start of each live stream, the ex-president's Twitter account ('@jairbolsonaro') shared the YouTube link of the broadcast as a tweet. We identified that the clusters with the highest positive rates usually contain tweets from supporters, in response to the official account's tweet, congratulating Bolsonaro for carrying out the live stream. We discovered the tweets source (in response to) by looking at the tweets metadata obtained from snscrape.
- **Most neutral clusters (Fig. 5b):** sorting and investigating clusters by their neutral percentage, we identified clusters consisting of many short tweets containing laughter ('kkkkkk') without much context. HDBSCAN was able to isolate the 'laugh' clusters, but we suspected that the amount of neutral examples may indicate unwanted bias in the BERTimbau sentiment classifier. Thus, by looking at the examples in the clusters, we identified many ironic or sarcastic (with negative sentiment) tweets incorrectly labeled as being neutral.

5   Conclusions

In this paper, controversial political topics were successfully identified in Twitter data by applying clustering-based topic modeling and analyzing the negative publications. To the best of our knowledge, this is the first study to combine clustering and sentiment analysis based on Transformers to identify controversial topics in social media. We identified one event for each date and validated the detected polemic by searching the news published during the week. We compared two different approaches to identify controversial clusters and obtained favorable results using UMAP and HDBSCAN method. Finally, we presented results investigating positive and neutral tweets in the obtained clusters.

As future works, we intend to replace the sentiment analysis step with aspect-based sentiment analysis to monitor social posts about other Brazilian politicians. The sentiment analysis could be improved by treating politician names as an aspect of the analysis. We also intend to replicate our methodology to analyze data from other social networks (such as Facebook and Instagram).

References

1. Aguiar, A., Silveira, R., Furtado, V., Pinheiro, V., Neto, J.A.M.: Using topic modeling in classification of Brazilian lawsuits. In: Pinheiro, V., et al. (eds.) PROPOR 2022. LNCS (LNAI), vol. 13208, pp. 233–242. Springer, Cham (2022). https://doi.org/10.1007/978-3-030-98305-5_22

2. Akhgari, Z., Malekimajd, M., Rahmani, H.: Sem-TED: semantic twitter event detection and adapting with news stories. In: 2022 8th International Conference on Web Research (ICWR), pp. 61–69. IEEE (2022)

3. Alhaj, F., Al-Haj, A., Sharieh, A., Jabri, R.: Improving Arabic cognitive distortion classification in twitter using bertopic. Int. J. Adv. Comput. Sci. Appl. **13**(1), 854–860 (2022)

4. Angelov, D.: Top2vec: distributed representations of topics. arXiv preprint arXiv:2008.09470 (2020)

5. Antypas, D., Preece, A., Collados, J.C.: Politics and virality in the time of twitter: a large-scale cross-party sentiment analysis in Greece, Spain and united kingdom. arXiv preprint arXiv:2202.00396 (2022)

6. Archivist, J.A.: Github - snscrape is a scraper for social networking services (SNS). It scrapes things like user profiles, hashtags, or searches and returns the discovered items, e.g. the relevant posts (2020). https://github.com/JustAnotherArchivist/snscrape. Accessed 15 May 2022

7. Boon-Itt, S., Skunkan, Y., et al.: Public perception of the COVID-19 pandemic on Twitter: sentiment analysis and topic modeling study. JMIR Public Health Surveill. **6**(4), e21978 (2020)

8. Bose, R., Dey, R.K., Roy, S., Sarddar, D.: Analyzing political sentiment using twitter data. In: Satapathy, S.C., Joshi, A. (eds.) Information and Communication Technology for Intelligent Systems. SIST, vol. 107, pp. 427–436. Springer, Singapore (2019). https://doi.org/10.1007/978-981-13-1747-7_41

9. Brum, H.B., Nunes, M.D.G.V.: Building a sentiment corpus of tweets in Brazilian Portuguese. arXiv preprint arXiv:1712.08917 (2017)

10. Chaudhary, J., Niveditha, S.: Twitter sentiment analysis using tweepy. Int. Res. J. EngTech **8**, 4512–6 (2021)
11. Egger, R., Yu, J.: A topic modeling comparison between LDA, NMF, Top2Vec, and BERTopic to demystify twitter posts. Front. Sociol. **7**, 886498 (2022)
12. Ester, M., Kriegel, H.P., Sander, J., Xu, X., et al.: A density-based algorithm for discovering clusters in large spatial databases with noise. In: KDD, pp. 226–231 (1996)
13. Feldman, R., et al.: Knowledge management: a text mining approach. In: Proceedings of the 2nd International Conference on Practical Aspects of Knowledge Management (PAKM 1998), pp. 9–1. No. CONF (1998)
14. Grootendorst, M.: Bertopic: neural topic modeling with a class-based TF-IDF procedure. arXiv preprint arXiv:2203.05794 (2022)
15. Hendry, D., et al.: Topic modeling for customer service chats. In: 2021 International Conference on Advanced Computer Science and Information Systems (ICACSIS), pp. 1–6. IEEE (2021)
16. Lorenzo-Rodríguez, J., Torcal, M.: Twitter and affective polarisation: following political leaders in Spain. South Eur. Soc. Polit. **27**, 1–27 (2022)
17. Lund, M.: Duplicate detection and text classification on simplified technical english (2019)
18. Marjanen, J., Zosa, E., Hengchen, S., Pivovarova, L., Tolonen, M.: Topic modelling discourse dynamics in historical newspapers. arXiv preprint arXiv:2011.10428 (2020)
19. McInnes, L., Healy, J., Astels, S.: HDBSCAN: hierarchical density based clustering. J. Open Source Softw. **2**(11), 205 (2017)
20. McInnes, L., Healy, J., Melville, J.: UMAP: uniform manifold approximation and projection for dimension reduction. arXiv preprint arXiv:1802.03426 (2018)
21. Moulavi, D., Jaskowiak, P.A., Campello, R.J., Zimek, A., Sander, J.: Density-based clustering validation. In: Proceedings of the 2014 SIAM International Conference on Data Mining, pp. 839–847. SIAM (2014)
22. Na, S., Xumin, L., Yong, G.: Research on k-means clustering algorithm: an improved k-means clustering algorithm. In: 2010 Third International Symposium on Intelligent Information Technology and Security Informatics, pp. 63–67. IEEE (2010)
23. Radovanovic, M., Ivanovic, M.: Text mining: approaches and applications. Novi Sad J. Math. **38**, 227–234 (2008)
24. Reimers, N., Gurevych, I.: Sentence-BERT: sentence embeddings using siamese BERT-networks. arXiv preprint arXiv:1908.10084 (2019)
25. Röder, M., Both, A., Hinneburg, A.: Exploring the space of topic coherence measures. In: Proceedings of the eighth ACM International Conference on Web Search and Data Mining, pp. 399–408 (2015)
26. Senado, I.D.: Datasenado - portal institucional do senado federal. https://www12. senado.leg.br/institucional/datasenado/publicacaodatasenado?id=panorama-politico-2022. Accessed 06 Oct 2022
27. Sheikha, H.: Text mining twitter social media for covid-19: comparing latent semantic analysis and latent dirichlet allocation (2020)
28. Silva, N.F.F., et al.: Evaluating topic models in Portuguese political comments about bills from brazil's chamber of deputies. In: Britto, A., Valdivia Delgado, K. (eds.) BRACIS 2021. LNCS (LNAI), vol. 13074, pp. 104–120. Springer, Cham (2021). https://doi.org/10.1007/978-3-030-91699-2_8

29. Souza, F., Nogueira, R., Lotufo, R.: BERTimbau: pretrained BERT models for Brazilian Portuguese. In: Cerri, R., Prati, R.C. (eds.) BRACIS 2020. LNCS (LNAI), vol. 12319, pp. 403–417. Springer, Cham (2020). https://doi.org/10.1007/978-3-030-61377-8_28

30. Souza, F.D., Filho, J.B.O.S.: BERT for sentiment analysis: pre-trained and fine-tuned alternatives. In: Pinheiro, V., et al. (eds.) PROPOR 2022. LNCS (LNAI), vol. 13208, pp. 209–218. Springer, Cham (2022). https://doi.org/10.1007/978-3-030-98305-5_20

31. Vaswani, A., et al.: Attention is all you need. In: Advances in Neural Information Processing Systems, vol. 30 (2017)

32. Wang, W., Wei, F., Dong, L., Bao, H., Yang, N., Zhou, M.: MiniLM: deep self-attention distillation for task-agnostic compression of pre-trained transformers. Adv. Neural. Inf. Process. Syst. **33**, 5776–5788 (2020)

Leveraging Sign Language Processing with Formal SignWriting and Deep Learning Architectures

Fernando de Almeida Freitas[(✉)] [ID], Sarajane Marques Peres[ID], Otávio de Paula Albuquerque[ID], and Marcelo Fantinato[ID]

School of Arts, Sciences and Humanities, University of São Paulo, São Paulo, Brazil
{fernandokorban,sarajane,otavioalbuquerque,m.fantinato}@usp.br

Abstract. Advances in sign language processing have not adequately kept pace with the tremendous progress that has been made in oral language processing. This fact serves as motivation for conducting research on the potential utilization of deep learning models within the domain of sign language processing. In this paper, we present a method that utilizes deep learning to build a latent and generalizable representation space for signs, leveraging Formal SignWriting notation and the concept of sentence-based representation to effectively address sign language tasks, such as sign classification. Extensive experiments demonstrate the potential of this method, achieving an average accuracy of 81% on a subset of 70 signs with only 889 training data and 69% on a subset of 338 signs with 3,871 training data.

Keywords: Sign Language Processing · Formal SignWriting · Deep Learning · Convolutional Neural Network · Transformer

1 Introduction

The World Health Organization has projected that approximately 700 million individuals will experience moderate-to-severe hearing loss by 2050, with a majority being elderly due to aging population [38]. Hearing loss brings challenges to the daily lives of individuals. The challenge is even more difficult for deaf babies born into families that are not familiar with sign language, as it is essential for them to gain literacy in sign language to facilitate appropriate cognitive and socioemotional development, as well as equitable and effective communication [38]. Besides acquiring sign language, access to diverse forms

The authors of this work would like to thank the Center for Artificial Intelligence (C4AI-USP) and the support from the São Paulo Research Foundation (FAPESP grant #2019/07665-4) and from the IBM Corporation. This study was financed in part by the Coordenação de Aperfeiçoamento de Pessoal de Nível Superior - Brasil (CAPES) - Finance Code 001.

M. C. Naldi and R. A. C. Bianchi (Eds.): BRACIS 2023, LNAI 14197, pp. 299–314, 2023.
https://doi.org/10.1007/978-3-031-45392-2_20

of knowledge throughout their lives is essential for their intellectual and civic development.

Deaf individuals are inevitably placed in social environments where oral language predominates. Although they can still use a gesture-visual language to communicate, acquiring information, whether routine or formal, is primarily accomplished through the use of the oral language writing system. This system is based on the transcription of the phonological aspects of oral language. Due to this discrepancy, the acquisition of written proficiency in the local spoken language by deaf individuals is not only complex, but it also does not promote linguistic reflection in their first language. Furthermore, this discrepancy prevents the complete dissemination of knowledge to this segment of the population.

The availability of automatic translation tools between sign languages and oral languages could facilitate the social inclusion of the deaf community and enhance their access to knowledge. However, despite recent advances in natural language processing, tasks related to the automatic processing of sign languages still remain a challenging problem. One of the major obstacles in this task is the scarcity of labeled datasets [26]. Although sign language videos are currently available in significant volume, labeling them is a laborious task and requires refined linguistic knowledge [25]. Furthermore, the most modern resources for automatic resolution of tasks related to linguistic systems are based on the principle of a word sequence-based language. This aspect prevents the use of modern techniques of language processing on sign language due its organization based in gestural elements superimposed in a three-dimensional space.

One alternative to overcome this gap is to use a writing system for sign languages, such as the well-known SignWriting. Developed by Valerie Sutton [33], SignWriting is a universal notation for recording sign languages that uses the International SignWriting Alphabet 2010 to represent the parameters of sign languages. As a result, it provides a basis for writing gesture-visual signs. Using a writing system for sign language opens up an avenue for using the Transformer architecture [23,35] - a neural network architecture widely used for language modeling and solving various natural (oral) language tasks.

In this paper we present a framework developed for using lexicons of sign language, described through SignWriting, as input for deep learning architectures. We trained deep learning models using public datasets that contain sign descriptions in SignWriting, and tested the resulting model on the classification task of signs from three sign languages: American, Brazilian, and German.

We contend that our work makes significant contributions to the field of automatic sign language processing. We propose a new conceptual model for sign language processing. This model creates a latent representation space for signs, and supports the development of solutions for tasks such as translation. The representation space is applied in classifier implemented with Convolutional Neural Network and Transformer architectures. We have made pre-trained model checkpoints publicly available[1] to the research community, which can accelerate the development of solutions for various sign language processing tasks. Addi-

---

[1] https://github.com/fernandoafreitas/bracis_2023_signwriting.

tionally, as part of the conceptual model development process, we devised three mappings for adequate SignWriting descriptions for using in deep learning architectures.

This paper is organized as follows: Sect. 2 provides a theoretical background; Sect. 3 presents an overview of the related work; Sect. 4 discusses the dataset design and the method proposal; Sect. 5 presents the experimental results along with a discussions about them; Sect. 6 presents the final remarks.

2 Background

2.1 Sign Language

Natural languages have structure based on small, meaningless building blocks (smallest units called phonemes) and bigger meaningful building blocks (morphemes and lexemes that make up words). Despite an unlimited number of meanings that a natural language can create, the limitation of the number of minimum units makes a descriptive study of languages possible [19].

Sign languages are natural languages assumed to be primary for deaf individuals. This languages employ signs composed of hand configurations and movements combined with facial expressions and body postures to communicate meaning, as opposed to the spoken word used in oral languages [31]. For instance, the signs for "little house", "house" and "big house" in Brazilian Sign Language are built using a specific hand configuration carried out in different locations and applying different facial expressions (cf. Fig. 1).

Fig. 1. Examples of signs for "little house", "house" and "big house" in Brazilian Sign Language [1].

Notwithstanding common misconceptions, sign languages are not word-for-word translations of spoken languages. For instance, the relative sentence in the oral language Portuguese "*A garota que caiu de bicicleta? ... Ela está no hospital!!*"[2] is uttered in Brazilian Sign Language by using appropriate signs in the following sequence: <MENINA CAIU BICICLETA> ... <ELA LÁ HOSPITAL>[3].

While there is a consensus among scholars regarding the notion of a finite number of minimal units generating an infinite number of meanings, there is

---

[2] Translation to English: "The girl who fell from bike? ... She is in the hospital!".

[3] Mapping for English words: <GIRL FELL BIKE> ... <SHE THERE HOSPITAL>.

no academic agreement on the standardized approach to describe and organize signs. This issue arises from the various SignWriting-based annotation practices applied worldwide [5].

2.2   Writing Systems for Sign Language and the SignWriting

The written recording of languages in a permanent and standardized form is essential for reflecting on communication content, recording memories, and thinking language itself, enabling the advancement and improvement of thought and cognitive processes. The absence of records makes it challenging to organize social and cultural activities within a society, leading to distortions and incomplete information about memories and traditions [8]. Thereby, sign writing systems were proposed over the years.

The oldest writing system was developed by Auguste Bébian in 1825. Such a system, known as "mimographie", has 190 symbols that describe five constituent parameters of a sign [4]. In 1965, Stokoe published a more concise writing system specific to the American Sing Language consisting of 50 symbols [31]. Based on Stokoe's notation, François Xavier Neve proposed, in 1996, a system that uses characters from Western writing as a way of representing the minimum units of sign languages [24]. The Hamburg Sign Language Notation System (HamNoSys), created in 2004, is also based on Stokoe's system, but with more iconic symbols to facilitate its use with computational techniques [16]. Finally, Writing in Sign Language (ELiS) was created by the Mariângela Estelita Barros in 2008, bringing the concept of finger configurations [3].

The writing system of interest in this paper is SignWriting, developed by Valerie Sutton [33], mainly because it has the widest availability of collaborative databases, with adherence to deaf communities around the world, among all notations [30]. This is a universal notation for sign languages that represents the parameters of sign languages through a set of visual symbols hierarchically classified. It comprises seven categories, with 30 groups of symbols, and 652 icons called "visographemes". Each icon has varying degrees of freedom to describe a sign. They can be rotated on their central axis, mirrored, and contextualized with other parts of the sign, generating 35.023 unique symbols. The categories of symbols used in SignWriting are:

1. hand shape: symbols to represent possibilities of finger articulation;
2. movements: symbols used to represent contact and small finger movements;
3. dynamics/timing: symbols use to describe intensity or movement cadence;
4. head & face: symbols used to represent head position, head/neck/facial parts movements, and facial expressions;
5. body: symbols used to represent torso/shoulders/limbs/hips movements;
6. location: contains 8 symbols to represent where, in the 3D-space, the sign is performed[4];
7. punctuation: symbol used when writing complete sentences.

---

[4] Position symbols were not found in the databases used in this paper.

Figure 2 depicts some examples of how to write the sign "rain" and "strong rain" in SignWriting, considering the composition of the sign in American Sign Language. Notice that the same sign can be written in multiple ways, and there exists the possibility of representing varying levels of intensity.

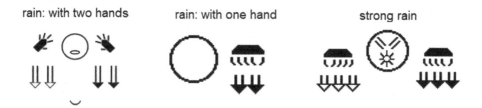

rain: with two hands   rain: with one hand   strong rain

Fig. 2. Examples of how to write "rain" and "strong rain" in SignWriting, considering the composition of the sign in American Sign Language [33].

Formal SignWriting. To facilitate the collection and dissemination of SignWriting content, Valerie Sutton developed the SignBank database[5], which serves as a collaborative platform for users worldwide. The database encourages users to contribute by adding new vocabulary entries and content written in SignWriting. Among the available contents, there are dictionaries that link signs written in SignWriting with their corresponding meanings utilizing words or expressions from the oral language of the respective country. Within this database, all inputs are encoded using Formal SignWriting, a standard symbol representation system. Formal SignWriting employs ASCII characters within a regular expression format, incorporating special tokens and $x, y-$coordinates to describe and position symbols within a designated area known as the *SignBox*. Tokens are organized in four groups: structural makers, the base symbol ranges, modifier indexes, and the numbers. The method for sign language processing introduced in this paper requires the use of Formal SignWriting for symbol representation. Below, we present the regular expression[6] employed in such formalism:

$$(S[123][0-9a-f]2[0-5][0-9a-f][0-9]3x[0-9]3)$$

The symbols have a unique identification, consisting of six characters, with the first character always being the letter S. The next three characters represent the base of a symbol; the fourth character identifies the filling of this symbol, which represents the plane in 3-dimensional space parallel to the palm of the hand, and the last character represents the rotation of the symbol. Each symbol has x, y coordinates within the SingBox. Optionally, the regular expression can be used to describe a sequence of symbols, meaning that signs have modifications to their structure during their execution. This notation uses the letter "A" in place of the letter "S" in the symbol key definition.

---

[5] https://www.signbank.org/.

[6] Detailed information about the regular expression is available at https://datatracker. ietf.org/doc/html/draft-slevinski-signwriting-text-05#section-2.3.

2.3 Deep Learning Basics

Convolutional Neural Network Architecture: According to [15], the term *convolutional neural network* signifies the network's utilization of the convolution operation, which proves particularly advantageous for processing data with grid-like structures like time series and images. The appeal of this neural network architecture stems from its capacity to extract meaningful features while disregarding the exact feature locations and focusing instead on their relative positions in relation to other features. The network comprises consecutive layers of convolution and pooling, culminating in a dense feedforward neural network. Each convolution layer consists of multiple feature maps generated by sets of synapses sharing the same weights. Maps are constructed through convolution followed by flattening carried out by sigmoid or ReLU functions [15,17]. Pooling layers follow each convolutional layer, reducing the feature map resolution through statistical operations. The final dense neural network is responsible for learning the mapping of features extracted by the preceding layers to a predefined set of classes corresponding to the input data labels.

Transformer Architecture: In the field of natural language processing, the *Transformer* [35] architecture has emerged as a prominent model renowned for its performance in text processing tasks. *Transformer* utilizes a neural network with an *encoder-decoder* [23] architecture. Considering the translation task, the network receives as input a sentence comprising words from the source natural language, encodes this information using attention layers, and subsequently decodes the encoded information, yielding a sentence translated into the target natural language. In this architecture, the so-called *attention layers* were introduced, which are responsible for identifying the importance of words as processing occurs. To accomplish this objective, every word is abstracted into vectors within the attention layer, and undergo processing through matrices of synaptic weights. Consequently, an internal parameterizable dimension encoding is obtained. Subsequently, a weighted normalization process is carried out using a softmax activation function. The weighting mechanism determines the significance of each encoding position.

3 Related Work

Previous attempts at automatic sign language processing for recognition, synthesis, translation, and application development relied on classification models using machine learning or hybrid approaches (using language parsers and rule-based reasoning) [2,9,12,14,26]. These efforts were restricted to specific languages and problems due to limitation of techniques and methods, and lack of labeled data [26]. Recent advances in natural language processing have addressed these challenges, enabling end-to-end solutions and efficient representation learning, mainly using writing systems for sign language, mitigating the scarcity of labeled data and promoting transfer learning. This section lists recent work that uses deep learning architectures in the context of sign language processing.

The initial endeavors aimed to enhance the automated processing of sign language by utilizing SignWriting descriptions primarily focused on German Sign Language signs [13,21], in 2012 and 2013. This initiative introduced an automated annotation process for sign subunits. Later, in 2015, the same research group employed the *RWTH-PHOENIX-Weather* dataset to train machine learning models for classification tasks [20].

In 2019, the description of signs was used to optimize BERT-type models, enabling the construction of a representation space [6]. Subsequently, this space was combined with visual information, encompassing hand and body movements, to facilitate classification tasks. This research involved the use of BERT models, convolutional neural networks, and Long Short-Term Memory networks for processing space-time information. In 2020, Camgoz *et al.* [7] introduced embeddings to represent space-time information in frame sequences, using data from the Montalbano II [11], MSR Daily Activity 3D [36], CAD-60 [32] and NTU-60 datasets [29]. Deep neural networks were employed to extract features, while textual information was provided as inputs in a BERT model. Both representations were used in a Transformer training for sign language recognition. In the same year, Li *et al.* [22] combined unstructured data (RGB images), BERT-based models, and spatial information from finger joints for solving the problem of classifying signs over the *Word-Level American Sign Language* dataset.

Several researchers employed efforts to apply deep learning to sign language processing in 2021 [10,18,27,34,39]. *Word-Level American Sign Language* dataset was again used in the study of Tunga, Nuthalapati and Wachs [34]. In this case, Convolutional Graph Network (GCN) and BERT model were combined to receive human poses as input and solve the glosses identification task. Deep neural networks were studied for feature extraction by Rastgoo *et al.* [27]. These authors combined the result of this representation learning with other traditionally used attributes (distances and angles). BERT attention layers were studied by Coster *et al.* [10] and proved to be effective in allowing the zero-shot classification process in the task of translating sign languages. Pre-trained models of *BERT-base* and *mBART-50* were applied to the task of translating sign language videos to German text. A new model called SignBERT was proposed by Zhou, Tam and Lam [39], using BERT in conjunction with ResNET to extract spatial information and apply it to the translation task. In this study, pose information in skeleton formats and hand configurations as inputs in the BERT strategy. The (SignBERT) was also used by Hu *et al.* [18] to compose another model, based on hand configurations and the sequence of gesture execution.

The method presented in this paper draws inspiration from related work but offers novel contributions in the following aspects: leveraging information from collaborative sources for training deep learning models, such as utilizing public data from platforms like Wikipedia to generate language models; introducing schema for mapping coding in Formal SignWriting for sentence-like representation, providing a strategy to create a versatile latent representation space applicable to various sign languages.

4   Method

Figure 3 illustrates the method proposed herein. The objective of this method is to establish a latent representation space for signs in a sign language, enabling efficient processing by deep learning models, such as for constructing classifier models. In this scenario, the classifier model takes a sign instance as input and outputs the corresponding concept (or word) represented by that sign.

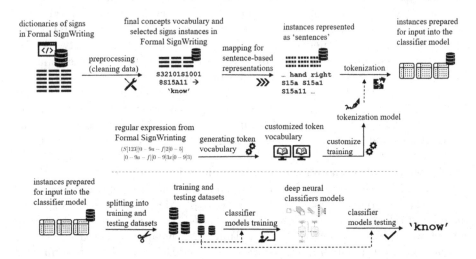

Fig. 3. Proposed method for sign language processing: generation of a representation space and its application for constructing classifier models.

The top section of Fig. 3 (above the dotted line) outlines the process of constructing the representation space. First, a collection of signs recorded in dictionaries is subjected to preprocessing to create a labeled dataset (sign representation → concept). Then, this representation is mapped to sentence-based structures and passed through a customized tokenizer. This tokenizer generates tensors for the sentences, which constitute the latent representation space. The custom tokenizer was trained using the regular expression from Formal Sign-Writing, which generates a customized vocabulary of tokens. The lower section of the figure illustrates the process of constructing classification models. Instances within the representation space are organized into datasets for both training and testing purposes, which are then utilized to build and evaluate classifier models.

The remainder of this section details the dataset used, the process of mapping the Formal SignWriting representation to a sentence-based representation, and implementation and evaluation strategies.

4.1   Datasets

The SignBank database portal has content related to more than 80 sign languages. For the purpose of our research, three dictionaries of signs were used[7]: German Sign Language with 24, 777 entries, American Sign Language with 11, 647 entries, and Brazilian Sign Language with 45, 412 entries. The dictionary entries form the vocabulary of signs being processed in our approach.

Each dictionary entry corresponds to a sign description representing a concept. Multiple signs can be associated with the same concept (cf. example shown in Fig. 2) and, not rarely, there are identical entries in the dictionary[8]. In addition, the dictionaries present some noise referring to data curation failures. During the data preprocessing stage, duplicated descriptions were removed, as well as noises were excluded.

Multiple signs associated with the same concept allowed for the creation of distinct subsets of data for model training and testing. These subsets are mutually exclusive in terms of descriptions and signs but share common concepts. This approach enables the evaluation of the deep learning model's ability to learn the patterns of concept descriptions. For instance, consider the formal and logographic descriptions illustrated in Fig. 4 for the concept "know". The training set can consist of descriptions (a) to (e), while descriptions (f) and (g) are allocated to the test set.

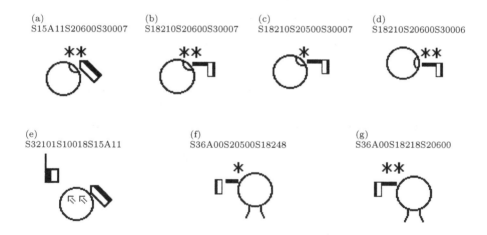

Fig. 4. Formal descriptions and logographic representations for the concept "know" using SignWriting [33], considering the signs used in American Sign Language.

---

[7] Accessed on 2023-01-31.

[8] The same sign can be represented in multiple ways, either due to slight variations in the execution of the gesture or by taking contextual factors into account during the signaling process.

Table 1 lists the number of minimum entries available in each dictionary when we consider the existence of a specific number of different descriptions for the same concept. Each row in the table represents a situation for training and testing a model. For example, the second row of the table shows the count of concepts represented and the number of entries selected in each dictionary when the condition is to have a minimum of two unique descriptions for each concept. In row number 12, the criterion is set to have at least 12 distinct descriptions. In the last scenario, the American Sign Language dictionary meets this requirement only for 13 concepts and only 197 entries are available for building training and testing datasets. The first line counts the total concepts and entries of each dictionary after the data preprocessing stage.

Table 1. Configuration of datasets for training and testing models. German - German Sign Language; ASL - American Sign Language; Libras - Brazilian Sign Language; #desc. - number of different descriptions for the same concept.

#desc.	German		ASL		Libras	
	#concept	#data	#concept	#data	#concept	#data
1	15985	23612	6239	10499	21627	36342
2	3943	11570	1987	6247	4835	19550
3	1599	6882	973	4219	2604	15088
4	777	4416	522	2866	1769	12583
5	446	3092	294	1954	1239	10463
6	273	2227	159	1279	929	8913
7	157	1531	104	949	713	7617
8	103	1153	61	648	557	6525
9	70	889	46	528	430	5509
10	50	709	35	429	338	4681
11	36	569	25	329	257	3871
12	30	503	13	197	199	3233

4.2 Data Mapping

Considering that Transformer-based solutions for natural language processing operate on word sequences (sentences) as input, we present three schema to map Formal SignWriting descriptions to sentence-based representations, involving sequences of meaningful units. We proposed three different mapping schema:

- **Mapping M1**: This scheme only represents the sequence of symbols, for instance, for representing a sign "know" in Fig. 4a, the mapping M1 generates the following sequence: *S15a11 S20600 S30007.*

- **Mapping M2**: the sentence provided by this mapping schema embeds an additional set of descriptive information about the sign. Such a set of information comprises:
  - what elements are involved in the sign description: hand, movement, dynamic, head, body or location;
  - indication if there is variation in the hand configuration of the right hand (notation: *staticityRTrue*) and of left hand (notation: *staticityLTrue*);
  - indication of the number of hands used during sign signaling (notation: *hands_qtt0, hands_qtt1* or *hands_qtt2*).

  The mapping M2 for a sign "know" generates the following sequence: *head S300 S30007. hand right S15a S15a11. movement S206 S20600. staticityRTrue. staticityLTrue. hands_qtt1.*
- **Mapping M3**: in this schema, the palm orientation and rotation information are presented separately. In accordance with the **Mapping M3** schema to describe the signs, we can represent the corresponding letters in Fig. 4 as follows:

(a) *head S300 S3000 S30007. hand right S15a S15a1 S15a11. movement S206 S2060 S20600. staticityRTrue. staticityLTrue. hands_qtt1.*
(b) *movement S206 S2060 S20600. hand right S182 S1821 S18210. head S300 S3000 S30007. staticityRTrue. staticityLTrue. hands_qtt1.*
(c) *head S300 S3000 S30007. hand right S182 S1821 S18210. movement S205 S2050 S20500. staticityRTrue. staticityLTrue. hands_qtt1.*
(d) *movement S206 S2060 S20600. hand right S182 S1821 S18210. head S300 S3000 S30006. staticityRTrue. staticityLTrue. hands_qtt1.*
(e) *head S321 S3210 S32101. hand left S100 S1001 S10018. hand right S15a S15a1 S15a11. staticityRTrue. staticityLTrue. hands_qtt2.*
(f) *head S36a S36a0 S36a00. movement S205 S2050 S20500. hand left S182 S1824 S18248. staticityRTrue. staticityLTrue. hands_qtt1.*
(g) *head S36a S36a0 S36a00. hand left S182 S1821 S18218. movement S206 S2060 S20600. staticityRTrue. staticityLTrue. hands_qtt1.*

4.3   Implementation Details

This project implemented the Tokenizer method from DistilBert [28] based-model using the HuggingFace Transformers library [37]. The vocabulary generated from the regular expression used in Formal SignWriting (cf. Sect. 2 and Fig. 3) is concatenated with the words used in the sentences (*hand, movement, dynamic, head, body, location, punctuation, stacityRTrue, stacityRFalse, stacityLTrue, stacityLFalse, hands_qtt0, hands_qtt1, hands_qtt2*), and with the special tokens [PAD], [UNK], [CLS], [SEP], and [MASK]. The custom vocabulary has 73, 743 tokens.

The Transformer architecture used for contruction of classifier models was also developed using the HuggingFace Transformers library [37]. The implementation of the convolutional neural network architecture was accomplished using libraries and frameworks provided by TensorFlow[9].

---

[9] https://www.tensorflow.org/.

4.4   Classifier Evaluation

The evaluation of the classifier models involved analyzing the impact of the number of concepts (classes) on their performance. This choice of analysis perspective is justified by the dynamic and distributed nature of sign languages. Due to the visual-spatial nature of these languages, signs can undergo numerous variations when used by different user communities. Moreover, the continuous development of new signs reflects the need to incorporate corresponding signs for emerging concepts in societal evolution. Thus, the models were generated using varying numbers of descriptions per concept (from one to eleven). We present the accuracy achieved through stratified cross-validation testing.

5   Experiments and Results

Table 2 presents the optimal outcomes achieved using the convolutional neural network architecture. The table provides the minimum, maximum, and average accuracies (along with their standard deviations) obtained through a comprehensive cross-validation process. Notably, the highest performance was observed in folds with smaller number of descriptions and concepts, indicating that while the experiment yielded promising, there exists an upper limit in terms of the number of concepts these models can effectively recognize. The Libras sign language models exhibited the lowest results, despite having a larger amount of available data. This outcome can be attributed to the experimental strategy employed, which involved a higher number of concepts for this language. Further experiments focusing on a smaller set of concepts are necessary to determine whether additional complexities exist in processing this specific sign language.

Table 2. Results for models using convolutional neural network. #d. - number of different descriptions for the same concept; μ - average; σ - standard deviation

Language	Mapping M1					Mapping M2					Mapping M3				
	#d	Min	Max	μ	σ	#d	Min	Max	μ	σ	#d	Min	Max	μ	σ
Germany	7	0.67	0.77	0.72	0.03	9	0.75	0.84	0.81	0.03	10	0.76	0.86	0.81	0.00
ASL	7	0.56	0.84	0.70	0.11	10	0.63	0.88	0.78	0.08	10	0.69	0.88	0.80	0.07
Libras	7	0.37	0.46	0.42	0.03	10	0.65	0.75	0.68	0.03	10	0.67	0.72	0.69	0.02

The results achieved using the Transformer architecture are presented in Table 3. On average, the results are slightly lower compared to the convolution-based architecture. While better performance in terms of maximum results was attainable for German and American sign languages, the model exhibited instabilities as indicated by the Brazilian sign language results.

Regarding the representation schemes, we observed a slight superiority for the **Mapping M3** schema when applying convolutional architecture, and for the **Mapping M1** and **Mapping M2** schemes when applying the Transformer

Table 3. Results for models using Transformer architecture. #d. - number of different descriptions for the same concept; μ - average; σ - standard deviation

Language	Mapping M1					Mapping M2					Mapping M3				
	#d.	Min	Max	μ	σ	#d.	Min	Max	μ	σ	#d.	Min	Max	μ	σ
Germany	12	0.70	0.84	0.78	0.05	12	0.63	0.91	0.77	0.05	12	0.63	0.88	0.74	0.07
ASL	12	0.79	0.94	0.83	0.05	12	0.76	0.94	0.84	0.05	12	0.70	0.91	0.80	0.07
Libras	12	0.48	0.55	0.51	0.02	11	0.40	0.48	0.42	0.03	12	0.04	0.46	0.24	0.16

architecture. These findings underscore the necessity for further investigations to harness the learning capabilities of each architecture, as they appear to encompass distinct features in the representation spaces.

6   Final Remarks

In this paper, we investigate a new sign language modeling approach based on SignWriting descriptions. The obtained classifier models demonstrate the ability to recognize signs based on detailed and unprecedented descriptions of their execution. Our method exhibits adaptability across sign languages from various countries, indicating the potential for utilizing a representation scheme that bridges the gap between SignWriting coding and sentence-based representation in automatic sign language processing research.

As future work, exploration of approaches involving texts and sentences in SignWriting could be conducted, enabling the incorporation of contextual information in sign execution. This would facilitate the development of models for various tasks within the same domain.

It is worth emphasizing that these models still require significant improvement, as their current capacity does not encompass a wide range of concepts. Therefore, conducting a detailed investigation into the errors that arise when introducing a larger set of concepts to the model can provide valuable insights for enhancing the latent representation space.

References

1. de Almeida Freitas, F.: Reconhecimento automático de expressões faciais gramaticais na língua brasileira de sinais. Master's thesis, Universidade de São Paulo, Brasil (2015)
2. de Almeida Freitas, F., Peres, S.M., de Moraes Lima, C.A., Barbosa, F.V.: Grammatical facial expressions recognition with machine learning. In: Proceedings of the 27th International Florida Artificial Intelligence Research Society Conference, pp. 180–185 (2014)
3. Barros, M.E.: ELis-Escrita das Línguas de Sinais: Proposta teórica e verificação prática. Ph.D. thesis, Tese (Doutorado em Linguística)-Universidade Federal de Santa Catarina (2008)

4. Bébian, A.: Mimographie, ou essai d'écriture mimique propre á régulariser le langage des sourds-muets. L. Colas (1825)
5. Bertoldi, N., et al.: On the creation and the annotation of a large-scale Italian-LIS parallel corpus. In: Proceedings of 7th International Conference on Language Resources and Evaluation, Valletta, Malta, pp. 19–22. European Language Resources Association (2010)
6. Bilge, Y.C., Ikizler-Cinbis, N., Cinbis, R.G.: Zero-shot sign language recognition: can textual data uncover sign languages? arXiv preprint arXiv:1907.10292 (2019)
7. Camgoz, N.C., Koller, O., Hadfield, S., Bowden, R.: Sign language transformers: joint end-to-end sign language recognition and translation. In: Proceedings of the IEEE Conference on Computer Vision and Pattern Recognition, pp. 10023–10033 (2020)
8. Capovilla, F., Raphael, W., Viggiano, K., Neves, S., Luz, R.: Sign writing: implicações psicológicas e sociológicas de uma escrita visual direta de sinais, e de seus usos na educação do surdo. Revista Espaço 33–39 (2000)
9. De Araújo Cardoso, M.E., Peres, S., De Almeida Freitas, F., Venância Barbosa, F., De Moraes Lima, C.A., Hung, P.: Automatic segmentation of grammatical facial expressions in sign language: towards an inclusive communication experience. In: Proceedings of the 53rd Hawaii International Conference on System Science, pp. 1499–1508 (2020)
10. De Coster, M., et al.: Frozen pretrained transformers for neural sign language translation. In: Proceedings of the 1st International Workshop on Automatic Translation for Signed and Spoken Languages, pp. 88–97. Association for Machine Translation in the Americas (2021)
11. Escalera, S., et al.: Multi-modal gesture recognition challenge 2013: dataset and results. In: Proceedings of the 15th ACM on International Conference on Multimodal Interaction, pp. 445–452. ACM, New York (2013)
12. Farooq, U., Rahim, M.S.M., Sabir, N., Hussain, A., Abid, A.: Advances in machine translation for sign language: approaches, limitations, and challenges. Neural Comput. Appl. **33**(21), 14357–14399 (2021)
13. Forster, J., et al.: RWTH-PHOENIX-weather: a large vocabulary sign language recognition and translation corpus. In: International Conference on Language Resources and Evaluation, Istanbul, Turkey, vol. 9, pp. 3785–3789. European Language Resources Association (2012)
14. Freitas, F.A., Peres, S.M., Lima, C.A., Barbosa, F.V.: Grammatical facial expression recognition in sign language discourse: a study at the syntax level. Inf. Syst. Front. **19**, 1243–1259 (2017)
15. Goodfellow, I., Bengio, Y., Courville, A.: Convolutional networks. In: Deep Learning, vol. 2016, pp. 330–372. MIT Press, Cambridge (2016)
16. Hanke, T.: Hamnosys-representing sign language data in language resources and language processing contexts. In: 4th International Conference on Language Resources and Evaluation, vol. 4, pp. 1–6 (2004)
17. Haykin, S.: Neural Networks and Learning Machines, 3rd edn. Pearson, London (2009)
18. Hu, H., Zhao, W., Zhou, W., Wang, Y., Li, H.: Signbert: pre-training of hand-model-aware representation for sign language recognition. In: Proceedings of the IEEE International Conference on Computer Vision, pp. 11087–11096 (2021)
19. Karnopp, L.B.: Aquisição fonológica na língua brasileira de sinais: estudo longitudinal de uma criança surda. Ph.D. thesis, Universidade Federal Do Rio Grande do Sul (UFRGS) (1999)

20. Koller, O., Forster, J., Ney, H.: Continuous sign language recognition: towards large vocabulary statistical recognition systems handling multiple signers. Comput. Vis. Image Underst. **141**, 108–125 (2015)
21. Koller, O., Ney, H., Bowden, R.: May the force be with you: force-aligned signwriting for automatic subunit annotation of corpora. In: 10th IEEE International Conference and Workshops on Automatic Face and Gesture Recognition, pp. 1–6. IEEE (2013)
22. Li, D., Opazo, C.R., Yu, X., Li, H.: Word-level deep sign language recognition from video: a new large-scale dataset and methods comparison. In: Proceedings of the IEEE Winter Conference on Applications of Computer Vision, pp. 1459–1469 (2020)
23. Luong, M.T., Pham, H., Manning, C.D.: Effective approaches to attention-based neural machine translation. In: Proceedings of the 2015 Conference on Empirical Methods in Natural Language Processing, Lisbon, Portugal. ACL (2015)
24. Nève, F.X.: Essai de grammaire de la langue des signes française, vol. 271. Librairie Droz (1996)
25. Polat, K., Saraclar, M.: Unsupervised term discovery for continuous sign language. In: Proceedings of the 9th Workshop on the Representation and Processing of Sign Languages: Sign Language Resources in the Service of the Language Community, Technological Challenges and Application Perspectives, Marseille, France, pp. 189–196. European Language Resources Association (2020)
26. Rastgoo, R., Kiani, K., Escalera, S.: Sign language recognition: a deep survey. Expert Syst. Appl. **164**, 113794 (2021)
27. Rastgoo, R., Kiani, K., Escalera, S., Sabokrou, M.: Multi-modal zero-shot sign language recognition. arXiv preprint arXiv:2109.00796 (2021)
28. Sanh, V., Debut, L., Chaumond, J., Wolf, T.: Distilbert, a distilled version of bert: smaller, faster, cheaper and lighter. arXiv preprint arXiv:1910.01108 (2019)
29. Shahroudy, A., Liu, J., Ng, T.T., Wang, G.: NTU RGB+ D: a large scale dataset for 3D human activity analysis. In: Proceedings of the IEEE Conference on Computer Vision and Pattern Recognition, pp. 1010–1019 (2016)
30. Stiehl, D., Addams, L., Oliveira, L.S., Guimarães, C., Britto, A.: Towards a signwriting recognition system. In: 13th International Conference on Document Analysis and Recognition, pp. 26–30. IEEE (2015)
31. Stokoe, W.C.: Sign language structure. Annu. Rev. Anthropol. **9**(1), 365–390 (1980)
32. Sung, J., Ponce, C., Selman, B., Saxena, A.: Unstructured human activity detection from RGBD images. In: IEEE International Conference on Robotics and Automation, pp. 842–849. IEEE (2012)
33. Sutton, V.: Signwriting. Sl: sn, p. 9 (2009)
34. Tunga, A., Nuthalapati, S.V., Wachs, J.P.: Pose-based sign language recognition using GCN and BERT. In: IEEE Winter Conference on Applications of Computer Vision Workshops, pp. 31–40 (2021)
35. Vaswani, A., et al.: Attention is all you need. In: Advances in Neural Information Processing Systems, vol. 30 (2017)
36. Wang, J., Liu, Z., Wu, Y., Yuan, J.: Mining actionlet ensemble for action recognition with depth cameras. In: IEEE Conference on Computer Vision and Pattern Recognition, pp. 1290–1297. IEEE (2012)
37. Wolf, T., et al.: Huggingface's transformers: state-of-the-art natural language processing. arXiv preprint arXiv:1910.03771 (2019)

38. World Health Organization: World report on hearing. Technical report, World Health Organization (2021)

39. Zhou, Z., Tam, V.W., Lam, E.Y.: SignBERT: a BERT-based deep learning framework for continuous sign language recognition. IEEE Access **9**, 161669–161682 (2021)

A Clustering Validation Index Based on Semantic Description

Roberto Douglas Guimarães de Aquino[1,2](✉) ⓘ, Vitor Venceslau Curtis[1] ⓘ, and Filipe Alves Neto Verri[1] ⓘ

[1] Aeronautics Institute of Technology, Sao Jose dos Campos, SP, Brazil
[2] Federal University of Sao Paulo, Sao Jose dos Campos, SP, Brazil
aquinordga@gmail.com

Abstract. In clustering problems where the objective is not based on specifically spatial proximity, but rather on feature patterns and the semantic description, traditional internal cluster validation indices might not be appropriate. This article proposes a novel validity index to suggest the most appropriate number of clusters based on a semantic description of categorical databases. To assess our index, we also propose a synthetic data generator specifically designed for this type of application. We tested data sets with different configurations to assess the performance of the proposed index compared to well-known indices in the literature. Thus, we demonstrate that the index has great potential for discovering the number of clusters for the type of application studied and the data generator is able to produce relevant data sets for the internal validation process.

Keywords: Clustering internal validation · Categorical data · Recommendation systems based on competence

1 Introduction

Clustering is a partitioning technique that aims to group data points so that points within a group have similar characteristics (or are close to each other) while points in other groups are different [11]. In this type of task, defining the most suitable technique for the type of data and the correct number of clusters can be a challenge, especially if little information about the data is known [12]. Despite the existence of several approaches to solve this problem, there is no optimal method, especially when considering categorical data. On this type of data the distance-based techniques may not be suitable [6,8].

In order to discover the adequate number of clusters, Cluster Validity Indices (CVI) [2] are generally used, which can have different approaches, such as distance between points [3,4,17], density [16], entropy [7], comparison between different indices [20], among others [18]. Thus, depending on the configuration of

Supported by Coordenação de Aperfeiçoamento de Pessoal de Nível Superior - Brasil (CAPES).

the data points and the purpose of the clustering certain indices may be more suitable.

Therefore, in scenarios where the objective is not based on specifically clustering nearby data points, but rather on obtaining clusters based on feature patterns or explaining them, using a CVI developed for this purpose may be more appropriate.

For example, consider a human resources data set in which each data point represents a person described by categorical features related to their professional characteristics, such as education, leadership, teamwork, creativity, etc. Given that a decision maker needs to identify the profiles present in this data set in order to select people for a given task, how could he know the number of potential professional profiles and their respective descriptions?

In this type of scenario, given the purpose of clustering and the type of features that describe the data, using a distance-based index may not be the best choice. An example of application of this type of scenario can be observed in the data sets of the competence system of the Department of Aerospace Science and Technology (DCTA)[1] of the Brazilian Air Force called *Hippocampus*. This system is restricted and stores competence levels in several areas – such as public administration, research, informatics, among others – of all DCTA's employees. DCTA's People Management sector uses the stored information to monitor the development of the skills of the workforce.

Thus, this article aims to propose an index that suggests the most appropriate number of clusters based on the semantic description potential, for applications in categorical data sets in binary form. Although the proposal is initially focused on a specific type of data, the index can also be applied to any type of data set, as long as the features are transformed into the categorical type [21].

In addition to the potential application in different types of data sets, the index can also be used to compare the quality of different partitions; it is independent of distance metrics and has significant performance when applied to competency-based data sets, where data points have features (competences) common to each other.

2   Related Works

In this section, we discuss various popular CVIs using different approaches to determine how good a particular clustering is. This variety, we think, ensures we have a solid set of CVIs to compare our proposed method against (for the actual comparison, see Sect. 6).

Average Silhouette Width (ASW) [17]

The Silhouette Width index scores how well a particular data point is clustered based on cohesion (the average within-cluster distance) and separation (how far

---

[1] https://dcta.mil.br/.

clusters are from each other). Let x_i be a data point in cluster S_l. Then, we can calculate the cohesion of x_i with

$$a(x_i) = \frac{1}{n_{S_l}} \sum_{x_j \in S_l, i \neq j} d(x_i, x_j),$$

and its cluster separation with

$$b(x_i) = \min_{S_t \in S, t \neq l} \left\{ \frac{1}{n_{S_t}} \sum_{x_j \in S_t} d(x_i, x_j) \right\}.$$

Thus, the ASW of a data set X under a partition S is given by

$$\text{ASW}(X, S) = \frac{1}{n} \sum_{x_i \in X} \frac{b(x_i) - a(x_i)}{\max\left\{a(x_i), b(x_i)\right\}}.$$

Notice that $-1 \leq ASW(X, S) \leq 1$, where a higher value indicates a better partition in terms of cohesion and separation. The literature certainly indicates there is no index that is the best in all cases, but the ASW has proven to be particularly successful (for a comprehensive comparison, see [2]).

Calinski-Harabasz (CH) [3]

The Calinski-Harabasz index is also rather popular and very much like ASW it can be easily found in many software packages used in data analysis, including MATLAB [10], scikit-learn [14], and R [15]. CH assesses the quality of a particular partition based on the ratio of the between-cluster-means and within-cluster sum of squares. Given a data set X and clustering S, we have that

$$CH(S) = \frac{(T - W_k)/(k - 1)}{W_k/(n - k)},$$

where T is the data scatter $T = \sum_{i=1}^{n} \sum_{v=1}^{m} (x_{iv} - \bar{x}_v)^2$, and $\bar{x}_v$ is the mean of feature v over all $x_i \in X$. The higher this index value is, the better the clustering is.

Davies-Bouldin (DB) [4]

The Davies-Bouldin index scores a clustering S by computing the cohesion of each cluster $S_l \in S$ as the average distance from each $x_i \in S_l$ and z_l (the centroid of S_l), and taking into account the pairwise distance between all centroids $z_l \in Z$. That is, one first calculates the cohesion for each cluster $S_l \in S$ as

$$C(S_l) = \frac{1}{N_{S_l}} \sum_{x_i \in S_l} d(x_i, z_l),$$

and then the value of this index is given by

$$\text{DB}(S) = \frac{1}{k} \sum_{S_l \in S} \max_{S_t \in S, l \neq t} \left\{ \frac{C(S_l) + C(S_t)}{d(z_l, z_t)} \right\}.$$

Unlike ASW and CH, a low value of DB indicates a good clustering.

Contiguous Density Region (CDR) [16]

The index is based on the uniformity measure of the clusters, obtained from the spatial pattern of the data points and the local density.

Initially, we define the local density of a data point x_i in a set of data points X as the distance to the nearest neighbor x_j, which belongs to the same cluster by the equation

$$\delta(x_i) = \min_{x_j \in S_l, i \neq j} \{d(x_i, x_j)\}.$$

The average density of a cluster S_l is calculated as the average of all the local densities of the cluster, defined as

$$\bar{\delta}(S_l) = \frac{1}{n_{S_l}} \sum_{x_i \in S_l} \delta(x_i).$$

Thus, the uniformity of each cluster is calculated as the difference between the local density of the cluster and the average of all the local densities of the cluster. It measures the degree of local density variation within a cluster. Given a cluster, its uniformity is defined by

$$v(S_l) = \frac{1}{\bar{\delta}(S_l)} \sum_{x_i \in S_l} |\delta(x_i) - \bar{\delta}(S_l)|,$$

for values of n_{S_l} greater than 1, otherwise a null uniformity value is assigned.

Given a partition S, the value of the CDR index is defined as

$$CDR(S) = \frac{1}{n} \sum_{S_l \in S} n_{S_l} v(S_l),$$

where n corresponds to the total number of data points in X. The CDR index minimization estimates the optimal data partition among all partitions generated by the clustering algorithm based on the structural characteristics of the data points.

CUBAGE [7]

The CUBAGE index (clustering utility based on the averaged information gain of isolating each cluster) uses the proposed *averaged information gain of isolating each cluster* (AGE) to measure the separation, and the reciprocal entropy of the set of data points conditioned on the partition to measure the compactness of the clusters.

Given that the data set X is described by a set of m independent features $V = \{v_1, \ldots, v_m\}$ and the value of the feature v_j can only be taken from the domain $D(v_j) = \{v_j^1, \ldots, v_j^{\eta_j}\}$ where η_j is the number of possible values of the respective feature.

Thus, starting from the concept of entropy, we initially calculate the information gain referring to the set of features $H(V)$ through the equation

$$H(V) = -\sum_{j}^{m}\sum_{i}^{\eta_j} p(\alpha_j^i) \log(p(\alpha_j^i)),$$

where $p(\alpha_j^i)$ is the probability of feature v_j taking the value α_j^i.

Given a partition S, the entropy of V conditioned on S, i.e., $H(V|S)$, and considered as *the whole entropy of the partition* is defined as

$$E(S) = H(V|S) = -\sum_{l}^{k}\sum_{j}^{m}\sum_{i}^{\eta_j} p(\alpha_j^i, S_l) \log(p(\alpha_j^i|S_l)),$$

where $p(\alpha_j^i, S_l)$ is the conditional probability of the value α_j^i, given cluster S_l. Finally, the information gain between variables and partitions is related through the equation

$$CUBAGE(S) = \frac{H(V) - E(S)}{E(S)}.$$

The index takes the form of the product of separation and compactness. The largest index value across partitions indicates the most suitable number of clusters.

Consensus Index (CI) [20]

This index aims to calculate the average pairwise similarity between different partitions. These partitions can be obtained by running different algorithms. Thus, suppose a set of w solutions of different clustering algorithms $S = \{S^1, \ldots, S^w\}$. The consensus index between the different partitions is given by

$$CI(S) = \sum_{p<q} \Phi(S^p, S^q),$$

where Φ is validation index based in similarity. In this study, we use the *Adjusted Rand Index* [9], also used in a similar study [19]. Hence, higher index values indicate higher quality in the partitions resulting from clustering.

Cluster Number Assisted K-means (CNAK) [18]

The CNAK algorithm is defined as a variant of k-means in which the proper cluster number is learned during the clustering process. The proposal is based on randomly sampled large-sized sets of data points having the same distribution as the original set, such that the number of generated cluster centroids from the sample is approximately the same as the original.

According to the results of the study, the algorithm presented satisfactory results, being able to successfully detect clusters, identify the hierarchy, but was not able to demonstrate robustness in dealing with noise and data overlap. In this study, we used the $\tilde{k}$ suggested by CNAK and compared it with other indices for the process of validating the results of the proposed method.

3   Validation Based on the Semantic Description of the Clusters

Given that validation indices with distance-based metrics can be a problem for applications on categorical data sets, since the spatial representation of data points is not adequate [6,8], we propose an index, based on the potential for semantic description of each cluster, for internal validation of clustering results.

For this, given the result of the clustering, frequent patterns within the clusters are identified and we calculate some indicators that are related to the semantic description potential of the set of frequent patterns. Thus, we consider an adequate clustering to be one in which the resulting clusters are adequately described, since the points within each of them were selected based on the similarity between the observed frequent patterns.

3.1   Proposed Index

Let $V = \{v_1, \ldots, v_m\}$ be a set of m features (item set) each representing a description of a categorical data set (transactional data set) $X = \{x_1, \ldots, x_n\}$ of n data points, where each point $x_i \in \mathbb{Z}_2^m$ represents the competency description of a person (transaction) from a subset of V items.

The support σ_{v_j} (or frequency of occurrence) of a v_j item is defined as the proportion of transactions that contain v_j. Even if σ_{v_j} is greater than a minimum support limit σ_{min}, we define v_j as *frequent pattern* [1].

Thus, through the partition $S = \{S_1, \ldots, S_k\}$ containing k clusters, obtained by a clustering algorithm, we define $F_1(S_l)$ as the collection of all $n_l \geq 0$ *frequent patterns* with length 1 contained in S_l, each with a support value σ_l assigned, which represent the semantic description of the respective cluster.

After that we evaluate the potential of the sets of items to perform the semantic description of the clusters. For this, we propose the *Support, Length, Exclusivity and Difference for Group Evaluation* (SLEDge) index, where the support values found, set size, exclusivity and support variability between items are evaluated, respectively. Thus, each of the indicators that are part of the index are presented in the subsections that follow.

Support. The indicator $S_{F_1(S_l)}$ is calculated from the average of support values σ_l of the collection of frequent patterns belonging to $F_1(S_l)$ by the equation

$$S_{F_1(S_l)} = \frac{1}{n_l} \sum_{\sigma_l \in F_1(S_l)} \sigma_l \tag{1}$$

when $n_l \geq 1$, otherwise the indicator is equal to zero.

We suggest that clusters described with frequent patterns of greater support between items are better, as this way we can infer more accurate information about the data points belonging to the cluster.

Length. To calculate this indicator we initially get the average cardinality based on the number of frequent patterns in all partitions by the equation

$$\bar{n}_{F_1(S)} = \frac{1}{k} \sum_l^k n_l,$$

where k represents the number of clusters. Then, we obtain the indicator through the equation

$$L_{F_1(S_l)} = \frac{1}{1 + \left| n_l - \bar{n}_{F_1(S)} \right|}. \tag{2}$$

We suggest that clusters are best described when they have an average cardinality of frequent patterns, given that either the presence of a large or small number of these should be insufficient to accurately define the semantic description of the cluster.

Exclusivity. The $E_{F_1(S_l)}$ indicator is related to the number of exclusive frequent patterns within the cluster and is calculated by the function

$$E_{F_1(S_l)} = 1 - \frac{\hat{e}_l}{n_l}, \tag{3}$$

where $\hat{e}_l$ is the number of non-exclusive frequent patterns of the cluster compared to others $k - 1$ clusters g such that $g \neq j$.

We suggest that clusters with exclusive frequent patterns are better defined in relation to the others that have more common patterns.

Difference. This indicator assess the maximum variability among the frequent patterns that describe each cluster. If there is more than 1 pattern, the indicator is calculated by taking the maximum difference δ_{max} between the support values of all *frequent patterns* belonging to $F_1(S_l)$. If there is only 1 pattern, the value of the indicator will be its own support value and if there are no patterns, the value will be null, as described by the function such

$$D_{F_1(S_l)} = \begin{cases} \sigma_l & \text{if } n_l = 1, \\ \sqrt{\delta_{max}\left\{F_1(S_l)\right\}} & \text{if } n_l > 1, \\ 0 & \text{otherwise.} \end{cases} \tag{4}$$

As the variation is small between the values of the supports, we decided to increase the influence by the square root. We assume that greater variation between support values within the cluster is better for a more accurate semantic description. Given that there are less representative frequent patterns within the cluster, we expect that the difference in proportion of these in relation to the largest values are as large as possible.

SLEDge Index. The index of each of the j clusters is calculated from the *median* of the indicators obtained by the Eqs. (1) to (4) and the global index value (SLEDge) is the *average* of the cluster indices.

The index evaluates the quality of each grouping and of the grouping in general based on the semantic description potential given from the identified frequent patterns. The higher the index value, the better the clustering quality. The index can be easily obtained from a specific function (`sledge_score`), which is part of a package (`sledge`) built in Python and available in a repository on GitHub[2].

4   Synthetic Data

Consider the following scenario. Each data sample represents a person. Each feature indicates whether such a person has a certain skill. We assume that people can be grouped in clusters such that:

- the presence/absence of a certain few features (called *dependent features*) are correlated; and
- other features may be present/absent independently.

Dependent features vary among clusters. The rationale is that, for a particular group of people, some features are more important than others. These important features appear more frequent than other features in the group. Moreover, they are positively or negatively correlated. For example, IT professionals usually share common skills: programming, data base skills, etc. On the other hand, the presence of some pairs of skills may be rare. For example, IT professionals who program rarely use the same programming language or development environment.

Thus, let x_{ij} be the indicator whether the i-th person, belonging to cluster $c(i)$, has the j-th feature. We want to generate a dataset such that

$$P(x_{ij} = 1) \approx \begin{cases} \lambda & \text{if } j \in \mathcal{F}_{c(i)} \\ \epsilon & \text{otherwise,} \end{cases} \tag{5}$$

where $0 < \epsilon < \lambda < 1$ and $\mathcal{F}_{c(i)}$ is the set of dependents features of the cluster $c(i)$.

We also want that the following statements to hold:

$$P(x_{ij}, x_{ij'}) = P(x_{ij})P(x_{ij'}) \forall j \notin \mathcal{F}_{c(i)}, \tag{6}$$

$$P(x_{ij}) = P(x_{i'j}) \text{ if } c(i) = c(i'), \text{ and} \tag{7}$$

$$P(x_{ij}, x_{ij'}) \neq P(x_{ij})P(x_{ij'}) \forall j, j' \in \mathcal{F}_{c(i)}. \tag{8}$$

To generate data with such properties, each cluster c in the dataset is associated with an m by m matrix W^c whose elements are

$$w_{pq}^c \sim \mathcal{N}(0, 1), \tag{9}$$

---

[2] https://github.com/verri/sledge.

and each sample $\boldsymbol{x}_i = (x_{i1}, x_{i2}, \ldots, x_{im})$ such that $c(i) = c$ is associated with a vector m-dimensional $\boldsymbol{a}^i$ whose elements are

$$a_p^i \sim \mathcal{N}(0, 1). \tag{10}$$

Then, let

$$\boldsymbol{b}^i = \frac{1}{\sqrt{m}} W^{c(i)} \times \boldsymbol{a}^i, \tag{11}$$

then, we generate the dependent features $(j \in \mathcal{F}^{c(i)})$

$$x_{ij} = \begin{cases} 1 & \text{if } \Phi(b_j^i) < \lambda \\ 0 & \text{otherwise,} \end{cases} \tag{12}$$

where the Φ is the cumulative distribution function of the standard normal distribution. The other features, $j \notin \mathcal{F}^{c(i)}$, are generated using

$$x_{ij} = \begin{cases} 1 & \text{if } u_{ij} \sim \mathcal{U}(0, 1) < \epsilon \\ 0 & \text{otherwise,} \end{cases} \tag{13}$$

Remark: note that the product XY of random variables $X, Y \sim \mathcal{N}(0, 1)$ is not a standard normal distribution. However, it is similar enough to use Φ to satisfy Eq. 5.

5   Set of Experiments

In this section we do some experiments to evaluate the performance of the proposed index, defined in Sect. 3, comparing with known indices described in Sect. 2. For this, we use synthetic data generated using the *Categorical Binary Random Data* (catbird) function, described in Sect. 4, which is part of the library called *Random Data Generator Algorithm for Clustering* (rdga_4k), built in Python and available in a GitHub repository[3].

For the generation of synthetic binary data, all possible combinations of the parameters listed in Table 1 are used. In the end, we obtain 3600 unique data sets from combinations of 2 parameters of ϵ, 2 of λ, 3 different numbers of both clusters and features, 2 types of data balance for each number of clusters and 50 seeds. The criteria for choosing the values of ϵ and λ are based on the different scattering scenarios of data points that the combinations of these parameters can simulate. In this way, we choose more extreme scenarios and two intermediate ones. For the K values, we try to evaluate three intervals of the same distance between 2 and 10 in order not to overload the tests, but still have interesting results to analyze. For the number of seeds, we believe that 50 seeds are enough for the analysis. In addition, during the tests, we identified that with half of the seeds the results did not change significantly.

---

[3] https://github.com/aquinordg/rdga_4k.

Table 1. Parameters used in the generation of synthetic binary data. In total, 3600 unique data sets are generated, obtained from combinations of 2 parameters of ϵ, 2 of λ, 3 different numbers of clusters, 3 of features, 2 types of data balance for each number of clusters and 50 seeds.

Parameter	Value
ϵ	0.1, 0.2
λ	0.8, 0.9
K	3, 5, 7
Number of features (relevants)	30 (20), 50 (33), 100 (66)
Balance	balanced, unbalanced
Number of seeds	50
Number of data set generated	3600

As clustering algorithms for the experiments, we use k-means and two hierarchical clusters with *average* (HACa) and *ward* (HACw) linkages. Although they are based on distance, they are considered suitable for use on binary data sets [5].

The algorithms are applied to each of the data sets for different values of k, starting at 2 and increasing in intervals from 1 to 10. Thus, using the obtained labels we calculate each of the indices, except CNAK, as it uses its own clustering algorithm (k-means++) to suggest the value of k. Given that the CI index uses different labels in a consensus approach, we named KAW the algorithm responsible for calculating the index, given that it uses the three algorithms.

6 Results and Discussion

In this section, we present and discuss some results obtained after the experiments performed, as described in Sect. 5.

As criteria to determine the quality of the indices, we used two approaches: one based on the proportion of times each index identifies the correct number of clusters (*Hit rate*) and another on the *Mean Relative Error* (MRE), defined by the equation

$$\text{MRE} = \frac{|k - \tilde{k}|}{k}, \tag{14}$$

where k represents the correct number of clusters in the data set and $\tilde{k}$ the predicted number of clusters. Thus, we use the average of the MRE to analyze each of the indices and calculate the Standard Deviation (STD). To improve the visualization of the results, we highlight the best performances in bold.

Initially, analyzing the overall performance of the indices, presented by Table 2, we can observe that SLEDge obtains the best performance, i.e. the smallest MRE. We use the standard deviation (STD) in order to verify how much the MRE values deviate from the average. The *Hit rate* presents the ratio of the number of times the MRE is zero, i.e. the index correctly identifies the number of clusters. On both measures, SLEDge also performs best, identifying the correct number of clusters in 85% of the data sets and the lowest STD among the rest. In second, we can observe that the ASW obtains a very similar performance, even with a similar STD, despite the proportion of hits being lower than expected.

Table 2. Overall performances of each index, presented using the MRE average, STD and Hit rate.

Score	Hit rate	MRE (STD)
ASW	0.76	0.07 (0.15)
CH	0.39	0.32 (0.29)
DB	0.57	0.30 (0.51)
CDR	0.00	0.55 (0.16)
CUBAGE	0.31	0.36 (0.28)
CI	0.35	0.23 (0.21)
CNAK	0.08	0.49 (0.21)
SLEDge	**0.85**	**0.05 (0.14)**

Given the different configurations of dispersion of the data points of the variables ϵ and λ, we expect that in the configuration with $\epsilon = 0.1$ and $\lambda = 0.9$ the MRE is smaller, since the clusters are more separated and the data points closer to each other. Unlike the setting $\epsilon = 0.2$ and $\lambda = 0.8$, where the opposite scenario is observed. As described by Saha and Mukherjee (2021) [18], as the proportion of overlapping between data points increases and the separation between clusters decreases, the performance of the indices tends to decrease as well. Despite varying the noise (ϵ) and overlapping (λ) parameters to produce different dispersion configurations, these configurations are not significantly different to the point that sudden changes are verified within the same index.

As for the results, according to Table 3, we can see that the SLEDge index achieves the best performance, despite being just a little higher than the ASW index. CNAK, given that it is especially sensitive to overlapping and noise in the data sets [18], presented a much lower performance, but expected, given the presence of a large number of these characteristics in the test data sets. As for CUBAGE [7], the index performs better in more compact and separate clusters, which can be observed when we analyze the different scenarios.

As for the number of clusters tested, we identified that for most indices the MRE tends to increase directly proportional to the number of k, except for the DB index. Some indices in particular, such as CNAK, CH and ASW for example, the smaller the number of clusters in the data set, the greater the chances of success of the indices [18]. Thus, given Table 4, we can observe that the SLEDge index had the best performance, followed by the ASW, but that it decreases a little when the number of clusters increases.

Table 3. Average MRE of each index given different ϵ and λ settings.

ϵ	λ	ASW	CH	DB	CDR	CUBAGE	CI	CNAK	SLEDge
0.1	0.9	0.03	0.26	0.29	0.55	0.28	0.23	0.47	**0.02**
0.1	0.8	0.07	0.32	0.29	0.55	0.36	0.23	0.50	**0.04**
0.2	0.9	0.07	0.30	0.31	0.55	0.35	0.22	0.49	**0.05**
0.2	0.8	0.13	0.39	0.31	0.55	0.46	0.23	0.49	**0.08**

Table 4. Average MRE of each index evaluated from different numbers of clusters.

k	ASW	CH	DB	CDR	CUBAGE	CI	CNAK	SLEDge
3	0.04	0.18	0.45	0.33	0.19	0.19	0.28	**0.02**
5	0.07	0.33	0.27	0.60	0.39	0.24	0.54	**0.05**
7	0.11	0.44	0.18	0.71	0.50	0.26	0.64	**0.07**

In the studies we use as a reference, given the average performance of the classic cluster validation indices, when different algorithms are used to evaluate the experiments, it is observed that ASW usually performs better, followed by CH and finally DB, as shows Table 2. The same is also observed for different degrees of data point dispersion. When the k-means algorithm is used, the ASW remains the best evaluated, but the DB usually presents a better result than the CH index, as demonstrated by studies by [5,16,18,19]. Thus, the indexes demonstrate to have influence on the type of clustering algorithm, except for the DB index that has better performance when k-means is applied instead of hierarchical, as shown in Table 5.

Table 5. Average MRE of each index evaluated from clustering algorithms.

Algorithm	ASW	CH	DB	CDR	CUBAGE	CI	CNAK	SLEDge
k-means	0.07	0.33	0.07	0.55	0.39	–	–	**0.04**
HACa	**0.10**	0.28	0.80	0.55	0.32	–	–	**0.10**
HACw	0.05	0.33	0.03	0.55	0.38	–	–	**0.01**
KAW	–	–	–	–	–	0.23	–	–
k-means++	–	–	–	–	—	–	0.49	–

In addition to verifying the performance of the indices for different values of k, we decided to do some tests in scenarios where the data points in the clusters are balanced and unbalanced. Thus, for the same number of k, we generate both types of data sets. As we can see, the SLEDge and ASW indices demonstrate significant and very similar performances, however, the SLEDge, given the proportions, continues to perform well in cases where the data is unbalanced. The results can be observed from the Table 6.

Table 6. Average MRE of each index applied to balanced and unbalanced data sets.

K	Distribution of data points	ASW	CH	DB	CDR	CUBAGE	CI	CNAK	SLEDge
3	1249,1249,1249	0.03	0.06	0.46	0.33	0.09	0.27	0.25	**0.02**
	2500,833,416	0.05	0.30	0.45	0.33	0.30	0.10	0.31	**0.02**
5	833,833,833,833,833	**0.03**	0.11	0.27	0.60	0.22	0.34	0.49	0.04
	2500,833,416,250,166	0.11	0.55	0.27	0.60	0.55	0.14	0.59	**0.05**
7	624,624,624,624,624,624,624	**0.03**	0.21	0.13	0.71	0.34	0.30	0.57	0.05
	2500,833,416,250,166,119,89	0.18	0.67	0.22	0.71	0.67	0.22	0.71	**0.10**

At the end, for the statistical significance test, we select the best performing indices in each data set, given the different algorithms, we rank them (using the average rank in the tie) and apply the Wilcoxon-Mann-Whitney test [13]. As a result, given a significance level of 0.05, we reject the null hypothesis that the performance of all indices has some similarity.

7 Conclusion

The objective of this article is to propose an index that suggests the most adequate number of clusters based on the potential of semantic description, for applications in categorical data sets, especially in clusters of profiles based on competences.

For index validation, we also propose a synthetic data generator suitable for the application, with a similar Gaussian distribution and apparently well-defined clusters, despite some overlapping points. As test experiments we use 3600 datasets with different configurations and clustered by different algorithms. Finally, we compare the proposed index with other well-known indices in the literature and with different approaches regarding the prediction of the number of clusters.

As for the results, the proposed index identifies with significant precision the correct number of clusters in relation to the others, although other indices present a similar performance. From the analysis of the results and comparing with other works, we identified that the behavior of some indices were the same of those works when applied in the synthetic data of the proposed generator. Thus, we believe that the generator can produce adequate data for the internal cluster validation task.

For future work we intend to test the index in real data sets for external validation and to develop a clustering algorithm to be applied in competence-based data sets that cluster data points in close profiles. We intend to apply our method in Hippocampus to increase the capabilities of the DCTA's People Management sector.

References

1. Agrawal, R., Srikant, R., et al.: Fast algorithms for mining association rules. In: Proceedings 20th International Conference on Very Large Data Bases, VLDB. vol. 1215, Santiago, Chile, pp. 487–499 (1994)
2. Arbelaitz, O., Gurrutxaga, I., Muguerza, J., Pérez, J.M., Perona, I.: An extensive comparative study of cluster validity indices. Pattern Recogn. **46**(1), 243–256 (2013)
3. Caliński, T., Harabasz, J.: A dendrite method for cluster analysis. Commun. Stat.-Theory Methods **3**(1), 1–27 (1974)
4. Davies, D.L., Bouldin, D.W.: A cluster separation measure. IEEE Trans. Pattern Anal. Mach. Intell. **2**, 224–227 (1979)
5. Dimitriadou, E., Dolničar, S., Weingessel, A.: An examination of indexes for determining the number of clusters in binary data sets. Psychometrika **67**(1), 137–159 (2002)
6. Dorman, K.S., Maitra, R.: An efficient k-modes algorithm for clustering categorical datasets. Stat. Anal. Data Mining ASA Data Sci. J. **15**(1), 83–97 (2022)
7. Gao, X., Yang, M.: Understanding and enhancement of internal clustering validation indexes for categorical data. Algorithms **11**(11), 177 (2018)
8. Guha, S., Rastogi, R., Shim, K.: Rock: a robust clustering algorithm for categorical attributes. Inf. Syst. **25**(5), 345–366 (2000)
9. Hubert, L., Arabie, P.: Comparing partitions. J. Classif. **2**, 193–218 (1985)
10. Inc., T.M.: Matlab version: 9.13.0 (r2022b) (2022). https://www.mathworks.com
11. Jain, A.K., Murty, M.N., Flynn, P.J.: Data clustering: a review. ACM Comput. Surv. (CSUR) **31**(3), 264–323 (1999)
12. Liu, Y., Li, Z., Xiong, H., Gao, X., Wu, J.: Understanding of internal clustering validation measures. In: 2010 IEEE International Conference on Data Mining, pp. 911–916. IEEE (2010)
13. Mann, H.B., Whitney, D.R.: On a test of whether one of two random variables is stochastically larger than the other. Ann. Math. Stat., 50–60 (1947)
14. Pedregosa, F., et al.: Scikit-learn: machine learning in python. J. Mach. Learn. Res. **12**, 2825–2830 (2011)
15. R Core Team: R: A language and environment for statistical computing (2021). https://www.R-project.org/
16. Rojas-Thomas, J.C., Santos, M.: New internal clustering validation measure for contiguous arbitrary-shape clusters. Int. J. Intell. Syst. **36**(10), 5506–5529 (2021)
17. Rousseeuw, P.J.: Silhouettes: a graphical aid to the interpretation and validation of cluster analysis. J. Comput. Appl. Math. **20**, 53–65 (1987)
18. Saha, J., Mukherjee, J.: Cnak: cluster number assisted k-means. Pattern Recogn. **110**, 107625 (2021)
19. Ünlü, R., Xanthopoulos, P.: Estimating the number of clusters in a dataset via consensus clustering. Expert Syst. Appl. **125**, 33–39 (2019)
20. Vinh, N.X., Epps, J.: A novel approach for automatic number of clusters detection in microarray data based on consensus clustering. In: 2009 Ninth IEEE International Conference on Bioinformatics and BioEngineering, pp. 84–91. IEEE (2009)
21. Witten, I.H., Frank, E., Hall, M.A., Pal, C.: Data Mining: Practical Machine Learning Tools and Techniques. Morgan Kaufmann, Burlington (2016)

Graph Neural Networks

Detecting Multiple Epidemic Sources in Network Epidemics Using Graph Neural Networks

Rodrigo Gonçalves Haddad$^{(\boxtimes)}$ and Daniel Ratton Figueiredo

System Engineering and Computer Science (PESC), Federal University of Rio de Janeiro (UFRJ), Rio de Janeiro, Brazil
{haddad,daniel}@cos.ufrj.br

Abstract. Epidemics start within a network because of the existence of epidemic sources that spread information over time to other nodes. Data about the exact contagion pattern among nodes is often not available, besides a simple snapshot characterizing nodes as infected, or not. Thus, a fundamental problem in network epidemic is identifying the set of source nodes after the epidemic has reached a significant fraction of the network. This work tackles the multiple source detection problem by using graph neural network model to classify nodes as being the source of the epidemic. The input to the model (node attributes) are novel epidemic information in the k-hop neighborhoods of the nodes. The proposed framework is trained and evaluated under different network models and real networks and different scenarios, and results indicate different trade-offs. In a direct comparison with prior works, the proposed framework outperformed them in all scenarios available for comparison.

Keywords: network epidemic · epidemic source identification · graph neural network · supervised learning

1 Introduction

Epidemics are among the most prominent and fundamental processes that take place in networks since they represent a diverse set of phenomena, e.g., the spread of information in social media or biological virus among the global human population [4,14,15]. In a network epidemic, nodes have an epidemic state (infected or susceptible) that may change over time while contagion occurs through network edges. It is not surprising that, over the past decades, many works have focused on developing models to predict real network epidemics sources for various kinds of phenomena. More importantly, the focus has also been on understanding the impact that different measures have on the epidemic process, such as quarantine and immunization, to cite a recent example largely debated during the coronavirus pandemic [7].

This work has received financial support through research grants from CNPq and FAPERJ (Brazil).

An epidemic starts with the infection of a single node or a few nodes, known as the epidemic sources. From this initial set of nodes, the epidemic unfolds through the edges of the network potentially reaching a large fraction of the network nodes. Precise information about epidemic process is often not available, such as timing or contagion information. Therefore, a fundamental problem in network epidemic is identifying the set of source nodes after the epidemic has reached a significant fraction of the network [11,20]. This problem is known to be hard for simple epidemics even when it starts on a single node and the network is infinite [18,19].

The difficulty in identifying the source nodes strongly depend on the observation process, i.e., what is observed from the network epidemic. A common snapshot model consists of a single and instantaneous snapshot that reveals the epidemic state of all network nodes [11,20]. If the snapshot is taken very early in the epidemic process, very few nodes will be infected and identifying the source nodes is intuitively easier. On the other hand, if it is taken very late, i.e., once all nodes have been infected, there is no information that can be used to identify the source nodes. The time the snapshot is taken can be translated to the moment the epidemic reaches a certain fraction of nodes, which is the model used in this paper.

The source identification problem can be posed as a classification problem as, given an epidemic snapshot, every node is to be classified as being an epidemic source, or not [3,17]. Besides the epidemic state of the nodes provided in the snapshot, the network structure is also available to the classification model. Therefore, a natural classification model is one that takes advantage of the network over which the epidemic unfolds. Graph neural networks (GNNs) seem well-suited for this task since this model integrates neighboring information. However, two nodes that have too many neighbors in common are likely to have similar representations under classic GNN models which would not be suited for source identification, since neighbors of the source node are often not a source (depending on how source nodes are chosen).

In order to tackle this issue, this work leverages the epidemic information within the k-hop neighborhood of the node. Two metrics are proposed: the fraction of nodes infected at the k-hop neighborhood and the average infection rate at the k-hop neighborhood (to be discussed in detail). These metrics are a function of the epidemic snapshot and are computed for every network node efficiently. The information, i.e., vector of rational numbers for each metric, is taken as node attribute, distinguishing neighboring nodes much beyond their epidemic state and hence providing better inputs to the GNN model.

The remainder of this paper is organized as follows. In Sect. 2, network epidemic and multiple source detection problem are introduced. A brief discussion of the related work is presented in Sect. 3. In Sect. 4, the proposed framework is presented. The evaluation of the proposed framework under different scenarios is shown in Sect. 5. Finally, Sect. 6 presents a brief conclusion for the paper.

2   Network Epidemics and Problem Statement

Network epidemics reflect different phenomena, such as influence models that describe "social contagion" [2] among individuals of a offline or online social network, and infection models that describe the contagion of a biological disease among a population. While social and biological epidemics have fundamentally different underlying contagion mechanisms, their essence can be represented by simple epidemic models.

The simplest epidemic model is the compartmental model where each individual is in one of only two possible epidemic states: susceptible or infected. This model is known as the classic SI epidemic model [12] where individuals can only change from the susceptible to infected. In the context of networks, individuals are represented by network nodes, and edges indicate the possibility of contagion. Thus, let $G = (V, E)$ denotes a undirected network (graph) where V and E denote the set of nodes and edges, respectively, and $n = |V|$ denotes the number of nodes.

The SI network epidemic model here considered is a discrete time model. Let $I(t)$ and $S(t)$ denote the set of infected and susceptible nodes at time t. Note that $I(t) \bigcup S(t) = V$ and $I(t) \bigcap S(t) = \emptyset$ for all $t = 0, 1, \ldots$ Let $I_0 \subset V$ denote the set of epidemic sources, namely, the set of nodes that are infected at time zero, thus, $I(0) = I_0$. At each time slot, an infected node infects each of its susceptible neighbors with probability p. Note that infection fails with probability $1 - p$, but the infected node is able to infect the same susceptible neighbor with probability p on subsequent time slots.

A susceptible node that has j infected neighbors in the current time step becomes infected in the next time step with probability $1 - (1 - p)^j$. Nodes become infected in the same time slot and never recover, causing $I(t)$ to be non-decreasing with t. Last, assuming the network is connected, it can be shown that there exists a finite large enough t^* such that $I(t^*) = V$ with high probability, and thus all nodes are infected.

As I_0 are the nodes chosen uniformly at random when the epidemic starts, $s = |I_0|$ is the parameter that determines the number of epidemic sources.

2.1   Epidemic Observation

The snapshot observation model which reveals the sets $I(t_o)$ and $S(t_o)$ is the network observation at a given time t_o. Note that G (the network) is also assumed to be known, and it is sufficient to observe $I(t)$, since $S(t) = V \setminus I(t)$.

The epidemic snapshot is taken when a predefined number of nodes have been infected. Let f_o denote that fraction of infected the nodes: the epidemic snapshot takes place the first time when $I(t)/n \geq f_o$. More precisely, $t_o = \min_t \{t | I(t)/n \geq f_o\}$. In what follows, f_o will be used to determine when the snapshot is taken.

2.2   Problem Statement

Consider a SI network epidemic with parameter p on the network $G = (V, E)$ having started with the epidemic sources I_0. A single epidemic snapshot is taken when a fraction of nodes infected is f_o. Design a model M that classifies nodes in $I(t_o)$ as epidemic sources, taking as input G and $I(t_o)$. Ideally, the model should recover I_0, namely

$$M_{G,I(t_o)}(v) = \begin{cases} 1 \text{ , if } v \in I_0 \\ 0 \text{ , otherwise} \end{cases} \tag{1}$$

Note that the model M has no prior information about p, the parameter for the SI epidemic. M is trained on different networks and epidemics and it should classify nodes for networks and epidemics never seen before. In fact, obtaining such generalization is the main challenge in this problem. In addition, in most scenarios the number of epidemic sources is very small with respect to the number of nodes infected in the snapshot, often by a factor of one hundred or more, hence, this classification task deals with very imbalanced classes.

3   Related Work

For over a decade, efforts have been made to address the problem of identifying the source of a network epidemic [11, 20]. The pioneering methods considered single source epidemics and simple networks while providing efficient algorithms to identify the source [19]. These approaches were based on simple probabilistic epidemic models and likelihood functions given the observation [5, 10, 23]. The rumor centrality algorithm proposed by Shah and Zaman is a prominent example that has been widely explored and extended [18, 19].

More recent works have tackled the multiple source detection (MSD) problem, where the epidemic simultaneously starts in different network nodes motivated by applications where the diffusion process starts in various nodes, e.g., information in an online social network, [3, 17, 21, 24]. *NetSleuth* [17] is an approach to infer the multiple epidemic sources in the probabilistic SI infection model using a likelihood function approach and Minimum Description Length (MDL) scores. *NetSleuth* requires prior knowledge of the number of epidemic sources. Another approach is LPSI (Label Propagation based Source Identification) [21] where labels assigned to nodes are propagated in iterations to build a weight matrix that is used to identify the multiple epidemic sources (the top ranked nodes according to a metric). The work of Zang et. al [24] considers SIR (Susceptible-Infected-Recovered) epidemics and a limited observation model by building "extended infected node sets" upon LPSI that are partitioned and ranked using a network centrality metric. Different number of epidemic sources are considered independently in order to determine the actual number of sources.

While the above approaches are based on likelihood functions and label propagation, learning-based approaches have been recently proposed to tackle the MSD problem. A prominent example is *Graph Convolutional Networks based Source Identification* (GCNSI) [3] where a GCN model is trained with supervision to generate node representations.

The approach uses LPSI as part of the node attributes and considers different epidemic models beyond SI (Susceptible-Infected), such as IC (Independent Cascade) and SIR. However, different configurations of GCN need to be trained for different networks. GCNSI is the closest approach to the framework designed in this paper, and results will be compared directly for scenarios that are available (see Sect. 5).

4 Proposed Framework

Recall that, in Graph Neural Networks, node attributes serve as the initial representation (input) to the neural network that generates the node representation. While the epidemic state of the node is the main node attribute, it is insufficient information to adequately distinguish nodes when it comes to latent representation generation, whereas sources representation should strongly distinguish from the other nodes. The framework here proposed as well as the evaluation scenarios are publicly available[1].

4.1 Metrics and Attributes

The idea behind the proposed attributes is to reflect the epidemic state around the node in their k-hop neighborhood. Intuitively, the epidemic state of the k-hop neighborhoods of source nodes should be similar but different from the epidemic state of k-hop neighborhoods of nodes infected late during the epidemic.

Therefore, two metrics are proposed to capture the epidemic state of k-hop neighborhoods: *Ring Infection* (RI) and *Depth Ring Infection* (DRI). The former leverages the average epidemic state of nodes at distance k, while the latter leverages the average neighboring epidemic state of nodes at distance k. Given the snapshot $I(t_o)$ and considering a node u, the RI metric is given by:

$$\alpha_k^u = \frac{\sum_{v \in N_k(u)} I_v}{|N_k(u)|}, \text{ for } k = 0, \ldots, K, \tag{2}$$

$N_k(u)$ denotes the k-hop neighborhood of node u, or equivalently, the set of nodes at distance k from node u, and I_v the epidemic state of node v in the snapshot $I(t_o)$. In particular,

$$I_v = \begin{cases} 1 \text{ , if } v \in I(t_o) \\ 0 \text{ , otherwise} \end{cases} \tag{3}$$

Note that $\alpha_0^u = I_u$ is simply the epidemic state of node u. Also, K is the parameter that determines the maximum distance to be considered in the attribute generation.

The DRI metric depends on the epidemic state of the neighbors of nodes that are at distance k from node u (and not just the epidemic state of the node). In

---

[1] Accessible on: https://github.com/rodrigohaddad/multiple-source-detector-gnn.

particular, DRI is the average of RI at distance 1 among the nodes that are at distance k from node u. DRI is given by:

$$\beta_k^u = \frac{\sum_{v \in N_k(v)} \alpha_1^u}{|N_k(u)|}, \text{ for } k = 1, \dots, K. \tag{4}$$

A single node can be counted multiple times in this metric, since it can be neighbor of many nodes at distance k from u. Thus, DRI provides information that is significantly different from RI.

The calculations of RI and DRI require determining the k-hop neighborhoods of all nodes in the network. For a given node u, a Breadth First Search (BFS) starting on u can efficiently determine the k-hop neighborhoods, in linear time $O(|V| + |E|)$. Since k is often small when compared to the network diameter, the BFS stops before reaching all network nodes. RI and DRI for different nodes are computed fully in parallel, as there are no dependencies. This allows for both time and memory efficient calculation.

Finally, RI and DRI are used to determine the attribute vector for each node. In particular, node u has an attribute vector that is the concatenation of α_u and β_u for all k whose dimension is $2K + 1$ (since α_u has $K + 1$ values and β_u has K values). This is the initial representation (input) to the GNN model.

4.2   Graph Neural Network Model

Graph Neural Networks (GNNs) integrate structural information with attribute information (node labels) in order to generate representations for network nodes. The GNN is a stack of neural network layers whose input is the node representation along with an aggregated representation of its neighbors, that can take several forms such as averaging. In practice, the input to the first layer corresponds to node attributes, while the output of the last layer corresponds to the node representation. The implementation discussed here adopts a small number of layers, e.g., three layers, to avoid diluting the signal of the node's neighborhood.

While being possible to make GNN models learn representations without supervision, GNNs may also be trained in a supervised fashion. In this case, the output of the last layer is used to compute a loss function that then drives the training of the model parameters. This work uses GNN in a classification task whose final goal is to classify an epidemic network node as being an epidemic source. Therefore, a *sigmoid* function is applied to the output of the last (third) layer in order to generate values between 0 and 1. The first and second layers use the *ReLU* as activation function together with a dropout layer to help avoid overfitting. The layers adopted are SAGE layers proposed by GraphSAGE, given its wide applicability and efficiency [9].

4.3   Loss Function

Since the problem consists of binary classification, the binary cross entropy function was adopted as the loss function. In particular, for each node u, it follows:

$$\ell_u(I_0) = c_u \log \sigma(y_u) + (1 - c_u) \log(1 - \sigma(y_u)), \tag{5}$$

where c_u indicates if node u is an epidemic source ($u \in I_0$, the ground truth), y_u is the output of the neural network for node u, and σ is the *sigmoid* function applied to the output of the neural network.

The average loss function across all nodes must consider the fact that classes are very imbalanced in this problem. For that matter, a weighted average is considered where w_0 and w_1 denote the weights for each class. The weights must consider the number of nodes in each class; $w_0 = r_0/(1 - |I_0|/n)$ and $w_1 = r_1/(|I_0|/n)$ where r_0 and r_1 are constants used to increase the class weight beyond the class balance ratio, and, in this work, $r_0 = 1$ and $r_1 = 100$ (chosen experimentally). This indicates to the model the importance of correctly classifying the epidemic source node. The average loss function is given by:

$$\ell(I_0) = \frac{w_0 \sum_{u \notin I_0} \ell_u(I_0) + w_1 \sum_{u \in I_0} \ell_u(I_0)}{(n - |I_0|)w_0 + |I_0|w_1} \tag{6}$$

4.4   Datasets and Training

An arbitrary large and diverse dataset with ground truth concerning epidemic sources is generated for this problem. For each sampled network, the set I_0 (epidemic sources) is chosen uniformly random, and a SI epidemic is simulated until a fraction of nodes f_o is infected, giving rise to one epidemic network sample. This entire process is repeated to generate independent samples of network epidemics. Real networks are also covered by the dataset, and in this case the network is always the same (but not I_0 nor the epidemic network). The dataset is then split for training, validation and testing.

This work made use of PyTorch Geometric [6], which employs PyTorch [16] and provides a number of GNN layers. The Adam optimizer was used with a low *learning rate* of 0.001 and all network nodes were considered as a single batch at each learning epoch. As for sampling neighboring nodes for GNN training, all nodes within a distance of 3 from the target node were considered. This sampling strategy is used to balance computational cost and network coverage as nodes closer to the target carry more important information.

The model was trained on 500 epochs and an early stopping criteria is adopted such that if no improvements in the loss function are observed in the validation network for 75 consecutive epochs, the training with the input network stops in order to avoid overfitting. The experiments were carried on a computer with an Intel Core i7-11800H, GPU NVIDIA GeForce 3070 8 GB and 16 GB RAM.

5   Evaluation

The performance of different approaches to identify epidemic sources depends on various factors, including network structure. In order to better assess the proposed framework, two random graph models and two real networks are considered. The performance of the proposed approach will be characterized using two different criteria: identifying epidemic source nodes and identifying neighbors of source nodes.

5.1   Network Models and Real Networks

The Erdős-Rényi (ER) random graph model, also known as $G(n, p)$, is a classic model that has been widely studied [8]. The model yields a very homogeneous graph, with no special structure, in which each possible edge is present in the n node graph with probability p. In contrast, the Barabási-Albert (BA) model follows a iterative process driven by preferential attachment and generates a scale free network, better representing real networks such as the web and the AS Graph [1]. Networks generated by these two models are strikingly different and both will be used in the evaluation. To allow for a direct comparison, the parameters of the two random graph models are chosen such that the generated networks are connected and have the same number of nodes and edges, on average.

As for networks modeled after real life data, the Facebook ego network is a social network consisting of the ego-networks of ten Facebook users [13]. The power grid network represents the electrical power grid of the western United States with generators, transformers and substation acting as network nodes [22]. These two real networks are also very different in terms of node degree and distances. Table 1 summarizes some details on the networks considered.

Table 1. Network information (* indicates average values).

Network	Nodes	Edges	Avg. Degree	Diameter
BA 5000	5000	61438	24.575	4
ER 5000	5000	61438*	24.575	4
Power Grid	4941	6594	2.669	46
Facebook Ego	4039	88234	43.691	8

For each random graph model and parameters determining a scenario (such as number of epidemic sources and infection ratio), 30 independent epidemic networks are generated and split equally into training, validation and testing sets. When considering real networks, 50 independent epidemic networks are generated, from which 1/5 is used for training, 1/5 for validation and 3/5 used for testing. The testing dataset is larger to ensure a better estimation of the average performance and allow a better comparison with prior works.

5.2 Evaluation Metrics

Accuracy is not an adequate metric to capture the performance of the framework, since there are very few epidemic sources (less than 0.3%) and thus a very unbalanced dataset. Metrics that capture the ability of correctly identifying the sources are more adequate, hence taking into account true positives, false positives and false negatives.

Recollect that the proposed model outputs the probability that a node is an epidemic source. Therefore, given an epidemic network, each node is used as input to the trained model and ranked according to their probability in decreasing order (most probable first). From this list, the top-k nodes are taken as epidemic sources, where k is parameter of the evaluation methodology. Using this methodology, the precision and recall values for the top-k are computed, and consequently the F-score metric. Note that k should be a function of the number of epidemic sources, which is given by $s = |I_0|$.

Two different sets are considered as true positives when assessing the performance of the framework: i) the set I_0, namely the epidemic source nodes; ii) the set $O = \{v|v \in I_0 \vee v \in N_1(u), u \in I_0\}$, namely the epidemic sources and the neighbors of the epidemic sources ($N_1(u)$ denotes the neighbors of node u). Note that O considers the neighbors of the epidemic source nodes as nodes classified correctly by the framework. Obviously, identifying nodes that are neighbors of the epidemic source has much more value than nodes that are not neighbors.

5.3 Results

Figure 1 shows the performance of the proposed framework on all four networks (one network for each subfigure). Each graph shows the F-score for the two scenarios, considering just the epidemic sources (set I_0, **w/o neighbors**) and epidemic sources and their neighbors (set O, **w/ neighbors**). Each curve corresponds to a different number of sources, $s = |I_0|$ and the x-axis is the value for k in the k-top nodes, as a multiple of s. In all scenarios, $f_o = 20\%$.

It also shows that increasing k improves performance on all networks on the w/ neighbors evaluation scenario in most cases. One exception being Facebook Ego, where the performance decreases when s is 5 or 10. Additionally, increasing the number of epidemic sources also improves performance on all networks on both scenarios, with the exception of BA networks. Note that for ER networks, performance clearly improves as s increases. In BA networks, having less epidemic sources yields better performance when considering w/o neighbors scenario. This occurs possibly because of the distinctive network topology where having a few epidemic sources allows for easier identification.

Interestingly, Fig. 1 shows that it is not always the case that performance is better in w/ neighbors scenario. On the Power Grid network, leveraging neighbors clearly improves performance, while for BA network the opposite is observed, and for ER networks results are mixed. Therefore, w/ neighbors has greater precision but potentially smaller recall values, as the set of true positives has increased significantly. This intuition is verified in Fig. 2.

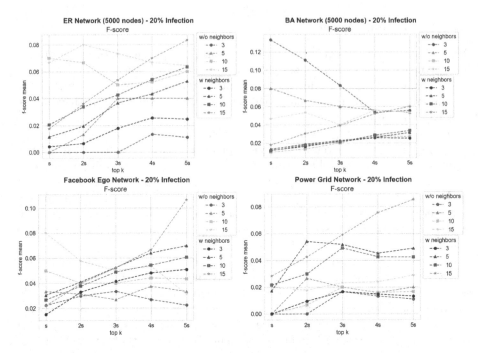

Fig. 1. F-score for all four networks for both evaluation scenarios (w/ and w/o neighbors) for different number of epidemic sources ($s \in \{3, 5, 10, 15\}$) and different top-k values (as a function of s). In all scenarios, $f_0 = 20\%$.

Figure 2 shows precision values for the same scenarios as Fig. 1. Note that in all networks, precision for a fixed number of sources is always higher when considering the set O (w/ neighbors) in comparison to the set I_0 (w/o neighbors). Moreover, for ER, Facebook and Power Grid networks, having more epidemic sources improves precision for the O set in all scenarios. When considering $s = 15$, precision for ER and Facebook network are above 20% and 45% for all top-k values considered, respectively, while for the Power Grid precision is always below 8%. This indicates the importance of the network structure on identifying epidemic sources.

5.4   Results with Mixed Datasets

The previous results considered specialized datasets where all samples were independent and identically distributed, all having the same parameters. Such samples were used for training, validation, and testing. However, it is also important to consider the performance of the framework when the datasets are more diverse and not composed by identically distributed samples.

A heterogeneous dataset was generated for each network model. The samples in each dataset have different networks (from the same model), different number of sources ($s \in \{3, 5, 10, 15\}$) and different infection rates ($f_o \in \{0.1, 0.2, 0.3\}$).

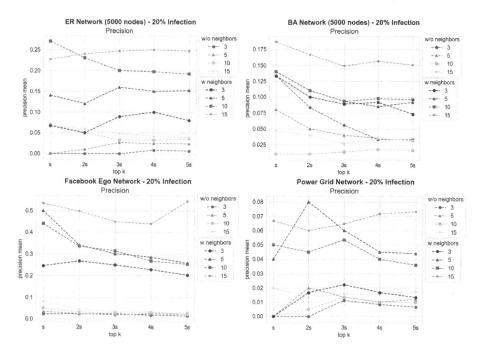

Fig. 2. Precision for all four networks for both evaluation scenarios (w/ and w/o neighbors) for different number of epidemic sources ($s \in \{3, 5, 10, 15\}$) and different top-k values. In all scenarios, $f_0 = 20\%$.

The exception are real networks where the network is always the same (but all else is simulated to generate an epidemic network). This heterogeneous dataset is used for training, validation and testing a general model.

Results comparing the performance of the framework on the homogeneous and heterogeneous datasets are shown in Fig. 3. Each subfigure shows a different infection rate, f_o and each couple of bars correspond to different number of sources, s. Performance on the homogeneous (specialized) dataset is almost always superior than performance on heterogeneous dataset. However, when the infection rate is small, e.g., $f_o = 0.1$, the performance of the models are comparable for both ER and BA networks for different number of sources. This indicates that the framework can be effectively trained on a more general dataset when the infection rate is small. Indeed, when the infection rate is small, increasing the number of sources will still allow them to have a distinctive signature in the epidemic network.

Figure 3 also shows that as the infection rate increases, performance degrades for both homogeneous and heterogeneous datasets, but falls faster for the heterogeneous. In fact, performance is zero for $f_o = 0.3$ and $s \in \{3, 5\}$) for ER networks, showing that heterogeneous dataset trained model cannot adequately identify source nodes in more difficult scenarios. A curious exception was BA

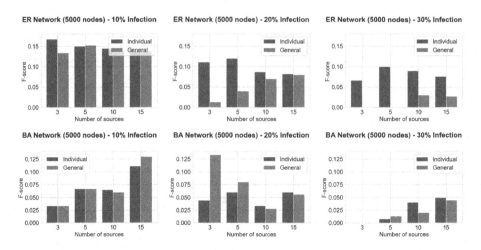

Fig. 3. Performance of network models using specialized/homogeneous dataset (Individual) and generalized/heterogeneous datasets (General) for different number of sources (s) and different infection rates (f_o).

for $f_o = 0.2$ where performance for the heterogeneous dataset was in some cases higher, $s \in \{3, 5\}$.

Figure 4 shows the performance of using the homogeneous and heterogeneous datasets when considering the two real networks. Results clearly indicate that using a heterogeneous dataset is significantly worst than using a homogeneous dataset, even in the case $f_o = 0.1$. Results also corroborate that for a given infection rate, increasing the number of sources improves performance, for both homogeneous and heterogeneous datasets. Moreover, results also corroborate that increasing the infection rate reduces the performance, and again for both datasets. Note that performance under the homogeneous dataset is in some cases twice larger than the heterogeneous dataset, indicating that generalization is much harder for real networks.

5.5 Comparison with Baselines

Only a few prior works have evaluated the multiple epidemic source detection (MSD) problem in scenarios that are reproducible and comparable. In fact, a contribution of this work is the evaluation of the proposed framework under well-studied network models and epidemic models that can be easily reproduced.

An exception is GCNSI [3] that performed numerical evaluations on real world networks and reported their scenario with sufficient details, using F-score metric, while also reporting results for LPSI and NetSleuth for the same scenarios. Table 2 compares F-score results for 3, 5 and 10 epidemic sources when 30% of the network is infected, $f_o = 0.3$, for GCNSI, LPSI and NetSleuth approaches. Recall that only GCNSI uses Graph Convolution Networks. Results for the proposed framework are shown in the row SAGE with maximum and average values across different values for k in the top-k ranking.

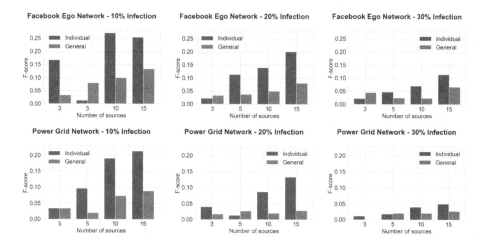

Fig. 4. Performance of real networks using specialized/homogeneous dataset (Individual) and generalized/heterogeneous datasets (General) for different number of sources (s) and different infection rates (f_o).

Table 2. F-score values of different frameworks on real networks with 3, 5 and 10 epidemic sources. GCNSI, LPSI and NetSleuth results extracted from [3].

$s = 3$		
	Power grid	Facebook ego
SAGE (Max)	0.011	0.022
SAGE (Average)	0.009	0.018
GCNSI	0.006	0.012
LPSI	0.003	0.009
NetSleuth	0.001	0.005
$s = 5$		
	Power grid	Facebook ego
SAGE (Max)	0.018	0.047
SAGE (Average)	0.014	0.039
GCNSI	0.009	0.017
LPSI	0.008	0.013
NetSleuth	0.003	0.006
$s = 10$		
	Power grid	Facebook ego
SAGE (Max)	0.066	0.138
SAGE (Average)	0.060	0.112
GCNSI	0.017	0.022
LPSI	0.013	0.020
NetSleuth	0.006	0.009

Note that results are generally poor across all table entries since reported F-score values are much closer to zero than to one, indicating that the MSD is rather a difficult problem. However, results in general improve with the increasing number of epidemic sources. Note that SAGE stands out showing the best performance across the table for all scenarios, for maximum and average performance. Also, the gap between SAGE and the other models increase along with the number of sources: SAGE's F-score on Facebook Ego with 10 sources are five times larger than the second best model.

6   Conclusion

Identifying multiple epidemics source nodes from the observation of an epidemic network is a challenging problem that has recently been explored in the literature. This work proposed a framework based graph neural networks that starts by generating features for nodes from the observed epidemic network. These features capture the infection state of k-hop neighborhoods around the node. Finally, a SAGE supervised trained model is used to rank nodes of a (never before seen) epidemic network with respect to their probabilities of being an epidemic source for that observation. The proposed model is generalizable and can be applied to any network, any SI contagion model, and any number of epidemic sources.

The performance of the proposed framework was evaluated on two random network models and two real networks, along with two evaluation scenarios (source nodes and source nodes with their neighbors). Results highlight the importance of the network structure on the framework's performance. Besides, results indicate that performance improves when the number of sources is larger, and decreases as the fraction of infected nodes increases in the observation. Results also indicate that a single model does not perform well when considering various scenarios (with the exception of low infection rates and random network models). Finally, the proposed framework significantly outperformed recent prior works in all scenarios where a direct comparison was possible.

In general, reported values for F1-score and other metrics (precision and recall) are low both here and in prior works, but this is an indication that detecting multiple epidemic sources is a challenging problem. Future works can explore node features and classification model in search of more promising approaches.

References

1. Barabási, A.L., Pósfai, M.: Network Science. Cambridge University Press, Cambridge (2016)
2. David, E., Jon, K.: Networks, Crowds, and Markets: Reasoning About a Highly Connected World. Cambridge University Press, New York (2010)
3. Dong, M., Zheng, B., Quoc Viet Hung, N., Su, H., Li, G.: Multiple rumor source detection with graph convolutional networks. In: ACM International Conference on Information and Knowledge Management (CIKM), pp. 569–578 (2019)

4. Draief, M., Massouli, L.: Epidemics and Rumours in Complex Networks. Cambridge University Press, Cambridge (2010)
5. Feizi, S., Médard, M., Quon, G., Kellis, M., Duffy, K.: Network infusion to infer information sources in networks. IEEE Trans. Netw. Sci. Eng. **6**(3), 402–417 (2019)
6. Fey, M., Lenssen, J.E.: Fast graph representation learning with PyTorch geometric. In: ICLR Workshop on Representation Learning on Graphs and Manifolds (2019)
7. Firth, J., Hellewell, J., Klepac, P., Kissler, S., Kucharski, A., Spurgin, L.: Using a real-world network to model localized covid-19 control strategies. Nat. Med. **26**, 1616–1622 (2020)
8. Gilbert, E.N.: Random graphs. Ann. Math. Stat. **30**(4), 1141–1144 (1959)
9. Hamilton, W.L., Ying, R., Leskovec, J.: Inductive representation learning on large graphs. In: NeurIPS, p. 1025–1035 (2017)
10. Jiang, J., Wen, S., Yu, S., Xiang, Y., Zhou, W.: K-center: an approach on the multi-source identification of information diffusion. IEEE Trans. Inf. Forensics Secur. **10**(12), 2616–2626 (2015)
11. Jiang, J., Wen, S., Yu, S., Xiang, Y., Zhou, W.: Identifying propagation sources in networks: state-of-the-art and comparative studies. IEEE Commun. Surv. Tutor. **19**(1), 465–481 (2017)
12. Kermack, W.O., McKendrick, A.G., Walker, G.T.: A contribution to the mathematical theory of epidemics. Roy. Soc. Lond. **115**(772), 700–721 (1927)
13. Leskovec, J., Mcauley, J.: Learning to discover social circles in ego networks. In: NIPS (2012)
14. Newman, M.E.: Spread of epidemic disease on networks. Phys. Rev. E **66**(1), 016128 (2002)
15. Pastor-Satorras, R., Castellano, C., Van Mieghem, P., Vespignani, A.: Epidemic processes in complex networks. Rev. Mod. Phys. **87**(3), 925 (2015)
16. Paszke, A., Gross, S., Chintala, S.: PyTorch: an imperative style, high-performance deep learning library. In: NeurIPS, pp. 8024–8035 (2019)
17. Prakash, B.A., Vreeken, J., Faloutsos, C.: Spotting culprits in epidemics: how many and which ones? In: IEEE International Conference on Data Mining (2012)
18. Shah, D., Zaman, T.: Rumors in a network: who's the culprit? IEEE Trans. Inf. Theory **57**(8), 5163–5181 (2011)
19. Shah, D., Zaman, T.: Rumor centrality: a universal source detector. In: ACM SIGMETRICS Performance Evaluation Review, vol. 40, pp. 199–210 (2012)
20. Shelke, S., Attar, V.: Source detection of rumor in social network-a review. Online Soc. Netw. Media **9**, 30–42 (2019)
21. Wang, Z., Wang, C., Pei, J., Ye, X.: Multiple source detection without knowing the underlying propagation model. In: AAAI Conference on Artificial Intelligence (2017)
22. Watts, D.J., Strogatz, S.H.: Collective dynamics of 'small-world' networks. Nature **393**(6684), 440–442 (1998)
23. Yu, P.D., Tan, C.W., Fu, H.L.: Epidemic source detection in contact tracing networks: epidemic centrality in graphs and message-passing algorithms. IEEE J. Sel. Topics Signal Process. **16**(2), 234–249 (2022)
24. Zang, W., Zhang, P., Zhou, C., Guo, L.: Locating multiple sources in social networks under the sir model: a divide-and-conquer approach. J. Comput. Sci. **10**, 278–287 (2015)

Prediction of Cancer-Related miRNA Targets Using an Integrative Heterogeneous Graph Neural Network-Based Method

Emanoel Aurelio Vianna Fabiano[1] and Mariana Recamonde-Mendoza[1,2(✉)]

[1] Institute of Informatics, Universidade Federal do Rio Grande do Sul,
Porto Alegre, RS, Brazil
[2] Bioinformatics Core, Hospital de Clínicas de Porto Alegre, Porto Alegre, RS, Brazil
mrmendoza@inf.ufrgs.br

Abstract. MicroRNAs (miRNAs) are crucial regulators of gene expression, including in diseases such as cancer. Although machine learning methods have shown promise in predicting miRNA-target interactions, they encounter challenges related to imbalanced classes and false positives. To tackle these issues, this study proposes a GNN-based model, using a variant of GraphSAGE algorithm named HinSAGE, which integrates validated miRNA-mRNA and mRNA-mRNA interactions with cancer-related gene expression data. Results show that our approach effectively learns miRNA-target interaction patterns from the graph structure and node features. The model achieves 77% precision, 80% recall, 78% F1-score, and 86% ROC AUC on the test data. It competes well with related approaches, reaching an F1-score of approximately 90% on a common test set. Thus, GNNs offer a promising avenue for studying miRNA-target interactions, providing balanced predictive power and improved precision through negative interaction sampling from the graph.

Keywords: microRNA target prediction · machine learning · bioinformatics · graph neural networks

1 Introduction

MicroRNAs (miRNAs) are small non-coding ribonucleic acid (RNA) molecules that regulate gene expression in post-transcriptional stages, inhibiting the expression of specific genes by messenger RNA (mRNA) silencing. Recent studies indicate that a single miRNA can target many different mRNAs, while a particular mRNA can be regulated by a set of miRNAs, simultaneously or in a context-dependent manner. Moreover, abnormal miRNA expression has been associated with the development and progression of different human diseases,

This study was financed in part by the Coordenação de Aperfeiçoamento de Pessoal de Nível Superior - Brasil (CAPES) - Finance Code 001, and by grants from FAPERGS [21/2551-0002052-0] and CNPq [308075/2021-8].

M. C. Naldi and R. A. C. Bianchi (Eds.): BRACIS 2023, LNAI 14197, pp. 346–360, 2023.
https://doi.org/10.1007/978-3-031-45392-2_23

including those related to the endocrine system and cancer [10]. Thus, miRNAs are important actors in the complex regulatory network governing gene expression in physiological and pathological processes.

Unraveling miRNA-mRNA target interactions is a fundamental step for discovering the regulatory network governed by miRNAs and their functional role within an organism. Due to the numerous potential target sites for a given miRNA, experimentally confirming every possible target candidate is impractical both in terms of cost and time. Machine learning (ML) has been increasingly explored to predict candidate targets using a series of descriptors (*i.e.*, features) of the miRNA-target mRNA interaction defined from experimentally characterized interactions and prior knowledge [13], filtering possible miRNA targets.

However, miRNA target prediction is still considered an open problem due to several challenges. First, manually engineered features based on biological assumptions capture only partially the numerous characteristics that influence the effectiveness of an interaction between miRNA and mRNA target [13]. The contribution of single features to the degree of repression of the miRNA target is still hard to completely determine. Second, the multiplicity of miRNA-target interaction patterns, often in a context-dependent manner, introduces large complexity into the problem. Third, there is an inherent class imbalance issue, as the majority of validated data refers to functional miRNA targets, *i.e.*, positive examples. This generates a bias towards large false positive predictions [11].

Over the past years, numerous deep learning (DL) methods have been developed, with various successful applications in biology and medicine. Among these, we highlight the DL algorithms in graphs, known as Graph Neural Networks (GNNs), and their capacity for solving complex problems where data requires a representation in non-Euclidean domains [1,16]. GNNs have found successful applications in identifying associations between miRNAs and diseases (*e.g.*, [4]); however, they are still little explored in the computational prediction of miRNA targets. Thus, in this work we aim to answer the question: *can GNNs precisely predict miRNA targets from known miRNA-mRNA target interactions and biological evidence regarding miRNAs and mRNAs expression?*

We propose an approach based on the HinSAGE algorithm and heterogeneous graphs containing prior knowledge about validated miRNA-mRNA target and mRNA-mRNA interactions. HinSAGE is a derivative of the GraphSAGE algorithm [3], an stochastic generalization of graph convolutions that can leverage node feature to efficiently generate embeddings on unseen data. Moreover, instead of using manually engineered features from known miRNA-target interactions, we use gene expression profiles as node features. Gene expression is the final product in the process of gene regulation, which is affected by molecular factors that are usually used as predictive features individually or combined, such as sequence complementarity, sequence conservation, favorable thermodynamics, and accessibility of the miRNA binding site. Here, we focus on miRNA-target interactions associated with cancer, motivated by the role of dysregulated miRNAs in cancer pathophysiology and as potential biomarkers [10].

Our main contributions are twofold. First, we propose and evaluate a prediction approach based on the GraphSAGE algorithm that integrates an heterogeneous graph and expression profiles to predict miRNA-target interactions. To the best of our knowlegde, this is the first work to explore GNNs for miRNA-target prediction in human cancer. Second, we conduct a detailed investigation of several hyperparameters involved in the algorithm and decisions regarding dataset construction, evaluating their impact on the addressed prediction problem.

2   Related Works

Several tools have been developed for miRNA target prediction, with ML and DL being recurrent among solutions. Since the related literature is vast [13], here we review the most relevant works for the proposed approach. TargetScan [9] is a widely used tool that employs linear regression and ML techniques for miRNA target prediction. However, it has limited sensitivity for targets without good evolutionary conservation. PITA [7] considers the accessibility of the binding site on the target, evaluating the scores for microRNA-target interactions by using energy-based secondary structure prediction algorithms. Both methods rely on specialist-designed features.

More recent approaches include miRAW [12], DeepTarget [8], and DeepMir-Tar [15], which are based on DL and aim to learn better representations directly from raw data, eliminating the need for manual feature extraction. They implement autoencoders and adopt distinct approaches for defining negative samples, such as generating mock examples or building negative examples based on experimentally verified data. We note that these approaches do not explore the graph structure and some of the works lack evaluation on independent test datasets.

Finally, we highlight comprehensive databases containing computationally predicted interactions: TarBase v8 [6] and mirDIP [14]. These databases integrate data from various sources and provide confidence scores for their predictions. We note that a main challenge faced was the difficulty of identifying, up to the time of writing this paper, a model for miRNA target prediction based on the concept of GNNs. It should be noted that GNNs have been previously applied in prediction tasks related to miRNAs, but with a special focus on predicting the association of miRNAs with human diseases (*e.g.,* [4]) - a task that has different objectives from the problem modeled in this study. Moreover, a previous study has explored GNNs in the miRNA-target prediction problem, although focusing on data from *Camellia sinensis* and without integrating gene expression profiles into their framework [2].

3   Methodology

GNNs are DL algorithms designed to operate on graph-structured data. The learning of node representations in the graph, *i.e.,* the generation of embeddings, is achieved through iterative updates that aggregate the representations of neighboring nodes and the node itself from previous iterations. This information can be used for tasks such as link prediction (our interest) or node classification. In this section, we detail how we collected and integrated the interactions to define the graph structure and the gene expression data used as node features in the model training process, as depicted in Fig. 1-a. We also explain the proposed learning approach based on the HinSAGE algorithm (Fig. 1-b).

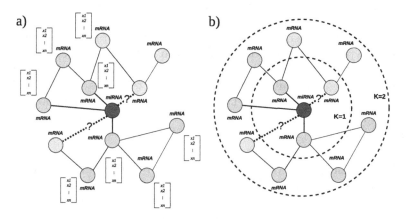

Fig. 1. Summary of the approach proposed. a) The graph consists of miRNAs (red node) and mRNAs (pink nodes) with their features vectors, and the known interactions among them (solid lines). b) The inductive learning process by the HinSAGE algorithm considering potential miRNA-mRNA interactions (dashed lines) using sampled 2-hop neighbors. (Color figure online)

3.1   Data Collection, Preprocessing and Integration

miRNA-mRNA and mRNA-mRNA Interaction Data. We obtained interactions from RNAInter v4.0 [5], a comprehensive database that integrates RNA interactome data, including experimentally validated and computationally predicted interactions. The current version has over 47 million annotated interactions for 156 species. RNAInter categorizes the type of experimental evidence for each interaction as strong or weak, and also calculates a confidence score per interaction based on characteristics such as experimental evidence confidence, scientific community confidence given by the number of article citations, and the number of different tissues or cells in which the interaction was identified – all integrated using a sigmoid function. We downloaded miRNA-mRNA and mRNA-mRNA interactions related to humans, removing those with inconsistent values, such as undefined miRNA or mRNA identifiers, and those derived from computational prediction to improve the quality of the training data.

Gene Expression Data in Cancer. Gene expression data for miRNAs and mRNAs was obtained from The Cancer Genome Atlas (TCGA) project, one of the leading consortia in cancer genomics. The data is publicly accessible through the GDAC FireBrowse portal[1]. We downloaded gene expression datasets for primary solid tumor samples (PT) and normal tissue (NT) samples (*i.e.,* non-tumor) preprocessed following well-established bioinformatics protocols. We only considered those cancers that did not exhibit significant imbalance between the number of tumor and normal tissue samples for miRNAs and mRNAs expression, resulting in 15 tumor types (see Supplementary Material[2]). The data was transformed to the log2 scale and a measure of differential expression, the log fold change (logFC), was calculated for each miRNA and mRNA in relation to each tumor type. The logFC expresses the ratio between the mean gene expression in PT vs. NT samples for a specific miRNA or mRNA, in a specific cancer type. A higher absolute value of logFC is assumed to indicate greater biological relevance in the studied domain. The step of summarizing expression patterns through logFC values generates a feature vector of length 15 for each miRNA and mRNA in the graph.

Data Integration for Graph Generation. We integrated the interaction data obtained from the RNAInter database with the gene expression data collected from FireBrowse using the unique identifier available for each miRNA and mRNA in the collected database. During this step, miRNA-target interactions without any identified expression records were also removed. This removal aimed to mitigate the induction of inconsistent results during the subsequent training of the proposed model. The final interaction network comprises over 100,000 miRNA-target interactions and nearly 50,000 mRNA-mRNA interactions. The network has a total of 2,617 miRNA nodes and 17,252 mRNA nodes. The number of interactions according to interaction type, evidence type, and score threshold is shown in Fig. 2.

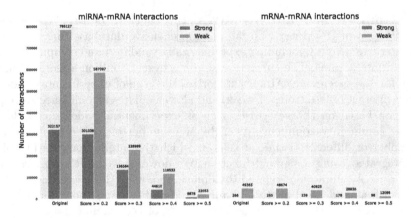

Fig. 2. Summary of collected miRNA-mRNA and mRNA-mRNA interactions.

---

[1] http://firebrowse.org/.
[2] https://encurtador.com.br/eilzA.

3.2   Model Training and Evaluation

In this work, we are interested in conducting a link prediction task handling a large amount of data, integrating node features into the learning process, and working with a heterogeneous graph. Based on the literature [16], we chose a variant of GraphSAGE [3] extended for heterogeneous graphs, named HinSAGE[3]. The algorithm generates embeddings by sampling and aggregating features from the local neighborhood of a node, instead of training individual embeddings focused solely on the node. The number of observed neighbors can be parameterized, both in terms of the number of hops (K) and the number of neighboring nodes to be sampled at each hop. This behavior can be observed in Fig. 1, where the value $K = 2$ represents the number of hops (*i.e.*, depth of search or number of HinSAGE layers). The neighborhood aggregation strategy is applied such that neighbors are randomly selected and fused together by edge type.

The predictive models were developed using the HinSAGE implementation provided by the StellarGraph library. The base graph was built as explained in Sect. 3.1. The hyperparameters were initially defined as follows: batch size equal to 200; 300 epochs of training; learning rate of 0.001 with Adam optimizer; two hops ($K = 2$); number of nodes sampled per hop equal to [8,4]; HinSAGE layer sizes setup as [32,32]; binary Cross-Entropy cost function; ReLU activation function; feature update rule based on the Graph Convolutional Network (GCN) aggregator. Variations in the hyperparameter values were tested and evaluated in a series of experiments that will be detailed later.

Our experiments were based on ten random divisions of the data into training, validation, and test sets. Considering our inductive domain, the training, validation, and test sets are subgraphs derived from reductions of the original graph, aiming to generate independent data for the different stages involved in model development (Table 1). We used the method EdgeSplitter provided by StellarGraph, which takes as input both the graph to be subdivided and a percentage p relative to the total number of edges in the provided graph, determining the number of positive and negative edge samples to be sampled from it. Positive edges are sampled from the actual set of edges present in the graph to maintain graph connectivity. Negative edges are randomly created by sampling pairs of miRNA-mRNA that are not connected by an interaction in the original graph. This sampling of negative examples aims to improve the model's ability to identify false positive edges, by mitigating class imbalance. The value of the hyperparameter p was adjust according to the set of interactions included in the creation of the graph, since it needs to be compatible with the number of negative interactions that can be randomly generated from the input graph. Model performance evaluation was based on the accuracy, precision, recall, F1-Score, and ROC AUC metrics.

4   Experiments

Two main sets of experiments were defined for the development of this study.

---

[3] https://stellargraph.readthedocs.io/en/stable/hinsage.html.

Table 1. Sizes of training, validation, and test sets.

Dataset	# miRNA-mRNA	# mRNA-mRNA	Hiperparameter p	# Positive Edges	# Negative Edges
Original data, containing all interactions					
Test	1040018	49601	0.004	4160	4160
Validation	1035858	49601	0.004	4143	4143
Training	1031715	49601	0.004	4126	4126
Dataset containing interactions with $score >= 0.2$					
Test	820906	48909	0.004	3283	3283
Validation	817623	48909	0.004	3269	3269
Training	814354	48909	0.004	3255	3255
Dataset containing interactions with $score >= 0.3$					
Test	309892	41039	0.01	3098	3098
Validation	306794	41039	0.01	3067	3067
Training	303727	41039	0.01	3036	3036
Dataset containing interactions with $score >= 0.4$					
Test	132992	28999	0.01	1329	1329
Validation	131663	28999	0.01	1316	1316
Training	130347	28999	0.01	1303	1303
Dataset containing interactions with $score >= 0.5$					
Test	22663	12176	0.01	226	226
Validation	22437	12176	0.01	223	223
Training	22214	12176	0.01	220	220

The first set, **S1**, is focused on analyzing the potential of the HinSAGE algorithm for miRNA target prediction using a graph-based learning approach. We evaluate the influence of aspects related to the construction of the interaction dataset and the configuration of the algorithms's hyperparameters on the results achieved by the HinSAGE model. We define 20 experimental scenarios organized into seven groups (Table 2), each of which with a specific experimental goal: **G1**: evaluation of the impact of using different datasets to define the base graph used in the model training, constructed by removing mRNA-mRNA interactions and with different score filters; **G2**: evaluation of the impact of removing mRNA-mRNA interactions from the base graph; **G3**: evaluation of the impact of reducing the number of training epochs; **G4**: evaluation of the impact of varying the batch size; **G5**: evaluation of the impact of varying the number of sampled neighboring nodes; **G6**: evaluation of the impact of varying the size of the hidden layers; **G7**: evaluation of the impact of varying the learning rate. Given the high computational cost involved in running each experiment, at the end of each experimental group, we identified the most promising configuration to be employed in the next group. Thus, best hyperparameter values found in previous experiments are carried over in the next group of experiments.

The second set, **S2**, aims to compare the approach developed in this study (based on the best model generated in the set **S1**) with other methods proposed in the literature. Due to lack of similar approaches, it was not possible to perform an equivalent comparison based on end-to-end graph analysis using GNNs. We conducted the comparison by focusing on the results of the models'

Table 2. Experimental scenarios investigated in the set of experiments **S1**.

G#	N².	Interactions	mRNA–mRNA	batch_size	epochs	num_sample	hinsage_layer_size	learning_rate
G1	00	All original	✓	200	300	[8, 4]	[32, 32]	0.001
	01	All original		200	300	[8, 4]	[32, 32]	0.001
	02	score >= 0.2	✓	200	300	[8, 4]	[32, 32]	0.001
	03	score >= 0.3	✓	200	300	[8, 4]	[32, 32]	0.001
	04	score >= 0.4	✓	200	300	[8, 4]	[32, 32]	0.001
	05	score >= 0.5	✓	200	300	[8, 4]	[32, 32]	0.001
G2	06	score >= 0.4		200	300	[8, 4]	[32, 32]	0.001
	07	score >= 0.5		200	300	[8, 4]	[32, 32]	0.001
G3	08	score >= 0.4	✓	200	100	[8, 4]	[32, 32]	0.001
G4	09	score >= 0.4	✓	1	100	[8, 4]	[32, 32]	0.001
	10	score >= 0.4	✓	50	100	[8, 4]	[32, 32]	0.001
	11	score >= 0.4	✓	100	100	[8, 4]	[32, 32]	0.001
	12	score >= 0.4	✓	300	100	[8, 4]	[32, 32]	0.001
G5	13	score >= 0.4	✓	300	100	[6, 3]	[32, 32]	0.001
	14	score >= 0.4	✓	300	100	[12, 6]	[32, 32]	0.001
G6	15	score >= 0.4	✓	300	100	[12, 6]	[64, 64]	0.001
	16	score >= 0.4	✓	300	100	[12, 6]	[128, 128]	0.001
	17	score >= 0.4	✓	300	100	[12, 6]	[16, 16]	0.001
G7	18	score >= 0.4	✓	300	100	[12, 6]	[32, 32]	0.005
	19	score >= 0.4	✓	300	100	[12, 6]	[32, 32]	0.01
	20	score >= 0.4	✓	300	100	[12, 6]	[32, 32]	0.0001

predictions for the test data, considering miRNA-target interactions found in common between the test set created in this study and the test sets used in the related works. We used the predictions provided by the authors of the original papers. To avoid greatly restricting the size of the common dataset due to the intersection of multiple methods, our comparison was performed pairwise with each selected related work. Since most of related works do not provide predicted probabilities, we did not use the ROC AUC metric in this analysis.

5   Results and Discussion

The next sections present and discuss the results for the two sets of experiments. We note that due to space constraints, we focus on the most meaningful findings. The complete results are provided as supplementary material[4].

5.1   S1: Exploring the Predictive Performance of HinSAGE

In the experiments of **G1**, we aimed to gain a better understanding of the model's behavior with different datasets used for training. We compared the original dataset, including both miRNA-mRNA and mRNA-mRNA interactions (Experiment 00), with a version excluding mRNA-mRNA interactions (Experiment 01) using the default hiperparameters values (Sect. 3.2). The results obtained during training showed promising numbers, with an accuracy close to 75% for both

---

[4] https://encurtador.com.br/eilzA.

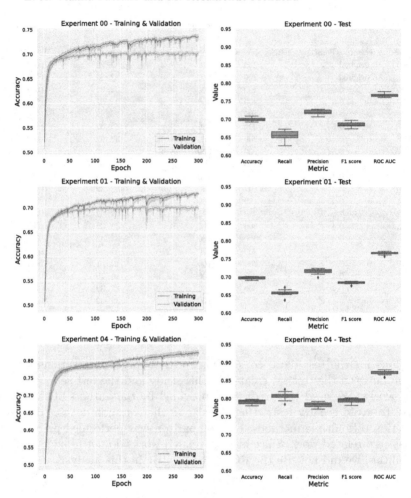

Fig. 3. Performance evaluation for selected experiments of **G1** in terms of mean and standard deviation for 10 random training, validation and test sets.

cases. However, upon examining the accuracy and loss throughout the training epochs, we identified characteristics of overfitting occurring from epoch 50 onwards. Moreover, we did not observe any significant performance change by removing mRNA-mRNA interactions, thus, we maintained them.

We also evaluated the influence of various filters applied to the confidence score associated to each interaction. Experiments 02 to 05 filtered interactions based on thresholds of 0.2, 0.3, 0.4, and 0.5 (Table 2), thus reducing the number of interactions in the graph (Table 1). We noticed a sequence of performance improvements for both training and testing data at the end of the 300 training epochs. The initial filter (score $>=$ 0.2 in Experiment 02) already showed improvements compared to Experiment 00. However, we also observed a highly negative overfitting effect in some cases, such as in Experiment 02 and, particularly, in Experiment 05. We highlight the results of Experiment 04, which proved to be the most promising in group **G1**, demonstrating an apparent improvement without a significant tendency towards overfitting. Figure 3 summarizes the training, validation, and test results for the main experiments of **G1**. It can be concluded in advance that applying the score filter does indeed influence the achieved results, and reducing the number of interactions described as strong (which decrease for larger thresholds) hinders the prediction of interactions, resulting in significant dispersion in the obtained results.

Next, in **G2**, we aimed to analyze whether the use of mRNA-mRNA interactions could influence the results obtained with filtered datasets. Therefore, Experiments 06 and 07 correspond to Experiments 04 and 05, respectively, but considering only miRNA-mRNA interactions. Once again, removing these interactions influenced the results, but the impact was not consistent between both experiments. We noticed that removing mRNA-mRNA interactions introduced overfitting in the experiment (Experiments 06 vs 04), whereas when comparing Experiments 07 vs 05, the training became more stable and showed less overfitting after removing these interactions, but without noticeable increase in the predictive performance. Hence, Experiment 04 (**G1**) remained the best scenario.

Since we observed some signs of overfitting, in **G3** we evaluated the reduction in the number of training epochs to mitigate overfitting and reduce the chances of predicting false positive interactions. Results are shown in Fig. 4 and they can be compared to Experiment 04, as the other configurations remain the same between the two experiments. We can observe that the training curves for 100 epochs exhibit more appropriate behavior, with close training and validation performances. This means that the confidence in the generalization power of the generated model becomes stronger. Moreover, the performances in the test set were quite satisfactory, although there was a slight reduction in values compared to the performance on the test data for the model generated in Experiment 04.

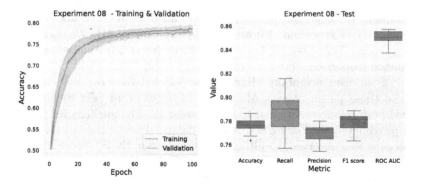

Fig. 4. Performance evaluation for **G3** in terms of mean and standard deviation for 10 random training, validation and test sets.

The precision varied mostly with values above 75% and the ROC AUC had a median value close to 85%. The following experiments will adopt 100 epochs.

Experiments of **G4** aimed to initiate the investigation of the hyperparameter values involved in the HinSAGE algorithm by exploring variations in the batch size, which was initially set to 200. Four distinct values were tested: 1 (Experiment 09), 50 (Experiment 10), 100 (Experiment 11), and 300 (Experiment 12). A batch size of 1 showed a training behavior described as stochastic, presenting overfitting within the first ten epochs of training. This characteristic was also present with a batch size of 50, which showed a large standard deviation for the first 25 epochs. Experiments 11 and 12, obtained more appropriate results. The latter, highlighted in Fig. 5, had a slight performance improvement when compared to Experiment 08, particularly in terms of performance variation across the 10 executions. The following experiments will apply a batch size of 300.

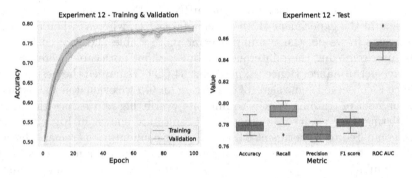

Fig. 5. Performance evaluation for selected experiment in **G4** in terms of mean and standard deviation for 10 random training, validation and test sets.

In **G5**, we investigated the impact of varying the number of sampled neighbor nodes in the HinSAGE algorithm on the generation of embeddings (*i.e.*, num_sample). We kept the number of hops fixed at $K = 2$ and tested two variations: [6,3] (Experiment 13) and [12,6] (Experiment 14). We observed that reducing the number of sampled neighbors (Experiment 13) resulted in increased performance dispersion between training and validation, while increasing the number of sampled neighbors (Experiment 14) led to a decrease in the dispersion of these performances. Moreover, Experiment 14 (Fig. 6) also showed improved performance compared to Experiment 12. The medians for all metrics were approximately 80% or higher. A more balanced performance between recall and precision was observed, resulting in an increase in the F1-Score metric. The ROC AUC metric had values very close to 90%. The better performances in the training, validation, and test datasets motivated the selection of Experiment 14 as the best scenario so far.

Finally, experiments in **G6** and **G7** aimed to compare variations in the size of hidden layers and in the learning rate, respectively. Our results indicate that increasing the dimensions of the hidden layers (Experiments 15 and 16) led to

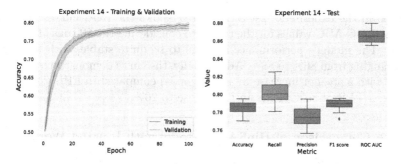

Fig. 6. Performance evaluation for selected experiment in **G5** in terms of mean and standard deviation for 10 random training, validation and test sets.

increased overfitting and had no positive effect on the maximum performance obtained by the model. On the other hand, reducing the size of the hidden layers (Experiment 17) caused the model to loose predictive power compared to the other evaluated scenarios, but mainly Experiment 14. Regarding the analysis for the learning rate, when its value was increased to 0.01 and 0.005, we observed more signs of overfitting without effective improvement in the achieved predictive performance. Conversely, decreasing the learning rate to 0.0001 had a detrimental effect on the predictive model's performance, causing underfitting of the model. In this case, the learning progress was very slow and the model failed to converge adequately for the fixed number of epochs (100).

Among all the evaluated scenarios in the set of experiments **S1** (Table 2), the best model was the one originated from Experiment 14 and it will be further

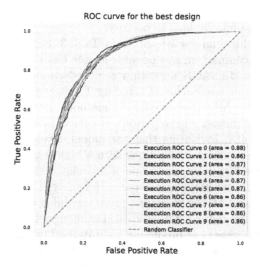

Fig. 7. ROC curves for the test sets in Experiment 14.

employed in the experiment set **S2**. Figure 7 shows the ROC curves and corresponding ROC AUC values for the test data considering the 10 runs of the 3-way holdout. The model's performance appears to be quite stable, with ROC AUC values ranging from 86% to 88%. Additionally, the curve demonstrates a suitable growth, with a steeper increase in TPR (y-axis) compared to FPR (x-axis). For example, a TPR of 80% yields an FPR close to 20%.

5.2   S2: Comparison of HinSAGE Model with Related Works

The related works for which a direct comparison was possible with our test data were miRAW [12], TargetScan [9], and PITA [7]. We also used computationally predicted interactions from TarBase v8 [6] and mirDIP [14]. To conduct the comparison, we intersected our test data with the pre-computed target prediction results provided by each tool or method. The prediction datasets were obtained from the official platforms of the respective works (accessed on January 10, 2023).

Some preprocessing steps were required in some cases. For TargetScan, predictions are associated to a *context++ score* and targets with lower *context++ scores* are considered more representative. However, there is no clearly defined cutoff point to determine whether a given miRNA-mRNA interaction should be classified as positive or negative. Therefore, we decided to apply three distinct thresholds: –0.5, –0.3, and –0.2. Interactions with *context++ scores* lower than the applied threshold were classified as positive. For TarBase, we only used interactions related to humans and classified as '*positive*' by the tool, indicating that they represent functional miRNA-mRNA interactions. For mirDIP, each interaction has a *score* assigned and described as: *Very High, High, Medium,* and *Low. Very High* indicates a high probability of the miRNA-mRNA interaction, while the probability decreases towards the *Low* designation. Due to the large volume of data, we obtained only the dataset with a *score* value described as *Very High,* inferring all resulting interactions as positives.

The results for this analysis are shown in Table 3. For all the related works with which a direct comparison was possible for the test instances, our proposed model based on the HinSAGE algorithm achieved results that are very close to or outperform other approaches. Excluding TarBase and mirDIP, which are databases rather than ML-based prediction methods, our work presents highly promising results. In almost all the performed comparisons, our model achieved an F1-score above 90%, indicating that the model developed in this work can predict positive and negative interactions with a high degree of confidence.

However, it should be noted that none of the related works focus on identifying miRNA-mRNA interactions associated with cancer. This limitation exists in comparing our results with the literature, given the difficulty of finding representative datasets for this context. Other critical challenges faced for this comparison were (i) the wide variety of predictive features used by previous works based on ML, turning the creation of the training dataset a slow and arduous process and (ii) the difficulty in running methods that are not made available through online tools, requiring local installation of the application and its dependencies.

Table 3. Performance comparison with related works

Method/Tool	# miRNAs	#Interactions	Recall	Precision	F1-Score	Accuracy
Comparison against miRAW						
Our model	11	12	0.9166	1.0000	0.9565	0.9166
miRAW	11	12	0.9166	1.0000	0.9565	0.9166
Comparison against TargetScan						
Our model	154	533	0.9640	0.9902	0.9769	0.9549
TargetScan (*context++ score* < -0.2)	154	533	0.5037	0.9888	0.6675	0.5028
TargetScan (*context++ score* < -0.3)	154	533	0.2518	0.9851	0.4012	0.2551
TargetScan (*context++ score* < -0.5)	154	533	0.0246	1.0000	0.0480	0.0337
Comparison against TarBase v8.0						
Our model	153	423	0.9607	0.9751	0.9679	0.9385
TarBase v.8	153	423	0.9877	0.9664	0.9769	0.9550
Comparison against PITA						
Our model	139	250	0.7175	0.8037	0.7582	0.6760
PITA	139	250	1.0000	0.7080	0.8290	0.7080
Comparison against mirDIP						
Our model	222	314	0.8175	0.9837	0.8929	0.8152
mirDIP	222	314	1.0000	0.9426	0.9704	0.9426

6   Conclusion

The premise that guided the present work is that the development of a computational strategy based on methods capable of learning miRNA regulation patterns from the analysis of heterogeneous interaction networks (*i.e.,* HinSAGE) could provide more robust predictions to noise and deficiencies in training data, as well as better capture the complexity involved in miRNA function. The use of graph-based DL algorithms not only introduces the advantage of exploring the graph structure in a much more comprehensive and robust way to discover patterns in miRNA-mRNA interactions, but also enables efficient handling of the data imbalance problem prevalent in this domain. Our experiments suggest that our approach was effective, reaching competitive performance and controlling false positive rates. Although there is no known equivalent work at the moment that can serve as a baseline for comparison to our work, the comparisons performed with computationally predicted interactions from other ML-based methods or deposited in databases have shown that the approach based on heterogeneous graphs and the application of the HinSAGE algorithm are highly promising. The performance achieved by our model stood out in various scenarios and demonstrated a suitable balance between recall and precision. Therefore, we believe that the objective of exploring the potential of the HinSAGE algorithm for miRNA target prediction in cancer has been successful, and the results obtained in this study serve as motivation for further exploration of this approach. Among possible directions for future work, we outline the application to different human diseases, the inclusion of other types of interactions in the heterogeneous networks, and the need to expand the comparison with other ML/DL algorithms or bioinformatics tools for miRNA-target prediction.

References

1. Cai, H., Zheng, V.W., Chang, K.C.C.: A comprehensive survey of graph embedding: problems, techniques, and applications. IEEE Trans. Knowl. Data Eng. **30**, 1616–1637 (2018)
2. Feng, H., Xiang, Y., Wang, X., Xue, W., Yue, Z.: Mtagcn: predicting mirna-target associations in camellia sinensis var. assamica through graph convolution neural network. BMC Bioinf. **23**(1), 1–18 (2022)
3. Hamilton, W., Ying, Z., Leskovec, J.: Inductive representation learning on large graphs. In: Advances in Neural Information Processing Systems, pp. 1024–1034 (2017)
4. Ji, C., Wang, Y., Ni, J., Zheng, C., Su, Y.: Predicting miRNA-disease associations based on heterogeneous graph attention networks. Front. Genet. **12**, 727744 (2021)
5. Kang, J., et al.: RNAInter v4.0: RNA interactome repository with redefined confidence scoring system and improved accessibility. Nucleic Acids Res. **50**, D326–D332 (2022)
6. Karagkouni, D., et al.: DIANA-TarBase v8: a decade-long collection of experimentally supported miRNA-gene interactions. Nucleic Acids Res. **46**(D1), D239–D245 (2018)
7. Kertesz, M., Iovino, N., Unnerstall, U., Gaul, U., Segal, E.: The role of site accessibility in microRNA target recognition. Nat. Genet. **39**(10), 1278–1284 (2007)
8. Lee, B., Baek, J., Park, S., Yoon, S.: deeptarget: end-to-end learning framework for microrna target prediction using deep recurrent neural networks. In: Proceedings of the 7th ACM International Conference on Bioinformatics, Computational Biology, and Health Informatics, pp. 434–442 (2016)
9. Lewis, B., Shih, I.H., Jones-Rhoades, M., Bartel, D., Burge, C.: Prediction of mammalian MicroRNA targets. Cell **115**, 787–798 (2004)
10. Peng, Y., Croce, C.M.: The role of MicroRNAs in human cancer. Signal Transd. Target. Tpherapy **1**(1), 1–9 (2016)
11. Pinzón, N., et al.: microRNA target prediction programs predict many false positives. Genome Res. **27**(2), 234–245 (2017)
12. Pla, A., Zhong, X., Rayner, S.: miRAW: a deep learning-based approach to predict microRNA targets by analyzing whole microRNA transcripts. PLOS Comput. Biol. **14**, e1006185 (2018)
13. Schäfer, M., Ciaudo, C.: Prediction of the miRNA interactome - established methods and upcoming perspectives. Comput. Struct. Biotechnol. J. **18**, 548–557 (2020)
14. Tokár, T., et al.: MirDIP 4.1 - Integrative database of human microRNA target predictions. Nucleic Acids Res. **46**, D360–D370 (2017)
15. Wen, M., Cong, P., Zhang, Z., Lu, H., Li, T.: Deepmirtar: a deep-learning approach for predicting human mirna targets. Bioinformatics **34**(22), 3781–3787 (2018)
16. Zhou, J., et al.: Graph neural networks: a review of methods and applications. AI Open **1**, 57–81 (2020)

Time Series Forecasting of COVID-19 Cases in Brazil with GNN and Mobility Networks

Fernando Henrique Oliveira Duarte[1]([✉]) [iD], Gladston J. P. Moreira[1] [iD],
Eduardo J. S. Luz[1] [iD], Leonardo B. L. Santos[2] [iD], and Vander L. S. Freitas[1] [iD]

[1] Universidade Federal de Ouro Preto, Ouro Preto, MG CEP 35402-163, Brazil
`fernando.hod@aluno.ufop.edu.br,`
`{gladston,eduluz,vander.freitas}@ufop.edu.br`
[2] Centro Nacional de Monitoramento e Alertas de Desastres Naturais (Cemaden),
São José Dos Campos, SP CEP 12247-016, Brazil
`leonardo.santos@cemaden.gov.br`

Abstract. In this study, we examine the impact of human mobility on the transmission of COVID-19, a highly contagious disease that has rapidly spread worldwide. To investigate this, we construct a mobility network that captures movement patterns between Brazilian cities and integrate it with time series data of COVID-19 infection records. Our approach considers the interplay between people's movements and the spread of the virus. We employ two neural networks based on Graph Convolutional Network (GCN), which leverage spatial and temporal data inputs, to predict time series at each city while accounting for the influence of neighboring cities. In comparison, we evaluate LSTM and Prophet models that do not capture time series dependencies. By utilizing RMSE (Root Mean Square Error), we quantify the discrepancy between the actual number of COVID-19 cases and the predicted number of cases by the model among the models. Prophet achieves the best average RMSE of 482.95 with a minimum of 1.49, while LSTM performs the least despite having a low minimum RMSE. The GCRN and GCLSTM models exhibit mean RMSE error values of 3059.5 and 3583.88, respectively, with the lowest standard deviation values for RMSE errors at 500.39 and 452.59. Although the Prophet model demonstrates superior performance, its maximum RMSE value of 52,058.21 is ten times higher than the highest value observed in the Graph Convolutional Networks (GCNs) models. Based on our findings, we conclude that GCNs models yield more stable results compared to the evaluated models.

Keywords: Time Series Forecasting · Graph Neural Networks · Mobility Networks · COVID-19

1 Introduction

The COVID-19 is an infectious disease that manifests itself in humans infected by the SARS-CoV-2 virus. The acronym SARS stands for Severe Acute Respiratory

M. C. Naldi and R. A. C. Bianchi (Eds.): BRACIS 2023, LNAI 14197, pp. 361–375, 2023.
https://doi.org/10.1007/978-3-031-45392-2_24

Syndrome, which belongs to the family of Coronaviruses who have been infecting humans for a long time. On December 31, 2019, the World Health Organization (WHO) alerted to a new strain (type) of coronavirus. On January 7, 2020, Chinese authorities confirmed the identification of a new type of coronavirus, that was initially named 2019-nCoV, and on February 11, 2020, it was given the name SARS-CoV-2 [15, 28].

Those infected transmit the virus through the mouth or nose, by expelling small particles when they cough, sneeze, talk, sing or breathe. Infection occurs by inhaling these particles through the airways or by touching a contaminated surface and then coming into contact with the eyes, nose or mouth [26]. Due to the nature of the virus, it spreads more easily indoors and in crowds. Those who become ill as a result of COVID-19 experience symptoms ranging from mild and moderate, where the patient recovers without special treatment, to severe, where the patient needs specialized medical care [2].

To this date (May 06, 2023), the virus that causes COVID-19 has infected more than 270 million people and caused more than 5.3 million deaths worldwide [5, 19]. In Brazil, the numbers reach more than 37 million infected and 701 thousand dead [1]. So far, the world has paid a heavy price for this pandemic in lost human lives, economic repercussions and increased poverty. The COVID-19 pandemic is still ongoing, and the virus continues to mutate, generating variants.

This work presents time series forecasting models to predict the number of COVID-19 cases in Brazilian cities [4], covering the period from February 2020 to December 2022. We aim to improve the accuracy of the predictions by incorporating Graph Convolutional Networks (GCN) models, specifically GCRN and GCLSTM. These models leverage a mobility network that captures the commuting patterns between cities, allowing us to account for the interdependencies among cities' time series and the influence of human mobility on disease transmission [7–9, 20].

The performance of our proposed GCN models is compared to two well-known forecasting methods, LSTM and Prophet. Through extensive experiments and evaluation, we obtained compelling results that highlight the effectiveness of the GCN models approach in capturing the complex dynamics of COVID-19 transmission and improving the accuracy of predictions.

This research contributes to the field of COVID-19 forecasting by demonstrating the value of incorporating mobility networks and considering the impact of human movement on disease spread. Our findings provide insights into the interplay between mobility patterns and the spread of COVID-19, and offer valuable implications for developing effective strategies to mitigate the impact of the disease.

2   Related Works

Xi et al. (2022) [27] simulate the propagation curve of COVID-19 in a university, based on the interaction between students through Graph Neural Network (GNN) and a SIR model (Susceptible - Infected - Recovered) [23]. They simulated three distancing strategies designed to keep the infection curve flat and help make the spread of COVID-19 controllable. Students and professors were considered as nodes and places or events as layers (face-to-face lectures, internal social activities, campus cafeterias, etc.). The GNN was used to validate and verify the effectiveness through the simulation of the infection curve, where the infected were randomly chosen. They provide a visualization of the dissemination process of COVID-19 and the comparison between the results of the forecast of dissemination for each of the control strategies.

With the aim of predicting the dynamics of the spread of the pandemic in the USA, the work carried out by Davahli et al. (2021) [6] developed two predictive models using GNN, fed by real data, obtaining results from the forecast of spread and a comparison between the models. The first model developed was based on graph theory neural networks (GTNN) and the second was based on neighborhood neural networks (NGNN). Each US state is considered a graph node, while the edges of the GTNN model document the functional connectivity between states, those of the NGNN model connect neighboring states with each other. The performance of these models was evaluated with existing real dataset that reflect conditions from January 22 to November 26, 2020 (prior to the start of COVID-19 vaccination in the United States). The results indicate that the GTNN model outperforms the NGNN model.

Malki et al. (2020) [17] estimate the spread of COVID-19 infection in several countries, using a supervised machine learning model developed based on a decision tree algorithm and linear regression, for time series forecasting. They show an R^2 of 0.99 in the prediction of confirmed cases at the global level.

Vaishya et al. (2020) [25] deal with the application of artificial intelligence (AI) in combating the COVID-19 pandemic. The authors presented several possibilities for the use of AI in different areas, such as diagnosis, triage, patient monitoring, forecasting disease trends, drug and vaccine development, and even to ensure compliance with social distancing measures. They emphasize the importance of international collaboration to ensure access to the data needed to train AI algorithms and highlight that technology can play a key role in the fight against the pandemic.

The related works presented in this section are all focused on predicting the spread of COVID-19 using different approaches. Xie et al. (2022) used a Graph Neural Network (GNN) and SIR model to simulate the propagation curve of COVID-19 in a university, while Davahli et al. (2021) used GNN to predict the dynamics of the spread of the pandemic in the USA. Malki et al. (2020) used a decision tree algorithm and linear regression for time series forecasting, and Vaishya et al. (2020) discussed the application of AI in combating the pandemic. In contrast, our study focuses on incorporating a mobility network that describes the commuting patterns between Brazilian cities into the forecasting task, leveraging Graph Convolutional Network (GCN) models. The main hypothesis is that this approach could improve the forecasting task by accounting for the dependencies among cities' time series. Therefore, our work contributes to the literature by emphasizing the importance of incorporating human mobility data into the forecasting task and highlighting the potential of GCN models for this purpose.

3 Methodology

3.1 Mathematical Preliminaries

$G = (V, E)$ is a graph, in which V is the set of nodes and E the set of edges, resulting in $N = |V|$ vertices and $|E|$ edges. The binary (or weighted) adjacency matrix $A \in \mathbb{R}^{N \times N}$ depicts the graph connections; $D_{ii} = \sum_j A_{ij}$ is the degree matrix, with zeros outside the main diagonal; and $X \in \mathbb{R}^{N \times C}$ is the feature vector of each vertex, where C is the number of vector dimensions.

3.2 GCN

The Graph Neural Networks (GNN) models allow us to extend the methods of neural networks, in order to process the data represented in the graph domain [21]. GNN models can process most graphs being some of them acyclic, cyclic, directed and undirected, or implement a particular function. The GCN proposed by [18] applies a convolution operation to each vertex of the graph. The approach has three steps: (i) Select a vertex, (ii) build a subgraph of its neighborhood and finally (iii) normalize the selected subgraph with the edge weights of neighboring vertices. Repeat the process for each vertex h^k, building the matrix H^k that contains all the vertices in the layer k (refer to Fig. 2 and Eq. (1)).

The subgraph is built using the adjacency matrix A and the diagonal degree matrix of the vertices D. The size of the subgraph is delimited by the parameter K (see Fig. 1). When $K = 1$ the vertex is not subjected to the influence of neighboring vertices. In the case of $K = 2$, first-degree vertices (direct neighbors) are considered, directly connected to the active vertex. For a $K = 3$, second-degree vertices are also considered, connected to first-degree vertices. The subgraph is normalized by the weight matrix.

When considering the subgraph of a node, we calculate the average feature vector by incorporating information from its neighbors. This process assigns greater importance to the less connected neighbors (lower degree) while also giving higher weight to the node's connections. Let $\widehat{A} = A + I_N$, A being the weighted adjacency matrix of the subgraph, I_N being the identity matrix of order N, and $\widehat{D}$ is the diagonal degree matrix of $\widehat{A}$. The output of the convolution layer $k + 1$ is as follows:

$$H^{k+1} = \sigma(\widehat{D}^{-1/2} \widehat{A} \widehat{D}^{-1/2} H^k W^{k+1}), \tag{1}$$

where W is the matrix of trainable parameters, and σ is the activation function.

In summary, each vertex in the graph is assigned a feature vector and convolution is applied at each node to compute a new feature vector that considers the neighborhood of the node.

3.3 GCN Models for Time Series Prediction

The Graph Convolutional Recurrent Networks (GCRN) [22] and Graph Convolutional LSTM (GCLSTM) [3] models are architectures based on GCN models

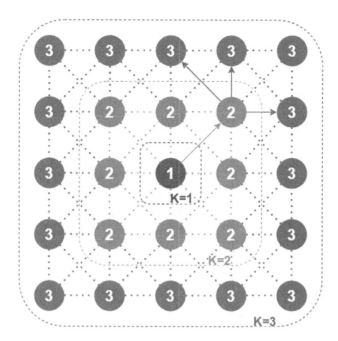

Fig. 1. Nearest neighbors parameter (K) in a Graph. In the purple color, $K = 1$ represents nodes without any neighbors. In the orange color, $K = 2$ represents nodes with first-degree neighbors, and in the red color, $K = 3$ represents nodes with second-degree neighbors. (Color figure online)

that are designed with graph processing and machine learning techniques for the purpose of prediction of data in time series. The convolutions in both architectures differ from a conventional convolution using the relationship between neighbors at each vertex of the graph.

To predict a future sequence, of length F, we use a sequence, of length L (lags), of observed data at previous timestamps:

$$\hat{x}_{t+1}, ..., \hat{x}_{t+F} = \underset{x_{t+1}, ..., x_{t+F}}{argmax} P(x_{t+1}, ..., x_{t+F} | x_{t-L+1}, ..., x_t) \qquad (2)$$

where x_t is a snapshot at time t of the features and $\hat{x}_t$ is the prediction (see Fig. 2).

Hence, x_t represents a value of the time series at time t for all nodes, defined on a directed, weighted static graph. On the other hand, a snapshot X can be interpreted as a matrix $X \in \mathbb{R}^{N \times C}$, which captures the temporal evolution of the time series over the interval of C observations.

The GCRN and GCLSTM models use the same design but with their own convolution blocks: in Fig. 2, the GCRN model uses GConvGRU blocks whereas the GCLSTM model uses GConvLSTM blocks. Both use deep learning techniques, but GCRN uses recurrent networks and GCLSTM uses LSTM cells, in

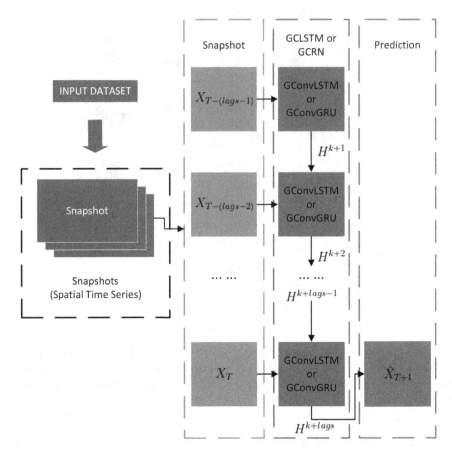

Fig. 2. Spatial and Temporal Snapshot Data Flow and Model Architecture: Processing GCNs with GCLSTM and GCRN Recurrences

order to capture patterns in time series data considering the influence of its neighbors through the graph.

The graph generated in this work has the spatial structure of the Brazilian mobility network (refer to Sect. 3.5), which has static edges, that is, fixed attributes (see Fig. 2).

3.4   Prophet and LSTM

LSTM (Long Short Term Memory) [12] and Prophet [24] are time series forecasting models, and each one has a different approach. In the LSTM model it resembles an RNN (Recurrent Neural Network) composed of 4 memory cells: Input, Forget, Cell, Hidden. These memory cells are responsible for retaining, forgetting or remembering selected information. LSTM is commonly used on data that exhibit complex patterns, sudden changes, trends, or seasonality.

The Prophet model, is based on an additive model that, through regression, models the linear or non-linear trend, in terms of independent variables such as trend, seasonality and holidays, components of the time series. The adjustments of these components are performed using an optimization algorithm such as MCMC [11] (Markov Chain Monte Carlo) or MAP [10] (Maximum A Posteriori) that seeks to find the values of the parameters that maximize the posterior probability of the data, considering the uncertainty of the parameters when making predictions.

3.5 Datasets: COVID-19 Cases and the Brazilian Mobility Network

Cota (2020) [4] provides a public dataset of COVID-19 records in Brazil. Among other data, it contains the numbers of infected and deaths related to municipal and federative units level in Brazil with official information from the Ministry of Health. The dataset is daily updated from official sources, providing reliable information. It also includes information on the cities' locations, which is useful for our mobility network.

Knowing that the spread of diseases like COVID-19 occurs from infected to susceptible people, one factor that may contribute to the construction of prediction models is to know the flow of people between cities. We thus use the dataset provided by the Brazilian Institute of Geography and Statistics (IBGE) [16] to build a mobility network whose weighted connections represent the number of vehicles from one city to another in an ordinary week. The connection identifies the cities directly accessible from a source, considering (1) the number of outgoing vehicles (frequency), (2) the type of transport (road, waterways, or alternative/informal), (3) travel time, and (4) cost of the traveling. In this study, the weight of each edge was represented solely by attribute 1. This weight was determined by the aggregation of frequencies associated with each mode of transportation (denoted as attribute 2) present on a given edge. It should be noted that attributes 3 and 4 were not incorporated into the calculations in this analysis. The dataset has $N = 5385$ vertices (cities) and $L = 65639$ edges (routes between the cities), as shown in Fig. 3. The subfigure on the right side gives the covered territory, in green, and the one on the left side presents the corresponding network.

3.6 Experimental Setup

The COVID-19 dataset has been structured to train models under four distinct scenarios: i) utilizing only the data from 2020; ii) using the 2021 data exclusively; iii) relying solely on the 2022 data; and finally, iv) a comprehensive version incorporating all years. The data undergoes standardization using Z-score normalization (Eq. (3)) to enable consistent and meaningful analysis.

$$z = \frac{(x - \mu)}{\sigma} \tag{3}$$

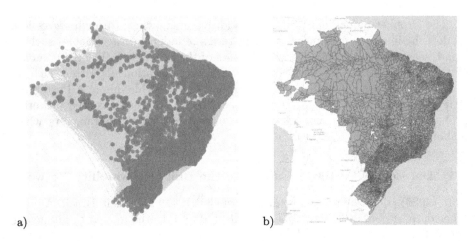

a)   b)

Fig. 3. Brazilian mobility network: a) Nodes representing cities and edges as the routes between them; b) The covered territory, in green. White areas inside the Brazilian map are cities without reported cases in our COVID-19 dataset. (Color figure online)

Here, x represents the value of the element, μ is the population mean, and σ is the population standard deviation. This normalization process ensures uniformity in the data and facilitates reliable comparisons. In each of the four scenarios, the dataset is partitioned into an 80% subset for training and a 20% subset designated for testing. Each city has 1,009 individual readings of COVID-19 case numbers, with each reading representing a single day's data.

These readings are grouped into sequences of 14 consecutive days (lags), which are utilized to predict the data for the 15th day. Consequently, each snapshot corresponds to a time series comprised of non-overlapping samples, each size 14. The subsequent day, the 15th, is the target prediction and assumes the first position in the succeeding time series (see Fig. 4). The entire 1009-day period, corresponding to February 2020 to November 2022, results in approximately 57 training snapshots and 15 test snapshots. Lauer et al. [14] provide empirical evidence supporting the recommendation of a 14-day active monitoring period by the US Centers for Disease Control and Prevention. Their study reinforces the importance of this time frame in effectively monitoring and controlling the spread of COVID-19. Regarding the training of the models is parameterized for a total of 100 epochs for the GCRN, GCLSTM, and LSTM models. In these models, an Adam optimizer [13] is used, which adjusts the model's weights during training, minimizing the Mean Square Error Loss.

As Adam is a stochastic gradient descent optimizer, it has an element of randomness, so we perform 10 rounds of experiments for each model. The Adam optimizer was chosen because it uses two moments to update the model weights: the gradient moment and the quadratic gradient moment [13]. At the gradient moment, it adapts the learning rate for each parameter based on how often the parameter is updated. Not all cities update cases daily, so the data does

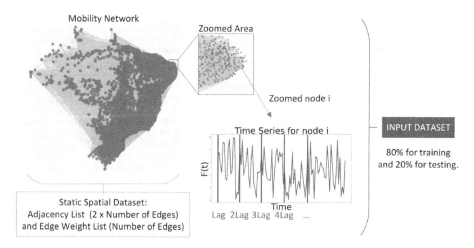

Fig. 4. Mobility Network Map and Spatial Time Series Data in Brazil: Input Dataset Overview

not have a unified update frequency. In the quadratic gradient moment, the weighting of the learning rate is performed by the moving average of the squares of the gradients of each parameter individually, adjusting the model weights.

To present and compare the results of the tests between the neural networks, we use the Root Mean Square Error (RMSE) and the R^2 score (Coefficient of Determination). The RMSE,

$$RMSE = sqrt(\frac{1}{n} \sum_{i=1}^{n} (y_i - \hat{y}_i)^2), \tag{4}$$

is a measure of dispersion that calculates the square root of mean square errors between predictions and actual values, i.e., it indicates how well the model's predictions fit the data. The lower the RMSE, the better the model performs against the test data.

The R^2 score,

$$R^2 = 1 - \frac{\sum (y_i - \hat{y}_i)^2}{\sum (y_i - \bar{y})^2}, \tag{5}$$

indicates how much the independent variables explain the variation of the dependent variable. It belongs to the interval $[0, 1]$, where 1 indicates that all variations in the dependent variable are explained by the independent variables, and it is 0 when there is no relationship between the variables. Therefore, the closer to 1, the better the model.

4   Results

The Tables 1 and 2 have references to those years, which we will comment on. As we train GCRN, GCLSTM, and LSTM models 10 times each, the results are presented in terms of mean, max, min, and std values of RMSE and R^2.

Table 1. GCRN model results in terms of RMSE and R^2 score

Time	RMSE (Cases)				R^2 score			
	Mean	Max	Min	Stand.	Mean	Max	Min	Stand.
2020	1495.18	1624.20	1364.51	114.70	0.9558	0.9630	0.9485	0.0064
2021	938.39	1389.69	**643.26**	250.44	0.9977	0.9990	0.9954	0.0012
2022	**819.62**	**1034.46**	722.62	**113.64**	**0.9993**	**0.9994**	**0.9989**	**0.0002**
All Years	3059.50	3699.74	2108.77	500.39	0.9889	0.9949	0.9840	0.0035

4.1 GCRN

The results of GCRN (see Table 1) in 2020, the first year of the COVID-19 pandemic, are: $RMSE = 1495.18$, while in 2021 and 2022, the values decreased significantly to 938.39 and 819.62, respectively. In the last two years, the mean value error became smaller, but the number of cases continued to grow.

The highest min error value was recorded in 2020 and the lowest min error value was recorded in 2021, with values of 1364.51 and 643.26, respectively, while in 2022 the max error value decreased to 1034.46 even with a greater variation than in the previous years. Note that the standard deviation in the years 2020 and 2022 are close, and in the year 2021 it doubles in relation to them. These factors are indicators that in the year 2021 there was a change in the dynamics of spreading the disease.

The highest RMSE for the three years combined suggests potential variations in the dynamics of COVID-19 spread over time, which could be attributed to various factors influencing the transmission patterns. These factors may include governmental interventions, changes in population behavior, the emergence of new variants, variations in testing strategies, healthcare capacity, public adherence to preventive measures, and other external factors that could impact the transmission dynamics of the virus.

Regarding the R^2 score, the models showed low variation in the standard deviation and a minimum value of 0.96 for 2020, which still represents a high prediction capacity of the model.

Figure 5b gives a visualization of the model predictions on the training set, with most points laying on the main diagonal as the high R^2 values suggest. Figure 5a presents a forecasting example, for November 20, with the model trained with all years. We highlight Sao Paulo, Manaus, and Brasilia as being state capitals in different regions of Brazil.

4.2 GCLSTM

Table 2 introduces the results for the GCLSTM model. As in the GCRN: the average RMSE for 2020 is the highest compared to 2021 and 2022; there is a decreasing trend from 2020 to 2022; and the error for all years is the largest. The R^2 score also decreases over the years and is better than the one obtained for 2020.

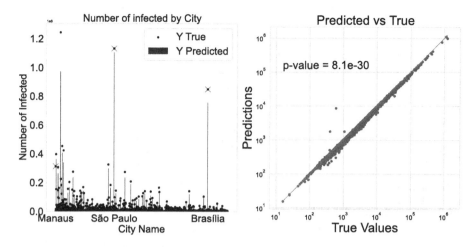

Fig. 5. GCRN trained with the dataset of all years (2020 to 2022): a) Result of the predicted and true number of cases in each city on November 29, 2022. The true values of Manaus, Sao Paulo and Brasilia are highlighted with red crosses; b) Linear regression between true and predicted values. (Color figure online)

Table 2. GCLSTM model results in terms of RMSE and R^2 score

Time	RMSE (Cases)				R^2 score			
	Mean	Max	Min	Stand.	Mean	Max	Min	Stand.
2020	1438.42	1892.34	1317.37	112.97	0.9590	0.9656	0.9506	0.0061
2021	1144.03	1472.16	919.75	178.72	0.9968	0.9980	0.9948	0.0010
2022	**750.71**	**1007.22**	**697.88**	**87.41**	**0.9994**	**0.9995**	**0.9989**	**0.0002**
All Years	3583.88	4569.97	2847.56	452.59	0.9849	0.9906	0.9767	0.0037

The experiment is repeated, but for the GCLSTM model, the results are shown for the R^2 score metric for Table 2 in Fig. 6b and also the results of prediction values compared to actual values for each city (see Fig. 6a).

4.3 Comparing Forecast Models Results

The lowest RMSE values are highlighted on Table 3, referring to the dataset of all years (from 2020 and 2022). Note that the GCRN and GCLSTM models present similar results concerning mean, max, min, and standard deviation. Interestingly, they achieved lower values of standard deviation, which shows their consistency.

The Prophet model stands out in the average and at the minimum, where it has the lowest values among the models. However, its standard deviation is 3 times greater than the lowest value found in the GCLSTM model and the maximum RMSE is 14 times greater than the maximum observed for GCRN. The model is incredibly good for most data points, but really bad in some cases, showing less stability than GCRN and GCLSTM.

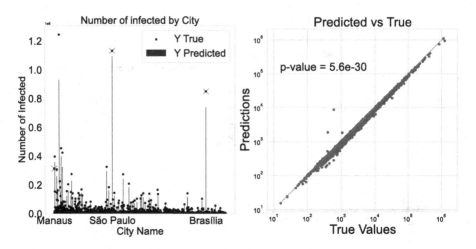

Fig. 6. GCLSTM trained with the dataset of all years (2020 to 2022): a) Result of the predicted and true number of cases in each city on November 29, 2022. The true values of Manaus, Sao Paulo, and Brasilia are highlighted with red crosses; b) Linear regression between true and predicted values. (Color figure online)

Table 3. RMSE values of the forecast models of COVID-19 cases in Brazil.

Modelo	RMSE (Cases)			
	Mean	Max	Min	Stand.
GCRN	3059.50	**3699.74**	2108.77	500.39
GCLSTM	3583.88	4569.97	2847.56	**452.59**
LSTM	13267.66	963263.36	115.91	34109.31
Prophet	**482.95**	52058.21	**1.49**	1758.85

The LSTM model did not obtain any statistics highlighted as the best and can be considered the model with the worst performance, although its minimum RMSE is the second best among all.

The results show that the mobility network indeed helps to forecast COVID-19 cases when embedded in machine learning models. LSTM alone presents the worst RMSE of all models, but when it is integrated with the GCN, via the GCLSTM, the predictions improve. Although Prophet gives the best mean RMSE, it fluctuates more than the graph-based models, showing that sometimes it outperforms them, but is not as stable.

5   Discussions and Conclusions

This study[1] investigates the application of GCNs for COVID-19 spread in Brazilian cities. Combining graphs with temporal forecasting models offers advantages. Integrating mobility networks, particularly GCNs with LSTM through GCLSTM, improves prediction accuracy over LSTM alone.

Among the models evaluated, LSTM performed the worst, with Prophet showing the best mean RMSE but higher fluctuations and less stability compared to the graph-based models (GCRN and GCLSTM). The GCN models achieved high R^2 score values above 0.95, representing the data with at least 95% certainty, the models also demonstrated similar results in terms of mean, maximum, minimum, and standard deviation of RMSE. Notably, they achieved lower standard deviation values, indicating higher consistency in their predictions.

In conclusion, the integration of mobility networks through GCNs in machine learning models has proven effective in forecasting COVID-19 cases. The GCN models (GCRN and GCLSTM) showed improved stability and consistency compared to Prophet, while LSTM alone exhibited the worst performance. The models' performance highlights the importance of considering spatial and temporal data in forecasting tasks, providing valuable insights for epidemiological surveillance and informing strategies to combat COVID-19.

One limitation of our analysis is that it only focuses on one-day predictions. However, it is crucial to extend the prediction window to longer time frames, considering the virus's incubation and transmission periods, as well as travel time between locations.

The strength of incorporating mobility data lies in our approach using GCNs, which enables us to capture the connectivity between cities and integrate changes in neighboring time series into the local predictions of each city. These changes in neighboring series have a time delay due to the incubation and transmission factors of the virus. By considering these factors and exploring longer prediction windows, we anticipate achieving more significant results in leveraging mobility networks for COVID-19 case prediction.

For future work, we will address the suggestion of conducting more in-depth analyses to assess the impact of including spatial data and applying GCNs for longer-term predictions. Additionally, we will investigate the convergence of the training process and explore ways to enhance the interpretability and explainability of the temporal and spatial data by the GCN models, particularly by examining the stabilization of the loss after 100 epochs, so that we can gain deeper insights into the underlying mechanisms driving the predictions of GCN models and enhance our understanding of the complex interplay between temporal and spatial factors in the spread of COVID-19.

---

[1] All data and source code are available at the repository link: https://github.com/hodfernando/BRACIS_GCN_2023.

Also, considering that GCN-based models may require more training time (epochs) compared to Prophet, further investigation into the impact of training duration on model performance could be valuable.

Lastly, exploring unexplored graph network regions is a promising research avenue. It can reveal COVID-19 spread insights and uncover hidden data patterns or relationships.

Acknowledgements. The authors thank the Brazilian Agencies for Research and Development (CNPq, grants 441016/2020-0, 307151/2022-0, 308400/2022-4), (FAPEMIG, grants APQ-01518-21, APQ-01647-22), (CAPES, grants 88887.506931 /2020-00) and Universidade Federal de Ouro Preto (UFOP).

References

1. Cumulative cases of covid-19 in Brazil (2023). https://covid.saude.gov.br//. Accessed 06 May 2023
2. Berlin, D.A., Gulick, R.M., Martinez, F.J.: Severe covid-19. N. Engl. J. Med. **383**(25), 2451–2460 (2020)
3. Chen, J., Wang, X., Xu, X.: Gc-lstm: graph convolution embedded lstm for dynamic network link prediction. Appl. Intell. **52**(7), 7513–7528 (2022)
4. Cota, W.: Monitoring the number of covid-19 cases and deaths in brazil at municipal and federative units level. SciELO Preprints (2020)
5. DASA: Cumulative cases of covid-19 in the world (2023). https://dadoscoronavirus. dasa.com.br/. Accessed 06 May 2023
6. Davahli, M.R., Fiok, K., Karwowski, W., Aljuaid, A.M., Taiar, R.: Predicting the dynamics of the covid-19 pandemic in the united states using graph theory-based neural networks. Int. J. Environ. Res. Public Health **18**(7), 3834 (2021)
7. Fanelli, D., Piazza, F.: Analysis and forecast of covid-19 spreading in China, Italy and France. Chaos Solitons Fract. **134**, 109761 (2020)
8. Freitas, V.L., Moreira, G.J., Santos, L.B.: Robustness analysis in an inter-cities mobility network: modeling municipal, state and federal initiatives as failures and attacks toward sars-cov-2 containment. PeerJ **8**, e10287 (2020)
9. Freitas, V.L.D.S., Konstantyner, T.C.R.D.O., Mendes, J.F., Sepetauskas, C.S.D.N., Santos, L.B.L.: The correspondence between the structure of the terrestrial mobility network and the spreading of covid-19 in Brazil. Cadernos de Saúde Pública **36**, e00184820 (2020)
10. Gauvain, J.L., Lee, C.H.: Maximum a posteriori estimation for multivariate gaussian mixture observations of markov chains. IEEE Trans. Speech Audio Process. **2**(2), 291–298 (1994)
11. Gilks, W.R., Richardson, S., Spiegelhalter, D.: Markov Chain Monte Carlo in Practice. CRC Press, Boca Raton (1995)
12. Hochreiter, S., Schmidhuber, J.: Long short-term memory. Neural Comput. **9**(8), 1735–1780 (1997)
13. Kingma, D.P., Ba, J.: Adam: a method for stochastic optimization. In: Bengio, Y., LeCun, Y. (eds.) 3rd International Conference on Learning Representations, ICLR 2015, San Diego, CA, USA, 7–9 May 2015, Conference Track Proceedings (2015)
14. Lauer, S.A., et al.: The incubation period of coronavirus disease 2019 (covid-19) from publicly reported confirmed cases: estimation and application. Ann. Intern. Med. **172**(9), 577–582 (2020)

15. Li, Q., et al.: Early transmission dynamics in Wuhan, China, of novel coronavirus-infected pneumonia. N. Engl. J. Med. **382**(13), 1199–1207 (2020)
16. Lulia, M.M.E.T., de Oliveira, D.H.: Instituto brasileiro de geografia e estatística-ibge (1936)
17. Malki, Z., et al.: The covid-19 pandemic: prediction study based on machine learning models. Environ. Sci. Pollut. Res. **28**, 40496–40506 (2021)
18. Niepert, M., Ahmed, M., Kutzkov, K.: Learning convolutional neural networks for graphs. In: International Conference on Machine Learning, pp. 2014–2023. PMLR (2016)
19. Organization, W.H.: Covid-19 weekly epidemiological update, 119 edn. (2022). Accessed 23 Nov 2022
20. Rothan, H.A., Byrareddy, S.N.: The epidemiology and pathogenesis of coronavirus disease (covid-19) outbreak. J. Autoimmun. **109**, 102433 (2020)
21. Scarselli, F., Gori, M., Tsoi, A.C., Hagenbuchner, M., Monfardini, G.: The graph neural network model. IEEE Trans. Neural Netw. **20**(1), 61–80 (2008)
22. Seo, Y., Defferrard, M., Vandergheynst, P., Bresson, X.: Structured sequence modeling with graph convolutional recurrent networks. In: Cheng, L., Leung, A.C.S., Ozawa, S. (eds.) ICONIP 2018. LNCS, vol. 11301, pp. 362–373. Springer, Cham (2018). https://doi.org/10.1007/978-3-030-04167-0_33
23. Smith, D., Moore, L., et al.: The sir model for spread of disease-the differential equation model. Convergence (2004)
24. Taylor, S.J., Letham, B.: Forecasting at scale. Am. Stat. **72**(1), 37–45 (2018)
25. Vaishya, R., Javaid, M., Khan, I.H., Haleem, A.: Artificial intelligence (ai) applications for covid-19 pandemic. Diab. Metabolic Syndr. Clin. Res. Rev. **14**(4), 337–339 (2020)
26. Velavan, T.P., Meyer, C.G.: The covid-19 epidemic. Trop. Med. Int. Health **25**(3), 278–280 (2020)
27. Xie, H., Li, D., Wang, Y., Kawai, Y.: Visualization method for the spreading curve of covid-19 in universities using gnn. In: 2022 IEEE International Conference on Big Data and Smart Computing (BigComp), pp. 121–128. IEEE (2022)
28. Zhu, N., et al.: A novel coronavirus from patients with pneumonia in China, 2019. N. Engl. J. Med. **382**(8), 727–733 (2020)

Pattern Recognition

Federated Learning
and Mel-Spectrograms for Physical
Violence Detection in Audio

Victor E. de S. Silva[1]([✉]) [iD], Tiago B. Lacerda[1] [iD], Péricles Miranda[2] [iD],
André Câmara[2] [iD], Amerson Riley Cabral Chagas[1] [iD],
and Ana Paula C. Furtado[2] [iD]

[1] Center for Advanced Studies and Systems of Recife, Recife, Brazil
{vess,tbl,arcc}@cesar.school
[2] Federal Rural University of Pernambuco, Recife, Brazil
{pericles.miranda,pericles.miranda,andre.camara,
anapaula.furtado}@ufrpe.br

Abstract. Domestic violence has increased globally as the COVID-19 pandemic combines with economic and social stresses. Some works have used traditional feature extractors to identify features from sound signals to detect physical violence. However, these extractors have not performed well at recognizing physical violence in audio. Besides, the use of Machine Learning is limited by the trade-off between collecting more data while keeping users privacy. Federated Learning (FL) is a technique that allows the creation of client-server networks, in which anonymized training result can be uploaded to a central model, responsible for aggregating and keeping the model up to date, and then distribute the updated model to the client nodes. In this paper, we proposed a FL approach to the violence detection problem in audio signals. The framework was evaluated on a newly proposed synthetic dataset, in which audio signals are represented as mel-spectrograms images, augmented with violence extracts. Thereby, it treats it as a problem of image classification using pre-trained Convolutional Neural Networks (CNN). Inception v3, MobileNet v2, ResNet152 v2 and VGG-16 architectures were evaluated, with the MobileNet architecture presenting the best performance, in terms of accuracy (71.9%), with a loss of 3.6% when compared to the non-FL setting.

Keywords: Federated Learning · Mel-spectrogram · Physical Violence Recognition

1 Introduction

Violence against women is widely recognized as a serious public health problem and a violation of women's human rights [21,22]. The COVID-19 pandemic combined with economic and social tensions, together with measures to restrict

M. C. Naldi and R. A. C. Bianchi (Eds.): BRACIS 2023, LNAI 14197, pp. 379–393, 2023.
https://doi.org/10.1007/978-3-031-45392-2_25

contact and movement is increasing violence against women and girls globally. Before the pandemic, there was an estimate that for every three women, one will experience violence throughout life. During the pandemic, these women are in their homes, cornered with their abusers, in which they exploit the inability of the woman to make a call asking for help or escape, on the other hand, health services are overloaded, non-governmental organizations and support houses are crowded, closed or were reused as a health unit [19].

In the face of this problem, the detection of physical violence is a challenging task, since it depends on the detection of changes in behavior caused by a disagreement of idea, injustice or serious disagreement [34]. A way to make such detection mechanism readily accessible, it can be embedded in a mobile application in order to help to detect and call the police to prevent the violence situation to go further through rapid intervention. Several papers in this topic were proposed recently [25,34], in which various techniques are used to detect violence in videos. However, there are few studies related to the detection of physical violence by audio. According to [3], only 2 works were found between the years 2015 and 2020 related to the theme, evidencing the complexity and lack of studies in the area.

However, it is necessary to respect the privacy of the data collected, stored and processed in a Machine Learning application, in accordance with European data protection laws (General Data Protection Regulation) and Brazilian (General Data Protection Law), where in some cases it is necessary authorization of the owner of the information for manipulation of information. In addition, it takes high computational power to keep Machine Learning (ML) models running, experiments, training, and even retraining with new data. Federated Learning [18] (FL) is gaining a lot of attention lately, since it allows the decentralization of the learning process to the user's own devices. Collective knowledge is then aggregated in a centralized model, built over several users' models on the federated network. Thus, the privacy of each user's data is maintained on the device, and there is no storage of sensitive information in a centralized location. In view of this, it raises the following research question that will guide the development of this work: *Is it possible to maintain similar results by using the FL approach [18], compared with the traditional approach, in identifying scenes of physical violence through audio mel-spectrograms and CNN's architectures?*

The experiments were conducted based on HEAR Dataset [15], a synthetic dataset that has 70,000 instances of 10-second audio clips, divided equally between two classes: presence or not of physical violence. In order to keep the experiments computationally feasible, only 12,500 records were used. The experiments conducted in this work considered the following CNN architectures: Inception [32], MobileNet [27], ResNet152 [8] and VGG-16 [29], and the results showed that CNN MobileNet was the best among the other models when evaluated in HEAR dataset, reaching 66.8% accuracy, with a loss of 8.6% when compared to non-FL experiments.

2   Background

In this Section, the concepts of Convolutional Neural Networks (CNN), FL and Mel-spectrogram to be used in this work will be presented.

2.1   Convolutional Neural Networks (CNN)

A Convolutional Neural Network [5] (CNN) is a deep neural network architecture, very popular in image classification tasks. It can learn to assign importance to relevant aspects of an image, allowing the learning of these characteristics and the distinction between them. For this, unlike traditional algorithms where filters/features needed to be implemented manually, CNN is able to learn a large amount of filters, allowing it to learn the best representation by itself. After the emergence of the AlexNet [14] architectures, exposed in the ImageNet challenge in 2012, and GoogleNet [31], presented in the ImageNet challenge in 2014, convolutional neural networks became popular influencing the emergence of architectures with upper 22 layers of GoogleNet, as is the case with ResNet [8] with 152 layers.

2.2   Federated Learning (FL)

Federated Learning [18] (FL) consists of the decentralization of the model learning process to customers (user devices), allowing a global model to be trained from various models of users of a network. In addition, the privacy of each user's data, often sensitive, is maintained on the device, with no sharing of raw data with the central server. Only the resulting trained models are shared on the federated network [7].

Furthermore, the FL technique can be applied when data for training is concentrated on other devices, allowing the centralizing server not to require a high computational power. Another scenario is when the data is sensitive and privacy needs to be respected, so only the training results are shared and the raw data remains on the source device. The FL technique also supports the constant updating of the model, since new data can be collected by the network of devices and used to improve the model.

The training process using FL works through learning rounds, where network devices receive the updated global model and start the learning process with data from their respective devices, and at the end, each device will send the training results to the centralizing server, which will aggregate these results into a global model using an aggregation algorithm, for example Federated Averaging (FedAvg) [18]. This process is repeated until the end of the defined training rounds or until a certain evaluation metric is achieved.

2.3   Mel-spectrogram

The problem addressed in this work belongs to the Sound Event Classification (Acoustic Scene Classification), which consists of adding markers to identify the

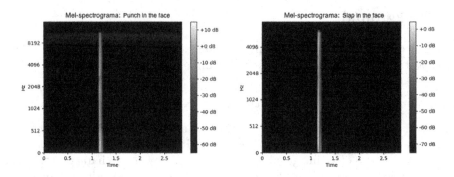

Fig. 1. Examples of two short-lived, high-intensity events of different types of physical violence.

existence or not of a particular characteristic present at any time in the audio. One instance of this problem is the detection of the presence of violence in audio samples. For this, HEAR Dataset uses the extraction of Mel-spectrogram from audios, which is a Mel scale spectogram [11] that represents sound components focused on frequencies accessible to human hearing, eliminating low frequency noise observed in the linear spectogram. From there, it can create a bi-dimencional image to be used in image classifiers. The Fig. 1 demonstrates the mel-spectrograms extracted from two short-lived audios containing physical violence ("punch in the face" and "slap in the face" respectively).

3   Related Works

The Acoustic Scene Classification problem is receiving a lot of attention lately. In [15], the authors introduce an application of Mel-spectrogram sound representation to be fed in a Deep Learning Neural Network for detecting physical violence in audio by converting audio into images. For that, the authors built a synthetic dataset with about 70,000 images, called HEAR dataset, resulting from the extraction of audio using the Mel-spectrogram technique. Finally, a comparison of the MobileNet, VGG16, ResNet-152 and InsceptionV3 architectures were realized, in which MobileNet achieved the best results, with an accuracy of 78.9% and f_1 score of 78.1%.

[28] presents an experiment to detect violence inside the car using audio signal. For that context, the author built the In-Car video dataset with the presence of ambient sounds and music to try to simulate the desired environment, and used Mel-spectrogram to generate images that represent each audio. The author then submitted the dataset to the ResNet, DenseNet, SqueezeNet, and Mobile architectures, in which he obtained the best result when applied to the ResNet-18 model with 92.95% accuracy.

[9] displays an emotion recognition system using deep learning approach from an emotional audiovisual big data. To this end, the audios were processed to obtain the Mel-spectrogram, allowing to treat the domain as an image problem,

and for the videos some frames were extracted to both feed a CNN each. Then, the output of the two CNNs were joined using two consecutive extreme learning machines (ELMs). The result of this union is an input source for a support vector machine (SVM) for final classification of emotions. Finally, the experiment demonstrates the efficacies of the proposed system involving CNNs and ELMs.

[20] demonstrates the content-based audio auto-tagging experiment using deep learning. To do that, it used deep neural network architectures such as Fully Convolutional Neural Network (FCN) and Convolutional Recurrent Neural Network (CRNN) in conjunction with MagnaTagATune (MTT) dataset. As input for the models, the author used the extraction of Mel-spectrogram, transforming the audios into images. As a result, the CRNN model obtained a better result with 88.56% of AUC-ROC (area under the curve - receiver operating characteristic) as opposed to 85.53% of the FCN model.

The authors in [3] presents two papers related to the detection of violence through audio published between 2015 and 2020, in addition to other works that used only video or audio and video, called multi-modal. The first work, produced by [26], depicts a scream detection system in public transport vehicles, in which it uses the characteristics of Mel Frequency Cepstral Coefficients (MFCC), energy, delta and delta-deltas. The author uses the Gaussian Model Mixture (GMM) and Support Vector Machine (SVM) models to construct the experiments from its own dataset with about 2500 s seconds of audio. The experiment demonstrated that SVM generated a low rate of false alarms and GMM approach had a better identification rate. It is observed that [3] exposes this work even though it is outside the established period.

[30] proposes the classification and detection of acoustic scenes evolving domestic violence using machine learning. To this end, the author proposes the use of the SVM classifier to detect scenes of domestic violence of a man against a woman through the audio parameters MFCC (Mel Frequency Cepstral Coefficients), ZCR (Zero Rate Crossing) and Energy, which are low-level acoustic parameters. As a result, achieved 73.14% of accuracy for the audio parameter MFCC against 71.3% and 66.54% of average was obtained for Energy and ZCR, respectively.

[24] also presents two papers on the processing of audio signals from deep learning using Mel-spectrogram published between 2016 and 2017. The first work, performed by [2], presents an automatic music tagging system using fully convolutional neural networks (FCNs) in conjunction with the MagnaTagATune and Million Song datasets. The author also used the Mel-spectrogram technique to extract characteristics from the audios and convert them to image. The experiment demonstrated that the use of Mel-spectrogram as input representation resulted in better performance compared to STFTs (Short-time Fourier Transform) and MFCCs.

[16] demonstrates in the experiment a comparison between Mel-spectrogram, Frame-level strided convolution layer and Sample-level strided convolution layer as input for models using the deep convolutional neural networks (DCNN) architecture. To this end, the author used the MagnaTagATune and Milion Song

datasets as the input source, and made adjustments to all audios, reducing to 29.1 s of duration and reaming to 22050 Hz when necessary. The results showed better results when compared to Mel-spectrogram with the other, obtaining 0.9059 of AUC (Area Under the Curve) against 0.8906 and 0.9055 of Frame-level strided convolution layer and Sample-level strided convolution layer respectively.

However, to the best of our knowledge, all previous works evaluated architectures that required either the upload of raw user data to train the classifiers, or trained the classifiers on public or synthetic datasets to produce a model to be distributed to users. Nevertheless, such models tend to become obsolete over time, due to changes in the examples, or to the lack of personalization to the users real usage.

Therefore, the goal of this experiment is to use the FL technique and HEAR dataset (audio mel-spectrograms) to train CNN's architectures to identify scenes of physical violence and compares the results with the traditional approach to identify whether or not similar results are maintained. The Table 1 summarizes everything that has been presented and positions our work in relation to others.

Table 1. Related works as well as our experiment describing Algorithm, Characteristics extractor, Data type and Dataset.

Ref	Algorithm	Characteristics extractor	Data type	Dataset
[15]	MobileNet	Mel-spectrogram	Audio	Google AudioSet and Freesound
[28]	ResNet-18	Mel-spectrogram	Video	Youtube and their videos
[9]	SVM	Mel-spectrogram	Audio	Author dataset
[20]	CRNN	Mel-spectrogram	Audio	MagnaTagATune dataset
[26]	GMM	MFCC, energy, delta and delta-deltas	Audio	Author dataset
[30]	SVM	MFCC, ZCR and Energy	Audio	Author dataset
This experiment	MobileNet	Mel-spectrogram	Audio	HEAR dataset

4   Materials and Methods

As demonstrated in the past sections, FL and CNN's architectures are able to decentralize model training from a clients network (devices), permitting a worldwide model to be trained by them, respecting the protection of every client's information, and lessening the need to have a major infrastructure to train it. Hence, this paper points, in light of the past works, to train four CNN based on a transfer learning approach utilizing FL and compare these results to standard Deep Learning models.

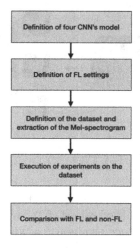

Fig. 2. The study workflow proposed.

The Fig. 2 shows the research process for this work. To detect violence in audios, we proposed to evaluate four CNN architectures under a transfer learning setting. We characterized the FL settings with three clients and the central server, which permits the decentralized training of models from a network. The HEAR dataset was set up as the data source for the experiments, and it was processed to extract mel-spectrograms from the audio. The dataset was parted into three equivalent parts for FL approach, one for each client. We used the Google Colab platform to run the experiments on the dataset, with a highly RAM-enabled environment. At last, we compare the results of the FL approach using Accuracy, f_1 score, Precision, Recall metrics. We also show the static (nonparametric) test of Friedman [4], Pairwise comparisons and compare the FL and non-FL results to analyze the metric results between both approaches.

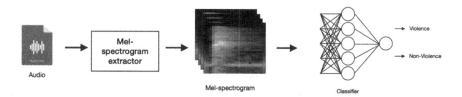

Fig. 3. The schematic of the proposed training procedure.

The experiment scheme shown in Fig. 3 describes the step to prepare the dataset as an input to the model in which it is the extraction of the mel-spectrogram from the audio. After that, we can handle the problem as image to be model inputs. For the FL approach, as displayed in Fig. 4, the mel-spectograms were split in the FL clients into three folders equally, and every

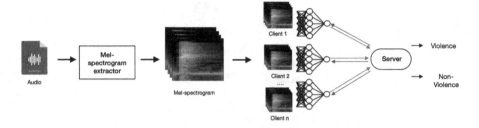

Fig. 4. The FL schematic of the training process.

client has its folder of the mel-spectrograms to train the models. Then, it would send the outcome to the centralized server after each round to build a worldwide model for classifying the presence or absence of violence.

4.1   Dataset

The HEAR Dataset [15], is a synthetic dataset that features 70,000 mel-spectrograms audio representations, constructed from transformations of 35,045 audios combined from three audio open databases from Google AudioSet and Freesound: 15,000 Google AudioSet foreground audios from the inside, conversation and speech classes; 20000 background audios also from Google AudioSet of the classes pets, television, toilet flush, door, radio, water, vacuum cleaner, sobbing, noise, sink and frying; and, 45 short-lasting physical violence sounds from Freesound.

With the set of mel-spectrograms, the dataset was divided into 60,000 training audios, 50% with the presence of physical violence and 50% without the presence of violence, 5000 test and 5000 for evaluation basis (50% of each class respectively). For the present experiment, only 12500 records were used, 80% for training and 20% for testing (1250 for validation and 1250 for testing), both with 50% of audios labeled violence and non-violence, for reasons of computational limitation.

4.2   Compared Methods

For the experiment, this paper compared the performance between FL, and non-FL approaches using four CNN's architectures that were used to verify which one presents the best result from the HEAR dataset. The selection criteria were: the most recent, from 2015, and representative according to the taxonomy performed by [12]. They are:

– **Inception v3** [32]: With a complex architecture, Inception v3 is the third edition of Google's Inception Convolutional Neural Network, which began as a module for GoogleNet. The architecture has a high quality in results and uses bottleneck layer and asymmetric filters to lower computational cost.

- **MobileNet v2** [27]: With the proposal of being a light and deep neural network, also published by Google, MobileNet v2 allows the construction of highly efficient models with high precision and low memory consumption, allowing use on mobile devices.
- **ResNet152 v2** [8]: Coming from a Residual Neural Network, ResNet v2 differs by introducing a connection called Identity, proposed to avoid escape gradient problems or vanishing gradient. In addition, it has its residual blocks called "skip connection", in which it allows the model to be extremely deep, having up to 152 layers.
- **VGG-16** [29]: convolutional neural network model with increased depth (up to 19 layers), of simple and homogeneous topology, which uses 3 connected layers, but of high computational cost, allowing to deal with large-scale images.

To perform the experiment, this paper used the pre-trained models and, through the technique of transfer learning, taking advantage of the weights trained in other contexts to perform this work, including only a few additional layers at the end of each architecture. First it was added a layer of Flatten, followed by a 20% layer of dropout, to prevent overfitting. Soon after, the experiment included one each of dense layer, with 1024 neurons and RELU activation function. Finally, a layer completely connected to output with 2 neurons with SOFTMAX activation function, according to the two classes (existence or not of violence), for the display of the probabilities of the classes.

4.3 Flower Framework

Flower is an open source FL framework that enables experiments to be run using machine learning frameworks such as PyTorch, TensorFlow, PyTorch Lightning, MXNet, scikit-learn, TFLite, among others, specifically designed to advance FL research, enabling heterogeneous FL workloads at scale [1]. Using Flower, you can run experiments simulating the existence of clients/devices that, connected to the federated network, can evolve their datasets and send the parameters of what has been learned to the centralizing server, which in turn will receive each update by aggregating into a single model.

The framework also allows the customization of different configurations for the experiment, such as waiting for a minimum number of clients to start a training cycle together, select assorted customers who participated in a cycle, perform the measurement of evolution both on the client and on the server, send the initial parameters to each client when starting an experiment. This flexibility allows for a diversification of scenarios in a given experiment.

Settings. To perform the experiment, this paper used the FedAvg [18] strategy, with 3 clients running 1 season per cycle, with 40 cycles, each client having 1/3 of the training dataset. In addition to these settings, it defined the existence of at least 3 clients for the beginning of each cycle, as well as the execution of validations, and also enabled the sending of the initial parameters for each client and set the validation configuration of the updated model using the test dataset, which is performed at each end of the cycle.

4.4   Experimental Setup

The experiment was conducted on the Google Colab[1] platform, Pro version, with 27 GB of RAM and an Nvidia Tesla P100 PCIe GPU with 16 GB. MLFlow[2] v1.20 was used to make annotations, log records, experiment tracking, metric storage and model in a central repository, being an open source platform enabling integration with several libraries used in the life cycle of machine learning solutions. The Python[3] v3 language and the TensorFlow[4] v2.6 framework were used to build the CNN models.

5   Results

The four chosen models were evaluated in the HEAR Dataset, and the results obtained are shown in the Table 2. The table also exposes the number of clients and rounds executed, as well as the total execution time (in hours). The MobileNet architecture performed better in terms of accuracy (71.9%), but Inception V3 showed better results in the f_1 score, precision and recall, 89.1%, 85.8%, 92.7% respectively.

Table 2. Results obtained along with the number of times and runtime.

Architecture	Clients × rounds	Execution time (hours)	Accuracy	f_1 score	Precision	Recall	Standard deviation acc.
MobileNet	3 × 40	3.4	0.719	0.407	0.356	0.476	0.069
VGG16	3 × 40	11.4	0.652	0.697	0.618	0.797	0.052
ResNet-152	3 × 40	13.2	0.526	0.003	0.003	0.003	0.013
Inception V3	3 × 40	4.4	0.587	0.891	0.858	0.927	0.029

The Table 2 presents that the MobileNet and Inception V3 architectures had a 1 h difference in run time, a much lower value compared to the time spent by the VGG16 and ResNet-152 architectures, which was an average of 12.3 h. An experiment was also performed with the same architectures without using FL techniques, presented in Table 3, to check for performance loss when using the technique.

Another important fact is the standard deviation, presented in the Table 2, where it indicates how close to the average the accuracy data is, they are: ResNet-152 has the lowest value with 0.013, followed by Inception V3 which obtained 0.029, VGG16 with 0.052 and MobileNet with 0.069.

---

[1] https://colab.research.google.com.
[2] https://mlflow.org.
[3] https://www.python.org.
[4] https://www.tensorflow.org.

Table 3. Comparison between training with and without FL with the number of times and lead time.

	Architecture	Clients × rounds	Execution time (hours)	Accuracy	f_1 score	Precision	Recall
Non-FL	MobileNet	1 × 40	3.3	0.755	0.750	0.763	0.790
	VGG16	1 × 40	3.3	0.684	0.683	0.710	0.894
	ResNet-152	1 × 40	4.5	0.637	0.708	0.594	0.814
	Inception V3	1 × 40	11.4	0.615	0.618	0.536	0.855
FL	MobileNet	3 × 40	3.4	0.719	0.407	0.356	0.476
	VGG16	3 × 40	11.4	0.652	0.697	0.618	0.797
	ResNet-152	3 × 40	13.2	0.526	0.003	0.003	0.003
	Inception V3	3 × 40	4.4	0.587	0.891	0.858	0.927

In addition, the Table 3 presents a loss of percentage points of almost 10% in the accuracy result of the MobileNet and ResNet-152 architectures when comparing the results between FL and non-FL. However, for the VGG16 and Inception V3 architectures the loss of percentage points is not as significant as in the MobileNet and ResNet-152 architectures, since the VGG16 presented 68.4% for training without FL and 65.2% with FL. InceptionV3 presented 61.5% for training without FL and 58.7% for FL. Another point to be highlighted is the execution time between the two approaches, where in the FL approach the MobileNet architecture presented shorter execution time and in the non-FL approach the MobileNet and VGG16 architectures presented the same execution time.

The static (non-parametric) test of Friedman [4] was applied for comparison of multiple models on a single dataset, in which the calculation of significance (p-value) was presented below the significance level $\alpha = 0.050$, concluding that the distributions differ and the null hypothesis can be rejected.

Another point is that the MobileNet architecture presents better performance in terms of accuracy (75.5%), f_1 score (75.0%), precision (76.3%) and recall (79.0%) when fl was not used (Table 4 and 5).

Table 4. Friedman Test Statistics for FL.

N	10
ChiSquare	30,000
df	3
Asymp. Sig	<,001

Wilcoxon's [35] *post hoc* method was applied to verify the null hypothesis for the samples, as a result critical difference diagrams were generated, illustrated in Fig. 5. Another way to visualize the data and identify the most critical differences is through pairwise comparisons, displayed in the Table 6 for Non-FL approach and the Table 7 for FL.

Table 5. Friedman Test Statistics for Non-FL.

N	10
ChiSquare	14,880
df	3
Asymp. Sig	<,002

Table 6. Pairwise Comparisons for Non-FL.

Sample 1–Sample 2	Test Statistic	Std. Error	Std Test Statistic	Sig.	Adj. Sig.
Resnet152V2 - VGG16	1,000	,577	1,732	,083	,500
Resnet152V2 - InceptionV3	−2,000	,577	3,464	<,001	,300
Resnet152V2 - MobilenetV2	−3,000	,577	5,196	<,001	,000
VGG16 - InceptionV3	−1,000	,577	1,732	,083	,500
VGG16 - MobilenetV2	−2,000	,577	3,464	<,001	,003
InceptionV3 - MobilenetV2	1,000	,577	1,732	,083	,500

Table 7. Pairwise Comparisons for FL.

Sample 1–Sample 2	Test Statistic	Std. Error	Std Test Statistic	Sig.	Adj. Sig.
Resnet152V2 - VGG16	1,200	,577	2,078	,038	,226
Resnet152V2 - InceptionV3	−1,400	,577	−2,425	,015	,092
Resnet152V2 - MobilenetV2	−2,200	,577	−3,811	<,001	,001
VGG16 - InceptionV3	−,200	,577	−,346	,729	1,000
VGG16 - MobilenetV2	−1,000	,577	1,732	,083	,500
InceptionV3 - MobilenetV2	,500	,577	1,386	,166	,995

Fig. 5. Critical difference diagram.

6   Conclusion and Future Work

This work aims to investigate the performance of CNNs architectures in the detection of physical violence using FL through HEAR Dataset [15], which is a synthetic dataset with 70,000 audios converted to mel-spectrograms (images). However, due to computational limitations, only 10,000 images divided equally into two classes were used: presence or not of physical violence. Thus, it was possible to apply consolidated techniques and tools to the audio context. [15] presented the investigation and application of the use of mel-spectrograms in the field of detection of audio violence, opening a gap for their application using

FL. In this experiment, the dataset used was divided among 3 clients, each running 1 season each round, in a 40-round cycle, considering four models of CNN: Inception v3, MobileNet v2, ResNet 152 v2, and VGG-16. Finally, the results showed that MobileNet obtained a better result when used with the FL technique, presenting a performance of 71.9% in the accuracy metric, with a loss of 3.6% when compared to the experiment without FL.

For future work, it is aimed at conducting tests with models in audios of real violence, application of techniques to provide privacy to data trafficked between client-servers [6,10,17,23,33], use the model in a larger sample of customers. In addition, it is intended to build comparisons through other models developed via Transfer Learning from a Large-Scale Pretrained Audio Neural Networks (PANNs) [13].

References

1. Beutel, D.J., Topal, T., Mathur, A., Qiu, X., Parcollet, T., Lane, N.D.: Flower: a friendly federated learning research framework. arXiv preprint arXiv:2007.14390 (2020)
2. Choi, K., Fazekas, G., Sandler, M.: Automatic tagging using deep convolutional neural networks (2016)
3. Durães, D., Marcondes, F.S., Gonçalves, F., Fonseca, J., Machado, J., Novais, P.: Detection violent behaviors: a survey. In: Novais, P., Vercelli, G., Larriba-Pey, J.L., Herrera, F., Chamoso, P. (eds.) ISAmI 2020. AISC, vol. 1239, pp. 106–116. Springer, Cham (2021). https://doi.org/10.1007/978-3-030-58356-9_11
4. Friedman, M.: The use of ranks to avoid the assumption of normality implicit in the analysis of variance. J. Am. Stat. Assoc. **32**(200), 675–701 (1937). https://doi.org/10.1080/01621459.1937.10503522. https://www.tandfonline.com/doi/abs/10.1080/01621459.1937.10503522
5. Fukushima, K.: Neocognitron: a self-organizing neural network model for a mechanism of pattern recognition unaffected by shift in position. Biol. Cybern. **36**(4), 193–202 (1980). https://doi.org/10.1007/BF00344251
6. Gu, B., Xu, A., Huo, Z., Deng, C., Huang, H.: Privacy-preserving asynchronous vertical federated learning algorithms for multiparty collaborative learning. IEEE Trans. Neural Netw. Learn. Syst. **33**, 1–13 (2021). https://doi.org/10.1109/TNNLS.2021.3072238
7. Hard, A., et al.: Training keyword spotting models on non-iid data with federated learning (2020). https://arxiv.org/abs/2005.10406
8. He, K., Zhang, X., Ren, S., Sun, J.: Deep residual learning for image recognition (2015). http://arxiv.org/abs/1512.03385
9. Hossain, M.S., Muhammad, G.: Emotion recognition using deep learning approach from audio–visual emotional big data. Inf. Fusion **49**, 69–78 (2019). https://doi.org/10.1016/j.inffus.2018.09.008, https://www.sciencedirect.com/science/article/pii/S1566253517307066
10. Hu, R., Guo, Y., Gong, Y.: Concentrated differentially private federated learning with performance analysis. IEEE Open J. Comput. Soc. **2**, 276–289 (2021). https://doi.org/10.1109/OJCS.2021.3099108
11. Volkmann, J., Stevens, S.S., Newman, E.B.: A scale for the measurement of the psychological magnitude pitch. J. Acoust. Soc. Am. **8**, 208 (1937). https://doi.org/10.1121/1.1901999

12. Khan, A., Sohail, A., Zahoora, U., Qureshi, A.S.: A survey of the recent architectures of deep convolutional neural networks. Artif. Intell. Rev. **53**(8), 5455–5516 (2020). https://doi.org/10.1007/s10462-020-09825-6. http://arxiv.org/abs/1901.06032

13. Kong, Q., Cao, Y., Iqbal, T., Wang, Y., Wang, W., Plumbley, M.D.: PANNs: large-scale pretrained audio neural networks for audio pattern recognition. IEEE/ACM Trans. Audio Speech Lang. Process. **28**, 2880–2894 (2020). https://doi.org/10.1109/TASLP.2020.3030497. https://ieeexplore.ieee.org/document/9229505/

14. Krizhevsky, A., Sutskever, I., Hinton, G.E.: ImageNet classification with deep convolutional neural networks. Commun. ACM **60**(6), 84–90 (2017). https://doi.org/10.1145/3065386

15. Lacerda, T. B., Miranda, P., Camara, A., Furtado, A.P.C.: Deep learning and mel-spectrograms for physical violence detection in audio. In: The 18th National Meeting on Artificial and Computational Intelligence, pp. 268–279 (2021). https://sol.sbc.org.br/index.php/eniac/article/view/18259/18093

16. Lee, J., Park, J., Kim, K.L., Nam, J.: Sample-level deep convolutional neural networks for music auto-tagging using raw waveforms (2017)

17. Lu, Y., Huang, X., Dai, Y., Maharjan, S., Zhang, Y.: Differentially private asynchronous federated learning for mobile edge computing in urban informatics. IEEE Trans. Ind. Inf. **16**(3), 2134–2143 (2020). https://doi.org/10.1109/TII.2019.2942179

18. McMahan, B., Moore, E., Ramage, D., Hampson, S., Arcas, B.A.y.: Communication-efficient learning of deep networks from decentralized data. In: Singh, A., Zhu, J. (eds.) Proceedings of the 20th International Conference on Artificial Intelligence and Statistics. Proceedings of Machine Learning Research, vol. 54, pp. 1273–1282. PMLR (2017). https://proceedings.mlr.press/v54/mcmahan17a.html

19. Nations, U.: Policy brief: the impact of covid-19 on women (2020). https://www.un.org/sexualviolenceinconflict/wp-content/uploads/2020/06/report/policy-brief-the-impact-of-covid-19-on-women/policy-brief-the-impact-of-covid-19-on-women-en-1.pdf

20. Nayyar, R.K., Nair, S., Patil, O., Pawar, R., Lolage, A.: Content-based auto-tagging of audios using deep learning. In: 2017 International Conference on Big Data, IoT and Data Science (BID), pp. 30–36 (2017). https://doi.org/10.1109/BID.2017.8336569

21. Organization, W.H.: Violence against women (2021). https://www.who.int/news-room/fact-sheets/detail/violence-against-women

22. Organization, W.H.: Violence against women prevalence estimates, 2018: global, regional and national prevalence estimates for intimate partner violence against women and global and regional prevalence estimates for non-partner sexual violence against women (2021). https://www.who.int/publications/i/item/9789240022256

23. Paul, S., Sengupta, P., Mishra, S.: Flaps: Federated learning and privately scaling. In: 2020 IEEE 17th International Conference on Mobile Ad Hoc and Sensor Systems (MASS), pp. 13–19 (2020). https://doi.org/10.1109/MASS50613.2020.00011

24. Purwins, H., Li, B., Virtanen, T., Schluter, J., Chang, S.Y., Sainath, T.: Deep learning for audio signal processing. IEEE J. Sel. Topics Signal Process. **13**(2), 206–219 (2019). https://doi.org/10.1109/jstsp.2019.2908700

25. Ramzan, M., et al.: A review on state-of-the-art violence detection techniques. IEEE Access **7**, 107560–107575 (2019). https://doi.org/10.1109/ACCESS.2019.2932114

26. Rouas, J.L., Louradour, J., Ambellouis, S.: Audio events detection in public transport vehicle. In: 2006 IEEE Intelligent Transportation Systems Conference, pp. 733–738. IEEE (2006). https://doi.org/10.1109/ITSC.2006.1706829. http://ieeexplore.ieee.org/document/1706829/

27. Sandler, M., Howard, A., Zhu, M., Zhmoginov, A., Chen, L.C.: Mobilenetv2: inverted residuals and linear bottlenecks. In: 2018 IEEE/CVF Conference on Computer Vision and Pattern Recognition, pp. 4510–4520. IEEE (2018). https://doi.org/10.1109/CVPR.2018.00474. https://ieeexplore.ieee.org/document/8578572/

28. Santos, F.: In-car violence detection based on the audio signal. In: Yin, H., et al. (eds.) IDEAL 2021. LNCS, vol. 13113, pp. 437–445. Springer, Cham (2021). https://doi.org/10.1007/978-3-030-91608-4_43

29. Simonyan, K., Zisserman, A.: Very deep convolutional networks for large-scale image recognition (2015). https://arxiv.org/abs/1409.1556

30. Souto, H., Mello, R., Furtado, A.: An acoustic scene classification approach involving domestic violence using machine learning. In: Anais do ENIAC, pp. 705–716 (2019). https://doi.org/10.5753/eniac.2019.9327. https://sol.sbc.org.br/index.php/eniac/article/view/9327

31. Szegedy, C., et al.: Going deeper with convolutions (2014)

32. Szegedy, C., Vanhoucke, V., Ioffe, S., Shlens, J., Wojna, Z.: Rethinking the inception architecture for computer vision (2015). http://arxiv.org/abs/1512.00567

33. Triastcyn, A., Faltings, B.: Federated learning with bayesian differential privacy. In: 2019 IEEE International Conference on Big Data (Big Data), pp. 2587–2596 (2019). https://doi.org/10.1109/BigData47090.2019.9005465

34. Tripathi, G., Singh, K.V.D.K.: Violence recognition using convolutional neural network: a survey. J. Intell. Fuzzy Syst. **39**, 7931–7952 (2020). https://doi.org/10.3233/JIFS-201400. https://content.iospress.com/articles/journal-of-intelligent-and-fuzzy-systems/ifs201400

35. Wilcoxon, F.: Individual comparisons by ranking methods. Biometr. Bull. **1**(6), 80–83 (1945). http://www.jstor.org/stable/3001968

Police Report Similarity Search: A Case Study

José Alan Firmiano Araújo[1], Ticiana L. Coelho da Silva[1,2(✉)],
Atslands Rego da Rocha[1], and Vinicius Cezar Monteiro de Lira[2]

[1] Federal University of Ceará, Fortaleza, Brazil
alanfirmiano@alu.ufc.br, atslands@ufc.br
[2] Insight Data Science Lab, Fortaleza, Brazil
{ticianalc,vinicius.monteiro}@insightlab.ufc.br

Abstract. Several crimes occur daily, and the initial investigation begins with a police report. In cities with high crime rates, it is impractical to expect the police to read and analyze every crime narrative. Some police reports may involve multiple victims or the same crime may be reported more than once. Additionally, police reports may exhibit similarities due to a shared *modus operandi*. This study addresses the challenge of providing a police report and searching for the most similar report in the database. A similar police report can be either another report with overlapping words or one that shares a similar *modus operandi*. One potential solution is to represent each police report as a feature vector and compare these vectors using a similarity function. Different methods can be employed to represent the narrative, including embedding vectors and count-based approaches such as TF-IDF. This research explores the use of pre-trained embedding representations at both the word and sentence levels, such as Universal Sentence Encoder, Word2Vec, RoBERTa, Doc2Vec, among others. We determine the most effective representation for capturing semantic and lexical similarities between police reports by comparing different embedding models. Furthermore, we compare the effectiveness of available pre-trained embedding models with a model trained specifically on a corpus of police reports. Another contribution of this work is the development of trained embedding models specifically tailored for the domain of police reports.

1 Introduction

Usually, victims and witnesses report the crime events on police reports for the police investigation. The police report also protects the police action, demonstrating where this series of investigative efforts by the police department began. A police report might even make it to the Supreme Court in severe circumstances. Consequently, a police report needs to be thorough and factual when describing an occurrence of crime (for example, accurate, concise, clear, objective, timely and complete).

© The Author(s), under exclusive license to Springer Nature Switzerland AG 2023
M. C. Naldi and R. A. C. Bianchi (Eds.): BRACIS 2023, LNAI 14197, pp. 394–409, 2023.
https://doi.org/10.1007/978-3-031-45392-2_26

This paper addresses the challenge of conducting a search for the most similar police report in a database. When it comes to similarity search in text data, the goal is to find the most similar police report(s) based not only on string similarity but also taking account semantic similarity. A similar police report can be a police report with overlapping words or one that describes an equivalent *modus operandi*, even if it does not use the exact words.

To illustrate both cases, let's consider the following two sentences: "Augusta's car was stolen yesterday at night" and "Yesterday night, Augusta had her car stolen." These sentences share almost identical wording. Now, let's examine two different sentences that exhibit the same *modus operandi*: "Two armed men on a motorbike robbed Patricia and took her mobile phone" and "Two men with a knife on a motorbike approached Mary and stole her personal belongings." These police reports involve the same type of crime (robbery), a female victim, and a similar criminal approach (two individuals on a motorbike). One possible approach to address this issue is to represent each police report as a feature vector and compare these vectors using cosine similarity.

Pre-trained word embeddings are one of the most commonly used representations for document vocabulary. There are various pre-trained word embeddings available, such as [13,16,17,19]. Word embeddings can capture a word's relationships with other words, its contextual meaning within a document, as well as its semantic and syntactic similarities. However, word embeddings alone may struggle to capture nuanced shifts in meaning when a sentence undergoes minor changes. For example, consider the sentences "Augusta's car was stolen" and "Augusta's car was not stolen." Despite the difference of only one word, these sentences have completely different meanings. Nonetheless, when using word embeddings, the cosine similarity between the vectors obtained from these sentences might still be relatively high, despite being semantically opposite. Embedding techniques should ideally be able to address this issue and provide more nuanced representations.

One alternative is to encode the entire sentence into embedding vectors. This method is known as sentence embedding, and there are many pre-trained sentence embedding techniques available, including Doc2Vec [15], SBert [3] and Universal Sentence Encoder [21]. These models take the text as input and generate a fixed-dimensional embedding representation of the sentence. Sentence embedding techniques aim to capture the full sentences and the semantic information they contain in the form of vectors. This helps the machine understand the context, intention, and other nuances of the text as a whole. The quality of the training data has a significant impact on the resulting sentence embedding vectors. For optimal results, the sentences in the training set should be semantically related [14].

This study focuses on addressing the challenge of conducting similarity searches on police report documents. The primary objective is to find similar police reports - with overlapping words or reports that share the same *modus operandi*. Police reports pose unique challenges for similarity search due to specialized language and terminology specific to law enforcement, requiring tailored approaches. Furthermore, the wealth of highly contextual information in these reports, encompassing location, time, individuals involved, and preceding events, adds complexity, necessitating context-aware techniques for accurate assessment. Additionally, the sensitivity of confidential police reports, containing personal details and classified materials, demands privacy protection measures during similarity analysis, adhering to legal and ethical guidelines through anonymization or aggregation techniques. To guide this research, we have formulated the following key research questions:

1. (**RQ1**) The first research question aims to identify the pre-trained sentence-level representation that effectively captures both the syntactic and semantic features of the vocabulary found in police reports while being able to differentiate between similar reports. In this study, we consider various approaches for input sentence representation, including word-level and sentence-level embeddings, as well as count-based text representation techniques such as TF-IDF. Additionally, we explore lexical-based similarities, such as Jaccard similarity, to analyze the effectiveness of different representation methods. The goal is to determine the most suitable approach that can accurately capture the essential characteristics of police reports and enable the detection of similar reports.
2. (**RQ2**) Our second research question delves into the effectiveness of embedding models that are trained from scratch with police report documents. We aim to determine if these models are better at representing sentences for similarity searches. Furthermore, we compare the pre-trained models used in RQ1 with our own data-trained embeddings. Our primary goal is to assess the performance of these various versions when it comes to searching for similar police reports.
3. (**RQ3**) The third research question explores whether combining embeddings can enhance the quality of searching for similar police reports. Building on the findings from [8], which highlights the benefits of combining different embeddings due to their complementarity, we investigate whether combining embeddings can lead to improved performance on the task at hand. Based on the analysis conducted in the previous research questions, we select the two best-performing embeddings. We then compare the effectiveness of combining these two embeddings using different aggregation techniques, such as taking the maximum, minimum, or average vector. By investigating the potential of combining models, we aim to determine whether it is possible to achieve a better sentence representation that captures the semantic and syntactic features crucial for searching similar police reports.

The remaining sections of this article are organized as follows. Section 2 presents the background and related works. Section 3 explains the data and the solutions to cope with our problem. Section 4 explains the evaluation metrics used in the comparisons and shows the analysis performed compared to different models. Finally, Sect. 5 summarizes this work and proposes future developments.

2   Backgound and Related Works

The background information and studies that are relevant to our problem are covered in this section. We divide this section into four main topics: Traditional text representation, Word Embedding, Sentence Embedding and Similarity search in text data.

Traditional Text Representation. The method of vectorizing text involves turning text into tensors of numbers. The most frequent and fundamental representation is one-hot encoding. Every word is given a distinct integer index, and this integer index i is then converted into a binary vector of size N (the vocabulary size). The vector is all zeros except for the $i - th$ entry, which is 1 [4]. Another typical representation of text data is the use of *bag-of-words* or *bag-of-n-words* [9] for simplicity. According to this paradigm, a sentence or text is typically depicted as a collection of words, ignoring grammar and word order but keeping multiplicity.

There are some alternatives to represent each word in a document for such a model: (i) a boolean indicating the presence or not; (ii) the frequency of the word in the text; (iii) TF-IDF (*Term Frequency, Inverse Document Frequency*), a statistical tool that assesses how pertinent a word is to a document within a collection of documents. The two metrics are multiplied: how many times a word appears in a document (TF) and the inverse document frequency of the word across a set of documents (IDF). However, these techniques have some drawbacks, such as the lack of semantics over the representation. In this paper, we experimented with the representation of each word in a sentence as its TF-IDF value.

Word Embeddings. Text of various lengths (such as a document, paragraph, sentence, or even a word) is transformed by embedding models into a fixed-length numeric vector that may be used to train machine learning algorithms like classification and clustering models. Pre-trained *word embeddings* have been widely used [1,16,17,19,24] due to their capacity to retain a word's context inside a document, as well as their semantic and syntactic similarity to other words. A framework for learning word vectors is proposed by [17] by training a linguistic model that predicts a word given other words in a context. *Word2Vec* is one specific implementation of such a structure. The main disadvantage is that [17] under-utilizes corpus statistics as the model is trained on a separate local context window rather than on global co-occurrence counts.

Bypassing this issue, [19] provide a model that creates a vector space of words by training it on the worldwide count of word co-occurrence, making effective use of statistical data. Another word embedding proposal is FLAIR [1], which abstracts away from the particular engineering difficulties that various word embeddings bring. It creates a uniform interface for all word, sentence, and arbitrary embedding combinations. Brazilian Portuguese BERT models are provided by BERTimbau [24]. Three NLP tasks-textual similarity of the sentence, textual linkage recognition, and named entity recognition-were used to assess the models. Compared to BERT multilingual and earlier single language techniques, BERTimbau advances state-of-the-art in these tasks, demonstrating the value of large pre-trained language models for Portuguese.

We can obtain document vectors from word embeddings. An approach is to average all word vectors collectively. However, this process weighs both significant and insignificant words equally. The fact that each word would be represented with the same vector, regardless of context, is another drawback of utilizing word embeddings to represent text. Sentence embedding is an improvement from word embedding.

Sentence Embedding. performs the encoding of sentences in an n-dimensional vector space, which might have several word representations based on context. The most widely used sentence embedding proposals are Doc2Vec [15], InferSent [5], SBert [3], Universal Sentence Encoder [21] and LaBSE [7], among others.

Similar to Word2Vec, Doc2Vec trains sentence embeddings or paragraph vectors to predict the following word, given a variety of contexts drawn from the paragraph. The sentence and word vectors are combined to forecast the next word in a context. The pre-trained BERT network is used by SBert [3], which adds a grouping operation to the output to create a fixed-length sentence embedding. To enhance BERT and produce semantically meaningful sentence embeddings that can be compared using cosine similarity, SBert introduces Siamese and triple lattice structures. Using the output of *CLS-token*, the developers of [3] experiment with various pooling strategies, such as finding the average of all output vectors (the default strategy) or computing a max-over-time of the output vectors (MAX-strategy).

The trained LaBSE model is proposed by [7] for sentence embeddings in several languages. Only pairs of bilingual utterances that are translations of one another are used exclusively in training and optimizing LaBSE to generate comparable representations. The source and destination phrases are independently encoded using a shared BERT-based encoder in the framework's LaBSE, then passed into a combination function. This is known as a *dual encoder*. The sentence embedding for each entry is obtained from the last layer representations [CLS]. The cosine over the sentence embedding created by BERT encoders is used to score how similar the source and destination phrases are to each other.

This paper assesses a variety of embeddings, including Word2Vec, Flair, and BERT as word embeddings, as well as Doc2Vec, the Universal Sentence Encoder, and SBert.

Similarity Search in Text Data. Several studies have been conducted to assess the effectiveness of different embeddings for text similarity. One such study, as suggested by the authors of [23], compares the semantic similarity between various techniques in the patent field. This study utilizes three different vector spaces, including a basic TF-IDF baseline, an LSI topic model, and a Doc2Vec neural model.

Another article, [18], compares different learning vector representations like Latent Semantic Analysis (LSA), Word2Vec, and GloVe to identify the most efficient technique in the subject segmentation space. [22] trains a supervised meta-embedding neural network and combines multiple sentences pre-trained embedding models for sentence similarity recognition. The meta-model outputs the distance between two given sentences as input. The approach for selecting the best embedding models in our study can be applied to the methodology proposed by [22] to choose the embedding models for effectively creating their meta-embedding model.

In addition, [12] tackles the issue of recognizing duplicate questions. This article proposes a method that vectorizes questions based on the combination and individual use of FastText mining subword integration, Google news vector integration, and FastText mining integration. These question word vectors are then used to determine how similar words are semantically. The proposed model is trained on each of the characteristics derived from each of the three words separately, and input is received by the MaLSTM neural network, which measures the Manhattan distance to assess the semantic similarity between questions.

Various studies have utilized embeddings to map input to low-dimensional vectors and perform similarity searches. For instance, [11] investigates embedding complex data objects like DNA sequences and images in a vector space to make their distances as close to their real distances as possible. This allows queries to be run on the embedded objects in similarity search applications. Similarly, [6] conducts similarity searches in Knowledge Graphs (KG) to help users identify the most significant entities to their query.

3   Data and Methods

This section discusses our datasets and methodology for investigating the research questions. We employ a two-step approach: generating similarity matrices for different sentence representations and validating the accuracy and Mean Reciprocal Rank (MRR) of police report representations. The first step involves data preprocessing, sentence representation, model training, and ranking of the top K most similar sentences using cosine and Jaccard similarity. In the second step, we evaluate the accuracy of each police report representation by determining if the most similar report is retrieved within the top K sentences. We also calculate the MRR to assess the retrieval performance. We provide more details about MRR as follows. Figure 1 presents the steps followed by us in this paper.

Data. Two Brazilian Department of Public Security organizations provided the datasets utilized in this study. To maintain anonymity during the review process, we have omitted their names. The datasets comprise two corpora: one with 1,089 police reports (1,065 reports are similar, containing at least one copy for each incident report), and another with 30,011 reports, both electronic and

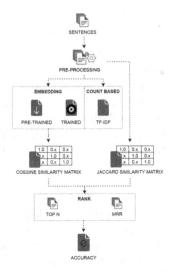

Fig. 1. Steps followed to assess the research questions in our work.

non-electronic. These corpora cover incidents that occurred between January 1, 2020, and March 29, 2020, encompassing various types of crimes such as theft, murder, robbery, attempted murder, and harassment. The first *corpus* with 1,089 police reports has 1,065 reports in all, in which each police report has a duplicate in this same set. Another 24 reports did not have similarities and were discarded from validation. Figure 2 presents some important information about the first dataset (with 1,065 police reports):

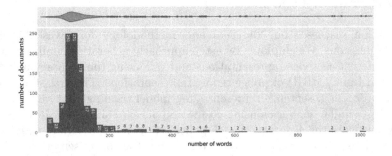

Fig. 2. Distribution of the number of documents by the word lenghh

Each column bin varies the number of words to 20; for instance, the first column contains 38 reports (for up to 20 words in each report). It is possible to observe that the number of documents decreases as we move to subsequent columns, i.e., fewer documents with a high number of words. In addition, Fig. 2

presents an asymmetry to the right, suggesting a greater concentration of documents with fewer words and a smaller number of documents with a more significant number of words in the documents. Some statistical measures from our dataset: the maximum document length in terms of the number of words is 1,019; on average, each document contains 119 words and around 597 characters.

To provide an example of a police report and its duplication:

- Original report: "The reporting party, identified above, informs via the Virtual Police Station that on the specified date and time, they were the victim of a robbery. The incident occurred as follows: Two individuals on a black motorcycle, wearing helmets and armed with a knife, approached and demanded the cell phone. They took a Samsung phone, model A205GT, with the serial number: 357621102968577. When asked about identifying or describing the suspects, the reporting party stated that it was not possible due to their helmets."
- Duplicated report: "The reporting party, identified as mentioned above, reports via the Virtual Police Station that on the mentioned date and time, they were the victim of a robbery. The incident occurred as follows: Two individuals on a black motorcycle, identified as POP, with a knife, threatened the reporting party and demanded the cell phone. When asked about identifying or describing the suspects, the reporting party stated that they were unable to do so as the suspects were wearing helmets."

Duplications of police reports can happen due to several factors: (i) the same occurrence may have happened to different victims, and thus each victim records the same event, however, with different words; (ii) electronic police reports can be made due to system or user failure, these police reports can be registered more than once, among other possibilities.

During the pre-processing phase, we conducted several steps. These included the removal of special characters, such as HTML tags and non-ASCII characters, as well as the elimination of terms containing digits. Additionally, we eliminated stopwords and converted all characters to lowercase. To represent the police reports using embedding models, we explored three potential solutions: utilizing pre-trained embeddings, training embeddings from scratch using our data, and combining different embeddings to represent the text.

Methods. In this paper, we utilize pre-trained neural embedding models, including Flair, Universal Sentence Encoder (USE), Doc2Vec, Word2Vec, and Sbert. Firstly, we employ word embeddings by computing the average of word representations to describe the narrative of each police report using models such as Word2Vec. Additionally, we investigate the performance of Word2Vec, Doc2Vec, and RoBERTa when trained from scratch using our dataset, which consists of approximately 30,011 reports obtained. Each word is assigned an initial representation, and the models refine these representations through training iterations to enhance their quality. To compare the performance of the embedding models and lexical approaches, we evaluate the use of a count-based method, TF-IDF, where each word in a sentence is represented by its TF-IDF value.

After representing each police report as a vector using either an embedding model or TF-IDF, we compare the reports to identify the most similar ones. For each representation approach investigated, we generate an N x N matrix to rank the top K most similar police reports for each report. For example, in the case of the dataset containing 1,065 police reports represented with Word2Vec, we calculate the cosine similarity between all 1,065 sentences, resulting in a $1,065 \times 1,065$ matrix that represents the similarity percentage. We disregard the diagonal of the matrix since the similarity between identical sentences would always be 100%. This similarity matrix is generated for each representation approach, including Word2Vec, Flair, USE, Doc2Vec, Sbert, RoBERTa, and TF-IDF.

Furthermore, we explore an alternative approach that retains the words in the sentences without representing them as vectors. In this case, we compute the Jaccard similarity matrix, which compares two sentences by assessing if they contain identical words. The Jaccard similarity ranges from 0% to 100%, with a higher percentage indicating a greater similarity between the sentences.

Evaluation Metrics. The ranking step in our evaluation involves analyzing the top K (top 1, top 5, and top 10) most similar sentences and calculating the Mean Reciprocal Rank (MRR). For each value of K, a list of the most similar police report IDs is generated. In the case of top 1, the list contains only the most similar sentence, while in the case of top 5, the list includes the five most similar sentences, and so on. If the list of most similar police report IDs includes the correct duplicate police report, it is considered a hit for the respective approach used for sentence representation. The approach's accuracy is calculated as the percentage of hits achieved by the approach across the entire dataset. The formula of MRR is shown below:

$$\text{MRR} = \frac{1}{|D|} \sum_{i=1}^{|D|} \frac{1}{\text{rank}_i}, \tag{1}$$

where D is the dataset of police reports and rank_i refers to the rank position of the first relevant document (most similar) for the $i-th$ police report. Additionally, the MRR is computed for each method, providing a harmonic mean of the ranks.

Comparative studies of word embedding vectors, such as those conducted by [2,25], play a vital role in ensuring the quality of word representation before their utilization in machine learning tasks. Evaluation methods for word vectors can be categorized as intrinsic or extrinsic [20,26].

Intrinsic evaluation focuses on assessing word relationships based on syntax or semantics and is independent of a particular NLP task. Human-curated benchmark datasets are used to compare words in terms of similarity and relatedness. On the other hand, extrinsic evaluation involves incorporating word vectors into an NLP task, such as sentiment analysis or natural language inference. The performance of the embedding models is then evaluated using task-specific metrics like accuracy, F1-score, or other relevant measures. This type of evaluation assesses the usefulness and effectiveness of the embeddings in real-world applications and can assist data scientists.

4   Experimental Results

In this section, we analyze and evaluate the results obtained using pre-trained embedding models as well as embedding models trained on a separate dataset of police reports. We investigate the impact of these models on identifying similar sentences, including the ones with the same *modus operandi*. Additionally, we examine the behavior of count-based methods, such as TF-IDF, in representing the text. Furthermore, we explore the potential benefits of combining different embeddings to improve the search for similar police reports.

The research questions defined in the Introduction guide our analysis in this section. By addressing these questions, we gain insights into the performance and effectiveness of various embedding models and representation techniques in the context of police reports. This evaluation allows us to understand the strengths and limitations of different approaches and provides valuable information for researchers and practitioners working in the field of natural language processing and law enforcement.

Study on the Results of RQ1. The first research question aims to determine the most effective representation method for capturing syntactic and semantic similarity between police reports. To investigate this question, we utilize a dataset containing 1,065 police reports. This dataset is accompanied by ground truth annotations, meaning that we have information about the most similar police report for each entry.

In our evaluation, we consider various pre-trained models, including Universal Sentence Encoder (USE), Word2Vec, Doc2Vec, BERT, SBert, and Flair. For each approach, we compare the sentences using cosine similarity. Additionally, we explore the option of keeping the words in the sentences without converting them into vector representations. In this case, we utilize Jaccard similarity to identify the most similar sentences.

Table 1 presents the accuracy and MRR results obtained by these approaches. These metrics allow us to assess the performance and effectiveness of each representation method in capturing the similarity between police reports.

The USE (Universal Sentence Encoder) model demonstrated the best performance among the embedding models evaluated. It is based on a Transformer encoder and achieved higher accuracy compared to the deep averaging network (DAN) model. The accuracy achieved by USE is similar to the results reported in the paper by [3] for the STS Bench (Semantic Textual Similarity Benchmark). The USE model incorporates sentences into a 512-dimensional vector, enabling a more robust semantic representation. Moreover, the Multilingual USE model allows for the representation of words from different languages.

Table 1. Results from the approaches for RQ1

Approach	Accuracy			
	Top 1	Top 5	Top 10	MRR
USE	70.47	80.62	82.61	75.19
BERT	65.58	74.46	76.99	69.85
Word2Vec	64.86	73.19	75.91	69.06
Doc2Vec	35.87	47.46	51.27	43.39
Flair	51.45	60.14	61.96	55.75
SBert	52.90	61.96	63.77	57.24
TF-IDF	**77.38**	**86.05**	**87.68**	**81.58**
Jaccard	75.18	83.15	85.51	79.14

BERT, another transformer-based model, obtained the second-best results among the models in capturing syntactic and semantic information. The pre-trained BERT model from [24][1] slightly outperformed the pre-trained Word2Vec model from [10][2]. SBert and Flair models showed similar performance, likely because they are both based on the same training base. However, the Doc2Vec model performed poorly compared to the other models and ranked last. This could be attributed to the use of Portuguese Wikipedia data for training, which may have introduced variations in phrase usage across different contexts.

TF-IDF was computed using the TfidfVectorizer[3] to represent the sentences. This involved creating a dictionary of words from the sentences and calculating the frequency of each term. Portuguese stopwords were removed to ensure consistency in similarity calculations. The TF-IDF representation yielded promising results. One key takeaway from this experiment: it is indeed expected that the embeddings used would be able to capture the context of the sentences and identify duplications as accurately as, if not better than, count-based and lexical methods. However, in the case of police reports, the text may not be exactly the same in terms of words, as demonstrated in the previous section. There may be repetitions of words, but they can occur in a different order within the text, which is a characteristic of this type of dataset.

Embeddings that are pre-trained on non-police data may not capture this particularity of the police reports as effectively as count-based and lexical methods. The pre-trained embeddings may have learned general language patterns, but they may not have specifically captured the nuances and variations present in police reports. This can result in lower accuracy when identifying similar sentences. In such cases, count-based and lexical methods, like TF-IDF and Jaccard similarity, tend to perform better because they directly compare the presence or

---

[1] Available at https://huggingface.co/neuralmind/bert-base-portuguese-cased.

[2] Available at https://nilc.icmc.usp.br/embeddings.

[3] https://shorturl.at/lmCXY.

absence of words in the sentences. They are not reliant on capturing contextual information and can handle cases where word order varies within the text.

Therefore, it is important to consider the specific characteristics of the dataset and the task at hand when choosing an appropriate representation method. While pre-trained embeddings offer many advantages, they may not always be the optimal choice for certain types of text, such as police reports with specific structural patterns and word repetitions.

Study on the Results of RQ2. The second research question investigates whether the embedding models are more effective in representing the sentences when trained with a police report dataset. Using an unsupervised approach, this study trained the RoBERTa (an improvement of the BERT model), Doc2Vec, and Word2Vec embedding models. The data for training is around 30,011 police reports provided by Brazilian Department of Public Security. To investigate **RQ2**, we use the dataset with 1,065 police reports (the same used for **RQ1**). The RoBERTa trained model was more assertive in identifying similar sentences; We trained the RoBERTa model for 13 epochs when the difference between the current and last epoch loss was not more significant than 0.01. Roberta embodied each word in 512 dimensions. The representation of the police reports for Word2Vec was made with vectors of 50 dimensions, trained for 100 epochs. For the Doc2Vec model, the representations of the sentences were with vectors of 150 dimensions, trained for 100 epochs, with a training window of size 10.

The findings of the study suggest that training embedding models with a specific dataset from the same context, in this case, the police report dataset, improves their effectiveness in representing sentences and capturing syntactic, semantic, and morphological relationships. The Word2Vec, RoBERTa, and Doc2Vec models trained with the 30,011 police reports achieved slightly better results compared to their pre-trained counterparts, which were trained on a much larger and diverse corpus.

The advantage of training embedding models with domain-specific data is that they can better capture the specific language patterns, vocabulary, and contextual relationships present in that particular domain. The embeddings trained on the police report dataset were able to maintain proximity between vectors representing syntactic, semantic, and morphological relationships relevant to police reports. This proximity allows for better identification of similar sentences and captures the nuances of language specific to the domain.

The results obtained with the trained models were considered satisfactory, considering that the pre-trained versions of Word2Vec, BERT, and Doc2Vec were trained on a much larger corpus with a significantly larger number of tokens. The pre-trained versions of Word2Vec, BERT, and Doc2Vec were trained with 1,395,926,282 tokens. Our 30K police reports are much smaller (with 2,579,815 tokens). Despite the smaller size of the police report dataset, it still proved to be effective in capturing the relevant patterns and relationships needed for the task at hand. This highlights the importance of training embedding models on data that is specifically relevant to the target domain, as it can lead to improved performance and accuracy in representing and analyzing text data (Table 2).

Table 2. Results from the approaches for RQ2

Model	Accuracy			
	Top 1	Top 5	Top 10	MRR
RoBERTa - trained	**68.66**	**77.17**	**79.89**	**72.75**
BERT	65.58	74.46	76.99	69.85
Word2Vec - trained	65.22	73.73	75.36	69.36
Word2Vec	64.86	73.19	75.91	69.06
Doc2Vec - trained	37.32	49.46	55.62	43.37
Doc2Vec	35.87	47.46	51.27	43.39

Study on the Results of RQ3. The third research question explores the potential of combining embeddings to enhance the search for similar police reports and improve the retrieval of similar reports. In this study, the embeddings from the USE model and Word2Vec-trained model, both trained with the 30,011 police reports, were combined to create a new embedding model called USEW2V.

To ensure compatibility in dimensionality, the Word2Vec model was adjusted to generate embeddings with 512 dimensions, matching the dimensionality of the embeddings generated by the USE model. By combining the embeddings of both models, the USEW2V model was created. The performance of the USEW2V model was compared against the individual USE and Word2Vec models. The evaluation focused on assessing the accuracy and effectiveness of the combined embedding model in identifying similar police reports and retrieving similar reports.

By combining the strengths and features of both models, it was expected that the USEW2V model could outperform the individual models in terms of accuracy and retrieval performance.

Table 3. Results from the approaches for RQ3

Model	Accuracy			
	Top 1	Top 5	Top 10	MRR
USE	**70.47**	**80.62**	**82.61**	**75.19**
Word2Vec - trained	65.22	73.73	75.36	69.36
USEW2V MEAN	69.93	80.25	82.07	75.05
USEW2V MAX	69.20	79.53	81.52	74.34
USEW2V MIN	68.66	78.99	81.16	74.32

Table 3 presents the results of the USEW2V models (max, min, mean) compared to the individual USE and Word2Vec models for the 1,065 police reports dataset. It can be observed that the combination of the two best embedding models, referred to as USEW2V models, yielded similar results to those obtained using only the USE model.

Since the 1,065 sentences in the dataset share a similar writing syntax, the improvement achieved by combining the two models to represent the sentences was not significant. However, it is worth noting that combining embeddings can sometimes yield results that are on par with or even better than the individual models.

This experiment demonstrates that combining different embedding models can be a viable approach, and there are various methods available for combining embeddings. Exploring these options and evaluating their effectiveness can be a valuable direction for future research in the field.

5   Conclusion and Future Works

This paper addresses the problem of finding the most similar police report in a database given a specific report. The goal is to consider not only lexical similarity but also syntactic and semantic similarity. The experiments conducted in this paper showed that the USE model, a pre-trained transformer-based sentence embedding model in its multilingual version, outperformed the other embedding models. Interestingly, even the raw comparison methods using TF-IDF and Jaccard similarity achieved similar results to the USE model.

Furthermore, the study found that training an embedding model with data from the same context as the target dataset is more effective than using pre-trained embeddings. The paper also explored the combination of different embeddings using various aggregation functions such as max, min, and average. However, it was observed that while combining embeddings can be competitive with using individual models, it might not surpass their performance as expected.

Future directions for research include evaluating the proposed approaches with different datasets, exploring embeddings from different architectures like graph embeddings, and investigating metrics to compare embeddings and assess their suitability for specific datasets and machine learning problems. These directions can contribute to further advancements in the field of similarity search and text representation. Another potential future direction is to explore the utilization of our dataset with 30,011 police reports for fine-tuning the models studied in the first research question. Another suggestion about combining embeddings would be to concatenate two embeddings generated by different models and assess whether it is possible to obtain a better representation.

Acknowledgments. The research reported in this work received support from the FUNCAP project titled "Big Data Platform to Accelerate the Digital Transformation of Ceará State" 04772551/2020. Part of the results presented in this work were also obtained through the UFC-FASTEF 31/2019.

References

1. Akbik, A., Bergmann, T., Blythe, D., Rasul, K., Schweter, S., Vollgraf, R.: FLAIR: an easy-to-use framework for state-of-the-art NLP. In: Proceedings of NAACL (Demonstrations), pp. 54–59 (2019)
2. Boggust, A., Carter, B., Satyanarayan, A.: Embedding comparator: visualizing differences in global structure and local neighborhoods via small multiples. In: 27th International Conference on Intelligent User Interfaces, pp. 746–766 (2022)
3. Cer, D., et al.: Universal sentence encoder. arXiv preprint arXiv:1803.11175 (2018)
4. Chollet, F.: Deep Learning with Python. Simon and Schuster (2021)
5. Conneau, A., Kiela, D., Schwenk, H., Barrault, L., Bordes, A.: Supervised learning of universal sentence representations from natural language inference data. In: Proceedings of the 2017 EMNLP, pp. 670–680 (2017)
6. Do, P., Pham, P.: W-KG2Vec: a weighted text-enhanced meta-path-based knowledge graph embedding for similarity search. Neural Comput. Appl. 33(23), 16533–16555 (2021)
7. Feng, F., Yang, Y., Cer, D., Arivazhagan, N., Wang, W.: Language-agnostic BERT sentence embedding. arXiv preprint arXiv:2007.01852 (2020)
8. Ghannay, S., Favre, B., Esteve, Y., Camelin, N.: Word embedding evaluation and combination. In: Proceedings of the Tenth International Conference on Language Resources and Evaluation (LREC 2016), pp. 300–305 (2016)
9. Harris, Z.S.: Distributional structure. Word 10(2–3), 146–162 (1954)
10. Hartmann, N., Fonseca, E., Shulby, C., Treviso, M., Silva, J., Aluísio, S.: Portuguese word embeddings: evaluating on word analogies and natural language tasks. In: Proceedings of the 11th STIL, pp. 122–131 (2017)
11. Hjaltason, G.R., Samet, H.: Properties of embedding methods for similarity searching in metric spaces. IEEE Trans. Pattern Anal. Mach. Intell. 25(5), 530–549 (2003)
12. Imtiaz, Z., Umer, M., Ahmad, M., Ullah, S., Choi, G.S., Mehmood, A.: Duplicate questions pair detection using siamese MaLSTM. IEEE Access 8, 21932–21942 (2020)
13. Kenton, J.D.M.W.C., Toutanova, L.K.: BERT: pre-training of deep bidirectional transformers for language understanding. In: Proceedings of NAACL-HLT, pp. 4171–4186 (2019)
14. Kiros, R., et al.: Skip-thought vectors. In: Advances in Neural Information Processing Systems (NIPS), pp. 3294–3302 (2015)
15. Le, Q., Mikolov, T.: Distributed representations of sentences and documents. In: Proceedings of the 31st International Conference on Machine Learning (ICML), pp. 1188–1196. PMLR (2014)
16. Liu, Y., et al.: RoBERTa: A robustly optimized BERT pretraining approach. arXiv preprint arXiv:1907.11692 (2019)
17. Mikolov, T., Sutskever, I., Chen, K., Corrado, G.S., Dean, J.: Distributed representations of words and phrases and their compositionality. In: Advances in Neural Information Processing Systems (NIPS), pp. 3111–3119 (2013)
18. Naili, M., Chaibi, A.H., Ghezala, H.H.B.: Comparative study of word embedding methods in topic segmentation. Procedia Comput. Sci. 112, 340–349 (2017)
19. Pennington, J., Socher, R., Manning, C.D.: GloVe: global vectors for word representation. In: Proceedings of the 2014 EMNLP, pp. 1532–1543 (2014)
20. Qiu, Y., Li, H., Li, S., Jiang, Y., Hu, R., Yang, L.: Revisiting correlations between intrinsic and extrinsic evaluations of word embeddings. In: Sun, M., Liu, T., Wang, X., Liu, Z., Liu, Y. (eds.) CCL/NLP-NABD -2018. LNCS (LNAI), vol. 11221, pp. 209–221. Springer, Cham (2018). https://doi.org/10.1007/978-3-030-01716-3_18

21. Reimers, N., Gurevych, I.: Sentence-BERT: Sentence embeddings using siamese BERT-networks. arXiv preprint arXiv:1908.10084 (2019)
22. Rodrigues, A.C., Marcacini, R.M.: Sentence similarity recognition in Portuguese from multiple embedding models. In: 2022 21st IEEE International Conference on Machine Learning and Applications (ICMLA), pp. 154–159. IEEE (2022)
23. Shahmirzadi, O., Lugowski, A., Younge, K.: Text similarity in vector space models: a comparative study. In: 2019 18th IEEE ICMLA, pp. 659–666. IEEE (2019)
24. Souza, F., Nogueira, R., Lotufo, R.: BERTimbau: pretrained BERT models for Brazilian Portuguese. In: 9th BRACIS (2020)
25. Toshevska, M., Stojanovska, F., Kalajdjieski, J.: Comparative analysis of word embeddings for capturing word similarities. arXiv preprint arXiv:2005.03812 (2020)
26. Zhai, M., Tan, J., Choi, J.: Intrinsic and extrinsic evaluations of word embeddings. In: Proceedings of the AAAI Conference on Artificial Intelligence, vol. 30 (2016)

Evaluating Contextualized Embeddings for Topic Modeling in Public Bidding Domain

Henrique R. Hott[1], Mariana O. Silva[1], Gabriel P. Oliveira[1],
Michele A. Brandão[1,2], Anisio Lacerda[1], and Gisele Pappa[1(✉)]

[1] Universidade Federal de Minas Gerais (UFMG), Belo Horizonte, MG, Brazil
{henriquehott,mariana.santos,gabrielpoliveira,anisio,glpappa}@dcc.ufmg.br
[2] Instituto Federal de Minas Gerais (IFMG), Ribeirão das Neves, MG, Brazil
michele.brandao@ifmg.edu.br

Abstract. Public procurement plays a crucial role in government operations by acquiring goods and services through competitive bidding processes. However, the increasing volume of procurement data has made manual analysis impractical and time-consuming. Therefore, text clustering and topic modeling techniques have been widely used to uncover hidden patterns in unstructured text data. This paper leverages the power of BERT-based models to overcome the challenges associated with analyzing public procurement data. Specifically, we employ BERTopic, a topic modeling technique based on BERT, to generate clusters that capture the underlying topics in procurement data. Additionally, we evaluate several sentence embedding models for representing procurement documents. By combining BERT-based models and advanced sentence embeddings, we aim to enhance the accuracy and interpretability of topic modeling in public procurement analysis. Our results provide valuable insights into the underlying topics within the data, aiding decision-making processes and improving the efficiency of procurement operations.

Keywords: public bids · topic modeling · contextualized embeddings

1 Introduction

Public procurement is a vital aspect of government and corporate operations, involving acquiring goods and services through competitive bidding processes. With the increasing volume of procurement data, manual analysis becomes an impractical and time-consuming task. Therefore, applying machine learning approaches, such as text clustering and topic modeling algorithms, has become increasingly essential to automate procurement analysis.

Text clustering and topic modeling techniques have proven effective in identifying hidden structures and patterns in unstructured text data [6,17]. Such methods aim to identify and extract hidden structures and patterns from a large corpus of unstructured text data. Topic modeling, in particular, has been widely

M. C. Naldi and R. A. C. Bianchi (Eds.): BRACIS 2023, LNAI 14197, pp. 410–426, 2023.
https://doi.org/10.1007/978-3-031-45392-2_27

used to identify prevalent topics in a corpus of documents. By identifying these topics, analysts can quickly identify common themes and trends in the data, which can inform decision-making processes.

However, traditional topic modeling algorithms such as Latent Dirichlet Allocation (LDA) [1] may not always produce accurate and interpretable results, especially when dealing with noisy and unstructured text data. In the context of public procurement data, characterized by its high technicality and specific jargon, processing and analyzing the data using traditional techniques can be challenging. Therefore, there is a need for advanced methods that can effectively capture and represent the underlying topics in such complex data.

In this paper, we leverage the power of BERT-based models to address the challenges associated with analyzing public procurement data. Specifically, we use BERTopic [10], a topic modeling framework based on BERT pre-trained model [5], to generate clusters that capture the underlying topics in procurement data. Additionally, we evaluate several sentence embedding models to represent procurement documents. Our main contributions are based on the following research questions (**RQs**):

RQ1. *What is the comparative performance of different sentence embedding models in representing procurement documents?* We conduct an internal evaluation based on topic coherence, topic diversity, and a weighted score (that combines both factors) to compare the performance of four different sentence embedding models and an additional model we trained, called LiBERT-SE. We aim to identify the most effective model for representing procurement documents' complex and specialized language by addressing this research question.

RQ2. *How well does BERTopic capture the underlying topics in procurement data?* We conduct an external evaluation to compare the BERTopic results with the true document class labels, assessing how well the topics identified in each cluster align with the document labels.

2   Related Work

Like many other government documents, public procurement documents are frequently stored in Portable Document Format (PDF) and consist of unstructured text [17]. The lack of attention to accessibility, usability, and data quality by governmental portals poses challenges in effectively accessing and extracting valuable information from these documents [13]. Consequently, there is a growing need to employ machine learning techniques, specifically text clustering and topic modeling algorithms, to automate the analysis of procurement data.

Text clustering techniques aim to group similar documents together based on their content, enabling the efficient organization and exploration of large document collections [7]. These methods employ various algorithms, such as k-means, hierarchical clustering, and density-based clustering, to identify patterns and similarities in the text data. By grouping related procurement documents, these techniques may facilitate the discovery of common themes, topics, and trends within the data.

On the other hand, topic modeling algorithms provide a means to uncover latent topics or thematic structures within a collection of documents. These algorithms, such as Latent Dirichlet Allocation (LDA), Probabilistic Latent Semantic Analysis (PLSA), and, more recently, BERTopic [10], assign topics to documents based on the distribution of words within the corpus. By extracting meaningful topics, topic modeling enables a deeper understanding of the content and allows for more targeted analysis and decision-making.

Previous research has explored the application of text clustering and topic modeling algorithms to unstructured text data [6,17]. For instance, Souza Jr. at al. [19] evaluate different pre-processing methodologies in topic modeling for Brazilian Portuguese political discussion extracted from Twitter and Reddit. In particular, the authors applied three document representation models, including two new proposals based on the *CluWords* model adapted to Portuguese. Also, Silva et al. [16] use topic models for analyzing and visualizing Brazilian comments about legislation. Specifically, the authors adapted the BERTopic topic mining tool to extract topics from political comments.

Furthermore, advancements in deep learning and language modeling have provided new opportunities to enhance the representation of text data [12]. Models like BERT (Bidirectional Encoder Representations from Transformers) [5] and its variants have shown remarkable performance in various NLP tasks, including document classification and named entity recognition. These models can capture intricate semantic relationships between words and provide context-aware embeddings, thereby improving the quality of text representations.

In this paper, we build upon these developments and leverage BERT-based models to address the challenges of analyzing public procurement data. By applying BERTopic, a topic modeling algorithm based on BERT, we aim to generate clusters that capture the underlying topics and themes within procurement documents. While previous studies have focused on topic modeling or document clustering alone, we extend the analysis by considering the performance of different sentence embedding models in capturing the semantics and contextual information of procurement documents.

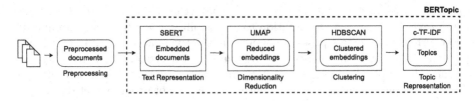

Fig. 1. Overview of the topic modeling methodology.

3   Methodology

The methodology followed in this work is summarized in Fig. 1 and consists of two main phases: text preprocessing and topic modeling. First, a set of unstructured documents undergoes a sequence of preprocessing steps to transform the text into a more structured format (Sect. 3.1). Then, we apply BERTopic, a topic modeling technique, to create dense clusters and uncover easily interpretable topics in our preprocessed documents (Sect. 3.2).

3.1   Preprocessing

Preprocessing plays a fundamental role in NLP tasks, as it generates a more structured representation of the text and reduces the final set of words (i.e., the vocabulary) that will be used as input for learning models. Indeed, applying a preprocessing step in a learning pipeline directly impacts the models' accuracy by reducing noise and improving the input quality. Therefore, in this work, we apply a sequence of three distinct preprocessing steps for each document: *text normalization*, *stop word removal*, and *numeral normalization*.

Considering our focus on Brazilian governmental documents, our preprocessing phase is specifically designed to handle texts in the Portuguese language. Hence, in this sequence, *text normalization* removes special characters, accents, city names, person names, and words with a sequence of repeated letters and lowercasing. City and person names are identified and eliminated based on an exact match against dictionaries containing over 5,000 Brazilian cities and 7,000 common Brazilian names. Next, *stop word removal* involves removing stop words, emails, URLs, and unit measures. Finally, *numeral normalization* means identifying and removing hours, numbers, and number symbols from the text.

3.2   Topic Modeling

The second phase of our methodology involves identifying the underlying topics within a set of documents. Such an approach is especially beneficial in documents that do not present clear structural patterns. In this work, we use BERTopic [10], an efficient topic model that leverages clustering techniques and a class-based variation of TF-IDF to create consistent topic representations.

Overall, BERTopic consists of four steps: text representation, dimensionality reduction, document clustering, and topic representation. First, a language model converts each preprocessed document to an embedding representation. Then, we reduce the dimensionality of these embeddings to avoid problems in the clustering process. The last step from these clusters is extracting topic representations, which is our primary goal. All such steps are detailed next.

Text Representation. Mapping sentences into numeric vector spaces is an efficient method to generate richer text representations in NLP tasks [20]. Such representations preserve semantic and syntactic information in sentences, leading to better performance in learning models that rely on the vector representation of the text as input. Here, we use the Sentence-BERT (SBERT) framework [14] to convert sentences and paragraphs into dense vector representations using pre-trained language models. One of the strengths of SBERT is its scalability since it can be executed over large volumes of data, allowing topic analysis in relevant domains that comprises several long documents, such as governmental processes.

Dimensionality Reduction. Reducing the dimensionality of the embeddings is an essential step in preparing the data for clustering. High-dimensional data is difficult for clustering algorithms, as they suffer from the curse of dimensionality. When the data is highly dimensional, the concept of spatial location becomes ill-defined, and distance measures differ little. Therefore, reducing the dimensionality of the embeddings is crucial to make the clustering process more efficient. UMAP (Uniform Manifold Approximation And Projection for Dimension Reduction) [11] is a popular dimensionality reduction method that stands out for preserving more local and global characteristics of high-dimensional data in lower projected dimensions. Using UMAP allows for representing the information more condensed using fewer dimensions, reducing the noise caused by highly correlated variables. As a result, this step increases the efficiency of the clustering algorithm since it will have a better capacity to group documents.

Clustering. After reducing the embeddings (i.e., document representation), the next step is to cluster them semantically. To do so, we use HDBSCAN [3], a density-based clustering algorithm, due to its superior performance compared to other density-based clustering algorithms, such as DBSCAN, in terms of both accuracy and efficiency. HDBSCAN can handle clusters of different shapes and sizes, making it ideal for identifying topics within a set of documents. Moreover, it can identify instances that do not belong to any cluster (i.e., outliers), reducing noise in the final topic representation. As a result, the detected topics are more coherent and representative of the underlying structure of the data.

Topic Representation. After generating the document clusters, we apply a traditional TF-IDF topic detection method variation to each cluster, called *c-TF-IDF* [10]. Such an adaptation considers that documents in a cluster share similar content and treats all documents in a cluster as a single document. Therefore, one can calculate the terms' importance and generate the topics' distribution in each cluster. The more important words are within a cluster, the more it represents that topic. In other words, if one extracts the most important words per cluster, the result is the topics' descriptions.

4   Experimental Design

This section describes the experimental design to evaluate the effectiveness of BERTopic and sentence embedding models. We first describe LiPSet, the dataset used in our experiments (Sect. 4.1). Then, we present the sentence embedding models used to represent procurement documents (Sect. 4.2). Finally, we detail the evaluation setup and metrics used in this work (Sect. 4.3).

4.1 Dataset

We consider LiPSet [15], a dataset with labeled public bidding documents from the Brazilian state of Minas Gerais, to evaluate the models. These documents are characterized by their technical language and specific jargon, making them an ideal candidate for evaluating the performance of the proposed clustering approach. The LiPSet comprises 9083 labeled documents, classified into four metaclasses: Minutes, Public Notice, Homologation, and Others (includes files belonging to other types of documents, including erratum, annexes, contracts, and descriptive memorials).

In the latest version of LiPSet, the documents were also assigned to classes using a hierarchical approach that consider their metaclasses. In other words, the documents were first labeled to a metaclass and then to one of the classes that belong to this metaclass. For example, minutes can be divided into price registration, waiver, and face-to-face auction. Overall, the four metaclasses are divided into 12 classes. The list of classes for all metaclasses and the number of documents for each one are presented in Table 1.

4.2 Sentence Embedding Models

In addition to evaluating the effectiveness of BERTopic, we also assess the performance of several sentence embedding models for representing procurement documents. Specifically, we consider four models as baselines: Multilingual Universal

Table 1. List of classes for each metaclass within LiPSet.

Metaclass	Class	Documents
Minutes	price registration	2200
	minutes of waiver	181
	face-to-face auction	160
	others	145
Public Notice	public notice	3589
Homologation	homologation	408
Others	others	1114
	contract	451
	notice	338
	amendment	211
	ratification	176
	erratum	110

Table 2. Transformer models used as a baseline in this work.

Model	Description
USE [21]	It is designed to generate universal sentence representations that capture the semantic meaning of text across multiple languages. The model leverages a deep neural network architecture with transformer layers to encode sentences into fixed-length dense vectors.
LaBSE [8]	It is a multilingual sentence embedding model that supports 109 languages, aiming to overcome the limitations of language-specific models. It is based on the popular BERT (Bidirectional Encoder Representations from Transformers) architecture. LaBSE incorporates language-agnostic training objectives to generate high-quality sentence representations that capture cross-lingual semantic information.
S-BERTimbau [14,18]	It is a Brazilian Portuguese sentence embedding model developed specifically for the Portuguese language. It is based on the weighting strategy proposed by the Sentence-BERT (SBERT) framework [14] to create a model based on BERTimbau [18].
LegalBERTPTbr [16]	It is a specialized sentence embedding model tailored for legal documents of Brazilian political comments. It is trained using SimCSE [9] coupled with BERTimbau. The data was extracted on two constitutional amendment projects from the official platforms

Sentence Encoder (USE) [21], Language-agnostic BERT Sentence Embedding [8] (LaBSE), S-BERTimbau [14,18], and Portuguese Legislative Sentence Embedding (LegalBERTPTbr) [16]. Table 2 summarizes these four language models. Additionally, we train a custom model called LiBERT-SE and explore a fine-tuned version using the SimCSE framework [9], described as follows.

LiBERT-SE. [1] Our model is adjusted in the Masked-Language Modeling (MLM) task for the domain of public procurements, using the BERTimbau [18] model checkpoint. Following the best practices of domain adaptation outlined in [5], we fine-tuned our model using a dataset of 300,000 official gazette segments sourced from various municipalities in Minas Gerais, as previously extracted in [4]. These segments encompass articles published in official gazettes, providing brief information about other documents related to the bidding process. To enhance the model's vocabulary, we employed a TF-IDF procedure, which allowed us to extract the most significant terms from the gazette segments. Additionally, we manually incorporated domain-specific jargon commonly found in public procurement contexts. To ensure compatibility with the BERTimbau

---

[1] https://huggingface.co/dccmpmgfinalisticas/LiBERT-SE.

model, we divided the segments into blocks of fixed sizes, ensuring that each block adhered to the maximum sentence length of 512 words supported by BERTimbau. For training, we followed the same hyperparameter settings employed by BERTimbau, except for the batch size, which we reduced to 16 due to computational resource limitations. By adhering to these settings, we aimed to maintain consistency and leverage the existing knowledge and insights gained from the BERTimbau model while tailoring it to the specific domain of public procurements.

LiBERT-SE + SimCSE. In addition to the initial fine-tuning step using the Masked-Language Modeling (MLM) task, we further optimized our model by performing a new fine-tuning step specifically tailored for the task of generating contextualized sentence embeddings. For this purpose, we use the SimCSE framework, which is a self-supervised sentence embedding model designed to learn rich sentence representations through contrastive learning. The SimCSE framework utilizes augmented positive and negative pairs to train the model. Augmented versions of the same sentence are considered positive pairs, and the model is trained to maximize their similarity. On the other hand, augmented versions of different sentences serve as negative pairs, and the model aims to minimize their similarity. By employing this contrastive learning approach, SimCSE effectively learns robust and discriminative sentence representations that capture essential semantic information. To fine-tune our model using the SimCSE framework, we used a dataset consisting of approximately 300,000 segments of official diaries, extracted and published in a previous work [4]. These segments were distinct from those used in the previous fine-tuning procedure for the MLM task.

4.3   Evaluation

Evaluating clustering and topic modeling results can be challenging due to their unsupervised nature. Popular approaches for evaluation involve either "internal" evaluation, where the results are summarized into a single quality score, or "external" evaluation, where the results are compared to an existing "ground truth" [7]. Here, we adopt both internal and external evaluation methods to assess the performance of the proposed approach.

Internal Evaluation. We used two primary metrics to evaluate the performance of models in the internal experiments. First, we use the topic coherence based on normalized pointwise mutual information (NPMI) [2], which ranges from -1 to 1, where a higher value indicates stronger coherence between words in a topic. We also evaluate the topic diversity, which measures the percentage of unique words across all topics [6]. The topic diversity score ranges from 0 to 1, where 0 indicates redundant topics and 1 indicates more varied topics.

   Additionally, to make a comprehensive evaluation, we combine the coherence and diversity scores by assigning weights to the normalized coherence (NC) and diversity (ND) based on their relative importance to our specific task. The weighted evaluation score is defined as the sum of the weighted coherence (WC_s) and weighted diversity (WD_s), where $WC_s = 0.8 \times NC$ and $WD_s = 0.2 \times ND$. By considering both coherence and diversity, we can better assess the overall performance of the topic modeling approach and ensure that the generated topics are not only coherent but also diverse.

External Evaluation. In external evaluation, results are evaluated based on data not used in the topic modeling, such as known class labels. To ensure optimal performance, we fine-tuned the hyperparameters of BERTopic using the language model that achieved the best results in the internal evaluation. During the evaluation, we manually compared the topics generated by BERTopic with the known class labels, which serve as ground truth for the evaluation. This comparison allowed us to gauge the accuracy and relevance of the generated topics in capturing the underlying themes and categories present in the procurement data. By conducting this manual comparison, we aimed to determine the extent to which the generated topics align with the known class labels, thereby providing insights into the effectiveness of BERTopic in capturing and representing the true structure and content of the procurement documents.

Experimental Setup. The **internal** evaluation metrics were calculated by averaging the results over 10 iterations, where we varied the number of topics from 10 to 50 with a step size of 10. This allowed us to assess the performance of BERTopic across a range of topic numbers and identify the optimal number of topics for our analysis. In the **external** evaluation, we fine-tuned two specific hyperparameters of BERTopic using the best-performing language model: *nr_topics_list*, which includes a list of potential numbers of topics,[2] and *min_topic_sizes*,[3] which represents the minimum size threshold for each topic. By fine-tuning these hyperparameters, we aimed to further optimize the performance of BERTopic in generating meaningful and coherent topics. Additionally, for the *n_gram_ranges* parameter, we set it to (1, 1), indicating that we considered only individual words (unigrams) during the topic modeling process. This choice focused on capturing the most important and distinct keywords within each topic without considering combinations of multiple words. The experiments were performed using 3 GPU models NVIDIA GeForce RTX 3090 Ti, NVIDIA GeForce RTX 4090, and NVIDIA Tesla GRID V100D-32C.

5   Experimental Results

In this section, we present the results of both internal and external evaluations, which aim to address the two research questions defined in Sect. 1. The internal evaluation focuses on assessing the performance and effectiveness of different sentence embedding models in representing procurement documents (Sect. 5.1),

---

[2] (10, 13, 14, 15, 16, 17, 19, 20, auto).
[3] (10, 20, 30, 40, 50, 60, 70, 80, 90, 100).

Table 3. Comparison of sentence embedding models based on internal evaluation metrics. The best result for each metric is underlined.

	Topic Coherence	Topic Diversity	Weighted Score
USE	0.073 ± 0.006	0.814 ± 0.008	0.414 ± 0.022
LaBSE	0.081 ± 0.007	0.833 ± 0.008	0.457 ± 0.026
S-BERTimbau	0.108 ± 0.005	0.815 ± 0.007	0.537 ± 0.021
LegalBERTPTbr	0.034 ± 0.008	0.737 ± 0.012	0.229 ± 0.034
LiBERT-SE	$\underline{0.139 \pm 0.008}$	0.843 ± 0.010	$\underline{0.664 \pm 0.032}$
LiBERT-SE + SimCSE	0.036 ± 0.006	0.842 ± 0.011	0.307 ± 0.024

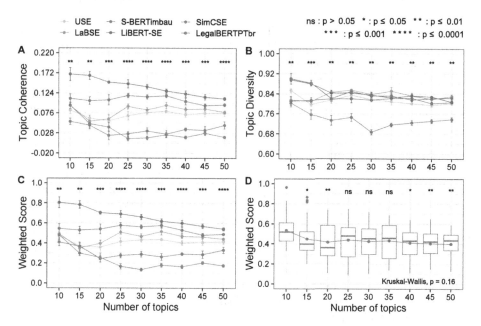

Fig. 2. Internal evaluation. (**A–C**) Distribution of internal evaluation metrics, varying across the different models and number of topics. The Kruskal-Wallis test is applied for the models' mean comparison. (**D**) Grouped distribution of the weighted score. The Paired Wilcoxon test is applied for grouped distributions' mean comparison, referring to the number of topics = 10.

while the external evaluation examines the quality of the generated clusters using BERTopic (Sect. 5.2).

5.1   Internal Evaluation

The internal evaluation results are presented in Table 3 and Fig. 2, providing a comparative analysis of the performance of the evaluated models. Table 3 summarizes the evaluation metrics, including topic coherence, diversity, and the weighted score. The results indicate that our trained model, LiBERT-SE, consistently outperforms the other five models across most internal evaluation metrics. Our model performs similarly to the fine-tuned SimCSE version and LaBSE regarding topic diversity, indicating that it can generate diverse and distinct topics. These findings underscore the efficacy of LiBERT-SE in representing procurement documents and capturing the underlying topics accurately.

Figure 2 complements the results' evaluation, depicting the distribution of topic coherence (**A**), topic diversity (**B**), weighted score (**C**), varying across the different models and number of topics. Furthermore, Fig. 2(**D**) shows the grouped distribution of the weighted score, providing a concise overview of the performance of each model across the number of topics range. Kruskal-Wallis and Paired Wilcoxon tests were conducted to determine the statistical significance of the observed differences. These tests allow us to verify whether there are significant differences between the models considered, providing additional insights into their relative performance.

Figures 2(**A**–**C**) corroborate the findings presented in Table 3, further reinforcing the superior performance of our model (LiBERT-SE) compared to the other models across most evaluated metrics. Our model consistently achieved higher scores in terms of coherence across the different numbers of topics. However, regarding topic diversity, the performance of our model was comparable to the SimCSE and LaBSE models. Therefore, while our model excelled in topic coherence, it also maintained a high level of topic diversity, which is crucial for generating meaningful and varied topics. This indicates that our model balances coherence and diversity, producing coherent topics while avoiding redundancy.

Figure 2(**D**) provides a comprehensive overview of the models' performance by showing the grouped distribution of the weighted score across the number of topics range. On average, considering ten topics resulted in a high general performance. However, as the number of topics increased, the performance varied across the models and slightly decreased. This finding suggests that the number of topics can impact the performance of the models. While a lower number of topics tends to yield higher overall performance, increasing the number of topics introduces more granularity but may result in a slight decrease in performance.

To conclude the internal evaluation, Fig. 3 provides further analysis explicitly focusing on (**A**) topic coherence, (**B**) topic diversity, and (**C**) the weighted score for each model when considering ten topics. By examining these metrics individually, we can assess the strengths and weaknesses of each model in capturing the underlying themes and generating diverse topic representations. Consistent with the previous findings, our model outperforms the other five evaluated models, and it consistently achieves higher topic coherence, indicating a more substantial alignment of words within each topic. Additionally, our model exhibits competitive performance regarding topic diversity, suggesting that it can generate varied and distinct topics.

Considering only the weighted score, which combines coherence and diversity, the models that perform well after our model are S-BERTimbau and LabSE. S-BERTimbau's high performance can be attributed to its design specifically

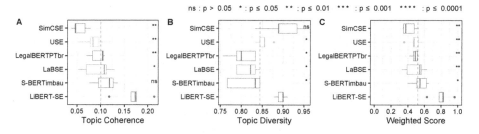

Fig. 3. Comparative of (**A**) topic coherence, (**B**) topic diversity, and (**C**) the weighted score for each model when considering ten topics. The vertical dashed line represents the median value for each internal evaluation metric. The Paired Wilcoxon test is applied for the models' mean comparison, referring to the LiBERT-SE model.

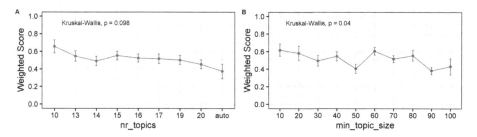

Fig. 4. Performance comparison based on the weighted score across different variations of *nr_topics_list* and *min_topic_sizes* hyperparameters.

for the Portuguese language. As public procurement data often contain technical terms, S-BERTimbau's domain-specific embeddings enable it to capture the nuances and intricacies of the text more effectively. LaBSE, on the other hand, is a language-agnostic model that leverages large-scale multilingual training data. Its competitive performance can be explained by its ability to handle various languages, including Portuguese. Finally, our model's superior performance can be attributed to its ability to leverage pre-training on large-scale datasets and capture the contextual understanding of the text, thereby enhancing the quality of the generated topic clusters.

5.2 External Evaluation

To perform the external evaluation, we fine-tuned two specific hyperparameters of BERTopic using the best-performing language model, i.e., LiBERT-SE. These hyperparameters are *nr_topics_list* and *min_topic_sizes*. Figure 4 shows the performance based on the weighted score across different variations of both hyperparameters. To determine the statistically significant combination of *nr_topics_list* and *min_topic_sizes*, we applied the Kruskal-Wallis test, allowing us to identify the combination that achieved the best performance: 10 and 60, respectively.

Fig. 5. Word clouds of the ten topics identified by BERTopic.

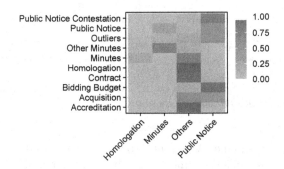

Fig. 6. Summary of the topic-metaclass associations.

Figure 5 shows the word clouds of the ten topics identified by BERTopic, using the optimal hyperparameters. Each word cloud visually represents the most significant keywords associated with a specific topic. The first topic (-1) represents the outlier documents that do not align with any particular theme or topic. These outlier documents may contain diverse or unrelated content that does not fit into the predefined categories.

The remaining topics capture distinct themes and content within the procurement documents. These topics range from specific domains or subfields within procurement, such as construction projects, supply chain management, legal regulations, or financial aspects. To compare the topics generated with the ground truth (i.e., metaclasses and class labels), we named each topic based on its top words. Table 4 presents the mapping between the topic ID and its given name, along with the top words and their frequency within the topic.

Table 4. Mapping between the topic ID and its given name.

ID	Topic Name	Top Words	Freq
−1	Outliers	licitacao, edital, ativo, efetivos, propostas, precos, publico, pregao	1629
0	Public Notice	centro, licitacao, rua, pregao, presencial, edital, situada, local	4979
1	Minutes	ata, precos, administracao, total, lei, fornecedor, federal, prazo	2991
2	Bidding Budget	bdi, sinapi, concreto, thome, total, letras, sao, financeiro	252
3	Contract	contrato, termo, clausula, aditivo, inscrito, doravante, sob	225
4	Homologation	prefeito, homologacao, moreti, aviso, filho, resultado, vencedoras	124
5	Accreditation	credenciamento, edital, publico, presente, secretaria, praca	107
6	Pub. Not. Contest	impugnacao, empresa, edital, impugnante, processo, nao, razoes	101
7	Acquisition	familiar, alimenticios, agricultura, generos, resolucao, fnde	75
8	Other Minutes	mg, br, ml, prati, cloridrato, ata, oral, donaduzzi	70

Using the mapped names in Table 4, we generate a heatmap to analyze each document's occurrence between topics and metaclass labels (Fig. 6). This heatmap provides a visual representation of the associations between the identified topics and the metaclass labels assigned to the procurement documents. The intensity of the colors in the heatmap reflects the frequency or occurrence of the topic-metaclass pair within the documents. Darker colors indicate a higher frequency, while lighter colors represent a lower frequency or absence of the topic-metaclass pair. Overall, the topic-metaclass associations do not perfectly align in the heatmap. While some topics show strong associations with specific metaclasses, indicating a clear thematic correspondence, others exhibit a more scattered or diverse distribution across different metaclasses.

This misalignment can be attributed to several factors. Firstly, the BERTopic is based on the underlying patterns and co-occurrences of words within the documents, which may not always align precisely with the metaclass labels assigned to the documents. Secondly, the complexity and diversity of the procurement domain can contribute to the variation and dispersion of topics across different metaclasses. Consequently, topics may overlap or span multiple metaclasses, leading to a less direct one-to-one correspondence between topics and metaclasses. Additionally, the quality and representativeness of the training data can influence the topic modeling results. If the training data does not fully cover the diversity of procurement documents or lacks sufficient labeled examples for each metaclass, it may affect the accuracy of the topic-metaclass associations.

6   Conclusion

In this paper, we presented a comprehensive methodology for topic modeling in Brazilian procurement documents using the BERTopic framework. Our approach involved preprocessing steps to enhance the quality of the text and utilized HDBSCAN for clustering the documents. The resulting clusters were then analyzed using c-TF-IDF to extract meaningful topic representations. Through evaluating several sentence embedding models, we found that our trained model, LiBERT-SE, consistently outperformed the other models in terms of topic coherence. However, regarding the efficiency of BERTopic in capturing the underlying topics in the purchase data, there is still room for improvement in aligning the topics with the predefined class labels. This finding highlights the inherent challenge of topic modeling in public procurement data, which is characterized by its complexity and diversity.

Limitations and Future Work. While our methodology provides valuable findings, it is important to acknowledge some limitations: The preprocessing steps and embedding models used in this study were specifically tailored for the Portuguese language. Applying the methodology to other languages may require adaptations and modifications. The focus of this work was on Brazilian governmental documents. The methodology's effectiveness and the models' performance may vary when applied to different domains or document types. The evaluation and comparison of sentence embedding models were conducted on a specific dataset and task. As future work, we aim to conduct experiments on diverse datasets across various languages and domains to validate the generalizability and robustness of the methodology, models, and evaluation metrics. Additionally, we plan to broaden our evaluation by testing a wider range of sentence embedding models, incorporating state-of-the-art models.

Acknowledgments.. This work was funded by the Prosecution Service of the State of Minas Gerais (*Ministério Público do Estado de Minas Gerais*) through the Analytical Capabilities Project (*Programa de Capacidades Analíticas*) and by CNPq, CAPES, and FAPEMIG.

References

1. Blei, D.M., Ng, A.Y., Jordan, M.I.: Latent dirichlet allocation. J. Mach. Learn. Res. **3**, 993–1022 (2003)
2. Bouma, G.: Normalized (pointwise) mutual information in collocation extraction. Proc. GSCL **30**, 31–40 (2009)

3. Campello, R.J.G.B., Moulavi, D., Sander, J.: Density-based clustering based on hierarchical density estimates. In: Pei, J., Tseng, V.S., Cao, L., Motoda, H., Xu, G. (eds.) PAKDD 2013. LNCS (LNAI), vol. 7819, pp. 160–172. Springer, Heidelberg (2013). https://doi.org/10.1007/978-3-642-37456-2_14

4. Constantino, K., et al.: Segmentação e classificação semântica de trechos de diários oficiais usando aprendizado ativo. In: SBBD, pp. 304–316. SBC (2022). https://doi.org/10.5753/sbbd.2022.224656

5. Devlin, J., et al.: BERT: pre-training of deep bidirectional transformers for language understanding. In: NAACL-HLT, pp. 4171–4186. Association for Computational Linguistics (2019). https://doi.org/10.18653/v1/n19-1423

6. Dieng, A.B., Ruiz, F.J.R., Blei, D.M.: Topic modeling in embedding spaces. Trans. Assoc. Comput. Linguistics **8**, 439–453 (2020). https://doi.org/10.1162/tacl_a_00325

7. Feldman, R., Sanger, J.: The Text Mining Handbook - Advanced Approaches in Analyzing Unstructured Data. Cambridge University Press (2007)

8. Feng, F., et al.: Language-agnostic BERT sentence embedding. In: Proceedings of the 60th Annual Meeting of the Association for Computational Linguistics (ACL), pp. 878–891. Association for Computational Linguistics (2022). https://doi.org/10.18653/v1/2022.acl-long.62

9. Gao, T., Yao, X., Chen, D.: SimCSE: simple contrastive learning of sentence embeddings. In: Proceedings of the 2021 Conference on Empirical Methods in Natural Language Processing (EMNLP), pp. 6894–6910. Association for Computational Linguistics (2021). https://doi.org/10.18653/v1/2021.emnlp-main.552

10. Grootendorst, M.: BERTopic: Neural topic modeling with a class-based TF-IDF procedure. arXiv preprint arXiv:2203.05794 (2022)

11. McInnes, L., et al.: UMAP: uniform manifold approximation and projection. J. Open Source Softw. **3**(29), 861 (2018). https://doi.org/10.21105/joss.00861

12. Naseem, U., et al.: A comprehensive survey on word representation models: from classical to state-of-the-art word representation language models. ACM Trans. Asian Low Resour. Lang. Inf. Process. **20**(5), 74:1–74:35 (2021). https://doi.org/10.1145/3434237

13. Nikiforova, A., McBride, K.: Open government data portal usability: a user-centred usability analysis of 41 open government data portals. Telematics Inform. **58**, 101539 (2021). https://doi.org/10.1016/j.tele.2020.101539

14. Reimers, N., Gurevych, I.: Sentence-BERT: sentence Embeddings using Siamese BERT-Networks. In: EMNLP-IJCNLP, pp. 3980–3990. Association for Computational Linguistics (2019). https://doi.org/10.18653/v1/D19-1410

15. Silva, M., et al.: LiPSet: um conjunto de dados com documentos rotulados de licitações públicas. In: Anais do IV Dataset Showcase Workshop, pp. 13–24. SBC, Porto Alegre, RS, Brasil (2022). https://doi.org/10.5753/dsw.2022.224925

16. Silva, N.F.F., et al.: Evaluating topic models in Portuguese political comments about bills from Brazil's chamber of deputies. In: Britto, A., Valdivia Delgado, K. (eds.) BRACIS 2021. LNCS (LNAI), vol. 13074, pp. 104–120. Springer, Cham (2021). https://doi.org/10.1007/978-3-030-91699-2_8

17. Silveira, R., et al.: Topic modelling of legal documents via legal-BERT. CEUR Workshop Proc. **1613**, 0073 (2021)

18. Souza, F., Nogueira, R., Lotufo, R.: BERTimbau: pretrained BERT models for Brazilian Portuguese. In: Cerri, R., Prati, R.C. (eds.) BRACIS 2020. LNCS (LNAI), vol. 12319, pp. 403–417. Springer, Cham (2020). https://doi.org/10.1007/978-3-030-61377-8_28

19. Souza Júnior, A.P., et al.: Evaluating topic modeling pre-processing pipelines for Portuguese texts. In: WebMedia, pp. 191–201. ACM (2022)
20. Turian, J.P., Ratinov, L., Bengio, Y.: Word representations: a simple and general method for semi-supervised learning. In: Proceedings of the 48th Annual Meeting of the Association for Computational Linguistics (ACL), pp. 384–394. The Association for Computer Linguistics (2010)
21. Yang, Y., et al.: Multilingual universal sentence encoder for semantic retrieval. In: Proceedings of the 58th Annual Meeting of the Association for Computational Linguistics: System Demonstrations (ACL), pp. 87–94. Association for Computational Linguistics (2020). https://doi.org/10.18653/v1/2020.acl-demos.12

AI Applications

A Tool for Measuring Energy Consumption in Data Stream Mining

Eric Kenzo Taniguchi Onuki⬤, Andreia Malucelli⬤,
and Jean Paul Barddal$^{(\boxtimes)}$⬤

Programa de Pós-Graduação em Informática (PPGIa),
Pontifícia Universidade Católica do Paraná (PUCPR),
Curitiba, Paraná, Brazil
{eric.kenzo,malu,jean.barddal}@ppgia.pucpr.br

Abstract. Energy consumption reduction is an increasing trend in machine learning given its relevance in socio-ecological importance. Consequently, it is important to quantify how real-time learning algorithms tailored for data streams and edge computing behave in terms of accuracy, processing time, memory usage, and energy consumption. In this work, we bring forward a tool for measuring energy consumption in the Massive Online Analysis (MOA). First, we analyze the energy consumption rates obtained by our tool against a gold-standard hardware solution, thus showing the robustness of our approach. Next, we experimentally analyze classification algorithms under different validation protocols and concept drift and highlight how such classifiers behave under such conditions. Results show that our tools enable the identification of different classifiers' energy consumption. In particular, it allows a better understanding of how energy consumption rates vary in drifting and non-drifting scenarios. Finally, given the insights obtained during experimentation on existing classifiers, we make our tool publicly available to the scientific community so that energy consumption is also accounted for in developing and comparing data stream mining algorithms.

Keywords: Data stream mining · Energy consumption · Green computing

1 Introduction

Data is consistently being generated, stored, and processed. Several applications provide data generated in large amounts, frequency, and speed. In this paper, we focus on such scenarios the so-called data streams. Data streams are, per definition, potentially unbounded and non-stationary data sequences [10]. Consequently, performing data mining to extract useful insights and patterns from such data requires algorithms that are tailored to such scenarios.

In contrast to traditional machine learning, the data stream mining area has shown concerns with the trade-off between accuracy and computational resources, i.e., processing time and memory consumption [5]. Nonetheless, recent research has shown that we still need more significant steps toward sustainability. For instance, authors in [28] depict that the CO2 emissions of training neural

M. C. Naldi and R. A. C. Bianchi (Eds.): BRACIS 2023, LNAI 14197, pp. 429–440, 2023.
https://doi.org/10.1007/978-3-031-45392-2_28

networks are larger than that of a car in its lifetime. First, there is no clear relationship between processing time, memory consumption, and energy consumption. Even though these components are tied to one another, multi-threading, compiler, and other low-level computational architecture have been shown to impact the entire process, and, consequently, energy consumption cannot be directly estimated from such components [13,18].

Having both sustainability and the lack of generic tools for quantifying energy consumption in streaming scenarios as motivation, we bring forward a tool for researchers and practitioners to investigate how data stream mining algorithms behave under different streaming settings, e.g., with and without concept drifts, different validation schema, etc. In opposition to previous works [15,16], our tool is generic because it can be coupled with any classifier and data stream available in the Massive Online Analysis (MOA) framework [6], which is the off-the-shelf solution for implementing and testing streaming methods. We experimentally evaluate our tool against a hardware solution and assess the energy consumption of different classifiers under different streaming settings. Finally, the tool is made publicly available to the scientific community as a byproduct of our research.

This paper is divided as follows. Section 2 describes data stream mining and brings forward the main concepts in energy consumption. Section 3 discusses related works that lie at the intersection of energy consumption and data stream mining. Section 4 describes our tool for energy consumption measurements and describes how it has been combined with the Massive Online Analysis (MOA) [6] framework. Section 5 discusses the analysis conducted to validate our tool and assess different classifiers under different experimental conditions. Finally, Sect. 6 concludes this work and states envisioned future works.

2   Data Stream Mining and Energy Consumption

Data streams are potentially unbounded data sequences made available over time, which may be non-stationary. Consequently, storing an entire data stream is unfeasible since it is not entirely available at once, and it would not fit in memory [10]. As a result, researchers and practitioners have devoted efforts towards developing efficient algorithms to process and mine data that arrive sequentially over time. Therefore, data stream mining is understood as the investigation of patterns, anomalies, and correlations in streaming data. In particular, in this work, we focus on classification, the most popular task in data stream mining that conveys the prediction of a discrete output given a set of input variables. More formally, we denote a data stream S to provide instances $i^t = (\boldsymbol{x}^t, y^t)$ at timestamps denoted as t. We also denote classification as the task of learning a predictive model $f : \boldsymbol{x} \rightarrow y$, where y is a discrete label in Y. In practice, we expect predictions $\hat{y}$ to be accurate given the ground-truth y values.

One of the main challenges in data stream mining is concept drift [31], which regards changes in the data distribution that may render a classifier obsolete. Formally, a concept $C = \bigcup_{y_i \in Y} \{(P[y_i], P[\boldsymbol{x}|y_i])\}$ is a set of class priors and class-conditional probability density functions [10]. Therefore, a concept drift is said to occur between two timestamps t_i and t_j if $C^{t_i} \neq C^{t_j}$ [12]. Consequently, classifiers for data streams must be adaptive, which means that f may be adjusted when newly labeled instances are made available.

2.1 Classifiers

Over the years, different approaches have been developed for the classification task in data streams. In practice, these classifiers are variants of traditional classifiers available for batch scenarios.

A popular approach for classification in streaming scenarios is the Incremental Naive Bayes [24]. As its batch counterpart, it assumes that input features are independent. With the arrival of a training instance, all probabilities are updated according. Since probabilities are based on counters, and there is no need to store instances, Naive Bayes has a constant memory consumption and processing time. Nonetheless, it does not present any traits to identify and adapt to concept drifts.

The most common approach for learning from data streams is decision trees. In particular, Hoeffding Trees [9] is the most popular approach as it branches over time when statistically enough data (grace period, $n_{\min}$) and evidence has been gathered, according to the Hoeffding Bound [22]. The Very Fast Decision Tree (VFDT) is a popular implementation of incremental Hoeffding Trees, meaning that it continuously branches as new data becomes available and does not revisit the quality of previously created split nodes. In contrast, the characteristic of revisiting split nodes is observed in Hoeffding Adaptive Trees [4], in which each split node is coupled with an ADWIN drift detector [3]. Whenever a drift is flagged, the corresponding split node is replaced by a leaf node, which can branch again if the Hoeffding inequality is met. Even though Hoeffding Adaptive Trees significantly improve accuracy rates compared to incremental trees, even better results are obtained when creating ensembles of Hoeffding Trees. A state-of-the-art exemplar of ensembles of Hoeffding Trees is the Adaptive Random Forest (ARF) [19], in which Randomized Hoeffding Trees are trained in parallel and coupled with drift detectors to identify and adapt to concept drifts rapidly. ARF adjusts the sampling process with Poisson($\lambda = 6$) so that instances have higher chances of being used during training, thus speeding up the drift adaptation process. In the test step, classifiers' votes are combined using weighted majority voting, i.e., classifiers with higher accuracy have a higher impact on the final prediction. Consequently, ARF is a strong learner that achieves state-of-the-art results in terms of accuracy, yet, at the high expense of computational resources.

2.2 Requirements

Throughout the training and test steps of data stream classification, streaming classifiers must meet certain requirements [5,6]:

- **Requirement #1:** Process an example at a time, and inspect it only once (at most);

- **Requirement #2**: Use a limited amount of memory;
- **Requirement #3**: Work in a limited amount of time;
- **Requirement #4**: Be ready to predict at any point; and
- **Requirement #5**: Detect and adapt to concept drifts.

This list has been incremented in the works of García-Martín [13,14,17], in which energy consumption is highlighted as a relevant aspect in data stream mining since several classifiers have been tailored focusing solely on accuracy and overlooking sustainability. This is one of the main drivers of our work, i.e., to allow researchers and practitioners to quantify the energy consumption of data stream classifiers and determine under which conditions they fail to meet energy sustainability criteria.

3 Related Works

Over the years, different approaches to measuring energy consumption have been proposed. This section highlights approaches for measuring energy consumption in data stream mining and decreasing energy usage. A first significant study is [16], in which authors used PowerAPI [30] to quantify the energy consumption of Hoeffding Trees despite acknowledging that it overlooks RAM consumption. The same authors have changed their approach in [15], in which Jalen (now called JourlarX) [26] has been used to quantify energy consumption of Hoeffding Trees on a function-basis. This allowed the authors to identify bottlenecks in the existing Hoeffding Tree implementation available in MOA.

Finally, authors in [13] and [14] have used Intel's RAPL [7] to quantify the energy consumed by the DRAM and the processor based on accesses to the processor performance counters. Even though Intel's RAPL code is not available, authors disclose that its accuracy has been checked in [21] and that it does not introduce processing overheads. It is also relevant to highlight that the work of [14] introduces a Hoeffding Tree variant in which the grace period (n_{min}) is adjusted so that branching is only attempted according to a user-given threshold. The results showed that the proposed Hoeffding Tree variant has accuracy convergence while approximately 65% less energy consumption rates. On the other hand, the work of [13] introduces a framework to quantify the energy consumption of Hoeffding Tree ensembles while accounting for decision tree learning, drift detectors, and tree replacement.

Regarding all of the works mentioned above, a significant drawback is that energy consumption has been tackled solely for Hoeffding Trees, and no general open-source tool makes energy consumption is easily available for researchers and practitioners. Our proposal is brought forward in the next section, and it circumvents such problems.

4 Proposal

In this section, we detail our tool to quantify the energy consumption of data stream mining algorithms. Our tool is embedded within the Massive Online

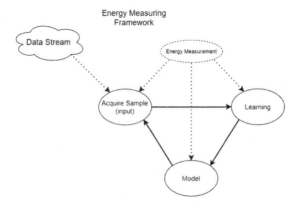

Fig. 1. Overview of the proposed energy measuring framework. The tool measures energy consumption during data acquisition and model training and testing.

Analysis (MOA) framework [6], yet, its rationale can be used in other tools like River [25].

Our tool uses Intel's RAPL [7] and can be seen as a plugin to the Massive Online Analysis (MOA) framework. The general idea of our tool is given in Fig. 1, in which we see that RAPL is used to quantify energy consumption in data acquisition and models' training and testing phases. Once an experiment starts, energy measurements are initialized. During data stream processing, arriving instances are used to determine processing and energy readings before and after processing, i.e., testing and training steps. These readings are used to compute energy consumption rates during the experiment. For each of the stages, there are different forms of measuring energy consumption. The framework controls the flow of the data stream model to start the energy measurement right before it begins processing the samples. At each cycle, the measurement is taken and presented in real-time to the user. At the end of the process, a graph showing the instantaneous measurement for each cycle is presented to the user.

Since our tool is based on Intel RAPL, we provide in Fig. 2 details on how our plugin interacts with MOA, RAPL, and the Linux Kernel. As new data becomes available for processing, the plugin requests measurements from the Linux Kernel, receiving the response of how much energy has been spent during the testing and training phases. These values are summed and made available whenever the evaluation interface (the so-called evaluation frequency parameter) requests an energy consumption rate. Figure 3 exemplifies the energy consumption rates measured by the proposed tool and how it is reported in MOA alongside other evaluation metrics.

The source code of our proposal and experimentation can be found at https://github.com/ericonuki/moa-bringing-awareness-green-ict.

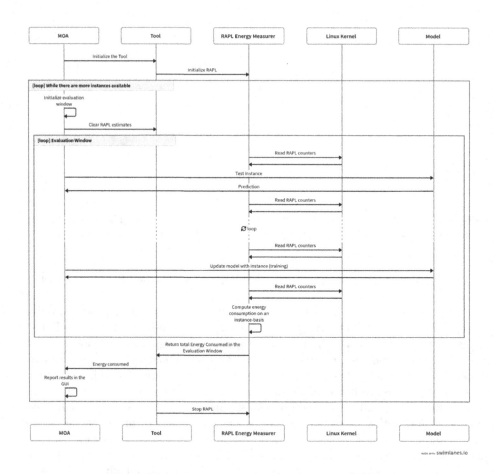

Fig. 2. Measuring energy consumption using RAPL.

Fig. 3. Screenshot of the Massive Online Analysis (MOA) framework with results provided by the proposed tool.

5   Experiments

This section presents the experiments conducted to assess our proposed tool to measure energy consumption in data streams. In particular, this section is divided into two experiments. First, we analyze our tool against a hardware solution, which serves as a gold standard for the energy consumed. Next, we analyze the energy consumed by different classifiers in Prequential test-then-train validation in stationary and non-stationary scenarios. All tests were performed on a desktop computer running Ubuntu Desktop 18.04 LTS; Intel Core i7-2600 Sandy Bridge CPU; 4GB RAM, and 250 GB HDD Hard Drive.

5.1   Experiment 1 - Validation Against a Hardware Solution

The first experiment conducted aimed to assess whether the energy consumption measurement via hardware and software solutions were equivalent or, at least, correlated. In this experiment, both software and hardware TP-LINK HS 110 wall plug solutions were connected to the computer, and a CPU stress called stress-ng [23] was used to quantify energy consumption rates.

Initially, it was deemed necessary to assess whether the measurement of energy consumption by hardware or software is equivalent and when using only one software tool (the plugin) would not compromise the results of this study. The stress was configured to generate a 10% stress level for 10 min and progressively perform increments by 10% until 100% stress was reached. Once 100% stress was reached, the test was incremented to use an extra processor core. This process was repeated until all four cores were allocated. The entire testbed encompassed five runs, and the average results are given in Fig. 4. Both lines indicate the various stress levels on the computer, with the blue line relating to the software power consumption measurement and the green line the hardware results.

These results depict that the software solution does not meet the hardware solution readings. These results are expected since the hardware readings account for the entire computer, i.e., the operational system and other software that is being run; thus, it is reasonable to assume that the stress occupies a part of the overall energy being consumed. Even though it is clear that the results do not match, they possess a 99.97% correlation, which depicts a strong correlation. Therefore, we verify that although the software solution does not accurately describe a computer's total energy consumption, its results correlate with the actual consumption measured from a hardware tool, as also observed in [8,27].

5.2   Experiment 2 - Analyzing Different Classifiers in Stationary and Non-Stationary Environments

In this experiment, we used our proposed tool to quantify the energy consumption of different classifiers in stationary and non-stationary environments. In particular, synthetic data streams were created using the Massive Online Analysis (MOA) framework using the Agrawal (AGR) [1], Assets Negotiation (AN)

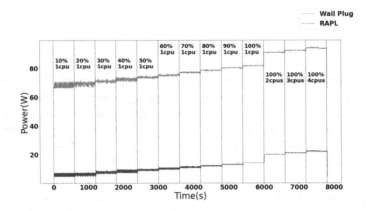

Fig. 4. Comparison of the energy consumption quantified by hardware and software.

[2], and SEA [29] generators. Each stream possessed 1 million instances, and two variants were created, one with concept drifts located at the middle of the experiment (drift position equal to 500,000) and another without concept drift. Experiment variants marked with a **-D** suffix stand for drifting experiments. The data streams mentioned above were used to assess Naive Bayes (NB), Hoeffding Tree (HT), Hoeffding Adaptive Tree (HAT), and Adaptive Random Forest (ARF) classifiers. All classifiers used the default parameters available in MOA, except for ARF, which used 100 ensemble members (each with an individual thread) and a grace period $n_{min} = 50$. All experiments were conducted using the Prequential validation scheme proposed in [11], i.e., each instance was retrieved and used for testing and training. The source code to reproduce this experimentation is also available in the code repository.

Discussion. The results obtained are given in Tables 1, 2, 3, and 4. These tables provides accuracy, processing time (in seconds), memory consumption (in GB-Hours), and energy consumption rates (in Watts), respectively. First, we highlight that Naive Bayes (NB) has the lowest accuracy rates in all scenarios. This result corroborates that decision trees and their ensembles are more interesting when this particular evaluation metric is pursuit. We highlight, for instance, the results obtained by ARF in drifting experiments, which are expected since it has multiple learners coupled with drift detectors to detect and adapt to such changes. Nonetheless, decision trees bring forward computational overheads that are quantified by the remainder of the metrics. First, we see that Hoeffding Tree (HT), Hoeffding Adaptive Tree (HAT), and the Adaptive Random dom Forest (ARF) are slower than Naive Bayes (NB). The processing times for HT, HAT, and ARF are 7, 11, and 44,751 times slower than NB, respectively. Similar results are observed for RAM consumption, in which HT, HAT, and ARF consume more RAM than NB. Again, we highlight the RAM consumption observed by ARF, which is approximately 10^8 times higher than its

Table 1. Accuracy results obtained during experimentation.

Experiment	Accuracy (%)			
	NB	HT	HAT	ARF
AGR	94.39	**94.97**	94.51	94.52
AGR-D	75.67	85.72	87.63	**90.66**
AN	92.36	**94.87**	94.83	94.86
AN-D	83.95	94.53	94.71	**94.73**
SEA	86.95	89.39	89.40	**89.67**
SEA-D	87.02	88.83	89.14	**89.55**

Table 2. Memory consumption results obtained during experimentation.

Experiment	Memory consumption (GB-Hours)			
	NB	HT	HAT	ARF
AGR	$\mathbf{1.60\times10^{-9}}$	6.64×10^{-6}	7.87×10^{-6}	7.72×10^{1}
AGR-D	$\mathbf{1.78\times10^{-9}}$	1.25×10^{-5}	7.85×10^{-6}	12.40×10^{1}
AN	$\mathbf{1.21\times10^{-8}}$	4.72×10^{-6}	2.02×10^{-5}	4.38×10^{1}
AN-D	$\mathbf{9.51\times10^{-9}}$	6.81×10^{-6}	8.13×10^{-6}	3.78×10^{1}
SEA	$\mathbf{3.59\times10^{-10}}$	6.11×10^{-7}	6.45×10^{-6}	5.99×10^{1}
SEA-D	$\mathbf{4.31\times10^{-10}}$	6.33×10^{-7}	2.27×10^{-6}	9.12×10^{1}

counterparts. Finally, the energy consumption values depict that NB is the less-consuming algorithm, which the exception of HAT in the AGR-D experiment. These results are expected since NB is much faster and less memory-consuming; thus, less energy is required to finalize an experiment. Regarding the AGR-D experiment, it is relevant to emphasize that HAT's energy consumption has decreased since it has a drift detector that restarted the entire tree learning process, i.e., it replaced the entire tree with a single decision stump, and thus, its computational cost after the drift has greatly decreased. This is a relevant scenario in which energy consumption is not directly related to processing time and memory consumption, and it allows a better analysis by researchers and practitioners on which classifier should be used in a specific scenario. Focusing on ARF, we also highlight that the energy consumption rates are not directly related to either processing time and memory consumption rates since, despite taking much more time and memory to run, its energy consumption was roughly twice when compared to its counterparts. This result can be explained due to ARF's implementation, which is multi-threaded, and even by combining 100 learners the energy consumption is not 100 times greater than its counterparts.

Table 3. Processing time results obtained during experimentation.

Experiment	Processing time (s)			
	NB	HT	HAT	ARF
AGR	**1.35**	16.57	19.24	74639.15
AGR-D	**1.50**	21.42	18.99	49600.75
AN	**16.28**	24.21	35.69	22538.36
AN-D	**12.80**	24.14	25.19	20149.68
SEA	**0.75**	5.72	16.94	96930.68
SEA-D	**0.90**	5.92	11.81	43513.95

Table 4. Energy consumption results obtained during experimentation.

Experiment	Energy (W)			
	NB	HT	HAT	ARF
AGR	**34334.04**	37452.53	35283.41	66811.31
AGR-D	37747.63	37435.13	**36280.75**	71646.43
AN	**32520.12**	33972.08	34425.10	84459.67
AN-D	**33414.73**	34277.01	34851.68	67507.25
SEA	**25749.19**	38479.07	35183.48	69747.71
SEA-D	**33910.95**	38363.17	35245.82	55118.25

6   Conclusion

In this work, we brought forward a tool for quantifying energy consumption in data stream mining. Our tool was embedded within the Massive Online Analysis (MOA) framework, thus allowing researchers to rapidly quantify the energy consumption of different classification methods under different streaming settings and validation processes. To validate our proposal, we first conducted a testbed using a processor stress to compare the proposed software readings against a hardware wall plug. Next, we tested different classifiers in stationary and drifting scenarios in a Prequential validation scheme. Results showed that our tool allows the identification of energy consumption rates of different classifiers under different scenarios, i.e., drifting and non-drifting data streams. The energy consumption rates allow a more fine-grained analysis of the classifiers as energy consumption is not directly tied with processing time and memory consumption, especially under concept drifting scenarios and multi-threading implementations.

In future works, we plan to extend our tool to encompass different classification, regression, and clustering validation schemes, including more datasets and scenarios. We also plan to port our tool to Python-based frameworks, such as River [25]. Finally, we also plan to make our tool available in AMD and Apple's ARM platforms and add support to GPU consumption since neural networks are increasingly used in streaming settings [20].

References

1. Agrawal, R., Imielinski, T., Swami, A.: Database mining: a performance perspective. IEEE Trans. Knowl. Data Eng. **5**(6), 914–925 (1993)
2. Barddal, J.P., Murilo Gomes, H., Enembreck, F., Pfahringer, B., Bifet, A.: On dynamic feature weighting for feature drifting data streams. In: Frasconi, P., Landwehr, N., Manco, G., Vreeken, J. (eds.) ECML PKDD 2016. LNCS (LNAI), vol. 9852, pp. 129–144. Springer, Cham (2016). https://doi.org/10.1007/978-3-319-46227-1_9
3. Bifet, A., Gavalda, R.: Learning from time-changing data with adaptive windowing. In: Proceedings of the 2007 SIAM International Conference on Data Mining, pp. 443–448. SIAM (2007)
4. Bifet, A., Gavaldà, R.: Adaptive learning from evolving data streams. In: Adams, N.M., Robardet, C., Siebes, A., Boulicaut, J.-F. (eds.) IDA 2009. LNCS, vol. 5772, pp. 249–260. Springer, Heidelberg (2009). https://doi.org/10.1007/978-3-642-03915-7_22
5. Bifet, A., Gavalda, R., Holmes, G., Pfahringer, B.: Machine Learning for Data Streams: With Practical Examples in MOA. MIT press, Cambridge (2023)
6. Bifet, A., Holmes, G., Kirkby, R., Pfahringer, B.: MOA: massive online analysis. J. Mach. Learn. Res. **11**(52), 1601–1604 (2010). http://jmlr.org/papers/v11/bifet10a.html
7. David, H., Gorbatov, E., Hanebutte, U.R., Khanna, R., Le, C.: RAPL: memory power estimation and capping. In: Proceedings of the 16th ACM/IEEE International Symposium on Low Power Electronics and Design. pp. 189–194. ISLPED 2010, Association for Computing Machinery, New York, NY, USA (2010). https://doi.org/10.1145/1840845.1840883
8. Desrochers, S., Paradis, C., Weaver, V.M.: A validation of dram RAPL power measurements. In: Proceedings of the Second International Symposium on Memory Systems, pp. 455–470. MEMSYS 2016, Association for Computing Machinery, New York, NY, USA (2016). https://doi.org/10.1145/2989081.2989088
9. Domingos, P., Hulten, G.: Mining high-speed data streams. In: Proceedings of the Sixth ACM SIGKDD International Conference on Knowledge Discovery and Data Mining, pp. 71–80. KDD 2000, Association for Computing Machinery, New York, NY, USA (2000). https://doi.org/10.1145/347090.347107
10. Gama, J.: Knowledge Discovery from Data Streams. Chapman & Hall/CRC, 1st edn. (2010)
11. Gama, J., Sebastião, R., Rodrigues, P.P.: On evaluating stream learning algorithms. Mach. Learn. **90**(3), 317–346 (2012). https://doi.org/10.1007/s10994-012-5320-9
12. Gama, J., Zliobaite, I., Bifet, A., Pechenizkiy, M., Bouchachia, A.: A survey on concept drift adaptation (2014). https://doi.org/10.1145/2523813
13. García-Martín, E., Bifet, A., Lavesson, N.: Energy modeling of Hoeffding tree ensembles. Intell. Data Anal. **25**(1), 81–104 (2020)
14. García-Martín, E., Bifet, A., Lavesson, N.: Green accelerated hoeffding tree (2020). http://urn.kb.se/resolve?urn=urn:nbn:se:bth-19152
15. Garcia-Martin, E., Lavesson, N., Grahn, H.: Energy efficiency analysis of the very fast decision tree algorithm. In: Trends in Social Network Analysis: Information Propagation, User Behavior Modeling, Forecasting, and Vulnerability Assessment, pp. 229–252 (2017)

16. Garcia-Martin, E., Lavesson, N., Grahn, H.: Identification of energy hotspots: a case study of the very fast decision tree. In: Au, M.H.A., Castiglione, A., Choo, K.-K.R., Palmieri, F., Li, K.-C. (eds.) GPC 2017. LNCS, vol. 10232, pp. 267–281. Springer, Cham (2017). https://doi.org/10.1007/978-3-319-57186-7_21

17. García-Martín, E., Lavesson, N., Grahn, H., Casalicchio, E., Boeva, V.: How to measure energy consumption in machine learning algorithms. In: Alzate, C., et al. (eds.) ECML PKDD 2018. LNCS (LNAI), vol. 11329, pp. 243–255. Springer, Cham (2019). https://doi.org/10.1007/978-3-030-13453-2_20

18. García-Martín, E., Rodrigues, C.F., Riley, G., Grahn, H.: Estimation of energy consumption in machine learning. J. Parallel Distrib. Comput. **134**, 75–88 (2019). https://doi.org/10.1016/j.jpdc.2019.07.007

19. Gomes, H.M., et al.: Adaptive random forests for evolving data stream classification. Mach. Learn. **106**(9), 1469–1495 (2017). https://doi.org/10.1007/s10994-017-5642-8, https://doi.org/10.1007/s10994-017-5642-8

20. Gunasekara, N., Gomes, H.M., Pfahringer, B., Bifet, A.: Online hyperparameter optimization for streaming neural networks. In: 2022 International Joint Conference on Neural Networks (IJCNN), pp. 1–9. IEEE (2022)

21. Hähnel, M., Döbel, B., Völp, M., Härtig, H.: Measuring energy consumption for short code paths using RAPL. ACM SIGMETRICS Perform. Eval. Rev. **40**(3), 13–17 (2012)

22. Hoeffding, W.: Probability inequalities for sums of bounded random variables. In: Fisher, N.I., Sen, P.K. (eds.) The Collected Works of Wassily Hoeffding. Springer Series in Statistics. Springer, New York, NY (1994). https://doi.org/10.1007/978-1-4612-0865-5_26

23. King, C.I.: Stress-ng. http://kernel.ubuntu.com/git/cking/stressng.git/. visited on 28/03/2018), p. 39 (2017)

24. Klawonn, F., Angelov, P.: Evolving extended Naive Bayes classifiers. In: Sixth IEEE International Conference on Data Mining-Workshops (ICDMW 2006), pp. 643–647. IEEE (2006)

25. Montiel, J.: River: machine learning for streaming data in python. J. Mach. Learn. Res. **22**(1), 4945–4952 (2021)

26. Noureddine, A.: PowerJoular and Joularjx: multi-platform software power monitoring tools. In: 2022 18th International Conference on Intelligent Environments (IE), pp. 1–4. IEEE (2022)

27. Phung, J., Lee, Y.C., Zomaya, A.Y.: Modeling system-level power consumption profiles using RAPL. In: NCA 2018–2018 IEEE 17th International Symposium on Network Computing and Applications. Institute of Electrical and Electronics Engineers Inc. (2018). https://doi.org/10.1109/NCA.2018.8548281

28. Shao, Y.S., Brooks, D.: Energy characterization and instruction-level energy model of intel's Xeon Phi processor. In: International Symposium on Low Power Electronics and Design (ISLPED), pp. 389–394. IEEE (2013)

29. Street, W.N., Kim, Y.: A streaming ensemble algorithm (SEA) for large-scale classification. In: Proceedings of the Seventh ACM SIGKDD International Conference on Knowledge Discovery and Data Mining, pp. 377–382 (2001)

30. Terpstra, D., Jagode, H., You, H., Dongarra, J.: Collecting performance data with PAPI-C. In: Müller, M., Resch, M., Schulz, A., Nagel, W. (eds.) Tools for High Performance Computing 2009. Springer, Berlin, Heidelberg (2010). https://doi.org/10.1007/978-3-642-11261-4_11

31. Widmer, G., Kubat, M.: Learning in the presence of concept drift and hidden contexts. Mach. Learn. **23**, 69–101 (1996)

Improved Fuzzy Decision System for Energy Bill Reduction in the Context of the Brazilian White Tariff Scenario

Raimunda Branco$^{(\boxtimes)}$ and Filipe Saraiva

Institute of Exacts and Natural Sciences, Federal University of Pará, Belém, Pará, Brazil
raimundancbranco@gmail.com, saraiva@ufpa.br

Abstract. The production of energy from renewable sources has become a more sustainable and environmentally correct alternative, where the use of solar energy through photovoltaic systems is evident. One of the main problems in the use of photovoltaic systems is the high cost of installation and maintenance of this system, in addition to the cost of the residential electricity tariff, which makes this technology expensive for most residential consumers in Brazil. An alternative for consumers to get around the high amounts paid on the energy bill is to opt for the white tariff modality, which is characterized by offering the variation of the energy value according to the day and time of consumption. The present work aims to develop a fuzzy system to manage the energy production from a photovoltaic system, optimizing the use of the produced energy between the consumer, the battery and the electric grid in a white tariff scenario for residential units in Brazil. Based on the simulations, the fuzzy system presented is efficient, with a significant economic reduction in the energy bill compared to a simple photovoltaic system without the ability to make intelligent decisions and used commercially in industries.

Keywords: Fuzzy System · Photovoltaic Production · Renewable energies

1 Introduction

A country's energy needs are an important indicator of its economic development. Electricity consumption has been identified as one of the most important factors for the well-being of society and a critical determinant of a country's progress [4]. As risk factors, increasing population demand combined with a country's energy base tied to a single or few energy sources are vulnerabilities that threaten economic growth.

The main sources of electrical energy in the world are hydro, biomass, wind, solar, natural gas, oil derivatives, nuclear and coal [6]. The demand for diversification of energy sources, the concern for sustainability and the reduction of costs have led several branches of science in the search for greater efficiency of electrical energy generation systems using renewable sources. Among the main sources of renewable energy, solar energy has become increasingly important, especially in countries with tropical climates. According

M. C. Naldi and R. A. C. Bianchi (Eds.): BRACIS 2023, LNAI 14197, pp. 441–454, 2023.
https://doi.org/10.1007/978-3-031-45392-2_29

to a report by [6], in Brazil, the photovoltaic solar energy generation has evolved from 3,287 MW in 2020 to 4,632 MW in 2021.

[15] assert that solar energy is the largest source of energy available on Earth, which is a renewable and inexhaustible source of energy. Also, according to the authors, the growing demand and research in this field will make solar energy become a significant part of the world's energy matrix in the future.

[1] explain that since the beginning of the 21st century, the traditional concept of centralized energy generation has been open to research and application of distributed generation (DG) or decentralized. A distributed system is mainly composed of small generator units connected directly to the consumer load and/or distribution grids, using mainly renewable energy sources such as photovoltaic [17]. In this system, each consumer can produce electricity and then connect to the distribution grid. In this way, the residential consumer can consume and also supply excess electricity to the local distribution center.

There are several advantages in the use of DG, among which the reduction of the electricity tariff paid by consumers to concessionaires stands out, since there is autonomy for the consumer to also produce energy. However, for [17], the study of which strategies could be used to increase the reduction of electrical energy consumption by concessionaires and consequently the use of DG to reduce the energy bill is still incipient.

When dealing with the management of photovoltaic systems, among the alternatives for a more efficient management, focusing on a higher quality and more economically viable energy distribution, computational intelligence (CI) has emerged as a promising option, since it seeks methods that possess or enhance the intelligent capacity of humans to solve problems, acquire and represent knowledge, in addition to recognizing patterns [11].

This article used as a reference the research by [17]. The idea at this first moment is to improve and update this work through adjustments in the fuzzy algorithm, mainly by improving the parameters of the linguistic variables, proposing other models for the system and adapting the configuration of photovoltaic technology to the present day.

In this sense, this research proposes to develop an intelligent photovoltaic system, using Fuzzy Logic as CI technique for decision-making about the best time to use or not to use the energy produced by DG, in a scenario where generators have an energy storage system coupled to them, with the aim of reducing the price of the energy bill for the final consumer. For this purpose, an environment with a different type of tariff modality is simulated, where the amount paid by the consumer depends on the time at which the energy is consumed, defined as a white tariff.

Finally, in addition to comparing the developed system with a previous work, the performance of a simple photovoltaic solar energy system (developed without any intelligent technique) is also simulated in order to compare it with the system developed through fuzzy logic.

The paper is presented as following: Sect. 2 proposes to present the literature review that supported this research. Section 3 describes the features of the proposed system, presenting the tariff modality used and the components of the photovoltaic system. Section 4 presents in detail the development of the fuzzy system. Section 5 describes the simulation created, explaining the database used, the specificities of the photovoltaic system components and how the fuzzy system is modeled. Section 6 describes the development of the research and presents the results obtained. Finally, Sect. 7 presents the conclusion about this research.

2 Literature Review

The application of artificial intelligence techniques in photovoltaic electrical systems is currently presented in a very frequent and promising way. There are several works that present fuzzy logic as an alternative to manage and optimize the distribution of energy in photovoltaic systems.

[12] presents an algorithm based on fuzzy logic to manage energy storage in a battery and also the energy demanded by consumers at the University of Naresuan, Thailand, in order to reduce the electricity bill. The main idea of the algorithm is to allow an intelligent switching between absorbing energy during periods of high solar irradiation and discharging energy to the load during times of high consumption. The fuzzy algorithm proved to be successful, as it reduced the annual energy bill by 17.58%.

The work by [5] addresses power management to improve network performance and generate smooth transitions between different power balance modes. This work presents a fuzzy system developed to provide a dynamic flow of energy from the grid, based on the network price, in order to decrease the energy rate and increase the useful life of the storage device. Experimental results were obtained which proved to be satisfactory.

Another relevant work found is the article by [14], which deals with energy management to maintain a balance between different energy sources, storage units and loads. In this article, fuzzy logic was used in order to maintain the balance between these two objectives. From the results exposed in this work, it was observed that this system demonstrated viability to be used as a solution to the problem.

[3] publishes a work in which an energy management system based on fuzzy logic is presented, where the objective is to smooth the profile of an electrothermal microgrid connected to the residential network. The proposed case study designs an energy management system to reduce the impact on the electrical grid when renewable energy sources are incorporated into pre-existing appliances connected to the grid. The simulation results used real data measured over a year and showed that the proposed management system design achieves a reduction of 11.4% of the maximum power absorbed from the network.

The publication by [10] proposes a residential smart microgrid topology linked to the distribution grid using fuzzy logic, which integrates a photovoltaic system, a fuel cell and a battery bank. To validate the functioning of the proposed energy management unit, a prototype of the system was developed and experimental tests were carried out. The management system was tested for three different residential load scenarios, both connected to the grid and isolated. The distribution and analysis of the energy cost provided for each management scenario showed the benefits of using this type of intelligent system for the consumer and for the utility.

[19] developed a coordinated control scheme of a battery energy storage system and DG units for a microgrid, based on fuzzy logic. The coordinated control scheme aims to mitigate the fluctuation of active power at the point of common coupling of the microgrid

when it is connected to the grid and also to keep the frequency of the microgrid within the defined range for operation when it is isolated. In the control scheme, the SoC of the battery energy storage system is used as input for the coordinated control based on fuzzy logic. The results of the case study were satisfactory and showed that the proposed coordinated control scheme is capable of mitigating the fluctuation of active power at the common coupling point for grid-connected control and performing efficient frequency control for isolated operation.

3 Features of the Proposed System

In addition to the use of a DG system, this research prioritized the use of the white hourly tariff modality. According to [2], the white tariff is characterized as consisting of differentiated tariffs for energy consumption depending on the time of day. This modality is divided into three tariff types: peak tariff, intermediate tariff, and off-peak tariff.

The days of the week and times stipulated for tariff posts are defined and implemented by each electricity company distributor in Brazil, provided that they are previously approved and ratified by ANEEL [2]. Equatorial Energia is one of the energy distributors present in Brazil that have made available the white tariff modality for all consumer units, since January 1st, 2020. The values of each tariff are also defined by the energy distributors. At Equatorial Energia, for example, the values and periods of tariff posts must be approved by ANEEL in each periodic tariff review, which must occur every four years [7]. The schedules and values of the tariff stations defined for the State of Pará – Brazil since the year 2019 are specified in the items below:

1. peak time (from 6 pm to 8:59 pm): R\$ 1.39;
2. intermediate time (from 4 pm to 5:59 pm and from 9 pm to 9:59 pm): R\$ 0.87;
3. off-peak time (from 10 pm to 3:59 pm): R\$ 0.49.

The benefits of using the white tariff can then be realized above all to consumers who consume very little between 4 pm and 10 pm, that is, during peak and intermediate time. The white tariff is a modality that, depending on the consumption profile and the habits of use of electric energy, presents itself as an interesting alternative for the residential consumer.

Another important issue for the system configuration of this application is the use of an integrated storage system connected to a hybrid photovoltaic system (both on-grid and off-grid). The presence of the battery in a photovoltaic production system is intended to accommodate the excess energy production of the residence. In this way, the extra energy can be used in case of power outages of the photovoltaic panel and/or the electrical network and to balance the load inserted in the grid [17]. In the context of the white tariff modality, it is possible to charge the battery when energy from the grid is cheap and to use it when energy is expensive. In this way, the use of the battery reduces the billed consumption of the consumer unit, especially during peak hours.

As a way of exemplifying this section, Fig. 1 presents the system components and the architecture proposed in this article for the residential photovoltaic installation.

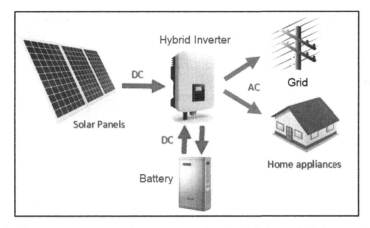

Fig. 1. System components and architecture of residential photovoltaic installation [8]

4 Development of the Fuzzy Algorithm

[18] defines a fuzzy system as a particular type of knowledge-based system where the knowledge base is built from a set of fuzzy rules. In this type of system, inputs are given in natural language, which, after going through the fuzzy inference process, are converted into a numerical format that is easy to manipulate. The steps for building a fuzzy system are described in Fig. 2.

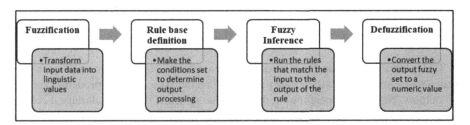

Fig. 2. Steps for implementing a fuzzy system

The fuzzy system developed in this research consists of 3 different fuzzy algorithms, one algorithm for each price range of the white tariff, as detailed previously. The fuzzy system is composed of 2 input variables (El and SoC) and an output variable (Eb). El represents the energy in kWh, that is, it is the value resulting from the subtraction of the energy produced by the DG (Eg) and the energy consumed by the consumer load (Ec). Thus, if El is positive, then there is more production than consumption, but if El is negative, then production is not enough to supply the demand for electricity.

SoC is the state of charge of electricity stored in the battery. If El is positive, it is possible to recharge the battery using the surplus electricity produced. If El is negative, the battery can be used to supply the energy demand.

Eb represents how much energy will be injected or utilized from the battery, that is, it represents the decision whether the battery will be charged or discharged. So, if Eb is positive, then energy can be inserted into the storage device. If Eb is negative, then the battery can be used by the customer load. However, these decisions will always depend on the time of day, El and SoC.

Table 1 presents the meaning of the linguistic sets of each variable.

Table 1. Definition of linguistic sets

Linguistic variable	Acronym	El	SoC	Eb
Positive Big	PB	big surplus photovoltaic production	battery fully charged	big injection of energy in the battery
Positive Small	PS	small surplus photovoltaic production	battery almost full	small injection of energy into the battery
Zero	ZE	no difference between production and consumption	battery half charged	no battery interaction
Negative Small	NS	small consumption surplus	battery almost discharged	small energy discharge in the battery
Negative Big	NB	big consumption surplus	battery very discharged	big energy discharge in the battery

Table 2 shows the rule bases defined for each tariff configuration. For each rule created, two conclusions will be specified, based on the association of predefined inputs and outputs. Note that depending on the grid energy price, the rule will have a certain behavior, resulting in an output with a different value. For example, if there is low consumption surplus (El = NS) and if the battery is half full (SoC = ZE), then 3 situations can result:

1. Cheap energy: then there is no interaction with the battery (Eb = ZE);
2. Intermediate energy: then there is a small energy discharge in the battery (Eb = NS);
3. Expensive energy: then there is a big energy discharge in the battery (Eb = NB).

Each configuration has a set of 25 rules, for a total base of 75 rules.

The membership functions used to represent the linguistic sets were the triangular and trapezoidal forms. The maximum and minimum limits of the input variables (El and SoC) were determined based on the dataset values, for El the minimum limit was set at −6 kW and the maximum limit 4 kW, while for the SoC it was agreed the value in percentage limits of the energy contained in the battery ranging from 0 to 100%. Regarding the output variable (Eb), the limits followed the ±6 kW battery charge flow.

Table 2. Rule base map for off-peak time (a), intermediate time (b) and peak time (c)

Eb		SoC				
		PB	PS	ZE	NS	NB
El	PB	PS	ZE	PS	PB	PB
	PS	PS	ZE	PB	PS	PS
	ZE	ZE	ZE	ZE	ZE	ZE
	NS	NS	NB	ZE	ZE	ZE
	NB	NS	NB	ZE	ZE	ZE

(a)

Eb		SoC				
		PB	PS	ZE	NS	NB
El	PB	PS	ZE	PS	PB	PB
	PS	PS	ZE	PB	PS	PS
	ZE	ZE	ZE	ZE	ZE	ZE
	NS	NB	NB	NS	ZE	ZE
	NB	NB	NB	NS	ZE	ZE

(b)

Eb		SoC				
		PB	PS	ZE	NS	NB
El	PB	PS	ZE	PS	PB	PB
	PS	PS	ZE	PS	PS	PS
	ZE	ZE	ZE	ZE	ZE	ZE
	NS	NB	NB	NB	ZE	ZE
	NB	NB	NB	NB	ZE	ZE

(c)

The following graphs show the membership functions of the input and output variables, with their respective linguistic sets (Fig. 3).

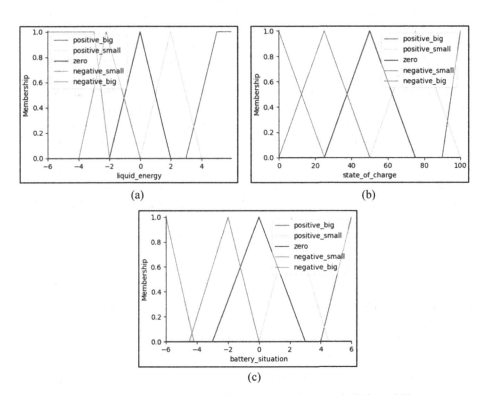

(a)   (b)

(c)

Fig. 3. Membership function of the output variable El (a), SoC (b) and Eb (c)

In addition, the fuzzy system uses the Mamdani inference method [18] and the defuzzification step is performed using the centroid method [18].

It is important to emphasize that the form used for the membership functions (tri-angular and trapezoidal) and the values used for the linguistic variables were defined empirically, based on a series of tests and adaptations, with the aim of finding the best configuration for the fuzzy system in question. In the same way, the configurations of the inference and defuzzification method were also defined both based on the used literature and based on system tests.

5 Simulation Specification

This paper improves the fuzzy system presented in [17] by changing the boundaries of the linguistic variables.

For the simulation of the system, real data on residential energy consumption and photovoltaic production over a period of one year were used. The public databases used were obtained from the Research Group in Knowledge Engineering and Decision Support (GECAD) at the Laboratory of Intelligent Electrical Systems (LASIE), located in Portugal at the Institute of Engineering - Polytecnica do Porto (ISEP /IPP) [9].

Data on residential consumption cover the period from December 2011 to March 2013, with a 15 min interval between one measurement and another. About photovoltaic production data, these include the months from January to December 2013, with an interval of 5 min between measurements. The measurement intervals were standardized in 5 min for residential consumption and photovoltaic production.

The simulation also adapted the composition of the photovoltaic electrical system to 13 photovoltaic panels, each one with a capacity of 340 W of power. Thus, the power of the installed system resulted in 4.42 kW. The residence has a maximum power demand of 6.8 kW and it is connected to a hybrid photovoltaic system, coupled to a LiFePO4 battery with a capacity of 13.5 kWh.

As a way of comparison, the system simulations were reproduced both for modeling considering a simple photovoltaic system, and for a photovoltaic system using fuzzy modeling. Both models are based on time of day, net energy and state of charge in the battery.

The simple photovoltaic system has a predetermined decision behavior, where, for this case, it follows the following procedures:

- Excess energy only charges the battery during the day. If the battery is full, the surplus is sent to the electrical distribution grid;
- The residence only uses energy from the distribution grid if the battery has a limit of 25% of the total;
- The residence only uses the energy stored in the battery during the hours from 4 pm to 10 pm, that is, when energy is generally more expensive in the white tariff mode.

Decision making in the fuzzy system behaves in a more complex way, but the idea is to make a more specific and intelligent assessment for each situation and moment, considering the white tariff modality:

- If the time of day is within the time range of the least expensive electricity rate, the system is more likely to store energy in the battery, depending on the state of charge;
- For intermediate electricity cost, the fuzzy system tends to use the battery moderately;

- When the cost of electricity is expensive, the system will try to use as much battery power as possible.

Equations (1) and (2) represent the punctual and annual electrical consumption P_c of the residence, respectively. When grid power P_{grid} is a positive value, it means that there was surplus energy produced by DG, so this extra energy must be transferred to the distribution grid. When P_{grid} is a negative value, it means that there is not enough energy in the residence, so the electrical grid must be used to supply this shortage. To arrive at the annual value of the residence's energy bill $B_{(total)}$, Eqs. (3) and (4) are used, where T_s is the time period used (5 min) and the value of the fee $f_{(t)}$ depends on the tariff value at that moment.

$$P_{c(n)} = P_{grid(n)} * T_s \quad (\text{if } P_{grid} < 0) \tag{1}$$

$$P_{c(total)} = \Sigma P_{c(n)} \tag{2}$$

$$B_{(n)} = P_{grid} * T_s * \text{fee}_{(t)} \quad (\text{if } P_{rede} < 0) \tag{3}$$

$$B_{(total)} = \Sigma B_{(n)} \tag{4}$$

For the development of the system, the Python programming language was used because it is known as a very intuitive syntax language and with a vast number of libraries and documentation available. For the implementation of fuzzy logic system in Python, the *scikit-fuzzy library* [16] was used. To analyze and manipulate data in a simpler way, the *pandas library* [13] was used. The graphical visualization of data in Python utilized was *matplotlib library* [13].

6 Research Results and Discussion

The total annual production of the photovoltaic system is 4426.37 kW, while the total annual residential consumption is 6801.28 kW. After developing the system and obtaining the results of the simulations, the resulting situations are observed and compared.

At first, the DG system without the fuzzy logic (commercially used by companies) consumes 4711.80 kW from the electrical distribution grid, which reduces the use of the electricity from the company by 30.72%, resulting in an annual invoice value of R$ 3312.20.

The second moment, which consists of adapting [17]'s research to current technologies, to the new data analysis and Python programming language, showed that the proposal consumes 3892.28 kW from the electrical distribution grid, with an annual cost of R$ 2590.25. Relative to the electricity distribution company's annual residential consumption, the reduction was 42.77%.

Finally, the simulation for the DG system using the fuzzy logic proposed in this paper resulted in a grid consumption of 3708.32 kW and an annual cost of R$ 2399.10, that is, a reduction of 45.48% compared to the annual consumption from the company. These data are better analyzed and compared in Table 3.

Table 3. Comparison of simulation results

	Simple System	Fuzzy System adapted from [17]	Fuzzy System
Total production (kWh)	4426.37	4426.37	4426.37
Total residential consumption (kWh)	6801.28	6801.28	6801.28
Total energy consumed from the distribution grid (kWh)	4711.80	3892,28	3708,32
Reduction based on total residential consumption (%)	30.72	42,77	45,48
Total annual bill (R$)	3312.20	2590,25	2399,10

The results presented are in line with the objective of this research. In Table 5, it is observed that there was a reduction in the total energy consumed by the electrical distribution grid, compared to the results obtained in the previous work. As a result of this reduction, the annual energy bill was reduced by R$ 191.15.

The simulation showed that the fuzzy system identifies the period in which there is greater energy production by the photovoltaic system and uses the battery more intensively, depending on the time of day, in order to use less the distribution grid. In contrast to the simple system that follows a linear pattern of energy storage in the battery. Figure 4 shows the behavior of the battery state of charge over a year, both for the simple system and for the fuzzy system.

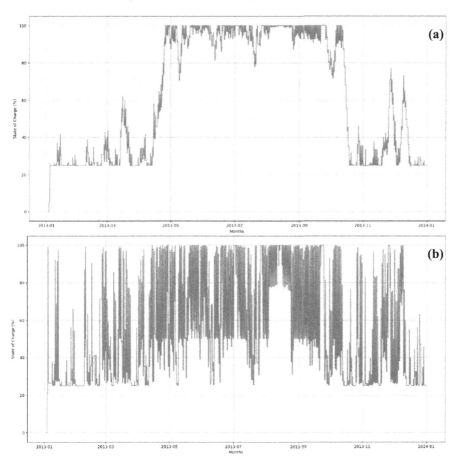

Fig. 4. SoC of the battery on simple (a) and fuzzy (b) systems over one year

It is observed that the fuzzy system requests less power from the distribution grid over time and injects power into it more uniformly. Positive values indicate injection of photovoltaic power into the grid and negative values indicate power consumption from the distribution grid throughout the residence. Figure 5 presents the behavior of the simple system and the fuzzy system in relation to the injection or consumption of power in the distribution grid.

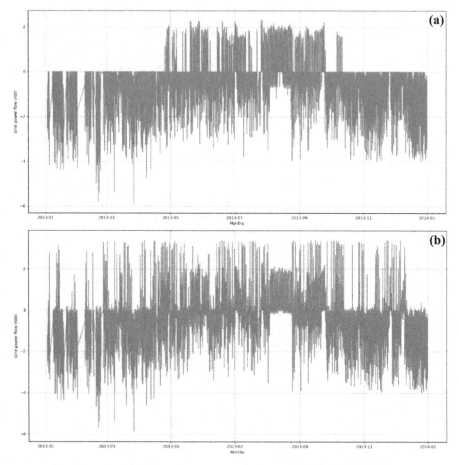

Fig. 5. Grid power flow on simple (a) and fuzzy (b) systems over one year

7   Conclusion

The main purpose of this article was to improve the use of photovoltaic energy in the context of white tariff in Brazil by using fuzzy systems. The proposed method was improved from previous work and the final setup could reduce the electricity bill for the residential consumer.

In order to adjust the system parameters, some limits of the input linguistic variables were modified. In the simulation, a battery sized in a more adjusted way and with a more current technology was also used to obtain a better cost benefit for a residence. A recent configuration of photovoltaic panels was used, with more efficient electrical power and better solar absorption technology. The modifications were satisfactory for the improvement in the results, allowing the continuation of the development of this research.

The results obtained brought a reduction of 14.05% in the energy consumed from the electrical distribution grid, when comparing the simple DG system developed without the use of IC with the proposed fuzzy system. This means a 31.73% reduction in the annual energy bill.

This work contributes to the CI area and to the academic community based on the expansion and progression of fuzzy logic studies applied to photovoltaic systems, with the aim of demonstrating the importance of these two subjects for improving decision-making and management of photovoltaic solar energy. In addition, the study and application of an intelligent system in a scenario that simulates the real world allows, among other contributions, the development of more innovative products for the energy market and products with better cost-benefits for the final consumer.

For future work, we intend to apply the proposed fuzzy systems for different energy production and consumption patterns from different cities of different Brazilian regions. We intend to verify if the proposal can be applied independently of the local, or if it requires small (or large) adjustments for better performance. It is also intended the study and application of other techniques of computational intelligence in order to compare with fuzzy logic.

References

1. Almeida, E., et al.: Energia Solar Fotovoltaica: revisão bibliográfica. Engenharias On-line **1**(2), 21–33 (2015). http://www.fumec.br/revistas/eol/article/view/3574/1911. Accessed 11 Apr 2023
2. ANEEL. Resolução Normativa N° 1000. Agência Nacional de Energia Elétrica. Brasília-DF, 07 de dezembro de 2021. Diário Oficial da União. https://www.in.gov.br/en/web/dou/-/resolu cao-normativa-aneel-n-1.000-de-7-de-dezembro-de-2021-368359651. Accessed 6 Apr 2023
3. Arcos-Aviles, D., et al.: An energy management system design using fuzzy logic control: smoothing the grid power profile of a residential electro-thermal microgrid. IEEE Access **9**, 25172–25188 (2021)
4. Balitskiy, S., Bilan, Y., Strielkowski, W., Streimikiene, D.: Energy efficiency and natural gas consumption in the context of economic development in the European Union. Renew. Sustain. Energy Rev. **55**, 156–168 (2016)
5. Dhar, R.K., et al.: Power balance modes and dynamic grid power flow in solar PV and battery storage experimental DC-Link microgrid. IEEE Access **8**, 219847–219858 (2020)
6. EPE, Empresa de Pesquisa Energética (Brasil). Balanço energético nacional: Relatório Final, ano base 2021, p. 264. Ministério de Minas e Energia, Rio de Janeiro (2022). https://www. epe.gov.br/pt/publicacoes-dados-abertos/publicacoes/balanco-energetico-nacional-2022. Accessed 7 Apr 2023
7. Equatorial Energia. Cartilha Online: Tarifa Branca (2019). https://pa.equatorialenergia.com. br/wp-content/uploads/2019/12/Cartilha-Online-PA.pdf. Accessed 6 Apr 2023
8. How Solar Power Works - On-grid, Off-grid and Hybrid Systems; Clean Energy Reviews. https://www.cleanenergyreviews.info/blog/2014/5/4/how-solar-works. Accessed 2 Apr 2023
9. IEEE-PES-ISS. Open Data Sets. http://sites.ieee.org/pes-iss/data-sets/. Accessed 6 Apr 2023
10. Jafari, M., et al.: Development of a fuzzy-logic-based energy management system for a multiport multioperation mode residential smart microgrid. IEEE Trans. Power Electron. **34**(4), 3283–3301 (2019)
11. Lopes, I.L., Santos, F.A.O., Pinheiro, C.A.M.: Inteligência Artificial, 1st edn. Elsevier Editora, Rio de Janeiro (2014)
12. Mansiri, K., Sukchai, S., Sirisamphanwong, C.: Fuzzy control algorithm for battery storage and demand side power management for economic operation of the smart grid system at Naresuan University, Thailand. IEEE Access **6**, 32440–32449 (2018)

13. McKinney, W.: Python for Data Analysis, 3rd. edn. O'Reilly Media, Inc., Sebastopol (2022)
14. Peña-Aguirre, J.C., et al.: Fuzzy logic power management strategy for a residential DC-microgrid. IEEE Access **8**, 116733–116743 (2020)
15. Simões-Moreira, J.R., et al.: Energias renováveis, geração distribuída e eficiência energética. [S.l: s.n.] (2017)
16. Singh, H., Lone, Y.A.: Deep Neuro-Fuzzy Systems with Python, 1st edn. Apress, Berlin (2019)
17. Sousa, M., Saraiva, F.: A fuzzy system applied to photovoltaic generator management aimed to reduce electricity bill. In: Moura Oliveira, P., Novais, P., Reis, L. (eds.) Progress in Artificial Intelligence, EPIA 2019. LNCS, vol. 11804, pp. 450–461. Springer, Cham (2017). https://doi.org/10.1007/978-3-030-30241-2_38
18. Wang, L.: A Course in Fuzzy Systems and Control. Prentice-Hall Inc., International Edition, Upper Saddle River (1997)
19. Zhao, H., et al.: Fuzzy logic based coordinated control of battery energy storage system and dispatchable distributed generation for microgrid. J. Mod. Power Syst. Clean Energy **3**(3), 422–428 (2015)

Exploring Artificial Intelligence Methods for the Automatic Measurement of a New Biomarker Aiming at Glaucoma Diagnosis

Gabriel C. Fernandes$^{(\boxtimes)}$, Fabio Lavinsky, Sandro José Rigo, and Henrique C. Bohn

Universidade do Vale do Rio dos Sinos (UNISINOS), São Leopoldo, Brazil
{gabcastro,fabiolavinsky,rigo,henriquecbohn}@edu.unisinos.br

Abstract. Using technologies capable of providing retina structure high-resolution images is one of the most widespread means of identifying structural changes that may indicate the onset or progression of visual impairment. Automated glaucoma detection using optical coherence tomography is still considered an area needing further research. Several manual analyzes are currently performed over the generated by imaging equipment. This work presents an approach to foster automatic glaucoma evaluation considering convolutional neural networks for semantic segmentation of retinal layers through optical coherence tomography images and image processing for measuring the cup region in the optic nerve head portion. We provide a quantitative evaluation comparing the results obtained by a specialist physician. The work's main contribution consists of the first approach supporting the automation of a new biomarker for diagnosing glaucoma.

Keywords: Deep Learning · Semantic Segmentation · Glaucoma

1 Introduction

Visual impairments and blindness can appear at any stage of life, but people over 50 are more likely to have some level of disability [6]. Monitoring through periodic examinations is very important as it makes it possible to identify any change in the structure of the retina. Follow-up through clinical examinations allows physicians to have an initial assessment, accompanied by intraocular pressure (IOP) measurements, a field of view testing, and structural imaging parameters. In cases of glaucoma, the disease may begin with a minimal disorder in the nerve fibers, almost invisible, and the loss of retinal ganglion cells (RGC). If it progresses to the intermediate stage, changes already become noticeable in the retinal nerve fiber layers (RNFL), along with peripheral vision loss. Finally, when fully installed, it presents serious damage and great vision loss [10].

Glaucoma is a chronic and progressive optic neuropathy characterized by the death of RGC associated with increased cupping of the optic nerve head

© The Author(s), under exclusive license to Springer Nature Switzerland AG 2023
M. C. Naldi and R. A. C. Bianchi (Eds.): BRACIS 2023, LNAI 14197, pp. 455–469, 2023.
https://doi.org/10.1007/978-3-031-45392-2_30

(ONH) and a corresponding loss of visual field (VF). Several studies have shown that the lamina cribosa (LC) has a role in the pathophysiology of glaucoma. The LC is the major area related to mechanical damage in glaucoma, and the detailed assessment of disease-related changes at that site is of the most relevant interest in understanding factors concerning the development and progression of glaucoma. Furthermore, Andrade and coworkers have also proposed that LC structural differences could be responsible for different effects of IOP on the tissue, contributing to individual susceptibility to glaucomatous damage mediated by the IOP [1].

The relationship between structural and functional damage is relevant for diagnosing glaucoma. The structural evaluation of the ONH, retinal nerve fiber layers (RNFL), and macula, when analyzed by optical coherence tomography (OCT), can provide important clinical information for the diagnosis of glaucoma and the evaluation of its progression [11]. Optical coherence tomography, a technology capable of providing high-resolution images of the structure of the retina, is one of the most commonly used means of inspecting it, providing access to the visualization of the retinal layers to monitor changes in thickness and possible deformations [12].

Detecting progress with visual field tests or optic disc imaging methods is limited due to the retrospective nature and high variability. Currently, functional and structural clinical tests suffer from the need for analysis and the significant time delay required to establish disease progression and response to treatment. So does the progress of the ganglion cell loss over that time. There is, therefore, a significant unmet need for glaucoma-related biomarkers to improve clinical trials, both for early diagnosis and detection of disease progression. Biomarkers then present a possibility to provide information that will eventually affect the decisions to be taken clinically [3].

Combining new technologies, such as imaging methods (OCT) and clinical evaluation based on parameters, is one way to assess glaucoma's criticality continuously. Optical coherence tomography provides relevant information such as cup-to-disc ratio (CDR), cup volume, rim area, and rim volume. Evaluations carried out by OCT demonstrated that regions such as Bruch's membrane opening (BMO) for the internal limiting membrane (ILM) are correlated with the thickness of the RFNL. As well as the posterior LC, a histopathological mark of glaucoma deformation was demonstrated in the deep part of the ONH structure. This information, made available by the imaging method, allowed the correlation of both regions, where Lavinsky and coworkers formulated the hypothesis that it would be helpful as a quantitative evaluation parameter of ONH remodeling in glaucoma [11]. BMO and LC represent the legs of a right triangle and result in the evaluation by discovering the hypotenuse of the vertical region of the ONH cup. The results are presented as a morphometric biomarker.

Studies regarding the automatization of OCT image analyses can foster automatic support for evaluating the retina's structure and identifying diseases such as glaucoma. These approaches aim to perform accurate analyses, especially with large volumes of data [5,17]. The studies of the retina structures are also supported by the rapid advancement in scanning technologies, which enable more researchers to obtain images with a high amount of information [16, 18].

Computational methods such as Computer Vision and Deep Learning can support the necessary OCT image analysis. They allow the resolution of problems such as segmentation, and pattern identification. This research aimed to contribute improving the way of extracting information related to a new biomarker on the progress of glaucoma. The importance of the study is directly related to the possibility of an accurate analysis of the retinal layers and the possibility of extracting information with automatic support.

The article is structured as follows: Section 2 discusses related work. Section 3 describes the approach and architecture performed and its relationship with the problem. Section 4 describes the evaluation carried out, and the results obtained. Finally, Sect. 5 concludes the work.

2 Related Work

This section summarizes the studied articles, presenting the objectives and techniques used. The articles were obtained through a non-systematic review, considering the research topics regarding Deep Learning applied to ophthalmology.

Automated diagnosis through OCT images has been the focus of many researchers in recent years, showing the relevance and improvements in the investigation for more accurate results. In [13], a review was carried out on several biomarkers that used OCT extractions as a basis and the potential to support the identification of diabetic retinopathy, age-related macular degeneration, retinitis pigmentosa, and vitreomacular interface. In conclusion, OCT-based imaging biomarkers help detect disease early, classify disease severity (qualitatively and quantitatively), and modify the treatment regimen accordingly.

Proposed by Fu and coworkers [4], the method automatically evaluates the optic disc in representations of an OCT slice through detection via low-level reconstruction dealing with noise and shadows of vessels in OCT slices. Training data were extracted from a single provided OCT slice using the anterior retina structure. The method is applied to discover the boundary of the optic disc that appears at the end of the retinal pigment epithelium (RPE). Results showed that it was possible to identify the RPE points that connect to the optic disc boundary through the low-level method. Similarly, the algorithm for calculating the cup disc ratio (CDR) [10] proposed a process for extracting and contouring the inner limiting membrane layer (ILM), and in calculating the disc diameter, the endpoints of the RPE layer were used to define the disc margin.

The automation of retinal segmentation in OCT [12] images for diagnosing glaucoma was developed using a convolutional neural network (CNN) architecture, so it was possible to identify the nine layers of the retina and the optic disc for the final segmentation. The following were applied: an optic disc detection neural network (NN), a neural network for the segmentation of retinal layers, and a fusion module. In another study [22], two separate neural networks were used to perform automated segmentation of the optic disc boundary and segmentation of the peripapillary retinal layer, where the final boundary of the peripapillary retina was calculated based on prediction and gradient map, using a multiweight graph search algorithm. The study mainly applied the U-Net architecture. A DL framework for segmenting multiple retinal layers and delineating fluid pockets in OCT images of the eye [15] was implemented using a Fully Convolutional Neural Network (FCN) adapted for semantic segmentation.

It is also possible to highlight works relevant to the segmentation of the optic disc's blood vessels or the segmentation of layers but related to other diseases, such as insufficiently corrected refractive error or age-related macular degeneration. The use of deep learning in these cases was relevant due to its possible generalization and application in cases of glaucoma. The U-Net [14] and FCN [19] inspire most architectures presented for clinical image segmentation.

There are U-Net extensions, which presented a densely connected convolutions [2], or demonstrating the possibility of having multiple U-Nets connected in a chain fashion, for the segmentation of blood vessels of the optic disc [23]. Further, the partially supervised form of retinal segmentation [18] was used in OCT images, which consists of a so-called "student-teacher" approach applied in a set with and without annotation. A new multi-prediction guided attention network (MPG-Net) [5] for automated segmentation of the retinal layer in OCT images was also created, consisting of two main steps to strengthen the discriminative power of a U-shaped FCN network and perform automated segmentation. Lastly, a framework was created to applied segmentation in OCT images of normal and pathological eyes, combining a CNN with Bidirectional Long Short-term Memory (BLSTM) [7].

Finally, it can be highlighted that OCT technology provides a range of information regarding the evaluation and progress of vision-related deformities, such as the evaluation of morphometry parameters and topography of the optic nerve head, calculation of the excavation proportion for the optic disc, cup volume, rim area, and rim volume [11]. These are essential concepts that help in investigations into the evolution of retinal damage.

3   Proposed Approach

Some studies demonstrate the importance of obtaining evaluation parameters (biomarkers). One of the ways observed was using the Pythagorean theorem [11] to assist in evaluating the vertical optic nerve head using SD-OCT enhanced depth imaging (EDI). The vertical B-scan with the largest cup seen in infrared is used for hypotenuse measurement. The depth of posterior displacement of the lamina cribrosa (LC), measured from the opening of Bruch's membrane level (BMO), and the length of the excavation between the BMO boundary and the ILM layer, form the sides of a right triangle, used in the calculation (Fig. 1 right). The discovery of the hypotenuse is a helpful parameter for the quantitative assessment of optic nerve head remodeling in patients with glaucoma. Therefore, the motivation of the work is associated with supporting the identification of the new biomarker, applying computational techniques for the measurement, and quantitative analysis of the identified method.

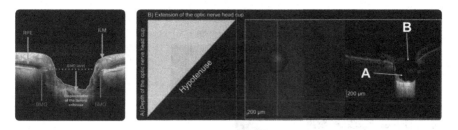

Fig. 1. Left figure shown each part of retina, in a B-Scan. Right figure shown the hypotenuse measurement [11].

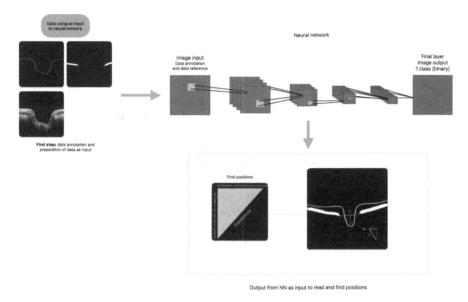

Fig. 2. General overview

In order to obtain analysis on the mentioned biomarker, the research was developed in three stages. The first stage includes collecting the OCT images with the physician to take notes and select the regions of interest. In the second stage, these images are used as input data in a convolutional neural network for semantic segmentation. In the last stage the output of the neural network, containing the segmentation images, a computer vision algorithm is applied to obtain coordinates that will result in the identification of the region of interest. This last stage is performed by implementing the biomarker approach defined by [11]. The results are compared with those analyzed by the specialist in the manual experiments in a previous study. The Fig. 2 shown an overview of each step.

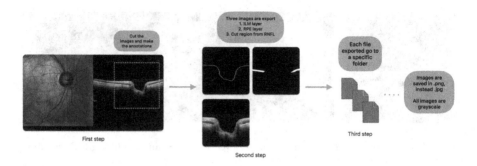

Fig. 3. Annotated data flow.

3.1 Data Annotation

This research developed a new annotated dataset to support the experiments. The overall context and main procedures are commented in this section.

All original research participants underwent SD-OCT (Spectralis OCT; Heidelberg Engineering GmbH, Dossenheim, Germany). Moreover, the scans (B-Scan and C-Scan) were exported in JPEG format files for the current research. Each image represents only one eye (right or left), with the union of the two scans in the same image. Finally, Photoshop image editing software was used to support the cuts, annotations, and export to the desired sizes.

As shown in Fig. 3, the first step refers to loading the images and selecting the region of interest to perform the cut. All images had the same size after cropping. Cropping all images with the same size is essential for input data in the neural network. In supervised neural networks, the image task must have the same pattern as the input data. The second step, shown in the exact figure, is the result of annotating the regions of interest after making the cuts. This step was conducted with a medical student to obtain the layers of the retina as accurately as possible, where files were obtained with the notes of each layer and later merged into the primary image. Each layer was exported separately in the annotation process: ILM layer annotation, RPE layer annotation, and the selected region of the retinal layer. The exported clippings were turned into PNG files because the format allows lossless compression and less loss of quality. The files are organized in different folders in the third step according to each export performed. This distinction is because the neural network requires actual (or ground truth) and annotated data as input. Therefore, the annotations images are to use with the image cropped under the same region.

The set of images obtained contains 80 folders with images of both eyes. However, some exclusions were made. Folders that did not contain one or both images, corrupted files, and folders with the same images were exclusion criteria. Therefore, from the total of 160 images, 134 images remained for use.

Throughout the experiments, the number of images was also modified. Tests were carried out with the entire set of images and new exclusions and separations. Tests without images with high contrast or noise were also applied. The same equipment can generate important differences in terms of noise presence or overall illumination in the image. This context fostered some experiments with image subsets, according to these main aspects.

We apply some different tests with the images obtained during the research period: test using all data; changing the process for choosing the images, where the images that seemed to present greater similarity in overall aspects were chosen manually; the way of annotating the RPE layer where outlines have been replaced by fills (this aspect was possible because to us it's essential the limit of RPE layer on level of BMO as shown on Fig. 1 left).

3.2   CNN Architecture

The present work used a neural network to perform tasks related to image segmentation. The use of the U-Net architecture [14] was defined as a starting point. Improvements were then included, in order to foster the results.

The architecture consists of a contraction path to capture context and a symmetric expansion path for localization. The contraction path consists of a typical convolutional network, with two convolutions followed by a ReLU (rectified linear unit) activation function and a max polling operation. The use of the regularization dropout technique to reduce overfitting was also considered. The number of resource channels doubles each reduction step (downsampling). The expansion path consists of an upsampling of the feature map, followed by an up-convolution that halves the number of feature channels, a concatenation with the corresponding clipped feature map from the contracting path, and two convolutions followed each by the ReLU function. Still, according to the authors, clipping is necessary due to the loss of edge pixels in each convolution. In the final layer, a 1×1 convolution maps each feature vector to the desired number of classes.

This is a network with no fully connected layers and which uses only the valid part of each convolution, i.e., the segmentation map only contains the pixels for which the entire context is available in the input image. It also allows the continuous segmentation of arbitrarily large images by a block overlay strategy, predicting the pixels in the region of the edge of the image that, due to the missing context, are extrapolated, mirroring the input image.

Changes were made to the architecture (Fig. 4) based on two articles, [8,21]. In both, there is the presence of additional blocks between the path of contraction and expansion. These modules allowed the extraction of information from the semantic context and the generation of high-level resource maps. The dense atrous convolution (DAC) module was used for the present work. DAC was designed to extract features from objects with different sizes using a set of dilated convolutions.

Dilated convolution was initially proposed for computing the wavelet transform. Allows insertion of zeros between consecutive values along the spatial dimension. Standard convolutions have a rate of change equal to 1, while atrous convolution allows changing this value. This type of operation supports the exponential expansion of receptive fields without loss of resolution or coverage [21]

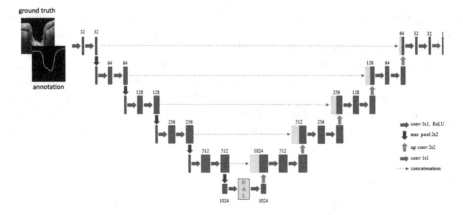

Fig. 4. U-Net Architecture.

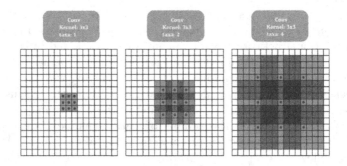

Fig. 5. Dilated convolution [21].

(Fig. 5). Thus, a DAC block was inserted between the contraction and extension paths to extract the features at different scales. Based on Li and coworkers, the same branch structure was used within the DAC module (Fig. 6), only changing the feature map from 512 to 1024.

Finally, the cross entropy loss function was initially used for model training due to its ability to calculate the difference between two probability distributions, which in this case are the pixels for the background and foreground. However, objects in medical images, such as the optic disc and retinal layers, occupy small regions compared to the whole. According to [8], cross-entropy is not performant in these cases and can be replaced by the loss function known as the Dice coefficient, which looks for the correlation between the prediction points and the annotations.

3.3 Data Augmentation

Some surveys often have the opportunity to use large amounts of data [22]. However, there are cases in which data collection could not be more voluminous

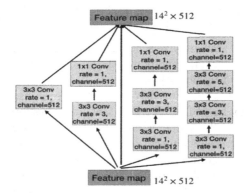

Fig. 6. Dense atrous convolution (DAC) [8].

due to access difficulties (restricted data), collection time, and variability to avoid creating bias, among other reasons. All this makes it impossible to have a more malleable set or with large variations. For this reason, the data augmentation technique allows the virtually simulating of new data based on the original input data, helping to increase the amount of data available for experiments.

Many modifications can be carried out using some specific API (e.g., Keras or PyTorch). However, it is also possible to carry out manually, implementing codes using libraries such as OpenCV, which allow greater flexibility in using functions. Operations as horizontal flip, rotation, scale and shift was used.

3.4 Measurement of Cup Portion on ONH Structure

The capture, annotation, and prediction process allows a flow to obtaining segmented images by a neural network. This result is then used as a final step to obtain values in the retinal excavation region. In this way, the calculation presented [11] can be carried out and evaluated computationally.

The first necessary step is to obtain the coordinates and excavation length value between the BMO boundary and the ILM layer. Next, the segmented images of the two layers are joined to be able to search the depth. The depth of posterior displacement of the lamina cribrosa (LC), measured from the line created at the level of opening of Bruch's membrane level (BMO), is then used as a source for obtaining the measurement.

The values obtained make it possible to assume that they are the legs in the Pythagorean equation. Moreover, in this way, obtaining the hypotenuse is discovered in pixel value. The last step is the conversion from pixel to micron, the unit of measurement used by the scanning equipment. This entire process is performed using the Python programming language.

4   Conversion to Micron Meter

The values found are still in pixels at the end of the search step for regions connected to the cup region. Therefore, there is a need for conversion to the measurement unit used by physicians.

The conversion process is currently manual since we use a scale available in the image obtained by EDI SD-OCT. This scale is always found in the lower part of the images and in an "L" shape, where the vertical bar represents a scale, and the horizontal bar represents another scale.

Using the ImageJ software, it is possible to convert the pixel distance of each orientation to the displayed reference (200 μm), as shown in the Fig. 7. The returned values are then used as fixed parameters in the flow presented in the previous section.

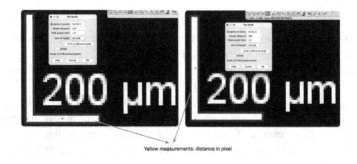

Yellow measurements: distance in pixel

Fig. 7. Measurements using ImageJ software.

5   Results

This section describes the experiments with the neural network architecture and the results obtained together with the results of the final stage, where the discovery of the hypotenuse over the disc region of the cup is performed and compared with the physicians' measurements.

Also, it's important mentioned that for the neural network, the implementation is based on the Keras API running under TensorFlow. All training was performed using Google Colab connected to GPUs.

The NN U-net trained images layers separately: one applied to images of the ILM layer and another to the RPE layer. For training and testing, we separated the images to leave 70% for training and 30% for validation and testing (half for each). Different hyperparameters and changes in the number of epochs were applied to train the model.

5.1   Evaluation of NN Segmentation

For the model evaluation, we decided to use mainly two metrics for segmentation tasks. The evaluation metrics were the Intersection Over Union (IoU) and F1-Score (F1). Other metrics, such as Recall and Precision, were also analyzed together, as they are part of the previous metrics' equation.

As mentioned in the work of Rezatofighi and coworkers, IoU, the Jaccard index, is the most commonly used metric for comparing the similarity between two arbitrary shapes. IoU encodes the shape properties of the objects under comparison, e.g., the widths, heights, and locations of two bounding boxes, into the region property and then calculates a normalized measure that focuses on their areas [9]. Also, according to another work, both metrics (IoU and F1) measure the same aspects and provide the same system classification. Therefore, the use of each one may depend on the case [20].

5.2   Results over Segmentation Using NN

It was possible to carry out four tests with the U-net network during the study period. The first one will not be included in the present study since the sample contains less than ten images. The other tests were performed with different numbers of images, changes in hyperparameters, re-annotation of images, and excluding some[1].

To outermost layer (ILM), it was possible to notice that there was a slight variation in the results obtained (Table 1) where most of the collected images there was little difference in contrast and variation in quality besides being a border region between the background and the scanned retina. In this way, it was possible to obtain results in this layer with less variation.

Table 1. Results obtained for the ILM layer.

	Epochs	T	Q1	Q2	F1	IoU	Recall	Precision
ILM (test 2)	90	134	96	38	0.69007	0.53283	0.77565	0.62636
ILM (test 3)	80	80	68	12	0.73776	0.58723	0.83270	0.66269
ILM (test 4)	60	77	65	12	0.76951	0.62975	0.83738	0.71587

Table 2. Results obtained for the RPE layer.

	Epochs	T	Q1	Q2	F1	IoU	Recall	Precision
RPE (test 2)	50	134	96	38	0.49687	0.34333	0.55054	0.45836
RPE (test 3)	50	80	68	12	0.53669	0.37704	0.58107	0.51507
RPE (test 4)	70	77	65	12	0.86320	0.77079	0.91495	0.84559

---

[1] To Table 1 and Table 2, was abbreviate some columns. Meaning of: **T**: Total images, **Q1**: Quantity of train images, and **Q2**: Quantity of test images.

In contrast, the RPE layer initially obtained lower values than the other region (Table 2). We noticed that the neural network did not perform well when detecting a more internal part with a thin line (annotation). Architectural changes were made but without success. The annotations for the last two tests were then modified, containing thicker annotations, focusing more on the boundary region between the RPE and the BMO since it is vital for the final calculation.

The results also show that adjusting the neural network and defining it as an isolated neural network for each image context is necessary. In this way, hyperparameters and configurations apply to each case.

5.3   Discovery of the Hypotenuse in the Excavation Region

Experiments were carried out at this stage to understand the feasibility of using computer vision libraries, such as OpenCV in Python, to identify measurements in the excavation region.

The algorithm created to identify what was described in the Sect. 3.4 obtained results that were compared with the values obtained by the physician in his study. In Table 3, it is possible to see the comparisons between each measured part and the final result, the hypotenuse. The final differences between the measurements taken by the doctor and those found by the algorithm can still be caused by manual measurement and the erroneous reading of some pixels in the final step. There is a need for a larger sample of data for comparison.

Table 3. Results obtained for hypotenuse findings (μm)

length	length-alg	depth	depth-alg	hyp	hyp-alg	diff: hyp
978	1062.5	288	384	1019.52	1129.762	110.239
517	662.5	150	212	538.320	695.593	157.273
699	887.5	313	424	765.878	983.581	217.703
1559	1725	385	476	1605.84	1789.469	183.634
985	1162.5	481	592	1096.17	1304.557	208.389
267	275	132	124	297.847	301.664	3.816
1377	1550	327	364	1415.29	1592.167	176.873
1338	1487.5	502	584	1429.07	1598.034	168.961
924	1037.5	486	572	1044.02	1184.732	140.715

As shown in the Table 3, columns: length, depth and hypo are the values obtained by the physician. Columns with the same name but with sufix "-alg" it is a reference to the algorithm created and described in the Sect. 3.4. The column "diff: hypo" show the difference between both results obtained by the physician and the algorithm. The results still are considere high, where it is necessary a new study to evaluate what may have affected this difference.

6   Conclusion

OCT is a modern and sophisticated image extraction technology that can perform excellent results in scanning the retina and inner layers. However, several studies still enable new ways for more accurate disease analysis. In the present study, deep learning focused on semantic segmentation was evaluated to enable the possible results and benefits of the technique and used to perform a mathematical calculation to reproduce the hypotenuse measurement over the optic nerve's head.

The retina structure presents very sensitive characteristics in which detailed analyses are necessary. In this study, a specialist physician and a medical student participated, enabling data availability and helping in the steps of notes taken.

It was possible to use neural networks to identify the layers of the retina and their segmentation. The results obtained, even though they need adjustments, were used to measure the region of the cup in the optic nerve head and calculate the hypotenuse. There are still improvements that need to be made, as well as new studies. It also noticed the need to separate the neural network in the future, applying different architectures for each layer of the annotated retina.

One limiting point that follows since the beginning of the research is the number of images. In which, as highlighted in Sect. 3.1, many images were excluded by quality criteria. There is still a significant challenge in working with variations in images, whether they are: contrast or the presence of noise. Different from the quality of the images, even with the cuts made, we use images in large sizes compared to other datasets of different themes.

The use of images from only one OCT device can be a factor that negatively contributes to the creation of a bias since we do not know the limitations and the pattern of the generated images. Still, there is more than one type of glaucoma, which can also affect results generation if the region does not have a similar pathology.

Moreover, finally, in order to calculate the hypotenuse, there is still a need to investigate how to convert the pixel aspect ratio to the micron unit (μm) without using secondary software. Currently, it is necessary to open some image through the ImageJ software and to read informations from scale bar, where it is possible know the distances of micron in pixels. This is necessary to convert the results from hypotenuse findings (Sect. 4).

From future adjustments, both in the neural network and the extraction algorithm, it will be possible to obtain even better results since the hypotenuse measurement over the excavation region is simple and reproducible. Moreover, it contributes to the possibility of making this approach viable as a support tool in diagnosing patients with glaucoma.

The ethics committee of the Hospital de Clinicas de Porto Alegre approved the study and it was conducted in accordance with the Declaration of Helsinki. Informed consent was obtained from all participants.

References

1. Andrade, J.C.F., Kanadani, F.N., Furlanetto, R.L., Lopes, F.S., Ritch, R., Prata, T.S.: Elucidation of the role of the lamina cribrosa in glaucoma using optical coherence tomography. Surv. Ophthalmol. **67**(1), 197–216) (2022). https://doi.org/10.1016/j.survophthal.2021.01.015

2. Azad, R., Asadi-Aghbolaghi, M., Fathy, M., Escalera, S.: Bi-directional ConvLSTM U-Net with Densley connected convolutions. In: Proceedings - 2019 International Conference on Computer Vision Workshop, ICCVW 2019, pp. 406–415 (2019). https://doi.org/10.1109/ICCVW.2019.00052

3. Beykin, G., Norcia, A.M., Srinivasan, V.J., Dubra, A., Goldberg, J.L.: Discovery and clinical translation of novel glaucoma biomarkers. Progress Retinal Eye Res. **80** (2021). https://doi.org/10.1016/j.preteyeres.2020.100875

4. Fu, H., Xu, D., Lin, S., Wong, D.W., Liu, J.: Automatic optic disc detection in OCT slices via low-rank reconstruction. IEEE Trans. Biomed. Eng. **62**(4), 1151–1158 (2015). https://doi.org/10.1109/TBME.2014.2375184

5. Fu, Z., et al.: MPG-Net: multi-prediction guided network for segmentation of retinal layers in OCT images. In: European Signal Processing Conference, pp. 1299–1303, January 2021. https://doi.org/10.23919/Eusipco47968.2020.9287561

6. GBD 2019 Blindness and Vision Impairment Collaborators: Vision Loss Expert Group of the Global Burden of Disease Study. Causes of blindness and vision impairment in 2020 and trends over 30 years, and prevalence of avoidable blindness in relation to VISION 2020: the Right to Sight: an analysis for the Global Burden of Disease Study [published correction appears in Lancet Glob Health. 2021 Apr; 9(4):e408]. Lancet Glob Health **9**(2), e144–e160 (2021). https://doi.org/10.1016/S2214-109X(20)30489-7

7. Gopinath, K., Rangrej, S.B., Sivaswamy, J.: A deep learning framework for segmentation of retinal layers from OCT images. In: Proceedings - 4th Asian Conference on Pattern Recognition, ACPR 2017, pp. 894–899 (2021). https://doi.org/10.1109/ACPR.2017.121

8. Gu, Z., et al.: CE-Net: context encoder network for 2D medical image segmentation. IEEE Trans. Med. Imaging **38**(10), 2281–2292 (2019). https://doi.org/10.1109/TMI.2019.2903562

9. Rezatofighi, H., Tsoi, N., Gwak, J., Sadeghian, A., Reid, I., Savarese, S.: Generalized intersection over union: a metric and a loss for bounding box regression. CoRR (2019). https://doi.org/10.48550/arXiv.1902.09630

10. Khalil, T., Akram, M.U., Raja, H., Jameel, A., Basit, I.: Detection of glaucoma using cup to disc ratio from spectral domain optical coherence tomography images. IEEE Access **6**, 4560–4576 (2018). https://doi.org/10.1109/ACCESS.2018.2791427

11. Khalil, T., Akram, M.U., Raja, H., Jameel, A., Basit, I.: Detection of glaucoma using cup to disc ratio from spectral domain optical coherence tomography images. IEEE Access **6**, 4560–4576 (2018). https://doi.org/10.1109/ACCESS.2018.2791427

12. Li, J., et al.: Multi-scale GCN-assisted two-stage network for joint segmentation of retinal layers and disc in peripapillary OCT images. Biomed. Opt. Express **12**, 2204–2220 (2021)

13. Phadikar, P., Saxena, S., Ruia, S., Lai, T.Y.Y., Meyer, C.H., Eliott, D.: The potential of spectral domain optical coherence tomography imaging based retinal biomarkers. Int. J. Retina Vitreous **3**(1), 1–10 (2017). https://doi.org/10.1186/s40942-016-0054-7

14. Ronneberger, O., Fischer, P., Brox, T.: U-Net: convolutional networks for biomedical image segmentation. IEEE Access **1**, 16591–16603 (2015). https://doi.org/10.1109/ACCESS.2021.3053408

15. Roy, A.G., et al.: ReLaynet: retinal layer and fluid segmentation of macular optical coherence tomography using fully convolutional networks. Biomed. Opt. Express **8**, 3627–3642 (2017). https://doi.org/10.1364/boe.8.003627

16. Sander, B., Larsen, M., Thrane, L., Hougaard, J.L., Jørgensen, T.M.: Enhanced optical coherence tomography imaging by multiple scan averaging. Br. J. Ophthalmol. **89**(2), 207–212 (2005). https://doi.org/10.1136/bjo.2004.045989

17. Schmidt-Erfurth, U., Sadeghipour, A., Gerendas, B.S., Waldstein, S.M., Bogunović, H.: Artificial intelligence in retina. Prog. Retin. Eye Res. **67**, 1–29 (2018). https://doi.org/10.1016/j.preteyeres.2018.07.004

18. Sedai, S., et al.: Uncertainty guided semi-supervised segmentation of retinal layers in OCT images. In: Shen, D., et al. (eds.) MICCAI 2019. LNCS, vol. 11764, pp. 282–290. Springer, Cham (2019). https://doi.org/10.1007/978-3-030-32239-7_32

19. Shelhamer, E., Long, J., Darrell, T.: Fully convolutional networks for semantic segmentation. IEEE Trans. Pattern Anal. Mach. Intell. **39**(4), 640–651 (2017). https://doi.org/10.1109/TPAMI.2016.2572683

20. Taha, A.A., Hanbury, A.: Metrics for evaluating 3D medical image segmentation: analysis, selection, and tool. BMC Med. Imaging (2015). https://doi.org/10.1186/s12880-015-0068-x

21. Yu, F., Koltun, V.: Multi-scale context aggregation by dilated convolutions. In: 4th International Conference on Learning Representations, ICLR 2016 - Conference Track Proceedings (2016)

22. Zang, P., Wang, J., Hormel, T.T., Liu, L., Huang, D., Jia, Y.: Automated segmentation of peripapillary retinal boundaries in OCT combining a convolutional neural network and a multi-weights graph search. Biomed. Opt. Express **10**(8), 4340 (2019). https://doi.org/10.1364/boe.10.004340

23. Zhuang, J.: LadderNet: multi-path networks based on U-Net for medical image segmentation, pp. 2–5 (2019)

Investigation of Deep Active Self-learning Algorithms Applied to Named Entity Recognition

José Reinaldo Cunha Santos A. V. Silva Neto[1]([⊠]) and Thiago de Paulo Faleiros[2]

[1] Osaka University, Osaka, Japan
josereinaldoneto@gmail.com
[2] University of Brasilia, Brasilia, DF, Brazil
thiagodepaulo@unb.br

Abstract. Active Self-Learning algorithms reduce the labeled data required to train a Machine Learning model through supervised training. This paper explores various Active Self-Learning algorithms for named entity recognition tasks. Firstly, we investigate the impact of different self-training techniques on Active Self-Learning algorithms. Secondly, we propose a novel token-level Active Self-Learning algorithm that achieves near-peak performance using fewer hand-annotated tokens compared to existing works. Through numerous experiments, we found that the sentence-level Active Self-Learning algorithm did not consistently yield significant results compared to pure active learning. However, our proposed token-level Active Self-Learning algorithm showed promising performance, training a neural model to nearly peak accuracy with fewer human-annotated tokens compared to state-of-the-art active learning baseline algorithms. The experimental results are presented and discussed, demonstrating the superior performance of the token-level Active Self-Learning algorithm

Keywords: Active learning · Active Self-Learning · Named entity recognition · Deep learning

1 Introduction

Active learning (AL) algorithms reduce the labeled data required to train a machine learning model through supervised training. These algorithms can be generally separated into three classes: (1) pool-based, (2) stream-based, and (3) query synthesis. For this paper, we focus on the pool-based class which thrives in scenarios where large amounts of data are available, but the annotation process for the complete pool of data is costly. The objective of the pool-based AL technique is to identify, given a considerable pool of unlabeled data, a smaller subset

J. R. C. S. A. V. S. Neto—Research performed during the author's masters undertaking at the University of Brasilia (UnB).

M. C. Naldi and R. A. C. Bianchi (Eds.): BRACIS 2023, LNAI 14197, pp. 470–484, 2023.
https://doi.org/10.1007/978-3-031-45392-2_31

of data samples that represent the whole data distribution well. By annotating this smaller subset of data samples and training a model through supervised training, it is possible to achieve good model performance with a significant decrease in data annotation costs.

In Active Learning for Named Entity Recognition (NER), most research proposes using sentence-level querying strategies. Sentence-level querying means that the whole unlabeled sentence is queried to the oracle, which is expected to provide labels for all tokens. It provides more context and can lead to more accurate annotation decisions, but it can be more computationally expensive and time-consuming. Some works propose alternative strategies such as token-level [7] or subsentence-level [16] querying. These strategies aim to make the annotation process more efficient by reducing the number of tokens that need to be manually labeled.

This paper proposes cooperative approaches for reducing the cost of annotation by an oracle and speeding up the annotation process. We focus on **Active Self-Learning (ASL)** algorithms for that. The most straightforward implementation of an Active Self-Learning algorithm uses the machine learning model to predict classes for the unlabeled data. The predicted labels are then split into high-confidence and low-confidence, depending on the model's confidence in its predictions. High-confidence samples are labeled automatically using the model's predictions, while low-confidence samples are queried for the human annotator to be labeled manually. Labeled samples are added to the labeled dataset, which is used to train the machine-learning model further. This process is repeated until the desired level of model performance is achieved, or a stopping criterion is reached. In Active Self-Learning algorithms, the active learning part corresponds to the query to the human annotator. In contrast, the self-learning part corresponds to the automatic labeling done by the trained model. The ASL creates a cooperative scenario where humans and models annotate data together and can potentially reduce annotation costs compared to using only active learning strategies [21].

Our 2 main contributions in this paper are: (1) We investigate the impact of different self-training techniques on Active Self-Learning algorithms; And (2) We propose a novel Active Self-Learning algorithm based on token-level querying that achieves peak performance with less hand-annotated data than previous works.

The overview of the paper is as follows. In Sect. 2, we present relevant works from the literature on both Active Learning and Active Self-Learning algorithms applied to named entity recognition tasks. Section 3 briefly reviews the Active Self-Learning algorithm from [2] based on sentence-level querying. In Sect. 4, we investigate the impact of different Self-Learning techniques on the proposed Active Self-Learning algorithm from [2]. In Sect. 5, we propose a novel Active Self-learning algorithm based on token-level querying, where both the human annotator and machine learning model cooperatively annotate tokens from the same sentence. Section 7 we present the results of our investigations on the ASL algorithms proposed in Sects. 4 and 5. Finally, in Sect. 8, we summarize the results from both sentence-level and token-level Active Self-learning algorithms.

2   Related Works

Many studies in the current literature focusing on Active Learning algorithms for NER utilize sentence-level querying. **Shen et al.** [18] proposes, to the best of our knowledge, the first Active Learning algorithm based on neural networks for tasks of sequence tagging (e.g. NER and Part-Of-Speech tagging). Their proposed neural model uses convolutional layers for character and word-level feature encoding and an LSTM layer with greedy decoding as the tag decoder. The authors also proposed the *Maximum Normalized Log-Probability (MNLP)* sentence-level query function, which normalizes the model's confidence for a given unlabeled sentence. Their experiments show that the proposed algorithm trains the model to peak performance using only 25% of the training set of the OntoNotes5.0 dataset [15]. **Siddhant and Lipton** [19] extend previous work by Shen et al. They use the Bayesian Active Learning through Disagreement [6] (BALD) framework by querying the unlabeled sentences that generate the most disagreement over multiple passes on neural models. To introduce stochastic behavior in the neural models, they propose two solutions. The first solution is the Monte Carlo dropout, where the model makes multiple predictions for the same sentence but with a different dropout mask at a time. The second solution is the Bayes by backpropagation, where some layers have their deterministic parameters replaced by stochastic parameters. Experiments showed that the proposed query strategies performed consistently better than the previous MNLP, but this came with more complex and costly compute query functions.

In the literature on Active Learning applied to named entity recognition, sub-sentence-level querying strategies are found less often, including token-level querying strategies. The main challenge is how to implement the model training routine with sentences that are partially labeled. **Kobayashi and Wakabayashi** [7] propose the use of a multi-class logistic regression model with point-wise predictions [12]. They can use token-level queries to train their models through this method. **Radmard et al.** [16] proposes a sub-sentence-based query strategy, where only the most essential, non-overlapping sub-sentence are queried instead of the whole sentence. The strategy is to query sub-sentence without their surrounding context and store the annotation in a dictionary that associates each sub-sentence with its corresponding accurate labels. Annotations for a sub-sentence are propagated onto the unlabeled dataset, meaning that all occurrences of a particular sub-sentence are assigned the same labels. They then train a neural model using a loss function computed only on labeled tokens.

Active Self-Learning algorithms have received considerably less attention in the NER literature when compared to purely Active Learning Algorithms. **Tran et al.** [21] proposes an Active Self-Learning algorithm based on Conditional Random Fields (CRF) models. The Active Learning process is related to using a diversity measure to query the most informative samples to the oracle. At the same time, the Self-Learning process is related to using the trained CRF model from the previous iteration to annotate unlabeled sentences with high confidence in its predictions. The dataset used was composed of sentences extracted from Twitter. The experiments compared uncertainty, diversity query strategies, and Active Learning algorithms with and without Self-Learning. Results have shown that Active Self-Learning algorithms achieved, in general, better results than the purely Active Learning algorithm. Inspired by the work of **Tran et al.** [21], **Cunha and Faleiros** [2] presents another Active Self-Learning algorithm. They made specific changes to address the use of deep neural models instead of CRF. The proposed ASL algorithm has been shown to be less sensitive to the quality of the labeled set in initial iterations, when compared to the previous Active Self-Learning algorithm from the literature [1].

3   Deep Active Self-learning Algorithm

This section briefly describes the active self-learning algorithm proposed by Cunha and Faleiros [2]. The Active Self-Learning algorithms described in existing literature require collaboration between a human annotator and a trained model to annotate samples from an unlabeled database. However, this process is sensitive to the initial labeled set used to train the machine learning model [1]. We argue that this sensitivity arises from poorly annotated samples selected by the model in the early rounds of the active self-learning algorithm. This poorly annotated data may introduce permanent bias to the labeled set. To address this issue, the work of [2] proposed an active Self-Learning algorithm that distinguishes between samples labeled by the model and those labeled by the human annotator. The former has less impact on the model's parameters during training. Additionally, samples labeled automatically by the model were returned to the unlabeled set after each iteration of the active self-learning algorithm. These modifications seem to mitigate the risk of introducing permanent bias to the trained model and labeled dataset.

Algorithm 1 presents a more detailed explanation of active self-learning algorithm.

Algorithm 1. Active self-learning algorithm

1: **procedure** ASL(U, m, Q, $min\_confidence$, $epoch$)
2:     $AL \leftarrow$ init_labeled_set(U)
3:     Train_model(m, $A.L.$, $epoch$)
4:     **while** Stopping Criterion not True **do**
5:         $A.L. \leftarrow$ Active_Learning_Query(U, m, Q)
6:         $S.L. \leftarrow$ Self_Learning_Query(U, m, $min\_confidence$)
7:         Train_model(m, $A.L.$, $S.L.$, $epoch$)
8:         $U \leftarrow S.L.$

Note that in Algorithm 1, m represents the machine learning model, Q is the query budget, $min\_confidence$ is the minimum confidence level required for

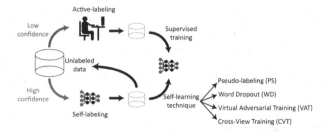

Fig. 1. Annotation process for a sentence-level active-self learning algorithm. (Color figure online)

the model to annotate an unlabeled sample, *A.L.* denotes the active labeled set that contains samples annotated by the oracle, *S.L.* contains samples labeled by the trained model, and U represents the set of unlabeled data. The active learning procedure, represented by the $Active\_Learning\_Query(\cdot)$ function, identifies the most informative unlabeled samples for annotation by the oracle. The self-learning procedure, represented by the $Self\_Learning\_Query(\cdot)$ function, corresponds to the different self-learning strategies that will be applied. We investigate the impact of different self-learning techniques based on sentence-level and word-level querying strategies.

4   Sentence-Level Active Self-learning Algorithm

This section introduces our proposed Active Self-Learning algorithms that utilize sentence-level querying. This algorithm is built upon the work of [2] by incorporating various Self-Learning techniques.

The diagram of the Active Self-Learning algorithm proposed is shown in Fig. 1. Two main characteristics make this algorithm different from previous algorithms in the literature. (1) The first change is that we now have two individual labeled sets (shown in orange in Fig. 1). One of these labeled datasets is responsible for keeping sentences hand-annotated by the oracle, thus having highly reliable labels. At the same time, the other is responsible for keeping sentences labeled by the machine learning model with less reliable labels. This separation of highly reliable and less reliable labeled data allows us to use them differently during training. We employ hand-annotated data for traditional supervised learning and utilize data labeled by the model for self-learning. (2) The second change is that after training the machine learning model in any given algorithm iteration, the data self-labeled by the machine learning model returns to the unlabeled data pool. These alterations allow a better-trained model to re-annotate samples in later iterations.

The algorithm proposed in [2] adopted pseudo-labeling as the self-learning strategy. Here, our alternate version replaces the pseudo-labeling technique with Self-Learning methods. We test three different Self-Learning techniques, namely: (1) Word dropout, (2) Virtual adversarial training, and (3) cross-view training.

Next, we will explain how our algorithm utilized pseudo-labeling and integrated these three techniques:

- The **Pseudo-Labeling (PL)** technique identifies highly reliable unlabeled samples to be automatically annotated by the model. In this case, highly reliable means that the model's confidence in its predictions is above a predefined threshold. Our strategy consists of using the machine learning model's predictions as labels. Then, apply supervised training with the pseudo-labeled and the hand-labeled data.

- The **Cross-View Training (CVT)** [3] aims to improve the representation capabilities of the model by forcing it to output similar predictions with different views of the same input data. The CVT algorithm uses neural models with auxiliary classification heads. Each head learns to predict tokens given a limited view of the input sentence (e.g., left-context only, right-context only). The auxiliary classification head comprises a fully-connected layer with ReLU [11] activation, followed by a softmax function to generate a distribution over predicted classes. The CVT loss is the Kullback-Leibler (KL) divergence between the outputs of the main classification head, which sees the complete input, and the auxiliary heads, which see partial views of the input.

- The **Word Dropout (WD)** technique [3,8] consists in replacing random words in a sentence with special tokens, such as $< removed >$ or $< UNK >$, and training the model to produce a similar output distribution as when the word was unmasked.

- The **Virtual Adversarial Technique (VAT)** is a Self-Learning technique proposed by Miyato et al. [10] for text classification and applied to NER by Clark et al. [3]. This technique extends the Adversarial Training [4], where a sample of input data is perturbed with specially crafted noise designed to fool the model. This helps the model to be more robust to small perturbations in the input data.

5   Token-Level Active-Self Learning Algorithm

The model's overall confidence for the entire sentence is considered when using sentence-level querying. This implies that the model assumes average confidence in its predictions for all the words/tokens within the sentence. However, hard-to-predict entities may be surrounded by various easy-to-predict tokens. We argue this may lead to an overall overconfidence in the model for complex tokens, leading to poor self-annotations that may hamper the Active Self-Learning process. We leverage this idea to propose a token-level Active Self-Learning algorithm.

The token-level Active Self-Learning algorithm proposed is a modified version of the Deep Active Learning (DAL) algorithm presented by Shen et al. [18]. The main difference between the original DAL and our proposed algorithm is its labeling process. The original algorithm queried the most uncertain sentences and asked the oracle to annotate them. Our proposed algorithm takes the queried sentences, identifies the tokens with low confidence, and asks the oracle to annotate them. The remaining unlabeled tokens from the queried sentence

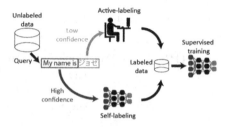

Fig. 2. An illustration of the collaborative configuration where an oracle and machine learning model annotate the same sentence jointly.

are automatically labeled using the model's output predictions. We also improve the accuracy of our self-labeling process by using hand-annotated labels to refine the predictions made by the model. This cooperative scenario can alleviate the cost of manual annotation for the oracle and speed up the annotation process by highlighting specific words in a sentence that must be hand-annotated. Our proposed token-level Active Self-Learning algorithm is illustrated in Fig. 2, which shows the annotation process.

Our proposed algorithm has three main procedures: 1) predict the highest confidence tokens to automatic annotation; 2) query low-confidence tokens to the oracle for manual annotation and 3) self-labeling refinement.

1. **Identifying low-confidence tokens**: Suppose a neural network produces a probability distribution for an input token, which represents the likelihood of the token belonging to one of the classes of named entities. In this context, we define low-confidence tokens as those for which the model has confidence in its prediction lower than a predefined threshold. We empirically selected a threshold confidence of 99%, implying that tokens with less than 99% of confidence in their predicted class will be labeled by the oracle (i.e., human annotator). In contrast, the remaining tokens will be self-labeled by the model.

2. **Query to the oracle:** It is the traditional active learning technique implemented at a token level, where only the low-confidence tokens in the selected sentences are queried to the oracle.

3. **Self-labeling refinement**: Once an oracle has labeled the low-confidence tokens, we can use self-labeling to label the remaining unlabeled tokens from the queried samples. A simple approach is to predict the classes for all tokens and use the predictions to label high-confidence tokens. However, we can leverage these labels to improve our predictions since we have the proper labels of the low-confidence tokens, which were labeled by the oracle. For example, the CNN-CNN-LSTM model [18] uses a greedy decoding approach where it receives the predicted label of the previous token as an additional input to help to predict the current token's class. If the previous token had low confidence and was manually labeled by the oracle, we could modify this approach by using the oracle-assigned label instead of the label previously predicted by the model. This process can be repeated a predefined number

of times, iteratively, with a decreasing number of tokens replacement in each iteration.

The refinement step was inspired by the iterative algorithm proposed by Park et al. [13]. They use iterative refinement to identify synonyms to substitute specific tokens from a sentence, with a low impact on its coherence. The idea is to identify potential synonyms for specific tokens, replace them, and verify if a masked language model predicts the synonyms with high confidence. Other synonyms replace the synonyms with the lowest masked language model scores. For our refinement step, however, we use reliable Oracle annotated tokens to enhance the model predictions for the unlabeled tokens.

6   Experimental Design

We use two consolidated English NER datasets, namely CoNLL03 [17] and OntoNotes5.0 [15], and one legal domain Portuguese NER dataset, named *Aposentadoria* [2]. *Aposentadoria* is a new legal domain NER dataset. It contains named entities from 10 classes associated with retirement acts of public employees from the *Diário Oficial do Distrito Federal* (Brazilian Federal District official gazette, in direct translation). The datasets chosen for use are listed in Table 1, along with relevant information such as their language, subject domain.

Table 1. Datasets description.

Dataset		CoNLL03	OntoNotes5.0	Aposentadoria
Domain		Reuters News	Variety	Brazilian legal texts
Language		English	English	Portuguese
Train set	Sent.	14,987	59,924	3,860
	Token	203,621	1,088,503	311,231
Valid set	Sent.	3,466	8,528	828
	Token	51,362	147,724	68,740
Test set	Sent.	3,684	8,262	827
	Token	46,435	152,728	65,912

We use two neural models for sentence-level and one for token-level Active Self-Learning. For sentence level, we use CNN-CNN-LSTM proposed by Shen et al. [18] and the CNN-biLSTM-CRF proposed by Ma and Hovy [9]. We could not use CNN-biLSTM-CRF in the token-level case because CRF classification layers output a distribution over the entire sentence instead of a distribution of classes for each token. More details about the model's and training algorithm's hyperparameters are presented in the following subsections.

We evaluate four versions of our proposed Active Self-Learning algorithm using self-training techniques at the sentence level (Sect. 4). We also compare the results with the deep active learning algorithm proposed by Shen et al. [18]. Moreover, we compare the token-level Active Self-Learning algorithm with the subsequence-based active learning algorithm by Radmard et al. [16]. This algorithm was chosen as a baseline because its queries use sub-sentences instead of whole sentences.

For all proposed algorithms, we compute the maximum normalized log-probability (MNLP) measure proposed by Shen et al. [18] to generate queries for new samples. The MNLP value for a sentence X of length n can be calculated as

$$MNLP(X) = \max_{y_1,...,y_{N-1}} \frac{1}{n} \sum_{i=0}^{n} log \ P(y_i|x_i,y_0,y_1,...,y_{i-1}). \tag{1}$$

where x_i is the i-th token of a sentence X, and y_i is the class predicted for the i-th token x_i.

The querying for the oracle and the trained model is based on the MNLP measure. We use the least confidence for the Oracle annotation, where unlabeled samples with the lowest MNLP scores in a given algorithm iteration will be queried to the Oracle. For automatic annotation, we use the exponentiated MNLP measure. The unlabeled samples with exponentiated MNLP measure higher than a predefined threshold will be used for self-training in the next iteration of the Active Self-Learning algorithm. For all experiments, the threshold selected empirically was 0.99. This threshold means that the model's prediction confidence must be equal to or higher than 0.99 for the sample to be used for self-training.

Similar to previous works in the literature [18,19], we apply early stopping of the model training. The early stopping is based on the model's performance on the validation set. We empirically chose a patience of 15 epochs, meaning that if the model's performance does not improve in 15 consecutive epochs, the model's training is interrupted.

6.1   CNN-CNN-LSTM Model

The CNN-CNN-LSTM model consists of a character-level convolutional encoder, a word-level convolutional encoder, and an LSTM tag decoder.

The character-level CNN is used to generate character-level vector representations of a word. This CNN works by first transforming each character of a word into a vector representation, with embeddings being initialized with uniform samples from $\left[-\sqrt{\frac{3}{dim}}, +\sqrt{\frac{3}{dim}}\right]$ as proposed by Ma and Hovy [9]. Dropout [20] is applied to the generated embeddings as presented by Ma and Hovy [9]. Then, a one-dimensional convolutional layer is used to extract information of neighboring characters, followed by ReLU activation [11] and max-pooling to generate fixed-size character-level word embeddings.

We generate a word representation by concatenating a character-level vector with a vector representation from a lookup table, which we initialize with pre-trained word embeddings that we update throughout the training process.

We used GloVe embeddings of 100 dimensions pre-trained on English newswire corpus [14], and GloVe embeddings of 300 dimensions pre-trained on multi-genre Portuguese corpus [5].

The LSTM tag decoder is responsible for the tagging of each word. The LSTM tag decoder uses a one-hot encoded vector of the previous tag, concatenates it to the encoded vector representation of the current word, and uses it as input.

A thorough explanation of the hyperparameter tuning routine and hyperparameter values used in our experiments is presented in [1].

6.2   CNN-biLSTM-CRF Model

Similarly to the CNN-CNN-LSTM model, the CNN-biLSTM-CRF model uses a CNN to generate a character-level representation for each word. The full embedding, formed by the concatenation of character-level and word-level embeddings, is fed to a biLSTM layer that generates vector representations for each word in a sentence. A fully-connected layer is then used to reduce the encoded vector's dimension to the number of possible tags. The reduced dimension vector is then fed to a CRF layer. Dropout layers are used before and after the biLSTM layer. All weight matrices for the biLSTM and fully-connected layers are randomly initialized using a uniform distribution to select samples from $[-\sqrt{\frac{6}{r+c}}, +\sqrt{\frac{6}{r+c}}$ as proposed by Ma and Hovy [9], where r and c are the numbers of rows and columns in the weight matrix. We initialize the bias parameters from the biLSTM layer to 0.0 and the forget gate bias to 1.0.

The model's hyperparameters for the English NER datasets are similar to those presented in the experiments of Siddhant and Lipton [19]. A grid search was employed for the Portuguese dataset to define the model's hyperparameters. A thorough explanation of the hyperparameter tuning routine and hyperparameter values used in our experiments is presented in [1].

7   Results

We present and discuss the experimental results focusing on the performance overview (f1-score) of the models trained on each of the three datasets (see Table 1). Additionally, we compare the performance of neural models trained using various Active and Active Self-Learning algorithms. In separate sections, we present the results of sentence-level and token-level querying strategies.

7.1   Results of Sentence-Level Strategy

The ASL algorithm that uses pseudo-labeling as the default Self-Learning strategy as proposed by [2], is referred to as DASL (Deep Active Self-Learning). To conduct an almost comprehensive experiment with various Self-Learning methods, we replace pseudo-labeling with three other techniques: DASL_WD (Deep Active Self-Learning with Word Dropout), DASL_VAT (Deep Active Self-Learning with Virtual Adversarial Technique), and DASL_CVT (Deep Active

Self-Learning with Cross-View Training). We also compare the results of Active Self-Learning strategies with Deep Active Learning (DAL). In Fig. 1, Supervised (i.e. dashed line) represents the best performance of the neural models when trained using the entire training set with true labels in a supervised fashion. DAL represents the baseline active learning algorithm by Shen et al. [18].

Figure 3 presents the performance (i.e., f1-score) achieved by the models and the percentage of oracle annotated tokens (human labeled tokens). It should be noted that Active Self-Learning algorithms use additional data for training in addition to the hand-annotated data. Note that the DASL_CVT technique is only implemented with the CNN-biLSTM-CRF model. This technique required multiple views of the input, which is more computationally costly to implement in neural models with CNN-only encoders such as the CNN-CNN-LSTM.

From the experiments performed in Fig. 3, we noticed that no technique performed consistently better than the others. However, for the *Aposentadoria* dataset, the DASL_VAT and DASV_CVT techniques help the model achieve its best performance faster than the baseline algorithms. We believe this behavior is because this dataset is less imbalanced than the other datasets, meaning that most NER entities are present in most sentences. As a result, the model generalizes faster with less hand-annotated data and more effectively utilizes unlabeled data through self-training in the early iterations of the Active Self-Learning algorithms.

7.2   Results of Token-Level Strategy

In this section, we present the results of the token-level Active Self-Learning algorithm proposed. Our algorithm, namely MDAL, is a modified version of Deep Active Learning (DAL).

Figure 4 compares the original and the modified DAL algorithms, both using the validation f1-score for early stopping. The graphs show that both the original and the modified DAL algorithms achieve similar performance with the same amount of labeled data but with our modified algorithm allowing for tokens to be self-labeled. From the graphs in the center, we observe that the models trained with the modified DAL algorithm reach peak performance with significantly less hand annotated tokens. This is justified by the fact that for the modified algorithm, most tokens are self-annotated by the trained model reliably, as shown in the graphs to the right in Fig. 4. Table 2 presents the percentage of hand-annotated tokens required for the trained model to reach 99% of its peak f1-score performance. We observe that the proposed method requires significantly fewer hand-annotated tokens. However, we do note that these results do not consider that many of the tokens in the queried sentences were not named entities and therefore were not required to be annotated. This is the reason why our algorithm is capable of training a model to peak performance with only 6.96% of human-labeled tokens in the CoNLL2003 dataset when compared to the 36.20% and 27% of the baseline algorithm. However, for fine-grained NER datasets, such as the OntoNotes5.0, and datasets with few tokens that do not have a named entity class, such as the Aposentadoria dataset, the presented results are convincing

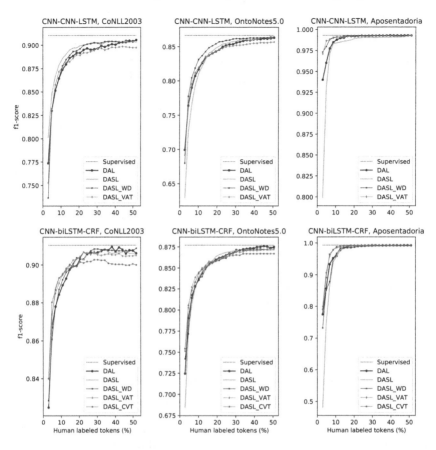

Fig. 3. The graph shows the performance (f1-score) of the neural models as a function of the percentage of hand-annotated tokens in each iteration of the Active and Active Self-Learning algorithms.

evidence that our method is capable of reducing the human annotation costs in the active learning process when compared to the baselines.

Table 2. Percentage of the training set that was annotated by the oracle in order to train a model that reaches 99% of its peak performance.

Algorithm	Dataset		
	CoNLL2003	OntoNotes5.0	Aposentadoria
Shen et al.	36.20%	25.91%	8.68%
Radmard et al.*	27%	13%	-
MDAL (ours)	6.96%	11.54%	4.73%

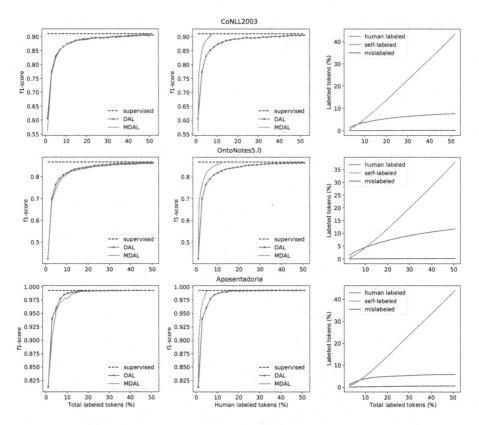

Fig. 4. Comparison between the original and modified DAL algorithms. Graphs on the left column compare the performance of the trained model (*y-axis*), by the total amount of labeled data (*x-axis*), including tokens annotated by the oracle and self-annotated by the trained model in the case of our modified algorithm. Graphs on the center column compare the f1-score of the trained model (*y-axis*) by the amount of data annotated by the Oracle (*x-axis*). Graphs on the right column compare the percentage of the training set that was annotated by a human, by the model through self-labeling, and the samples that were mislabeled. In all graphs, DAL represents the original deep active learning algorithm from the literature, while MDAL indicates our modified algorithm with token-level self-labeling.

8   Conclusion

In this work, we presented an investigation into different types of active-self learning algorithms for named entity recognition tasks. From the many experiments performed, the sentence-level Active Self-Learning algorithm could not consistently achieve significant results compared to pure active learning. However, the proposed token-level Active Self-Learning could train a neural model to near-peak performance using fewer human-annotated tokens compared to the state-of-the-art algorithms used as baselines. Our proposed token-level algorithm is particularly effective for fine-grained datasets where most tokens are assigned a named entity class.

While we failed at creating a sentence-level Active Self-Learning that overcomes the current state-of-the-art, future research may investigate how pretrained transformer models may impact these algorithms. Bridging the current gaps in sentence-level Active Self-Learning algorithms research using models with substantial a priori information may be possible.

Acknowledgements. The authors were supported by the *Fundação de Apoio a Pesquisa do Distritio Federal (FAP-DF)* as members of the *Knowledge Extraction from Documents of Legal content (KnEDLe)* project from the University of Brasilia.

References

1. Neto, J.R.C.S.A.V.S.: Deep active learning approaches to the task of named entity recognition. Masters Dissertation [University of Brasilia] (2021)
2. Neto, J.R.C.S.A.V.S., Faleiros, T.P.: Deep active-self learning applied to named entity recognition. In: Britto, A., Valdivia Delgado, K. (eds.) BRACIS 2021. LNCS (LNAI), vol. 13074, pp. 405–418. Springer, Cham (2021). https://doi.org/10.1007/978-3-030-91699-2_28
3. Clark, K., Luong, M.T., Manning, C.D., Le, Q.: Semi-supervised sequence modeling with cross-view training. In: Proceedings of the 2018 Conference on Empirical Methods in Natural Language Processing, pp. 1914–1925. Association for Computational Linguistics, Brussels, Belgium, October–November 2018
4. Goodfellow, I.J., Shlens, J., Szegedy, C.: Explaining and harnessing adversarial examples (2015)
5. Hartmann, N.S., Fonseca, E.R., Shulby, C.D., Treviso, M.V., Rodrigues, J.S., Aluísio, S.M.: Portuguese word embeddings: evaluating on word analogies and natural language tasks. In: Anais do XI Simpósio Brasileiro de Tecnologia da Informação e da Linguagem Humana, pp. 122–131. SBC, Porto Alegre, RS, Brasil (2017)
6. Houlsby, N., Huszár, F., Ghahramani, Z., Lengyel, M.: Bayesian active learning for classification and preference learning (2011)
7. Kobayashi, K., Wakabayashi, K.: Named entity recognition using point prediction and active learning. In: Proceedings of the 21st International Conference on Information Integration and Web-Based Applications and Services, iiWAS2019, pp. 287–293. Association for Computing Machinery, New York, NY, USA (2019)
8. Lakshmi Narayan, P., Nagesh, A., Surdeanu, M.: Exploration of noise strategies in semi-supervised named entity classification. In: Proceedings of the Eighth Joint Conference on Lexical and Computational Semantics (*SEM 2019), pp. 186–191. Association for Computational Linguistics, Minneapolis, Minnesota, June 2019
9. Ma, X., Hovy, E.: End-to-end sequence labeling via bi-directional LSTM-CNNs-CRF. In: Proceedings of the 54th Annual Meeting of the Association for Computational Linguistics (Volume 1: Long Papers), pp. 1064–1074. Association for Computational Linguistics, Berlin, Germany, August 2016
10. Miyato, T., Dai, A.M., Goodfellow, I.: Adversarial training methods for semi-supervised text classification. In: International Conference on Learning Representations (ICLR) (2017)

11. Nair, V., Hinton, G.E.: Rectified linear units improve restricted Boltzmann machines. In: Proceedings of the 27th International Conference on International Conference on Machine Learning, ICML 2010, pp. 807–814. Omnipress, Madison, WI, USA (2010)

12. Neubig, G., Nakata, Y., Mori, S.: Pointwise prediction for robust, adaptable Japanese morphological analysis. In: Proceedings of the 49th Annual Meeting of the Association for Computational Linguistics: Human Language Technologies, pp. 529–533. Association for Computational Linguistics, Portland, Oregon, USA, June 2011

13. Park, J., Kim, G., Kang, J.: Consistency training with virtual adversarial discrete perturbation (2021)

14. Pennington, J., Socher, R., Manning, C.: GloVe: global vectors for word representation. In: Proceedings of the 2014 Conference on Empirical Methods in Natural Language Processing (EMNLP), pp. 1532–1543. Association for Computational Linguistics, Doha, Qatar, October 2014. https://doi.org/10.3115/v1/D14-1162

15. Pradhan, S., et al.: Towards robust linguistic analysis using OntoNotes. In: Proceedings of the Seventeenth Conference on Computational Natural Language Learning, pp. 143–152. Association for Computational Linguistics, Sofia, Bulgaria, August 2013

16. Radmard, P., Fathullah, Y., Lipani, A.: Subsequence based deep active learning for named entity recognition. In: Proceedings of the 59th Annual Meeting of the Association for Computational Linguistics and the 11th International Joint Conference on Natural Language Processing (Volume 1: Long Papers), pp. 4310–4321. Association for Computational Linguistics, Online, August 2021

17. Sang, E.F.T.K., Meulder, F.D.: Introduction to the CoNLL-2003 shared task: language-independent named entity recognition. In: Proceedings of the Seventh Conference on Natural Language Learning at HLT-NAACL 2003, pp. 142–147 (2003)

18. Shen, Y., Yun, H., Lipton, Z., Kronrod, Y., Anandkumar, A.: Deep active learning for named entity recognition. In: Proceedings of the 2nd Workshop on Representation Learning for NLP, pp. 252–256. Association for Computational Linguistics, Vancouver, Canada, August 2017. https://doi.org/10.18653/v1/W17-2630

19. Siddhant, A., Lipton, Z.C.: Deep Bayesian active learning for natural language processing: results of a large-scale empirical study. In: Proceedings of the 2018 Conference on Empirical Methods in Natural Language Processing, pp. 2904–2909. Association for Computational Linguistics, Brussels, Belgium, October–November 2018. https://doi.org/10.18653/v1/D18-1318

20. Srivastava, N., Hinton, G.E., Krizhevsky, A., Sutskever, I., Salakhutdinov, R.: Dropout: a simple way to prevent neural networks from overfitting. J. Mach. Learn. Res. **15**(1), 1929–1958 (2014)

21. Tran, V.C., Nguyen, N.T., Fujita, H., Hoang, D.T., Hwang, D.: A combination of active learning and self-learning for named entity recognition on Twitter using conditional random fields. Knowl.-Based Syst. **132**, 179–187 (2017)

Author Index

Printed in the United States
by Baker & Taylor Publisher Services